Nonlinear dynamics, chaotic and complex systems constitute some of the most fascinating developments of late twentieth century mathematics and physics. It transpires that chaotic behaviour can be understood, and sometimes even utilized, to a far greater degree than hitherto suspected. Surprisingly, universal constants have been discovered. The implications have changed our understanding of important phenomena in physics, biology, chemistry, economics, medicine and numerous other fields of human endeavor.

In this book, two dozen people largely responsible for the 'nonlinear revolution' cover most of the basic aspects of the field in 15–20 page articles. The book is subdivided into five parts: dynamical systems, bifurcation theory and chaos; spatially extended systems; dynamical chaos, quantum physics and foundations of statistical mechanics; evolutionary and cognitive systems; and complex systems as an interface between the sciences.

This book is expected to become a standard text on the subject. It grew out of an EC sponsored conference held in Zakopane, a Polish mountain resort, in November 1995. This conference was attended by many of the founders of the field and was considered to be one of the most important gatherings of 1995.

NONLINEAR DYNAMICS, CHAOTIC AND COMPLEX SYSTEMS

Nonlinear Dynamics Chaotic and Complex Systems

Proceedings of an International Conference
held in Zakopane, Poland
November 7–12, 1995
Plenary Invited Lectures

Edited by

E. INFELD
Sołtan Institute for Nuclear Studies, Warsaw, Poland

R. ŻELANZNY
Institute of Plasma Physics and Laser Microfusion, Warsaw, Poland

A. GAŁKOWSKI
Institute of Plasma Physics and Laser Microfusion, Warsaw, Poland

CAMBRIDGE UNIVERSITY PRESS
Cambridge, New York, Melbourne, Madrid, Cape Town, Singapore,
São Paulo, Delhi, Dubai, Tokyo, Mexico City

Cambridge University Press
The Edinburgh Building, Cambridge CB2 8RU, UK

Published in the United States of America by Cambridge University Press, New York

www.cambridge.org
Information on this title: www.cambridge.org/9780521152945

First published 1997
First paperback edition 2010

A catalogue record for this publication is available from the British Library

Library of Congress Cataloguing in Publication Data

Nonlinear dynamics, chaotic and complex systems: proceedings of an international conference held in Zakopane, Poland 7-12, 1995: plenary invited lectures/edited by E. Infeld, R. Żelazny, A. Ga łkowski.
p. cm.
ISBN 0 521 58201 6 (hardcover)
1. Differentiable dynamical systems – Congresses. 2. Chaotic behavior in systems – Congresses. 3. Nonlinear theories – Congresses. I. Infeld. E. (Eryk) II. Żelazny. Roman. III. Ga łkowki A. (Andr zej). 1951-
QA614.8 N66 1997
003'.85–dc21 97-4012 CIP

ISBN 978-0-521-58201-8 Hardback
ISBN 978-0-521-15294-5 Paperback

Contents

Evolutionary and Cognitive Systems

Complex Systems as an Interface between Natural Sciences and Environmental, Social, and Economic Sciences

Conference Banquet Speech

Participants

Pavel G. Akishin	Join Institute for Nuclear Research, Russia
Peter Allen	Cranfield University, United Kingdom
Ioannis Antoniou	Université Libre de Bruxelles, Belgium
Roberto Artuso	Istituto di Scienze Matem. Fisiche e Chimiche, Italy
Erik Aurell	Royal Institute of Technology, Sweden
Jan Awrejcewicz	Technical University of Łódź, Poland
Agnes Babloyantz	Université Libre de Bruxelles, Belgium
L. Bacry	Centre Nationale de la Recherche Scientifique, France
Radu Balescu	Université Libre de Bruxelles, Belgium
Pavol Baňacký	Comenius University, Slovak Republic
Piotr Banat	Warsaw University, Poland
Bogdan Baranowski	Polish Academy of Sciences, Poland
Julius Bene	Eötvös University, Hungary
Iwo Białynicki-Birula	Polish Academy of Sciences, Poland
Silviu Birauas	University of Timişoara, Romania
Stefano Boccaletti	Istituto Nazionale di Ottica, Italy
Valentin Boju	University of Craiova, Romania
Catalin Borcia	Al. I. Cuza University, Romania
Aleksander B. Borisov	Russian Academy of Sciences, Russia
Skaidra Bumelienė	Semiconductor Physics Institute, Lithuania
Leonid A. Bunimovich	Georgia Institute of Technology, United States
Friedrich H. Busse	Universität Bayreuth, Germany
Adrian Cârstea	Institute of Atomic Physics, Romania
G. Casati	Universita' Degli Studi di Milano, Italy
A. Cenian	Polish Academy of Sciences, Poland
Antanas Čenys	Semiconductor Physics Institute, Lithuania
George Chechin	Rostov State University, Russia
Leonid Chekhov	Steklov Mathematical Institute, Russia
Yury Cherkashin	Russian Academy of Sciences, Russia
Boris V. Chirikov	Budker Institute of Nuclear Physics, Russia
Steliana Codreanu	Babes-Bolyai University, Romania
Constana D. Constantinescu	University of Craiova, Romania
Maurice Courbage	Université Paris 7, France
Per Dahlqvist	Royal Institute of Technology, Sweden
Vladimir N. Damgov	Bulgarian Academy of Sciences, Bulgaria
Huyen Dang-Vu	Université Pierre et Marie Curie, France
Claudine Delcarte	LIMSI-Université de Paris-Sud, France
Stanisław Dembiński	Nicholas Copernicus University, Poland
Florin Despa	Institute of Atomic Physics, Romania
Lioudmila Dmitrieva	St Petersburg State University, Russia
Vladimir A. Dobrynskiy	Kiev University, Ukraine
John P. Dougherty	University of Cambridge, United Kingdom
Stanisław Drożdż	Institute of Nuclear Physics, Poland
Danuta Dudek	Polish Academy of Sciences, Poland
Mihaela Dumitru	'Politehnica' University of Bucharest, Romania
Werner Ebeling	Humboldt-Universität zu Berlin, Germany
Hans T. Elze	University of Arizona, United States
Bengt Enflo	Royal Institute of Technology, Sweden
Juri Engelbrecht	Estonian Academy of Sciences, Estonia

Victor Eremenko	Russian Academy of Sciences, Russia
Mitchell J. Feigenbaum	Rockefeller University, United States
Lachezar Filipov	Bulgarian Academy of Science, Bulgaria
Evald Fradkin	St Petersburg University, Russia
Mats Fredriksson	Royal Institute of Technology, Sweden
Wojciech Gadomski	University of Warsaw, Poland
Sangeeta D. Gadre	University of Delhi, India
Ryszard Gajewski	Physical Optics Corporation, United States
Andrzej Gałkowski	Inst. Plasma Physics and Laser Microfusion, Poland
Piotr Garbaczewski	University of Wrocław, Poland
Oleg Gerasimov	Odessa State University, Ukraine
Reiner Gerold	European Commission, DG XII, Belgium
Eugen Gheorghiu	Biophysics Laboratory, Romania
Alexander V. Glushkov	Applied Mathematics Department, OHM, Ukraine
Sergey V. Gonchenko	Inst. Applied Math. and Cybernetics, Russia
Vladimir A. Gordin	Russian Hydrometeocentre, Russia
Dan Grecu	Institute of Atomic Physics, Romania
Ioan Grosu	Al. I. Cuza University, Romania
Volker Matthias Gundlach	Universität Bremen, Germany
Zbigniew Haba	Wrocław University, Poland
Krzysztof Haman	Warsaw University, Poland
Xavier de Hemptinne	Catholic University Leuven, Belgium
Janusz Hołyst	Warsaw University of Technology, Poland
William G. Hoover	Lawrence Livermore Nat. Lab., United States
Richard Hsieh	Royal Institute of Technology, Sweden
Eryk Infeld	Soltan Institute for Nuclear Studies, Poland
Cristian Ioana	Romanian Academy, Romania
Anzelm Iwanik	Wrocław Technical University, Poland
Krzysztof Janicki	Polish Academy of Sciences, Poland
Krzysztof Kacperski	Warsaw University of Technology, Poland
Brunon Kamiński	Nicholas Copernicus University, Poland
Tomasz Kapitaniak	Technical University of Łódź, Poland
Marina T. Kapoustina	NASci of Ukraine, Ukraine
Bronislavas Kaulakys	Inst. Theoretical Physics and Astronomy, Lithuania
Bogdan Kaźmierczak	Polish Academy of Sciences, Poland
Valery N. Kharkyanen	NASci of Ukraine, Ukraine
Olga Kocharovskaya	Institute of Applied Physics, Russia
Vitaly Kocharovsky	Institute of Applied Physics, Russia
Vladimir Kocharovsky	Institute of Applied Physics, Russia
Alexander Kovalev	B. Verkin Institute for Low Temperature, Ukraine
Zbigniew S. Kowalski	Wrocław University of Technology, Poland
Ladislav Krlin	Academy of Science, Czech Republic
Juraj Kumicak	Technical University, Slovak Republic
Maciej Kuna	Pedagogical University of Słupsk, Poland
Yuri Kuperin	St Petersburg State University, Russia
Pavel Kurasov	Ruhr University Bochum, Germany
Jacek Kurzyna	Polish Academy of Sciences, Poland
Marek Kuś	Polish Academy of Sciences, Poland
Jarosław Kwapień	Institute of Nuclear Physics, Poland
Muthusamy Lakshmanan	Bharathidasan University, India

Polina Landa	Moscow State University, Russia
Lev Lerman	Inst. Applied Mathematics and Cybernetics, Russia
Ignace Loris	Vrije Universiteit Brussel, Belgium
Vladimir Maistrenko	Kiev University, Ukraine
Yuri Maistrenko	Kiev University, Ukraine
Eva Majernikova	Slovak Academy of Sciences, Slovak Republic
Hanna Makaruk	Polish Academy of Sciences, Poland
Vladimir G. Makhankov	Los Alamos National Laboratory, United States
Adam Makowski	Nicholas Copernicus University, Poland
Svetlana V. Malinovskaya	Odessa Institue of Spectroscopy, Ukraine
Szymon Malinowski	University of Warsaw, Poland
Mikhail Malkin	Nizhny Novgorod Pedagogical University, Russia
Natalia Manaenkova	Institute of Terrestial Magnetism, Russia
Konstantin Mardanov	St. Petersburg University, Russia
Milos Marek	Institute of Chemical Technology, Czech Republic
Mária Markosová	Institute of Measurement, Slovak Republic
Mihaela Mehedintu	Biophysics Laboratory Biotechnos S.A., Romania
Leonid Melnikov	Chernyshevsky State University, Russia
Yuri Melnikov	Institute for Physics, Russia
Viktor Mel'nikov	Laboratory of Theoretical Physics, Russia
Corina M. Mihai	Biophysics Laboratory Biotechnos S.A., Romania
Jacques Misguich	Centre d'Etudes de Cadarache, France
Marian Mrozek	Jagiellonian University, Poland
Florin Munteanu	Romanian Academy, Romania
Renata Murat	Maria Curie Skłodowska University, Poland
Audrius Namajunas	Semiconductor Physics Institute, Lithuania
Wojciech Nasalski	Polish Academy of Sciences, Poland
Gregoire Nicolis	Université Libre de Bruxelles, Belgium
Valerica Ninulescu	'Politehnica' University of Bucharest, Romania
Arne Nordmark	Royal Institute of Technology, Sweden
Dagmar Novotná	Academy of Science, Czech Republic
Andrzej Nowak	Warsaw University, Poland
Christer Nyberg	Royal Institute of Technology, Sweden
Andrzej Okniński	Kielce Technical University, Poland
Sorinel Adrian Oprişan	Al. I. Cuza University, Romania
Marek Orlik	Warsaw University, Poland
Edward Ott	University of Maryland, United States
Robert Owczarek	Polish Academy of Sciences, Poland
Boris S. Pavlov	St Petersburg University, Russia
Rafał Pawlikowski	Polish Academy of Sciences, Poland
Piotr Pepłowski	Nicholas Copernicus University, Poland
Zbigniew Peradzyński	Polish Academy of Sciences, Poland
Elvira Perekhodtseva	Hydrometeorological Centre of Russia, Russia
Laurent Pezard	Hopital de la Salpetriere, France
Jarosław Piasecki	Warsaw University, Poland
Jean Paul Pique	Laboratoire de Spectrometrie Physique, France
Mihai Piscureanu	'Politehnica' University of Bucharest, Romania
Juri Pismak	St Petersburg State University, Russia
Andrzej Plonka	Technical University of Łódź, Poland
Pavel Pokorny	Institute of Chemical Technology, Czech Republic

Andrzej Posiewnik	Inst. Theoretical Physics and Astrophysics, Poland
Jiři Pospichal	Slovak Technical University, Slovak Republic
Ilya Prigogine	Université Libre de Bruxelles, Belgium
Itamar Procaccia	The Weizmann Institute of Science, Israel
Feliks Przytycki	Polish Academy of Sciences, Poland
Mircea Puta	University of Timişoara, Romania
Kestutis Pyragas	Semiconductor Physics Institute, Lithuania
Victor L. Rapoport	St Petersburg University, Russia
James Reineck	SUNY at Buffalo, United States
Roman Romanov	Institute for Physics, Russia
Vitali A. Rostovtsev	Joint Institute for Nuclear Research, Russia
Piotr Rozmej	Maria Curie Skłodowska University, Poland
Konstantin G. Rubinstein	Hydrometeocentre of Russia, Russia
Vladislav Ruchlia	State University of Belarus, Republic Belarus
German Rudin	St Petersburg State University, Russia
Ryszard Rudnicki	Polish Academy of Sciences, Poland
Czesław Rymarz	Military University of Technology, Poland
Aurel Salabas	Institute of Atomic Physics, Romania
Vladimir Samuilov	State University of Belarus, Republic Belarus
Mircea Sanduloviciu	Al. I. Cuza University, Romania
Alexander N. Sharkovsky	Ukrainian Academy of Sciences, Ukraine
Vladimir Sepman	St Petersburg University, Russia
Arkadi G. Shagalov	Ural Branch of Russian Academy of Sciences, Russia
Leonid P. Shilnikov	Inst. Applied Mathematics and Cybernetics, Russia
Clement Sire	Universite Paul Sabatier, France
Andrei Sivak	Ukrainian Academy of Sciences, Ukraine
Zygmunt Składanowski	Inst. Plasma Physics and Laser Microfusion, Poland
Jan J. Sławianowski	Polish Academy of Sciences, Poland
Wojciech Słomczyński	Jagiellonian University, Poland
Artyom Smirnov	Institute for Physics, Russia
Leonard Smith	University of Oxford, United Kingdom
Tatiana Soboleva	Ukrainian Academy of Sciences, Ukraine
Lars Söderholm	Royal Institute of Technology, Sweden
Dmitry M. Sonechkin	Hydrometeorological Research Centre, Russia
G. Sonnino	European Commission, Belgium
Florin Spineanu	Institute of Atomic Physics, Romania
Roman Srzednicki	Jagiellonian University, Poland
Angela Stevens	Universität Heidelberg, Germany
Andrzej Sukiennicki	Warsaw University of Technology, Poland
Cristian Suteanu	Romanian Academy, Romania
Janusz Szczepański	Polish Academy of Sciences, Poland
Wanda Szemplińska-Stupnicka	Polish Academy of Sciences, Poland
Harold Szu	University of Louisiana, United States
Arunas Tamaševičius	Semiconductor Physics Institute, Lithuania
Tamaś Tél	Eötvös University, Hungary
Dmitry Turaev	Inst. Applied Math. and Cybernetics, Russia
Andrzej Turski	Polish Academy of Sciences, Poland
Juras Ulbikas	Semiconductor Physics Institute, Lithuania
Sergei Vakulenko	Inst. Problems of Mechanical Engineering, Russia
Chris Van den Broeck	Limburgs Universitair Centrum, Belgium

Antonin Vaněček	Academy of Science, Czech Republic
Svetlana Vikul	Ukrainian Academy of Sciences, Ukraine
Vilutis Gediminas	Institute of Theoretical Physics and Astronomy, Russia
Sergue I. Vinitsky	Joint Institute for Nuclear Research, Russia
Andrei Vladimirov	St Petersburg State University, Russia
Gabriel E. Weinreb	NASci of Ukraine, Ukraine
Marek Wójcik	Institute of Nuclear Physics, Poland
Marek Wolf	Institute of Theoretical Physics, Poland
Henryk Woźniakowski	Warsaw University, Poland
Jerzy Zagrodziński	Polish Academy of Sciences, Poland
Piotr Zgliczyński	Jagiellonian University, Poland
George Zhuvikin	St Petersburg University, Russia
Irina Zhuvikin	St Petersburg University, Russia
Witold Zieliński	University of Wrocław, Poland
Vladimir Zolotarev	University of Kharkov, Ukraine
Henryk Zorski	Polish Academy of Sciences, Poland
Jan J. Żebrowski	Warsaw University of Technology, Poland
Roman Żelazny	Inst. Plasma Physics and Laser Microfusion, Poland
Karol Życzkowski	Jagiellonian University, Poland

Preface

The last thirty years have witnessed a powerful new development across the board of scientific research. The insight gained has changed our whole understanding of natural phenomena. It involves the complex behaviour in which many different states of a system occur. These states have wholly different stability properties and can exhibit abrupt transitions from one to the other. The seemingly erratic motion involved defies prediction and is extremely difficult to control. We usually refer to this phenomenon as *deterministic chaos*. It coexists with the classical, organized behaviour we learned about at school. It now seems that an understanding of both is required in a wide range of situations.

Chaotic behaviour is observed in most fields of science (e.g. subatomic to molecular physics, fluid dynamics, plasma physics, chemistry, molecular biology). It appears in environmental, social and economic phenomena, as well as in technological processes.

The important discovery is that there are universal mechanisms underlying the complex behaviours involved. Fortunately, these mechanisms can be studied by looking at simplified theoretical models. Once these two facts were grasped, people working in theory, experiments, and numerics inevitably came together to develop the field.

There is an interesting political aspect to all this. The human potential of people active in the field in Central and Eastern Europe is still immense (especially in theory). How can we avoid dispersion of these invaluable human resources? Does the EC wish to stand by while these scientists follow their colleagues working in other fields in fleeing our Continent? If not, what steps should be taken? With all this in mind, a group of scientists from EC countries joined forces with their colleagues in Central and Eastern Europe. They gained support from the Twelfth Directorate (DG XII) of the EC in Brussels. The idea was to convene a conference on nonlinear dynamics and complex systems. This conference would unite the two parts of Europe (EC and non-EC), but with strong North American participation. Largely due to personal contacts between the Brussels and Warsaw schools, the conference took place in Poland. The venue was Zakopane, a mountain resort south of Kraków. Forbidding weather notwithstanding, the turnout was excellent. Twenty three invited lectures were given. Mitchell Feigenbaum gave a fascinating banquet talk, a transcript of which is included in this volume under the heading 'Where will the future go?'. There were 212 contributed talks (to be published in the Journal of Technical Physics) and four very uneven panel discussions. The panels were on: Mathematics (coordinated by E. Ott and L. Bunimovich); Chemistry and Biology (G. Nicolis and M. Marek); Physics (R. Balescu, I. Białynicki-Birula, M. Feigenbaum, V. G. Makhankov and H. Woźniakowski); and on the future of nonlinear dynamics research (R. Żelazny). This last-mentioned panel might merit a comment here. The Local Organizing Committee put forward an idea for establishing a nonlinear dynamics center in Poland. At the time of writing, this idea is still being persued.

The invited talks

A short glance at the talks indicates just how diverse and dynamic the field is proving to be.

Professor Gundlach spoke about describing chaos in mechanical systems by using symbolic dynamics. His procedure allows us to describe quite general systems evolving under the influence of noise and exhibiting stochastic features. He is able to cope with fractal structures, at least formally.

Professor Shilnikov talked about the occurrence of homoclinic chaos. The main feature of the attractors involved here is the presence of homoclinic tangencies. He pointed out the possibility of the coexistence of infinitely many stable periodic orbits. There can also be infinitely many arbitrarily degenerate periodic orbits. It should be stressed that an infinity of non-periodic attractors can coexist in the systems concerned (with each other and with the periodic orbits).

Modern ergodic theory is the cornerstone of the formalism of classical chaos. However, in quantum systems, the energy and frequency spectra are discrete. To overcome the resulting difficulties, B. Chirikov introduced the concept of pseudochaos. However, examples of this phenomenon in classical systems were also given. As an aside, we were offered a warning. Computers can in fact exhibit pseudochaos, thus leading to a dynamic trajectory becoming a periodic one (an artefact).

Dr Dougherty gave a lecture on the foundations of non-equilibrium statistical mechanics. He shed some light on the otherwise difficult concept of subdynamics as introduced by Ilya Prigogine's school.

His talk was a rare example of a survey of the Brussels school's activity from the outside.

The Prigogine lecture was a survey of the Nobel Prize winner's results over his later period.† During the last hundred years, we have been faced with the fact that practically all interactive particle systems are nonintegrable. An important example of such a system is the Large Poincaré System having a continuous spectrum. This leads to the appearance of diffusive terms. It also leads to limitations on classical trajectory dynamics and wave functions. As a consequence, two important aspects appear in the fundamental dynamics description. They are probability and irreversibility. This is so because we must abandon the trajectory description and use the probabilistic, distribution function, approach. This in turn leads to the Liouville operator and its complex spectral decomposition. We are thus forced to go beyond the usual Hilbert space to a 'rigged' Hilbert space. In the case of chaotic maps, this procedure shows that the eigenvalues of an evolution operator are directly related to the Lyapunov exponents. Application of this idea to other Hamiltonian operators was presented.

Professor Woźniakowski spoke on the notion of computational complexity, which concerns the intrinsic difficulty of solving mathematically posed problems. Information based complexity is a branch of computational complexity that deals with continuous problems defined on spaces of multivariate functions. For such problems, only approximate solutions are possible. The complexity is defined as a minimal cost needed to compute an approximation with error at most ϵ.

Microscopic simulations along with applications in thermomechanical problems connected with heat transfer far from equilibrium were reported by Professor Hoover. The least action principle can be applied far from equilibrium, where it is necessary to control the temperature or internal energy or else other dynamical variables. The resulting equations of motion can be solved.

Professor Ott talked on the control of chaos by the selection and maintenance of an optimum or near optimum unstable periodic orbit from the point of view of a specified goal. Analogous considerations based on different models were discussed by Professor Vaněček. The point is that chaos can even be desirable. The control of chaos was also addressed by other contributors. Chaotic systems can often be managed. This is of immense importance in applications (medicine, industry, etc.).

Driven spatially extended systems, consisting of many almost identical elements, are quite frequent in nature. Some aspects of their complexity can be modelled with the help of sandpile-like cellular automata. Their dynamical properties are interesting from both

† Although the talk was given by Dr Antoniou, we were sent a version co-authored by a different collaborator.

a theoretical and a practical point of view. The question arises of whether scale invariant structures in nature are the consequences of the self-organized critical states that arise. The characteristic properties of these states have been studied both numerically and analytically. Analytical calculations of the characteristics of these states are done mainly for Abelian sandpile models on cubic and Bethe lattices. Abelian sandpile calculations were presented by M. Markošová.

Professor Procaccia spoke on scenarios for anomalous scaling in hydrodynamic turbulence. The scaling behaviour of turbulent hydrodynamic systems is often anomalous; the scaling exponents not being simply related to each other. Nevertheless, it was possible to formulate a theory yielding these exponents. Comparisons with experiments were given. Similar problems, in particular as related to transport in the von Karman sheet, were considered by Professor Tél.

Further important modelling devices are furnished by networks with learning capabilities.

Professor Van den Broeck presented problems connected with the understanding of the mechanism of learning. The learning process can be described as a competition between entropy (log of the number of configurations) and error (of a given configuration with respect to the target) and it is not surprising that a formalism, similar to that of equilibrium statistical mechanics, can be set up.

Professor Szu presented the concept of an Artificial Neuron Network defined to have fixed layers of neurons with threshold logic and capable of learning by the 'small perturbation' Hebb rules. Mathematical attempts to formulate such a self-organization architecture were given.

One way of studying evolutionary processes is by investigating sequences generated by evolution. The basic idea is that the evolution generates criticality structures which occur on the border between regular and chaotic regimes or between two different regular or chaotic critical structures. There is also an interesting class of systems showing 'self-organized criticality' or 'self-tuned criticalities'. Those criticality structures should exhibit very long memory effects, and thus long correlations. By using the methods of symbolic dynamics, any trajectory of a dynamic system may be mapped onto a string of letters in an alphabet.

Professor Ebeling presented studies on sequences generated by dynamic models of evolutionary processes, as in the Fibonacci model, the logistic map, and in strings carrying information such as a book, a piece of music, or DNA. The results show that, in particular, the low frequency spectra and the scaling of the mean square deviations are appropriate measures for long range correlations in symbolic sequences.

There is hope that the analysis of entropies, power spectra, and scaling exponents could be useful for studies of the large scale structures of series and information sequences.

Magnetic fields in plasma confining devices (e.g. tokamaks) or in astrophysics can have a broad range of behaviour ranging from integrable structures to erratic wandering in space. This is caused primarily by the nonlinearities of ideal magnetohydrodynamics. In addition, the presence of small dissipation (resistivity) induces reconnection of magnetic lines with the generation of magnetic islands. These are surrounded by regions where the field lines exhibit a high sensitivity to initial conditions. This means exponential instability, which is a signature of chaos. In confining devices, simplified configurations of the magnetic fields are being studied. These problems were examined by Dr Spineanu.

Excitability is a property shared by some chemical systems and a large number of biological systems. In an excitable medium, suprathreshold stimuli are propagated as a pulse or wave front by a mechanism combining local excitatory kinetics with transport. The Belousov–Zhabotinski reaction has often been used in studies of excitability. Typical examples of oscillatory and excitable phenomena in biology are Ca^{2+} oscillations and waves in cells. Intracellular Ca^{2+} release is connected with many receptor induced cell signals, controlling processes ranging from secretion and heart rate frequency control to transcription and cell division. The introduction of digital Ca^{2+} imaging techniques made it possible to visualize the spatiotemporal patterns of the ion. Ca^{2+} waves in biological systems are mostly treated within the framework of excitable media with ion transport. These waves can often be either provoked by electric stimulation or else influenced by an electric field. A talk on this was given by Professor Marek.

Interesting applications of stochastic differential geometry in financial studies were presented by Professor Makhankov. The idea was to develop a generalized pricing method for over the counter derivatives.

Analytical estimates and computer experiments reveal a variety of possible market developments:

(*a*) explosive instability of the solution which leads to unpredictability of the bond market,

(*b*) a number of stationary solutions: rising, falling and humped curves.

The form of the solution is completely determined by boundary conditions and hence is an exogenously defined feature of the system rather than an intrinsic one.

The difficulty in predicting the evolution of a nonlinear system varies with the state of the system. However, this variation is highly organized within the attractor. The point is to exploit this structure to improve predictions.

Panels

Four discussion panels augmented these talks and endeavoured to formulate some specific problems. All participants stressed the fact that research in the field is just emerging from its infancy and still has a wide range of problems to clarify. Further efforts are necessary to apply this knowledge to solve specific applicational problems. The tendency is to treat chaos in a more positive way (not just as a nuisance).

Formulating future lines of development can be risky. Science develops in an unpredictable and chaotic way resembling patterns and behaviours considered by the Conference. A surprise solution or idea that may emerge in one field is often used elsewhere. Nevertheless, we tried to look at where future development could go.

As for a **mathematical** way of looking at our problems, the following directions were formulated:

The development of new computer algorithms and software.

Consideration of, to what extent computer results reproduce the correct qualitative behaviour of the phenomena simulated ('artifacts', 'ghosts', etc.).

The development of new techniques for the analysis of experimental data (e.g. in the past the introduction of embedding techniques has had an enormous effect on experiments, and the prediction of chaotic system time evolutions promises wide applications).

The development of new bifurcation theory, software, etc., for higher dimensional dynamical systems.

The development of techniques for influencing chaotic systems.

The development of methods for the study of extended systems (e.g. cellular automata, lattice dynamical systems, fluids, reaction–diffusion systems, systems with nonlinear boundary conditions, self-organized criticality, etc.). It would be useful to create a collection of exactly solvable model problems.

The development of a theory for spatial patterns and spatial and temporal chaos.

Studies of chaos as an underpinning of statistical mechanics and transport properties.

Studies of quantum problems (e.g. quantum chaos and quantum dynamics in a Markovian background).

The views expressed on the future directions of research as seen from a **physical** angle were:

We need a well formulated statistical mechanics on three consecutive levels: microscopic, mesoscopic and macroscopic.

Due to the nonintegrability of their Hamilton or Liouville equations many-body problems must be described on a probabilistic level. The extension of Hilbert space to a 'rigged' one may open a new perspective here.

The many-body problem for open systems which can be described as systems with a Markovian background. Can noise be used for controlling chaos?

Measurements in quantum many-body systems. Preparation of the systems by measurements.

Simulation from the point of view of: describing multifractal objects; the understanding of ergodic strange attractors; developing smooth boundaries; excluding unwanted fluctuations; improving the description of turbulence; relating Nosé's mechanics to mechanical variational principles.

Further research on the development of chaos in fluids, the onset of turbulence and its perseverance, the control of turbulence.

The theory of nonlinear fluids and solids (stress tensors depending quadratically on strain and/or velocity gradient). Coefficients may exhibit additional dependence on dynamical variables.

Chaotic phenomena in plasmas, particularly in thermonuclear plasmas, understanding of anomalous and 'strange' transport processes in turbulent plasmas, the control and stabilization of plasma behaviour.

Improvement in the reliability of meteorological prognoses on the basis of experimental data and numerical simulations.

Capital markets are not well described by the traditional Efficient Market Hypothesis. By applying the theory of chaos to investments and economics, a new Fractal Market Hypothesis was recently presented by E. E. Peters. It gives an economic and mathematical structure to fractal market analysis. Through this approach, we can understand why self-similar statistical structures exist, as well as how risk is distributed among investors. All in all, results originally obtained for nonlinear and complex physical systems could be useful for an analysis of social and economic models.

The panel on **chemistry and biology** formulated the following recommendations on directions of nonlinear research:

The most common evidence of complex behaviour in chemistry and biology comes from phenomena on the macroscopic level. Here emphasis is on the origin of collective behaviour in multi-unit systems giving rise to new emergent properties.

There is increasing awareness that complexity also appears at the microscopic and mesoscopic levels, for example in the form of complex distributions of molecular spectra, anomalous transport and kinetics in dispersed media and fractal structures, or of supramolecular assemblies endowed with recognitional and other unusual properties, traditionally attributed exclusively to biomolecules. Bridging the gap between these three levels is one of the priorities. Potential implications are, among others, the control of chemical reactions by ultra-short light pulses and a better understanding of the structure and function of enzymes.

By and large, complex system studies were hitherto based on nonlinear dynamical systems of few degrees of freedom. The limitations of this approach are becoming evident. For instance, the behaviour of spatially extended systems described by partial differential equations is not amenable to conventional dynamic system analysis. In view of the above, a special research effort is required for developing a qualitative theory of spatially extended systems. Special emphasis should be put on the classification of the types of behaviour predicted by the normal form equations.

In the **application area**, the following recommendations were sketched by the panels:

• *Fusion.* Nonlinear problems appear to be crucial in almost all aspects of the physics of hot plasma confinement.

• *Chemical processing.* The manufacturing of both traditional and new materials relies to a large extent on the process of heterogeneous catalysis. It has recently been established that chemical transformations do not happen synchronously on the catalytic surface, but instead give rise to instabilities and complex spatiotemporal patterns. In a similar vein, chemical synthesis in dispersed media (films, micelles) can benefit from studying regimes in which traditional equilibrium-like states are unstable and replaced by self-organizing regimes.

• *Energy technology.* At the root of energy technology is the process of combustion, which gives rise to a variety of complex behaviours, from multistability to oscillations and the appearance of wave fronts. Problems involving flame stabilization are crucial for the efficiency and performance of heat engines. Their evolution could benefit from cross-fertilization with research on complex systems. Here microscopic simulation methods offer a promising tool.

• *Nonlinear dynamics in biology and medicine.* Chemical instabilities provide a natural prototype of a number of important biological phenomena from biological rhythms to morphogenesis and embryonic development. Evidence of wave-like behaviour in cardiac tissues and of spatiotemporal chaos in the brain is mounting. Nonlinear dynamics offers new ways of analyzing data which could culminate in the development of new diagnostic tools and mathematical models for certain types of disease.

• *Evolutionary and cognitive systems.* The emergent adaptive properties observed in biological evolution, animal interaction and cognition furnish striking examples of complex behaviour in nature. Insight gained from such biological phenomena can be applied to techniques for artificial self-organizing and computational devices. Decentralized interactions of simple autonomous units may lead to structures with properties complementary to those of conventional machines. Among some systems and issues currently studied and worthy of support in the future are:

- Evolutionary biotechnology;
- Genetic algorithms and evolutionary programming applied successfully to optimization problems;
- Neural networks, pattern recognition and information compression;
- Multiagent systems and robotics.

• *Ecosystem stability and biodiversity.* Hierarchic mixed populations which coexist and compete in various niches are ubiquitous in nature. Studies of the development of spatiotemporal patterns of such populations and of their stability in distributed systems are of primary importance. Such studies can benefit from the development of a common methodology as mentioned above. They should include a description of man-made chemical agents on populations and could be done in collaboration with biologists. Such investigations could also be of importance in improving the performance of bioreactors, or controlling the spreading of infections and diseases.

These remarks, however incoherent, do show impressive research and applicational potential of the field. Their possible impact on science, technology, and also industrial, environmental, socio-economical processes in the future are impressive. An integrated Europe should meet this challange. As mentioned above, something should be done to utilize resources in Central and Eastern Europe. This could perhaps be achieved through a centre for nonlinear dynamics, chaotic and complex systems as proposed by the Local Organizing Committee.

The Editors

Acknowledgements

The authors would like to thank the DG XII of the European Commission and Dr R. Gerold personally for making this conference possible. We would also like to thank the Polish State Agencies involved: Science Research Committee (KBN), Polish Academy of Sciences and Atomic Energy Agency. Dr A. A. Skorupski was extremely helpful in correcting the texts. If this book is now a uniform entity, it is largely thanks to him. Finally, Dr S. Capelin of CUP was most helpful and patient.

Chaos in random dynamical systems

By VOLKER MATTHIAS GUNDLACH

Institut für Dynamische Systeme, Universität Bremen, Bremen, Germany

In the investigations of chaos in dynamical systems a major role is played by symbolic dynamics, i.e. the description of the system by a shift on a symbol space via conjugation. We examine whether any kind of noise can strengthen the stochastic behaviour of chaotic systems dramatically and what the consequences for the symbolic description are. This leads to the introduction of random subshifts of finite type which are appropriate for the description of quite general dynamical systems evolving under the influence of noise and showing internal stochastic features. We investigate some of the ergodic and stochastic properties of these shifts and show situations when they behave dynamically like the common shifts. In particular we want to present examples where such random shift systems appear as symbolic descriptions.

1. Introduction

Randomness and chaos are notions with rather strong links. Even intuitively, chaos is associated with the exhibition of random features. Mathematicians call a dynamical system with strong mixing properties, chaotic. This is done in particular when the system is measure-theoretically isomorphic to a Bernoulli shift, that is the full (left-) shift transformation on the space of infinite sequences with finite alphabets equipped with the product measure of the equidistribution on the alphabet. Shift systems also arise in the ergodic-theoretical description of stationary processes and the Bernoulli shift in particular for i.i.d. processes. The latter provide the standard examples for randomness like throwing dices or flipping coins.

Chaos is an inner property of dynamical systems, causing the unpredictability of the long-term behaviour of the system. This is particularly crucial in the presence of noise – and when can the presence of noise really be excluded? In this paper we do not only want to rely on the observation that an external stochastic influence brings out and underlines the stochastic properties of a chaotic system, but we want to describe rigorously the effect of the external influence on the system. We want to do it in a rather expressive way in a general setup, the theory of random dynamical systems. The latter present a useful generalization of dynamical systems by allowing the dependence on an additional parameter evolving in time and describing a stochastic influence.

Chaotic properties could be proved, either directly or via a conjugation to a symbolic dynamical system. In the case of random dynamical systems the first way is not very expressive, but was already executed under quite weak conditions on the hyperbolicity of the system (see Kifer (1991) or Khanin & Kifer (1994)). In Kifer (1991) it was also stated that the common random shift, also called the Ledrappier-shift, see Ledrappier (1974), is not suited to serve as symbolic dynamics for random dynamical systems. A more intrinsic shift system was introduced by Bogenschütz & Gundlach (1992) to describe the dynamics of random expanding maps. The use of such random subshifts of finite type was extended by Gundlach (1995) and by Arnoux & Fisher (1995) to the description of random homoclinic orbits and random Anosov diffeomorphisms, respectively. Moreover, a transfer operator method for random subshifts of finite type was developed by Bogenschütz & Gundlach (1995).

While the approach of Kifer works quite well, under weak hyperbolicity conditions, for random systems, where the stochastic influence is modelled by i.i.d. stochastic processes, the approach of Bogenschütz and Gundlach has the advantage of illustrating a real time picture of the dynamics of random dynamical systems exhibiting some uniform hyperbolicity conditions, and being driven by some less restrictive noise. In particular a benefit of this procedure is the possibility of investigating fractal structures, see Kifer (1994b).

In this paper we are extending some of the previous results on random subshifts of finite type, and trying to reconcile the approach of Kifer with that of Bogenschütz and Gundlach. Namely, we extend the class of hyperbolic random dynamical systems for which random symbolics can be constructed after some random scaling of time, and we show how random subshifts of finite type are isomorphic to common Bernoulli shifts.

The paper is organized as follows. In § 2 we give a short introduction to the theory of random dynamical systems. In particular we present the notion of a bundle random dynamical system, which is crucial for the definition of random subshifts of finite type, the main topic of § 3. Then we move on to give, in § 4, a review of smooth random dynamical systems that can be conjugated to random subshifts of finite type. We also consider an even wider class of systems that can be related to symbolic dynamics via a change of the system causing the external randomness. In the final section we present an extension of the transfer operator method for random subshifts of finite type, in order to equip the random shifts with suitable invariant measures, i.e. the Gibbs measures and the equilibrium states. We prove some new properties of these measures connected with the chaotic features of the systems, in particular we prove a strong mixing property and the crucial results for a relativized isomorphism theory for Bernoulli shifts and the natural extension from one- to two-sided random shifts.

2. Random dynamical systems

Random dynamical systems (RDS) form a special class of non-autonomous systems. The autonomy is violated in a stationary way, the dependence on time being explicitly given via a stationary process, the standard example of which being the Wiener process. In the case of a continuous time, random dynamical systems can be generated by random or stochastic differential equations, see Arnold (1995). Here we only want to consider the case of a discrete time, where RDS correspond to products of random transformations.

In order to give a rigorous definition we have to start with the law for the choice of random transformations. For this purpose we consider an abstract dynamical system $(\Omega, \mathcal{F}, \mathsf{P}, \vartheta)$, where $(\Omega, \mathcal{F}, \mathsf{P})$ is a probability space and ϑ is a measure-preserving transformation on that space (see also Arnold & Crauel (1991)). We will assume that ϑ is invertible and ergodic. In the case of Ω being a Lebesgue space these two properties may be assumed without loss of generality, the first one due to the existence of natural extensions (see Cornfeld *et al.* (1982), p. 239), the second one due to the decomposition into ergodic components (see Cornfeld *et al.* (1982), Chapter 1, § 3). We think of $(\Omega, \mathcal{F}, \mathsf{P}, \vartheta)$ being a model system for some stochastic influence like noise. Thus the standard example for such an abstract dynamical system in our context is provided by the canonical dynamical system for the discrete version of the Wiener process.

In the following we will denote by T the group or semigroup of time parameters, and assume that $T = \mathsf{Z}$ or N. Then we can give the following definition.

DEFINITION 2.1. *A (measurable) random dynamical system (RDS) on a measurable space* $(X, \mathcal{B})$ *over* $(\Omega, \mathcal{F}, \mathsf{P}, \vartheta)$, *for the time set* T, *is a mapping*

$$\varphi : T \times \Omega \times X \to X, \quad (n, \omega, x) \mapsto \varphi(n, \omega)x$$

satisfying

$$\varphi(0,\omega) = \mathrm{id} \quad \textit{for all } \omega \in \Omega \tag{2.1}$$

and the cocycle property

$$\varphi(n+m,\omega) = \varphi(n,\vartheta^m\omega) \circ \varphi(m,\omega) \quad \textit{for all } n,m \in T,\ \omega \in \Omega, \tag{2.2}$$

such that for all $n \in T$ the mapping $\varphi(n,.) : \Omega \times X \to X$ is $\mathcal{F} \otimes \mathcal{B}, \mathcal{B}$ measurable. If X is equipped with some topology and the corresponding Borel-σ-algebra, we call the RDS continuous, if for all $\omega \in \Omega$ the mapping $\varphi(.,\omega) : T \times X \to X$ is continuous. If furthermore X is equipped with some differentiable structure (manifold), then we call the RDS smooth, if for all $n \in T$ and all $\omega \in \Omega$ the mapping $\varphi(n,\omega) : X \to X$ is smooth.

Note that in the case of $T = \mathsf{Z}$, conditions (2.1) and (2.2) imply the invertiblity of $\varphi(n,\omega)$ with

$$\varphi(n,\omega)^{-1} = \varphi(-n,\vartheta^n\omega). \tag{2.3}$$

If furthermore the RDS is smooth, i.e. at least C^1, each mapping $\varphi(n,\omega)$ is C^1. It follows from (2.3) that all these mappings are diffeomorphisms. Another consequence of the cocycle property of φ is the fact that $\varphi(\omega) := \varphi(1,\omega)$ for all $\omega \in \Omega$ generates the RDS such that

$$\varphi(n,\omega) = \begin{cases} \varphi(\vartheta^{n-1}\omega)\ldots\varphi(\vartheta\omega)\varphi(\omega) & \text{for } n \geq 1, \\ \mathrm{id} & \text{for } n = 0, \\ \varphi(\vartheta^n\omega)^{-1}\ldots\varphi(\vartheta^{-1}\omega)^{-1} & \text{for } n \leq -1, \text{ if } T = \mathsf{Z}. \end{cases} \tag{2.4}$$

So the RDS is indeed given by products of random maps and, in the case of smooth RDS with $T = \mathsf{Z}$, by products of random diffeomorphisms.

In connection to the random dynamical system φ we can introduce a skew-product transformation $\Theta : \Omega \times X \to \Omega \times X$ by

$$\Theta : (\omega, x) \mapsto (\vartheta\omega, \varphi(\omega)x).$$

We can regard Θ as a bundle map on the trivial bundle $\Omega \times X$ (equipped with the projection onto Ω) and can draw the following picture:

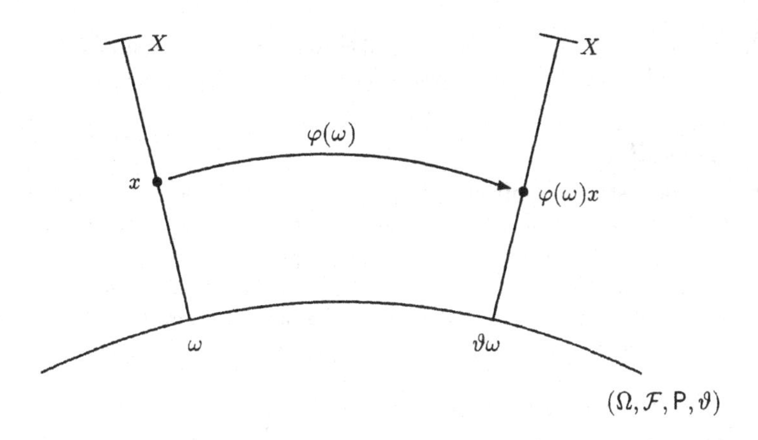

Also we would like to point out that it is sufficient to require the properties (2.1) and (2.2) to hold only on a subset $\tilde{\Omega} \subset \Omega$ of full P-measure. Nevertheless, without loss of generality, we assume here (and will assume throughout this article) that all properties hold for all and not just P-almost all $\omega \in \Omega$.

To study the ergodic properties of random dynamical systems, it is necessary to introduce a suitable invariant measure. As the abstract dynamical system plays an important role in the definition of the RDS, while its measure P is not at our disposal, the following definition is accepted as a natural one.

DEFINITION 2.2. *A probability measure μ on $(\Omega \times X, \mathcal{F} \otimes \mathcal{B})$ with marginal* P *on Ω is called φ-invariant, if it is invariant under the induced skew-product transformation, i.e. if μ satisfies $\Theta\mu = \mu$ and $\mathrm{pr}_\Omega \mu = \mathsf{P}$, where pr_Ω denotes the projection from $\Omega \times X$ onto X.*

It is well known that this kind of invariance can be described in terms of the disintegration $\{\mu_\omega\}$ with respect to P, i.e. $d\mu(\omega, x) = d\mu_\omega(x) d\mathsf{P}(\omega)$, as $\varphi(\omega)\mu_\omega = \mu_{\vartheta\omega}$. This is the reason why the invariance property is also called *equivariance*. In this context it is of importance that ϑ is invertible.

If the support of these disintegrations is non-atomic and μ_ω exhibits strong mixing properties, the behaviour of the system is rather complicated. For deterministic systems (where μ_ω has to be independent of ω) this phenomenon is called *chaos*. Of particular interest are those systems that are conjugated to shifts, especially the Bernoulli-shift. We will be confronted with such cases throughout this article.

The notion of a random dynamical system can be generalized, if the idea of a trivial bundle $\Omega \times X$ is abandoned in favour of a non-trivial subbundle E of $\Omega \times X$ such that even more randomness might be introduced via the phase space to the RDS φ. In the next section we will get to know a situation where this extension is crucial. We will consider here only the special case where (X, d) is a compact metric space with corresponding Borel-σ-algebra $\mathcal{B}$ and where $\mathcal{F}$ is countably generated and P-complete. This condition on $\mathcal{F}$ is sufficient to define a so-called random compact bundle over Ω, which is a set $E \in \mathcal{F} \otimes \mathcal{B}$ with closed sets $E_\omega := \{x \in X : (\omega, x) \in E\} \subset X$, called fibres, such that the distance function $(\omega, x) \mapsto d(x, E_\omega)$ is $\mathcal{F} \otimes \mathcal{B}$, $\mathcal{B}(\mathsf{R}^+)$ measurable for all $x \in X$.

DEFINITION 2.3. *Let E be a random compact bundle over Ω. Then a bundle random dynamical system on E is a family $\varphi := \{\varphi(n, \omega) : E_\omega \to E_{\vartheta^n\omega}\}$ of continuous mappings such that $(\omega, x) \mapsto \varphi(n, \omega)x$ maps E measurably to X and conditions* (2.1) *and* (2.2) *are satisfied.*

All the properties and notions, in particular the invariant measures take over from common RDS to bundle RDS, see Bogenschütz (1993).

3. Random shifts

The first random shift was introduced by Ledrappier (1974) as a relativized version of a standard (left-) shift σ defined on the sequence space $\mathcal{S}_k^+ = \{1, \ldots, k\}^{\mathsf{N}}$. Just consider an abstract dynamical system $(\Omega, \mathcal{F}, \mathsf{P}, \vartheta)$ and the product $(\Omega \times \mathcal{S}_k^+, \mathcal{F} \otimes \mathcal{B}^+)$ together with the product mapping $\Theta = \vartheta \times \sigma$, where $\mathcal{B}^+$ denotes the Borel-σ-algebra with respect to the product of the discrete topology on $\mathcal{S}_k^+$. We can equip now $(\Omega \times \mathcal{S}_k^+, \mathcal{F} \otimes \mathcal{B}^+, \Theta)$ with a σ-invariant measure in the sense of the last section. If this invariant measure is nontrivial, i.e. is not a product measure, so that the disintegrations are not constant, this RDS is called a random (Ledrappier) shift.

In the Ledrappier shift, all the information about the stochastic influence is carried by the random measures. The shift mapping itself is of a deterministic nature and this fact cannot be changed. Nevertheless, there is a chance to leave some additional noise-induced informations in the shift space. This can be done, as proposed by Bogenschütz and Gundlach (see for example Bogenschütz & Gundlach (1995)), by turning Θ into a bundle map on a non-trivial subbundle in the product of Ω with a sequence space X. To ensure compactness it was suggested to choose $X = \bar{\mathsf{N}}^{\mathsf{N}}$ where $\bar{\mathsf{N}}$ is the one point compactification of N and X is endowed with the product of the discrete topology on $\bar{\mathsf{N}}$. Thus X is compact, and in fact it can even be made a metric space. Moreover, the (left-) shift σ on X defined by $(\sigma x)_n = x_{n+1}$ for $x = (x_i)$, $i \in \mathsf{Z}$, is continuous.

For an N-valued random variable k we put

$$\Sigma_k^+(\omega) := \{x \in X : x_i \leq k(\vartheta^i\omega) \text{ for all } i \in \mathsf{N}\} = \prod_{i=0}^{\infty} \{0, \dots, k(\vartheta^i\omega)\}. \tag{3.5}$$

It was shown by Bogenschütz & Gundlach (1995) that $\{\sigma : \Sigma_k^+(\omega) \to \Sigma_k^+(\vartheta\omega)\}$ defines a *bundle random dynamical system* on the compact bundle $\Sigma_k^+ = \{(\omega, x) : x \in \Sigma_k^+(\omega)\}$ over Ω.

DEFINITION 3.1. *Let* $A = \{A(\omega) = (a_{ij}(\omega)) \in \{0,1\}^{(k(\omega)+1)\times k(\vartheta\omega)+1)} : \omega \in \Omega\}$ *be a random transition matrix such that* $\omega \mapsto a_{ij}(\omega)$ *is measurable for all* $(i,j) \in \mathsf{N} \times \mathsf{N}$ *and each* $A(\omega)$ *has at least one non-zero entry in each row and each column. Then the bundle random dynamical system determined by the family* $\{\sigma : \Sigma_A^+(\omega) \to \Sigma_A^+(\vartheta\omega)\}$ *where*

$$\Sigma_A^+(\omega) := \{x \in \Sigma_k^+(\omega) : a_{x_i, x_{i+1}}(\vartheta^i\omega) = 1 \text{ for all } i \in \mathsf{N}\} \tag{3.6}$$

is called a random subshift of finite type. For the choice $a_{ij}(\omega) \equiv 1$ *we obtain the random k-shift* $\{\sigma : \Sigma_k^+(\omega) \to \Sigma_k^+(\vartheta\omega)\}$.

One of the main features of these random shifts is that they act between different sequence spaces such that we can draw the following picture:

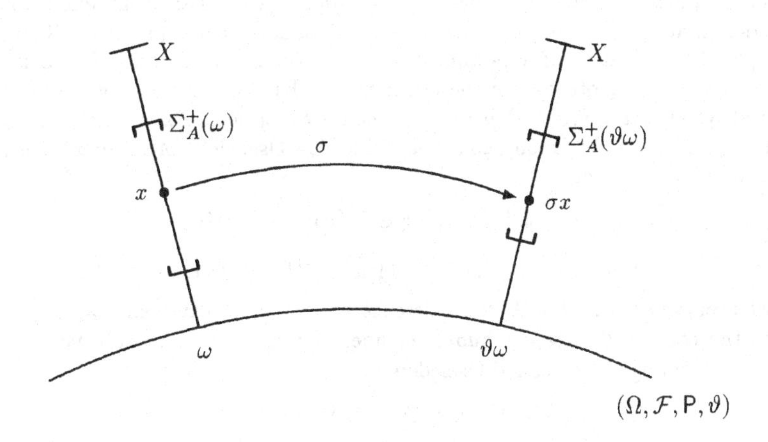

The analogous definitions are available for the two-sided random k-shift on Σ_k and the two-sided random subshift of finite type on Σ_A, which are obtained by replacing N in the definition of X and in (3.5) and (3.6), respectively, by Z, see Steinkamp (1995).

There exist a few natural candidates for invariant measures for random shifts, namely random Markov measures determinded by a family $P = \{P(\omega)\}$ of stochastic matrices and a family $p = \{p(\omega)\}$ of stationary probability vectors yielding the random (p, P)-Markov-shift (see Bogenschütz (1993)), and random Bernoulli measures for random k-shifts. The latter are determined on finite cylinders $C_\omega(i_0, \dots, i_n) = \{x \in \Sigma_A^+(\omega) : x_0 =$

$i_0, \ldots, x_n = i_n\}$ for $i_j \in \mathbb{N}$, $i_j \leq k(\vartheta^j\omega)$ by

$$\mu_\omega(C_\omega(i_0, \ldots, i_n)) = \prod_{j=0}^{n} p_{i_j}(\vartheta^j\omega)$$

for random probability vectors $\{p(\omega) = (p_1(\omega), \ldots, p_{k(\omega)}(\omega))\}$, in particular with $p_l(\omega) = 1/k(\omega)$ for all l.

While the standard Bernoulli shift is the ergodic-theoretical description of the event that a dice is thrown infinitely often, the random Bernoulli shift corresponds to the following scenario. We have got a reservoire of dices characterized by different numbers of sides k. We choose a die at a random time corresponding to the distribution $k\mathbb{P}$ and then we throw this die. Note that the drawing of the dice in this scenario is not necessarily an i.i.d. process. Anyway, if this procedure is executed infinitely often, the outcome is described by our random Bernoulli shift. In fact the stochastic process just described consists of two processes (which obviously are not independent of each other), but only the outcome of the second one is represented explicitly by the Bernoulli shift.

Although the random k-shifts and random subshifts of finite type might be regarded as interesting by themselves, their importance arises through their conjugations to other random dynamical systems. In the next section we will present several situations in which they arise.

4. Random symbolic dynamics

Chaotic properties are typical for dynamical systems with hyperbolic features. The basis for hyperbolicity, a property connected with the linearization of the system, in the theory of smooth random dynamical systems φ on a manifold X is provided by the famous Multiplicative Ergodic Theorem of Oseledets (1968). Let us assume throughout this section that X is a compact, connected and smooth manifold with a Riemannian metric $\|.\|$, the derivative of φ is almost surely non-degenerate and μ is a φ-invariant measure. Then we denote by $T\varphi$ the linear bundle RDS on $\Omega \times TX$ over the abstract dynamical system $(\Omega \times X, \mathcal{F} \otimes \mathcal{B}, \mu, \Theta)$ with the fibre mappings $T\varphi(n, \omega, x) : (\omega, T_xX) \to (\vartheta^n\omega, T_{\varphi(n,\omega)x}X)$, Arnold (1995), equation (3.8). Now Oseledets' MET guarantees, under the integrability conditions

$$\log^+ \|T\varphi(1, \omega, x)\| \in L^1(\mu) \quad \text{if } T = \mathbb{N},$$

$$\log^+ \|T\varphi(1, \omega, x)\| + \log^+ \|T\varphi(1, \omega, x)^{-1}\| \in L^1(\mu) \quad \text{if } T = \mathbb{Z},$$

μ-almost everywhere on $\Omega \times X$ the existence of Lyapunov exponents $\lambda_1 > \ldots > \lambda_d$ and – in the case of $T = \mathbb{N}$ – a bundle of linear forward invariant subspaces $V_i(\omega, x)$, $i = 1, \ldots, d$, forming the so-called Oseledets flag

$$T_xX = V_1(\omega, x) \supset \ldots \supset V_d(\omega, x) \supset \{0\}$$

such that

$$\lim_{n\to\infty} \frac{1}{n} \log \|T\varphi(n, \omega, x)v\| = \lambda_i \Leftrightarrow v \in V_i(\omega, x) \setminus V_{i+1}(\omega, x),$$

and – in the case of $T = \mathbb{Z}$ – an invariant splitting, called the Oseledets splitting,

$$T_xX = E_1(\omega, x) \oplus \ldots \oplus E_d(\omega, x)$$

such that

$$\lim_{n\to\pm\infty} \frac{1}{n} \log \|T\varphi(n, \omega, x)v\| = \lambda_i \Leftrightarrow v \in E_i(\omega, x).$$

Hyperbolicity could now be defined in terms of non-vanishing Lyapunov exponents. Such a procedure would require the existence and knowledge of the invariant measure μ. Therefore we define hyperbolicity directly in terms of the asymptotic behaviour rather than via Lyapunov exponents.

DEFINITION 4.1. *Let φ be invertible and $\Lambda \subset \Omega \times X$ be a φ-invariant random compact bundle. This set is called hyperbolic, if there exists a constant $\lambda > 1$, a random variable $c > 0$ and for all $(\omega, x) \in \Lambda$ a splitting*

$$T_x X = E^s(\omega, x) \otimes E^u(\omega, x)$$

which depends measurably on (ω, x) and for every fixed $\omega \in \Omega$ continuously on $x \in \Lambda_\omega$, is invariant under $T\varphi$ in the sense that for all $n > 0$

$$T\varphi(n, \omega, x) E^{s/u}(\omega, x) = E^{s/u}(\Theta^n(\omega, x)),$$

and satisfies

$$\|T\varphi(n, \omega, x)|_{E^u(\omega,x)}\| \geq c(\omega)\lambda^n, \quad \|T\varphi(n, \omega, x)|_{E^s(\omega,x)}\| \leq c(\omega)\lambda^{-n}$$

for all $n \geq 0$; Λ is called uniformly hyperbolic if c is constant, and it is called strictly uniformly hyperbolic if $c \equiv 1$.

This definition requires a comment. First note that the condition of the continuity of the splitting for every fixed ω is indeed a necessary condition, unlike the case of a deterministic system, where it is fulfilled automatically. Furthermore, it is not necessary to distinguish between the uniform and strictly uniform hyperbolicity, as the norm can be so adapted as to get $c \equiv 1$. A similar technique can be used in the random case, see Arnold (1995) or Gundlach (1995), but only on the expense of a so-called random or Lyapunov norm $\|.\|_{(\omega,x)}$, depending measurably on (ω, x). Such a norm was first introduced by Pesin (1976) for non-uniformly hyperbolic systems.

The case of the uniform hyperbolicity was extensively investigated by Gundlach (1995) for the case of manifolds embedded in R^n, and later by Arnoux & Fisher (1995) without working directly in the Euclidean space. In both cases, under certain restrictive conditions, a stable manifold theory for random dynamical systems is developped as well as a shadowing lemma and a theory of Markov partitions. We do not want to present this extensive machinery and the restrictions in detail here, but rather focus on a short review of the results on symbolic dynamics.

The simplest example is provided by expanding the RDS in which φ is non-invertible, and the unstable subbundle E^u coincides with the tangent bundle. If φ is uniformly expanding, there exists a random norm $\|.\|_\omega$ and random variables $K > 1$ and $c > 1$ such that

$$\|\xi\| \leq \|\xi\|_\omega \leq K(\omega)\|\xi\|, \quad \|T\varphi(1, \omega, x)\xi\|_{\vartheta\omega} \geq c(\omega)\|\xi\|_\omega.$$

In Bogenschütz & Gundlach (1992) the following result was proved.

THEOREM 4.1. *If c varies slowly so that the random variables $\varepsilon > 0$ satisfying*

$$\varepsilon(\omega)c(\omega) > \varepsilon(\vartheta\omega)$$

can be found, then φ is conjugate to a random subshift of finite type (Σ_A^+, σ), i.e. there exist homeomorphisms $h(\omega)$ from open and dense sets $L(\omega)$ in X to Σ_A^+ such that $h(\vartheta\omega)\varphi(\omega) = \sigma h(\omega)$.

The assumption of theorem 4.1 is satisfied if e.g. the support of $\varphi\mathsf{P}$ is compact. It is also fulfilled in the simplest examples like the following one, correspoding to random base expansions.

Example 1. Let X be the unit circle, and consider the mapping $(\omega, z) \mapsto (\vartheta\omega, z^{n(\omega)})$ on $\Omega \times X$ for a random variable $n \geq 2$ taking values from N. For this example theorem 4.1 is applicable.

As pointed out by Kifer, there exist a lot of 'expanding in average' RDS, where this result cannot be used. Let us show now a way how to obtain symbolic dynamics also in the case where the expansion is not uniform. In this case define a new random variable M such that $c(\omega)\lambda^{M(\omega)} > 1$ and then consider, for $k \in \mathsf{N}$, the set $D_k := \{\omega \in \Omega : M(\omega) \leq k\}$. If k is sufficiently large, we obtain $\mathsf{P}(D_k) > 0$, and we can define an induced transformation $\hat{\vartheta}_k$ on D_k, which still is ergodic, see Cornfeld *et al.* (1982). Corresponding to this induced transformation the resulting random dynamical system $\hat{\varphi}_k$ is expanding uniformly, and the construction of a conjugation to a standard shift with finite number of elements is a trivial particular case of the general procedure for expanding RDS.

PROPOSITION 4.1. *If φ is an expanding RDS and k is sufficiently large so that the corresponding set $D_k \subset \Omega$ satisfies $\mathsf{P}(D_k) > 0$, then $\hat{\varphi}_k$ is on D_k conjugate to a subshift of finite type.*

Note that for the investigation of chaotic properties of φ it is of no importance that the conjugation exists only on D_k and for $\hat{\varphi}_k$, as D_k, $k \in \mathsf{N}$, is exhaustive for Ω, and the time scaling by the transition to the induced transformation is not relevant for mixing properties of the system. Nevertheless, it has to be said that such symbolic dynamics does not present a real picture of the dynamics of the RDS, as much of the information produced by the random source is now hidden in the induced transformation, i.e. in the new time scale.

Let us move on now to the case of invertible RDS with hyperbolic sets. Here we can do the same construction as above for the non-uniform situation to obtain conjugations to standard shifts over induced transformations. Therefore from now on we want to consider only the case that the hyperbolicity is uniform. Here only partial results exist. The first one we want to present is concerned with random Anosov systems and is due to Arnoux & Fisher (1995). These results are obtained for $\Lambda \equiv \Omega \times X$ in the definition 4.1. A particular example of a uniformly random Anosov system is provided by small random perturbations of Anosov diffeomorphisms. It was shown by Arnoux & Fisher (1995) how to construct a conjugation to a random subshift of finite type (Σ_A, σ). We want to present here another example from that paper which can be regarded as an extension of example 1.

Example 2. Let X be the 2-torus. Assume that the random variable $n \geq 1$ taking values from N is given. Then consider linear random mappings $\tilde{\varphi}_i(\omega) : \mathsf{R}^2 \to \mathsf{R}^2$, $i = 1, 2$, defined by

$$\tilde{\varphi}_1(\omega) = \begin{bmatrix} 1 & n(\omega) \\ 0 & 1 \end{bmatrix}, \quad \tilde{\varphi}_2(\omega) = \begin{bmatrix} 1 & 0 \\ n(\omega) & 1 \end{bmatrix},$$

as lifts of maps φ_i on X. Then a RDS φ on X over $(\Omega, \theta, \rho, \vartheta^2)$ is generated by $\varphi_2(\vartheta\omega) \circ \varphi_1(\omega)$, which was shown by Arnoux & Fisher (1995) to be uniformly random Anosov and conjugate to a random subshift of finite type (Σ_A, σ).

The last example we want to present is concerned with random homoclinic points for smooth invertible RDS on $X = \mathsf{R}^d$. We consider stationary points $x : \Omega \to \mathsf{R}^d$ for φ, i.e. random variables x which satisfy $\varphi(n, \omega)x(\omega) = x(\vartheta^n\omega)$ for all n and ω. Without loss of generality we can take $x \equiv 0$. Then we assume that 0 is hyperbolic in the sense that no Lyapunov exponent corresponding to the random Dirac measure $\mathsf{P} \times \delta_0$ vanishes. If the fibred stable and unstable manifolds of 0 for φ, which due to the random nature of φ

also depend on ω, intersect transversally for all ω, then we call the random intersections $y_0 : \Omega \to \mathbb{R}^d$ random homoclinic points. Iterating these random homoclinic points y_n leads to a random hyperbolic set Λ with $y_n(\omega) = \varphi(n,\omega)y_0(\omega) \in \Lambda_{\vartheta^n\omega}$, if $\log \|y_0\|$ is integrable with respect to P. On this random hyperbolic set a shadowing lemma can be proved, if the derivative of φ satisfies an integrability condition on a small random neighbourhood of Λ. In particular the following result holds, see Gundlach (1995).

THEOREM 4.2 (RANDOM BIRKHOFF-SMALE THEOREM). *Let y_0 be a transversal random homoclinic point with respect to the hyperbolic fixed point 0 of the C^1 RDS φ on $\mathbb{R}^d$ with P-integrable $\|T\varphi(1,.,x)\|$ for all $x \in \mathbb{R}^d$. Then for each $n \in \mathbb{Z}^+$ there exist*

(i) a $\varphi(n,.)$-invariant random compact set $S^{(n)}$,

(ii) a random variable $k_n : \Omega \to \mathbb{N}$,

(iii) a measurable mapping $h_n : \Sigma_{k_n} \to S^{(n)}$ such that each $h_n(\omega) := h_n|_{\Sigma_{k_n}(\omega)} : \Sigma_{k_n}(\omega) \to S^{(n)}_\omega$ is a homeomorphism satisfying, for all $\omega \in \Omega$,

$$\varphi(n,\omega) \circ h_n(\omega) = h_n(\vartheta^n\omega) \circ \sigma \quad \text{on } \Sigma_{k_n}(\omega).$$

The following interesting cases can occur.

(i) If n is sufficiently small, then $S^{(n)}$ is trivial, i.e. $S^{(n)}_\omega = \{0\}$ for all $\omega \in \Omega$.

(ii) If n is sufficiently large, then $S^{(n)}$ is nontrivial, i.e. for all ω out of a set of positive measure there exists $m = m(\omega) \in \mathbb{Z}$ with $\{0, y_m(\vartheta^{-m}\omega)\} \subset S^{(n)}_\omega$ and $k_n(\omega) \geq 1$.

(iii) If φ and its hyperbolic set have compact support we can construct an $N \in \mathbb{N}$ such that $k_N \geq 1$ and $S^{(N)}$ is non-trivial with $\Lambda \subset S^{(N)}$.

Examples to which this theorem can be applied are provided by small random perturbations of systems with homoclinic points.

5. Chaos in smooth RDS

In the last section we have constructed conjugations from RDS to random shifts in a topological set-up. Now we face the problem to construct an appropriate invariant measure for these shifts and, due to the conjugation, also for the original RDS. A standard technique for constructing natural invariant measures, in particular SRB-measures, was carried over to random dynamical systems by Bogenschütz & Gundlach (1995). There a transfer operator theorem was proved under some rather strong primitivity condition on the random transition matrix defining the random subshift of finite type. Following a suggestion by Kifer (1994a) we are going to present now an extension of this result.

For that purpose let us denote by $\mathcal{C}(\Sigma_A^+(\omega))$ the space of all continuous functions on $\Sigma_A^+(\omega)$ and by $\mathsf{L}_A^0(\Omega, \mathcal{C}(X))$ and $\mathsf{L}_A^1(\Omega, \mathcal{C}(X))$ the spaces of random and integrable random variables adopting values in $\mathcal{C}(\Sigma_A^+(\omega))$, $\omega \in \Omega$. Integrability has to be understood here as corresponding to the sup-norm of the random function on fibres. The structure of these two spaces is rather intricate and we refer the interested reader for details to Bogenschütz (1993). We call the members of these spaces *random functions*. Of special interest will be the functions satisfying certain Hölder conditions. Let us denote for a function $\phi \in \mathcal{C}(\Sigma_A^+(\omega))$

$$\mathrm{var}_n\phi := \sup\left\{|\phi(x) - \phi(y)| : x, y \in \Sigma_A^+(\omega) \text{ with } x \overset{n}{\sim} y\right\},$$

where $x \overset{n}{\sim} y :\Leftrightarrow x_i = y_i$ for $0 \leq i \leq n-1$. If there exist positive real numbers b and $\alpha \in (0,1)$ such that for all $\omega \in \Omega$ the random function $\phi \in \mathsf{L}_A^0(\Omega, \mathcal{C}(X))$ satisfies

$$\mathrm{var}_n\phi(\omega) \leq b\alpha^n \quad \text{for all } n \in \mathbb{N}, \tag{5.7}$$

then we call ϕ equi-Hölder continuous, and denote the family of all such random functions in $\mathsf{L}^0_A(\Omega, \mathcal{C}(X))$ by F_A. Though this might look like a rather strong restriction, it was shown in Meyer (1995) that $\mathsf{F}^1_A = \mathsf{F}_A \cap \mathsf{L}^1_A(\Omega, \mathcal{C}(X))$ is dense in $\mathsf{L}^1_A(\Omega, \mathcal{C}(X))$.

Let us fix now a random function $\phi \in \mathsf{L}^0_A(\Omega, \mathcal{C}(X))$. Then we introduce, for $f \in \mathcal{C}(\Sigma^+_A(\omega))$,

$$(\mathcal{L}_\phi(\omega)f)(x) = \sum_{y \in \Sigma^+_A(\omega): \sigma y = x} \mathrm{e}^{\phi(\omega,y)} f(y),$$

where $x \in \Sigma^+_A(\vartheta\omega)$. Thus we have defined a mapping $\mathcal{L}_\phi(\omega)$ from $\mathcal{C}(\Sigma^+_A(\omega))$ to $\mathcal{C}(\Sigma^+_A(\vartheta\omega))$. We call the operators $\mathcal{L}_\phi(\omega) : \mathcal{C}(\Sigma^+_A(\omega)) \to \mathcal{C}(\Sigma^+_A(\vartheta\omega))$ *random transfer operators* associated with ϕ. They are positive, linear and continuous operators, as $\|\mathcal{L}_\phi(\omega)f\|_\infty \leq \|\mathcal{L}_\phi(\omega)1\|_\infty \|f\|_\infty$.

Let us denote by $\mathcal{M}(\Sigma^+_A(\omega))$ the space of finite signed Borel measures on $\Sigma^+_A(\omega)$ and by $\mathcal{L}^*_\phi(\omega)$ the dual of $\mathcal{L}_\phi(\omega)$. Then $\mathcal{L}^*_\phi(\omega) : \mathcal{M}(\Sigma^+_A(\vartheta\omega)) \to \mathcal{M}(\Sigma^+_A(\omega))$, i.e. for $\rho \in \mathcal{M}(\Sigma^+_A(\vartheta\omega))$, $\mathcal{L}^*_\phi(\omega)\rho$ is determined by

$$\int f \mathrm{d}\mathcal{L}^*_\phi(\omega)\rho = \int \mathcal{L}_\phi(\omega) f \mathrm{d}\rho \quad \text{for all } f \in \mathcal{C}(\Sigma^+_A(\omega)).$$

For the transfer operator theorem we need a condition concerning aperiodicity of the random transition matrix A.

DEFINITION 5.1. *The random transition matrix A is called aperiodic, if there exists a random variable M adopting values in N such that the $k(\omega) \times k(\vartheta^{M(\omega)}\omega)$-matrix*

$$A(\omega)A(\vartheta\omega)\ldots A(\vartheta^{M(\omega)-1}\omega)$$

has nonzero elements for all $\omega \in \Omega$; A is called uniformly aperiodic, if M is constant.

In Bogenschütz & Gundlach (1995) a transfer operator theorem was proved under the uniform aperiodicity condition satisfied by A. Let us give now an extension of this result for aperiodic random transition matrices. This can be done with the help of two lemmas which we are going to present now. These results are concerned with the denseness of orbits in the random shift space (suggested by Kifer (1994a)) as well as with the asymptotic properties of the transfer operators, and make use of induced transformations. For $k \in \mathsf{N}$ let us introduce the sets

$$D_k := \{\omega : M(\omega) \leq k\}.$$

If k is sufficiently large, then $\mathsf{P}(D_k) > 0$ and we can define a return time for D_k in the form

$$t_k(\omega) = k + r_k(\omega), \quad r_k(\omega) > 0.$$

Moreover, we can define an induced transformation $\hat{\vartheta}_k : D_k \to D_k$, which preserves the P-induced measure on D_k, such that the resulting abstract dynamical system is ergodic and the random variables t_k and r_k are integrable.

LEMMA 5.1. *For any $\varepsilon > 0$ and P-almost all $\omega \in \Omega$ there exists $N(\varepsilon, \omega)$ such that for every $x \in \Sigma^+_A(\omega)$ and every $n > N(\varepsilon, \omega)$ the set $\sigma^{-n}x$ is ε-dense in $\Sigma^+_A(\vartheta^{-n}\omega)$.*

Proof. Choose $k \in \mathsf{N}$ so large that $\mathsf{P}(D_k) > 0$ and

$$p(\omega) = \min\{\ell \geq k : M(\vartheta^{-\ell}\omega) \leq k\}.$$

Then for any $n \in \mathsf{N}$ in the form $n = j + p(\omega)$, $j \in \mathsf{N}$, and any given $y \in \Sigma^+_A(\vartheta^{-n}\omega)$, $x \in \Sigma^+_A(\omega)$ there exists some $z \in \Sigma^+_A(\vartheta^{-n}\omega)$ with $z_i = y_i$, $i = 0, 1, \ldots, j$ and $z_l = x_{l-n}$

for $l \geq n+1$. As with increasing j the natural distance (the shift space is metrizable) between z and y decreases exponentially, the assertion is proved. □

In order to study the asymptotic behaviour of functions under the action of transfer operators let us denote, for $n > 0$, the operator cocycle $\mathcal{L}_\phi(\vartheta^{n-1}\omega) \circ \ldots \circ \mathcal{L}_\phi(\omega)$ by $\mathcal{L}_\phi(n,\omega)$, and define the induced one by $\hat{\mathcal{L}}_{\phi,k}(n,\omega) = \hat{\mathcal{L}}_{\phi,k}(\hat{\vartheta}_k^{n-1}\omega) \circ \ldots \circ \hat{\mathcal{L}}_{\phi,k}(\omega)$ and $\hat{\mathcal{L}}_{\phi,k}(\omega) = \mathcal{L}_\phi(t_k(\omega),\omega)$.

LEMMA 5.2. *Let k be sufficiently large so that $\mathsf{P}(D_k) > 0$, and assume that there exists a metric d_ω on $\mathcal{C}(\Sigma_A^+(\omega))$ for all $\omega \in \Omega$ such that for some $\omega \in D_k$*

$$\limsup_{n\to\infty} \frac{1}{n} \log d_{\hat{\vartheta}^n\omega}(\hat{\mathcal{L}}_{\phi,k}(n,\omega)g_1, \hat{\mathcal{L}}_{\phi,k}(n,\omega)g_2) \leq \kappa < 0$$

for $g_1, g_2 \in \mathcal{C}(\Sigma_A^+(\omega))$. Then there exists some $\rho < 0$ such that

$$\limsup_{n\to\infty} \frac{1}{n} \log d_{\vartheta^n\omega}(\mathcal{L}_\phi(n,\omega)g_1, \mathcal{L}_\phi(n,\omega)g_2) \leq \rho < 0.$$

Proof. Let us write

$$n = kq(n,\omega) + \sum_{i=0}^{q(n,\omega)-1} r_k(\hat{\vartheta}^i\omega) + R(n,\omega), \qquad \text{where } R(n,\omega) < k + r_k(\hat{\vartheta}^{q(n,\omega)}\omega).$$

Then

$$\mathcal{L}_\phi(n,\omega) = \mathcal{L}_\phi(R(n,\omega), \hat{\vartheta}^{q(n,\omega)}\omega)\hat{\mathcal{L}}_{\phi,k}(q(n,\omega),\omega)$$

and therefore

$$\begin{aligned} &\limsup_{n\to\infty} \frac{1}{n} \log d_{\vartheta^n\omega}(\mathcal{L}_\phi(n,\omega)g_1, \mathcal{L}_\phi(n,\omega)g_2) \\ &\leq \limsup_{n\to\infty} \frac{q(n,\omega)}{n} \frac{1}{q(n,\omega)} \log d_{\hat{\vartheta}^{q(n,\omega)}\omega}(\hat{\mathcal{L}}_{\phi,k}(q(n,\omega),\omega)g_1, \hat{\mathcal{L}}_{\phi,k}(q(n,\omega),\omega)g_2) \\ &\leq \limsup_{n\to\infty} \frac{1}{k + \frac{1}{q(n,\omega)}\sum_{i=0}^{q(n,\omega)-1} r_k(\hat{\vartheta}^i\omega)}\kappa. \end{aligned}$$

The limit on the right hand side exists due to Birkhoff's Ergodic Theorem. So due to the ergodicity of the induced transformation, Poincaré's Recurrence Theorem and the integrability of r_k the result follows immediately. □

On the basis of lemma 5.1 and lemma 5.2 the following extension of the transfer operator theorem of Bogenschütz & Gundlach (1995) is obvious.

THEOREM 5.1 (RANDOM TRANSFER OPERATOR THEOREM). *Assume that the random transition matrix is aperiodic, $\phi \in \mathsf{F}_A$ and $\log \mathcal{L}_\phi(M(.),.)1 \in \mathsf{L}_A^1(\Omega, \mathcal{C}(X))$, where $\mathcal{L}_\phi(M(.),.)1 := \{\mathcal{L}_\phi(M(\omega),\omega)1\} \in \mathsf{L}_A^0(\Omega,\mathcal{C}(X))$. Then there exist a random variable λ satisfying $\lambda > 0$ and $\log\lambda \in \mathsf{L}^1(\Omega,\mathsf{P})$, $g \in \mathsf{L}_A^0(\Omega,\mathcal{C}(X))$ such that $g > 0$ and $\log g \in \mathsf{F}_A$, $\Phi \in \mathsf{F}_A$ such that*

$$\Phi(\omega,x) = \phi(\omega,x) + \log g(\omega,x) - \log g(\vartheta\omega,\sigma x) - \log\lambda(\omega) \quad \text{for all } (\omega,x) \in \Sigma_A^+,$$

and $\nu,\mu \in \mathcal{M}_\mathsf{P}^1(\Sigma_A)$ such that the following holds P-a.s.

(i) $\mathcal{L}_\phi(\omega)g(\omega) = \lambda(\omega)g(\vartheta\omega)$.

(ii) $\mathcal{L}_\phi^*(\omega)\nu_{\vartheta\omega} = \lambda(\omega)\nu_\omega$.

(iii) $\displaystyle\int g(\omega)\mathrm{d}\nu_\omega = 1.$

(iv) $$\lim_{n\to\infty}\left\|\frac{1}{\lambda(\vartheta^{n-1}\omega)\dots\lambda(\omega)}\mathcal{L}_\phi(n,\omega)f - g(\vartheta^n\omega)\int f d\nu_\omega\right\|_\infty = 0$$

for all $f \in \mathcal{C}(\Sigma_A^+(\omega))$ and with exponential speed of convergence for f in a dense subset Λ_ω of $\mathcal{C}(\Sigma_A^+(\omega))$.

(v) $\mathcal{L}_\Phi^*(\omega)\mu_{\vartheta\omega} = \mu_\omega$.

(vi) $$\int f d\mu_\omega = \int f g(\omega) d\nu_\omega.$$

(vii) $$\lim_{n\to\infty}\left\|\mathcal{L}_\Phi(n,\omega)f - \int f d\mu_\omega\right\|_\infty = 0$$

for all $f \in \mathcal{C}(\Sigma_A^+(\omega))$ and with exponential speed of convergence for f in a dense subset Λ_ω of $\mathcal{C}(\Sigma_A^+(\omega))$.
Here $\|.\|_\infty$ denoted the supremum-norm. The triple (λ, g, ν) is P-a. s. uniquely determined and so are Φ and μ.

This theorem has a lot of consequences. First of all it contains as a special case a random version of the Perron–Frobenius Theorem for products of rectangular non-negative aperiodic matrices, see Gundlach & Steinkamp (1996), which can be used for investigations of random Markov chains in random environments. Furthermore it follows that μ is invariant, and is an equilibrium state for ϕ and Φ, a Gibbs state for Φ and a quasi-Gibbs state for ϕ, see Bogenschütz & Gundlach (1995). These results form a basis for a thermodynamic formalism for random subshifts of finite type, see Meyer (1995), Gundlach (1996) and Khanin & Kifer (1994).

We are going to consider now the consequences for chaotic features of the random shifts equipped with such measures. In the following μ will denote any measure arising from theorem 5.1.

COROLLARY 5.1. *The random shift (σ, μ) is strong-mixing, and the fibrewise partition of Σ_A^+ into 1-cylinders is weak-Bernoulli.*

Proof. Assume that $f \in \Lambda_\omega$ satisfies $\int f d\mu_\omega = 0$. Then

$$\lim_{n\to\infty}\|\mathcal{L}_\Phi(n,\omega)f\|_\infty = 0,$$

where the convergence is exponential. As $g_{\vartheta^n\omega}$, for $g \in \mathsf{L}_A^2(\Omega, \mathcal{C}(X))$, and $\mu_{\vartheta^n\omega}$ are integrable, they define stationary processes with Lyapunov exponents 0, which leads to

$$\lim_{n\to\infty}\int g_{\vartheta^n\omega}\mathcal{L}_\Phi(n,\omega) f d\mu_{\vartheta^n\omega} = 0,$$

with exponential speed of convergence. As on the basis of theorem 5.1 the left hand side can be rewritten as $\int g_{\vartheta^n\omega}\circ\sigma^n \cdot f d\mu_\omega$, we obtain the strong-mixing property for a special choice of f. For more general $f \in \Lambda_\omega$ consider $f - \int f d\mu_\omega$ in order to get back to the first case and obtain

$$\lim_{n\to\infty}\left|\int g_{\vartheta^n\omega}\circ\sigma^n \cdot f d\mu_\omega - \int g_{\vartheta^n\omega} d\mu_{\vartheta^n\omega}\int f d\mu_\omega\right| = 0,$$

with exponential speed of convergence. This has two consequences. Firstly we can use the denseness of Λ to extend this convergence (in general without the exponential speed of convergence) to the case $f \in \mathsf{L}^2(\mu_\omega)$, which finishes the proof of the strong-mixing property. On the other hand we can choose f and $g_{\vartheta^n\omega}$ as characteristic functions for 1-cylinders which are indeed contained in Λ_ω and $\Lambda_{\vartheta^n\omega}$, respectively. This yields the weak-Bernoulli property as follows. For any $k, s, n \in \mathsf{N}$ and any cylinders $C_\omega(i_0, \dots, i_k)$ and $C_{\vartheta^n\omega}(j_k, \dots, j_{k+s})$ we have

$$|\mu_\omega(C_\omega(i_0, \dots, i_k) \cap \sigma^{-n}C_{\vartheta^n\omega}(j_k, \dots, j_{k+s}))$$

$$- \mu_\omega(C_\omega(i_0,\ldots,i_k))\mu_{\vartheta^n\omega}(C_{\vartheta^n\omega}(j_k,\ldots,j_{k+s}))|$$
$$= |\mu_{\vartheta^n\omega}(\chi_{C_{\vartheta^n\omega}(j_k,\ldots,j_{k+s})}\mathcal{L}_\Phi(n,\omega)\chi_{C_\omega(i_0,\ldots,i_k)})$$
$$-\mu_\omega(\chi_{C_\omega(i_0,\ldots,i_k)})\mu_{\vartheta^n\omega}(\chi_{C_{\vartheta^n\omega}(j_k,\ldots,j_{k+s})})|$$
$$\leq \|\mathcal{L}_\Phi(n,\omega)\chi_{C_\omega(i_0,\ldots,i_k)} - \mu_\omega(C_\omega(i_0,\ldots,i_k))\|_\infty \mu_{\vartheta^n\omega}(C_{\vartheta^n\omega}(j_k,\ldots,j_{k+s}))$$
$$= \left\|\mathcal{L}_\Phi(n,\omega)\frac{\chi_{C_\omega(i_0,\ldots,i_k)}}{\mu_\omega(C_\omega(i_0,\ldots,i_k))} - 1\right\|_\infty \mu_\omega(C_\omega(i_0,\ldots,i_k))\mu_{\vartheta^n\omega}(C_{\vartheta^n\omega}(j_k,\ldots,j_{k+s})).$$

Due to the exponential convergence of the first term on the right hand side, as $n \to \infty$, we can conclude that the sum of the terms on the left hand side over all possible k-cylinders $C_\omega(i_0,\ldots,i_k)$ and all possible s-cylinders $C_{\vartheta^n\omega}(j_k,\ldots,j_{k+s})$ decreases and tends to zero exponentially as $n \to \infty$, which yields the weak-Bernoulli property of the partition into cylinder sets. □

Notice that the transfer operator theory developed in this section only works for the one-sided shifts, as the non-invertibility of the shifts is essential in deriving the results. Nevertheless, this theory can be made applicable also to the two-sided shifts via cohomology (see Steinkamp (1995)): equi-Hölder continuous functions on Σ_A are cohomologous to equi-Hölder continuous functions that could be regarded as functions on Σ_A^+ leading to the same equilibrium states. For these functions a measure μ with all the properties mentioned above could be constructed on Σ_A^+, and it remains to transfer this measure back to Σ_A. This could be done with the help of natural extensions. For this procedure the exactness of (σ, μ) is needed, see Steinkamp (1995). This can also be derived from the transfer operator theorem.

COROLLARY 5.2. *(σ, μ) is exact.*

Proof. According to the proof of corollary 5.1 we can assume without loss of generality that the funtions f in Λ_ω satisfy $\int f d\mu_\omega = 0$. If we denote by U_ω the Koopman operator induced by σ, i.e. $U_\omega f = f \circ \sigma$ for $f \in \mathsf{L}^0(\mu_\omega)$, then any $f \in \bigcap_{i=1}^n U_{\vartheta\omega} \circ \ldots \circ U_{\vartheta^i\omega}\Lambda_{\vartheta^i\omega}$ has necessarily the form $f = f_i \circ \sigma^n$ with $f_i \in \mathsf{L}^0(\mu_{\vartheta^i\omega})$ and $\mathcal{L}_\Phi(i,\omega)f = f_i$. Thus for $f \in \bigcap_{n\in\mathsf{N}_+} U_{\vartheta\omega} \circ \ldots \circ U_{\vartheta^i\omega}\Lambda_{\vartheta^i\omega}$ we can deduce for the L^2-norms $\|.\|_2$ on $\mathsf{L}^2(\mu_\omega)$ that (due to the σ-invariance of μ)

$$\|f\|_2 = \|f_n\|_2 = \|\mathcal{L}_\Phi(n,\omega)f\|_2 \to 0 \quad \text{as } n \to \infty,$$

from which we conclude that $f = 0$ μ_ω-a.s. For general $f \in \Lambda_\omega$ it follows that f is constant μ_ω-a.s., and for the special choice of characteristic functions we obtain

$$A \in \bigcap_{n\in\mathsf{N}} \sigma^{-n}\mathcal{B} \Rightarrow \chi_A = 0 \text{ or } 1 \text{ a.s.}$$

Thus $\bigcap_{n\in\mathsf{N}} \sigma^{-n}\mathcal{B}$ is modulo null-sets the trivial σ-algebra, and therefore (σ, μ) is exact. □

The weak-Bernoulli property can be carried over from one-sided shifts and is useful to establish a relative isomorphism theory for random dynamical systems. In an extension of the relative isomorphism theory of Thouvenot (1975), given in the Appendix to Lind (1977), it was shown that the weak-Bernoulli property is sufficient for the relative entropy to be a complete invariant for two-sided shifts. This means that any two two-sided random subshifts of a finite type, (Σ_{A_1}, μ_1) and (Σ_{A_2}, μ_2), which satisfy the weak-Bernoulli property and have the same entropy, $h(\mu_1) = h(\mu_2)$, for RDS – that is the same relative entropy in the set-up of random dynamical systems (see Bogenschütz (1993)) –

are relatively isomorphic, i.e. the induced skew-products Θ_1 and Θ_2 are isomorphic relative to ϑ. Examples of random subshifts of finite type satisfying the weak-Bernoulli property are the two-sided random Bernoulli shifts. Every positive value can be realized as entropy of an RDS by a random Bernoulli shift. This is also true for standard Bernoulli shifts, which can be viewed as random Bernoulli shifts where the noise has been decoupled. Thus we have the following result for two-sided shifts.

COROLLARY 5.3. *(σ, μ) is relatively isomorphic to a standard Bernoulli shift and every random Bernoulli shift is relatively isomorphic to a standard Bernoulli shift.*

This result shows implicitly that random shifts do not generate any new dynamical features. Nevertheless, they are useful in analyzing RDS with hyperbolic properties and in investigating random fractal structures (see Kifer (1994b)), as they draw a real-time picture of the dynamics in contrary to other symbolic descriptions, in particular to those obtained via induced transformations.

I would like to thank the Banach Center in Warsaw for hospitality during the Symposium on Ergodic Theory and Dynamical Systems in June 1995, where I had the chance to learn from Jean-Paul Thouvenot about Lind's extension of the relative isomorphism theory. I am very grateful to him as well as to my students, Wolfgang Meyer, and Marcus and Oliver Steinkamp, who contributed to my current understanding of random shifts.

REFERENCES

ARNOLD, L. 1995 *Random Dynamical Systems.* Preliminary Version 2.

ARNOLD, L. & CRAUEL, H. 1991 Random dynamical systems. In *Lyapunov Exponents Proceedings,* Oberwolfach 1990 (ed. L. Arnold, H. Crauel & J.-P. Eckmann). Lecture Notes in Mathematics, vol. 1486, pp. 1–22. Springer.

ARNOUX, P. & FISHER, A. 1995 Anosov families, renormalization and the Teichmüller flow. Preprint, Université Marseilles.

THOMAS BOGENSCHÜTZ 1993 Equilibrium states for random dynamical systems. PhD thesis, Universität Bremen.

BOGENSCHÜTZ, T. & GUNDLACH, V. M. 1992 Symbolic dynamics for expanding random dynamical systems. *Random & Computational Dynamics* **1**, 219–227.

BOGENSCHÜTZ, T & GUNDLACH, V. M. 1995 Ruelle's transfer operator for random subshifts of finite type. *Ergod. Th. & Dynam. Sys.* **15**, 413–447.

CORNFELD, I. P., FOMIN, S. V. & SINAI, YA. G. 1982 Ergodic theory. Grundlehren der Mathematischen Wissenschaften, vol. 245. Springer.

GUNDLACH, V. M. 1996 Thermodynamic formalism for random subshifts of finite type. Preprint.

GUNDLACH, V. M. 1995 Random homoclinic orbits. *Random & Computational Dynamics* (1 & 2) **3**, 1–33.

GUNDLACH, V. M. & STEINKAMP, O. 1996 Products of random rectangular matrices. Preprint.

KHANIN, K. & KIFER, YU. 1994 Thermodynamic formalism for random transformations and statistical mechanics. Preprint, Hebrew University, Jerusalem.

KIFER, Y. 1991 Equilibrium states for random expanding transformations. *Random & Computational Dynamics* **1**, 1–31.

KIFER, YU. 1994a Large deviations for paths and configurations counting. Preprint, Hebrew University, Jerusalem.

KIFER, YU. 1994b Fractal dimensions and random transformations. Preprint, Hebrew University, Jerusalem.

LEDRAPPIER, F. 1974 Principe variationnel et systèmes dynamiques symboliques. *Z. Wahrscheinlichkeitstheorie verw. Gebiete* **30**, 185–202.

LIND, D. A. 1977 The structure of skew products with ergodic group automorphisms. *Israel J. Math.* **28**, 205–248.

MEYER, W. 1995 Thermodynamische Beschreibung des topologischen Drucks für zufällige Shift-Systeme. Diplomarbeit, Universität Bremen.

OSELEDETS, V. I. 1968 A multiplicative ergodic theorem. Lyapunov characteristic numbers for dynamical systems. *Trans. Moscow Math. Soc.* **9**, 197–231.

PESIN, YA. 1976 Families of invariant manifolds corresponding to non-zero characteristic exponents. *Math. USSR IZV.* **10**, 1261–1305.

STEINKAMP, M. 1995 Zweiseitige zufällige Shifts. Diplomarbeit, Universität Bremen.

THOUVENOT, J.-P. 1975 Quelques propriétés des systèmes dynamiques qui se décomposent en un produit de deux systèmes dont l'un est un schéma de Bernoulli. *Israel J. Math.* **21**, 177–207.

Controlling chaos using embedded unstable periodic orbits: the problem of optimal periodic orbits

By BRIAN R. HUNT[1] AND EDWARD OTT[2]

[1]Institute for Physical Science and Technology, University of Maryland, College Park, USA

[2]Departments of Electrical Engineering and of Physics
and
Institutes for Plasma Research and for Systems Research, University of Maryland, College Park, USA

1. Introduction

The presence of chaos in physical systems has been extensively demonstrated and is very common. In practice, however, it is often desired that chaos be avoided and/or that the system performance be improved or changed in some way. Given a chaotic attractor, one approach might be to make some large and possibly costly alteration in the system which completely changes its dynamics in such a way as to achieve the desired behavior. Here we assume that this avenue is not available. Thus, we address the following question: Given a chaotic attractor, how can one obtain improved performance and a desired attracting time-periodic motion by making only *small* time-dependent perturbations.

The key observation is that a chaotic attractor typically has embedded within it an infinite number of unstable periodic orbits. Since we wish to make only small perturbations to the system, we do not envision creating new orbits with very different properties from the existing ones. Thus, we seek to exploit the already existing unstable periodic orbits. The approach is as follows: we first determine some of the unstable low-period periodic orbits that are embedded in the chaotic attractor. We then examine these orbits and choose the one which yields the best system performance. Finally, we tailor our small time-dependent parameter perturbations so as to stabilize this already existing orbit.

The above strategy, originally proposed in Ott *et al.* (1990) and commonly referred to as the OGY method, has been implemented in a wide variety of situations. These include optical systems, biology, solid state devices, mechanical systems, etc. (See Shinbrot *et al.* (1993), Ditto & Pecora (1993), Chen & Dong (1993), Ott & Spano (1995) and Ott *et al.* (1994) for reviews.)

In any implementation of the method a technique for stabilizing a periodic orbit is necessary. Here there are many possible choices, and many of these have been used in experiments. One technique is to apply perturbations to put the orbit on the stable manifold of the desired periodic orbit, Ott *et al.* (1990), another uses the pole placement technique of control theory, Romeiras *et al.* (1992), another, Pyragas (1993), uses an ad hoc feedback of the form $k(\mathbf{x}(t) - \mathbf{x}_0(t))$ where $\mathbf{x}_0$ is the desired periodic orbit and k is empirically adjusted seeking values that suitably stabilize the periodic orbit, still another, Pyragas (1993), empirical method uses feedback of the form $k(\mathbf{x}(t) - \mathbf{x}(t-T))$ where T is the period of the desired periodic orbit, etc. A key point in the utility of the OGY method is that it can be used in situations where the equations of motion of the system are not known (this is crucial to many of the experiments that have been carried out). In this case, one uses embedding techniques, Ott *et al.* (1994), to find the periodic orbits and, if desired, to examine their properties. Periodic orbit stabilization techniques have also been developed for the case where one uses a delay coordinate reconstruction of the system state, see Dressler & Nitsche (1992) and So & Ott (1995). Often system parameters drift or are purposely changed with time, and tracking techniques for following and maintaining control of the desired periodic orbit have been formulated and experimentally implemented, see Schwartz & Triandaf (1992) and Gills *et al.* (1992).

A question that naturally arises is which periodic orbit to use. Ideally one would like to have information on all the periodic orbits embedded in an attractor, and to choose the best one. Since there is an infinite number of such orbits, this is not an option. One might also wish to test non-periodic orbits. Say I test several low period orbits and choose the best. Could I do much better if I had information on all the orbits? Recent numerical and analytical work suggests that the optimal orbit is typically of low period, see Hunt & Ott (1996). The rest of this paper will be devoted to summarizing this work.

2. The problem of optimal periodic orbits

Many questions concerning dynamical behavior are addressed by consideration of the long-time average of a function F of the state vector x,

$$\langle F \rangle = \lim_{t\to\infty} \frac{1}{t} \int_0^t F(x(t'))\mathrm{d}t', \tag{2.1a}$$

$$\langle F \rangle = \lim_{t\to\infty} \frac{1}{t} \sum_{t'=1}^{t} F(x_{t'}), \tag{2.1b}$$

where t denotes time and is either continuous, (2.1a), or discrete, (2.1b).

We consider systems, such that, for *typical* choices of the initial x, the trajectory generated by the dynamical system is chaotic, and has a well-defined long-time average (2.1). (Here 'typical' is with respect to the Lebesgue measure of initial conditions in state space.) We note, however, that atypical initial conditions may generate orbits embedded in the chaotic attractor that have different values for $\langle F \rangle$ than typical orbits. For example, consider a chaotic attractor with a basin of attraction B. Even though there is a set of initial conditions in B all yielding the *same* value for $\langle F \rangle$, and the state space volume (Lebesgue measure) of these initial conditions is equal to the entire volume of B, there is still a zero volume set of initial conditions ('atypical' initial conditions) whose orbits asymptote to sets within the chaotic attractor but for which $\langle F \rangle$ is different from the average attained by typical orbits. In particular, this happens when the initial condition is placed exactly on an unstable periodic orbit embedded in a chaotic attractor (or on the stable manifold of the unstable periodic orbit).

The question we address is the following. *Which (atypical) orbit on the attractor yields the largest value of* $\langle F \rangle$*?* To our knowledge this question has not been previously addressed, yet it is fundamental to the controlling chaos methods discussed in § 1.

As discussed in § 1, the strategy is to first identify several low period unstable periodic orbits embedded in the chaotic attractor. One then determines the system performance that would apply if each of the various determined unstable periodic orbits were actually followed by the system. In many cases the system performance can be quantified as the value of some time average $\langle F \rangle$, as in (2.1). One then selects an orbit yielding performance that is best and feedback stabilizes that orbit. The question is whether one can obtain much better performance by looking exhaustively at higher period orbits or by considering stabilization of atypical *nonperiodic* orbits embedded in the chaotic attractor.

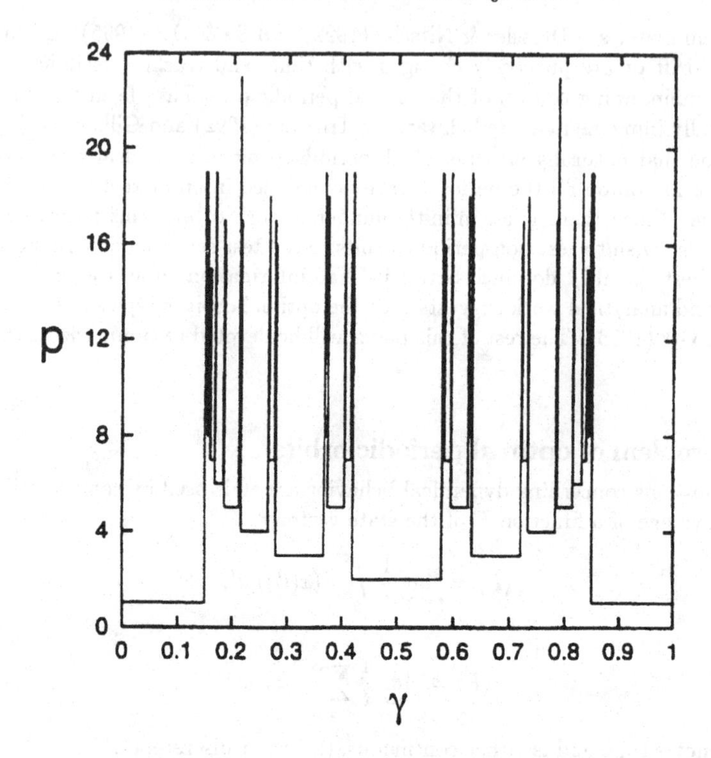

FIGURE 1. Period that optimizes $\langle F_\gamma \rangle$ as a function of γ for the doubling map (3.2) and function (3.3).

3. A simple example

To begin we consider a simple example, namely the doubling transformation,

$$x_{t+1} = 2x_t \pmod 1), \tag{3.2}$$

and for F we take

$$F_\gamma(x) = \cos[2\pi(x - \gamma)]. \tag{3.3}$$

Although some of the results we observe for (3.2) and (3.3) are model-specific, we claim that (3.2) and (3.3) also yield essential behaviors that should be expected in general for low dimensional chaotic systems. A main point will be that the optimal average is typically achieved by a low period periodic orbit.†

For each of 10^5 evenly spaced values of γ, we tested the value of $\langle F_\gamma \rangle$ for all periodic orbits of the map (3.2) with periods 1 to 24. There are on the order of 10^6 such orbits. Figure 1 shows the period of the orbit that maximizes $\langle F_\gamma \rangle$ for (3.2) and (3.3) as a function of the phase angle γ. The third column of table 1 gives the fraction $f(p)$ of phase values γ for which a period p orbit maximizes $\langle F_\gamma \rangle$. For example, if γ is chosen at random in $[0, 1]$, then over 93% of the time, the optimal periodic orbit does not exceed

† A different but related result has been obtained by Hunt & Yorke (1991) who argue that a certain type of crisis bifurcation of a chaotic attractor is typically mediated by a low period unstable periodic orbit. Also Yorke (private communication) has conjectured that the bifurcation to a riddled basin of attraction is mediated by a low period periodic orbit, and our results imply confirmation of this.

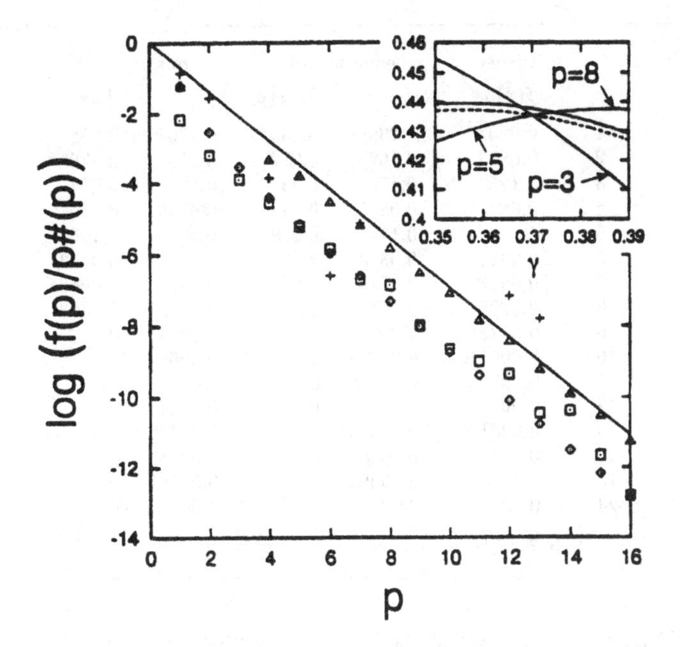

FIGURE 2. Graph of $\log(f(p)/p\#(p))$ versus p. The straight line has slope $-\log 2$. Inset shows $\langle F_\gamma \rangle_p$ versus γ for $p = 3, 5, 8$.

7 in period, and more than half the time the optimal orbit's period is 1, 2, or 3. The second column in table 1 gives a conjectured approximate asymptotic prediction of the fraction $f(p)$ of the time a period p orbit maximizes $\langle F_\gamma \rangle$ if γ is chosen at random in $[0, 1]$,

$$f(p) = Kp2^{-p}\phi(p). \tag{3.4}$$

Here $\phi(p)$ is the Euler function, which is defined as the number of integers between 1 and p (inclusive) that are relatively prime to p (e.g., the numbers 1, 5, 7, and 11 are relatively prime to 12, and so $\phi(12) = 4$). Thus $\phi(p) \leq p - 1$ for $p \geq 2$, and $\phi(p) = p - 1$ if p is a prime. The factor K is a fitting parameter, which we choose to be 1/6. We see from table 1 and the data plotted as diamonds in figure 2 that (3.4) agrees very well with the numerical results for large p (the straight line in figure 2 has slope $-\log 2$).

From table 1, the agreement with (3.4) is better than 5% for $p > 5$. Note that (3.4) apparently has nothing to do with the precise choice of the function F_γ in (3.3). We believe that the result (3.4) is often a good approximation for smooth functions with a single maximum. Tests using other quadratic maximum, single humped functions (e.g. $F_\gamma(x) = -(x - \gamma)^2$) in place of (3.3) confirm this.

Not only are low period orbits most often optimal, but, even when a somewhat higher period orbit is optimal, it apparently only leads to a relatively small increase in $\langle F_\gamma \rangle$ as compared to a lower period orbit. This point is emphasized by the fourth column in table 1, which gives the fraction of the γ values such that the lowest period orbit that yields a value of $\langle F_\gamma \rangle$ within 90% of the maximum value has period p. Thus, for this example, if one is willing to settle for 90% of optimal, one *never* has to go above period 5. Also for over 83% of the γ values it suffices to consider only period 1, 2, and 3. The relatively small increase of $\langle F_\gamma \rangle$ achieved by going to higher period is also evident from

	theory	equation (3.3)		equation (4.5)	
p	$f(p)$	$f(p)$	$f_{90\%}(p)$	$f(p)$	$f_{90\%}(p)$
1	0.0833	0.299	0.333	0.230	0.258
2	0.0833	0.160	0.212	0.163	0.175
3	0.125	0.176	0.294	0.186	0.234
4	0.0833	0.0985	0.143	0.0850	0.110
5	0.104	0.116	0.0180	0.136	0.169
6	0.0313	0.0310	0	0.0350	0.0473
7	0.0547	0.0573	0	0.0427	0.00664
8	0.0208	0.0211	0	0.0583	0.00031
9	0.0176	0.0178	0	0.0244	0
10	0.00651	0.00644	0	0.00697	0
11	0.00895	0.00918	0	0.0164	0
12	0.00195	0.00196	0	0.00516	0
13	0.00317	0.00324	0	0.00446	0
14	0.00085	0.00084	0	0.00389	0
15	0.00061	0.00062	0	0.00105	0
16–24	0.00091	0.00092	0	0.00167	0

TABLE 1. Numerical results for doubling map (3.2)

the plots of $\langle F_\gamma \rangle_p$ versus γ, where $\langle F_\gamma \rangle_p$ denotes the average of F_γ over the period p orbit that is optimal from among all period p orbits. For example, the γ region near $\gamma \approx 0.37$ (see figure 1) has $p = 3$ and $p = 5$ intervals with a smaller $p = 8$ interval in between. A plot of $\langle F_\gamma \rangle_p$ for $p = 3, 5, 8$ in this region is shown in the inset in figure 2. The weighted average $(5\langle F_\gamma \rangle_5 + 3\langle F_\gamma \rangle_3)/(5+3)$ is shown as a dashed curve. Note that $\langle F_\gamma \rangle_8$ closely follows this average but is slightly above it. Thus $\langle F_\gamma \rangle_8$ is slightly larger than both $\langle F_\gamma \rangle_3$ and $\langle F_\gamma \rangle_5$ in a small region near $\gamma \approx 0.37$.

It is also interesting to note the Farey tree structure present in figure 1; the periods follow the pattern of the denominators in the Farey construction of the rational numbers. That is, between any two γ-intervals with optimal orbits of periods p_a and p_b and only higher periods associated with any intervening γ-intervals, there is a smaller γ-interval of period $p_a + p_b$ in between, and all other γ-intervals in between have period higher than $p_a + p_b$. This is illustrated by figure 1. For example, consider the γ-interval $[0.35, 0.45]$. Between the period 3 interval and the period 2 interval there is a period 5 interval. Between the 3 and the 5 there is an 8, between the 5 and the 2 there is a 7, and so on. Numerically we find an exponential decrease, as p increases, of the total length of the γ-intervals with period at least p (this can be discerned from the data in table 1). Noting this and thinking of optimal nonperiodic orbits as being created in the limit as the Farey tree level approaches infinity,† we infer that optimal nonperiodic orbits typically do not occur on a positive Lebesgue measure set of γ.

The form of equation (3.4) is obtained as follows. The factor $\phi(p)$ is the number of times the integer p appears in the complete Farey tree (starting at the lowest level with $p_a = p_b = 1$). The factor $p2^{-p}$ is obtained from our numerical observations (and by direct analytical calculation in a special case) of how the width of an interval scales with the period p.

What is the character of the set S_γ of γ values for which the optimizing orbit is non-

† This line of argument is supported by the work of Sigmund (1972) who proved that for hyperbolic systems, every invariant measure can be approximated arbitrarily well by the δ-function measure on a periodic orbit.

periodic? From the above discussion, S_γ has zero Lebesgue measure. On the basis of our numerical evidence, we can show that S_γ is a Cantor set (in particular, S_γ is uncountable) whose fractal dimension is zero. Also, based on the Farey structure, we can show that when $\gamma \in S_\gamma$, the nonperiodic orbit that maximizes $\langle F_\gamma \rangle$ has topological entropy zero. The above arguments are deferred to a future, longer publication, Hunt & Ott (1996).

Based on our numerical results we make a general conjecture concerning typical maps with chaotic attractors and typical smooth optimization functions F with a parameter dependence.

Conjecture. Nonperiodic optimizing orbits occur on a set of zero Lebesgue measure in the parameter space of F.

In the remainder of this paper we present some further numerical results involving different choices of the optimization function F and different dynamical systems, in support of the above conjecture and the principle that for most parameters, $\langle F \rangle$ is maximized by a low period orbit. Other cases appear in Hunt & Ott (1996).

4. Other examples

The fifth column of table 1 shows the fraction of 10^5 evenly spaced values of γ for which a period p orbit of the map (3.2) maximizes the average of a different function

$$F_\gamma(x) = \cos[2\pi(x-\gamma)] + \sin[6\pi(x-\gamma)]. \tag{4.5}$$

The sixth column of table 1 gives the corresponding fraction for the lowest p within 90% of optimal. The function in (4.5) has three local maxima and three local minima. This increases the likelihood of a higher period orbit maximizing $\langle F_\gamma \rangle$, as is reflected in the data. The Farey structure, present for smooth functions with a single maximum (e.g. equation (3.3)), is found only partially in this case (and in the examples with two-dimensional maps that follow). Thus the number of intervals $\#(p)$ for which a period p orbit maximizes $\langle F_\gamma \rangle$ is in general not equal to the Euler function $\phi(p)$. However, we find that the size of each period p interval still tends to scale like $p2^{-p}$; if we replace the Euler functions $\phi(p)$ in (3.4) by the numerically observed number of γ intervals $\#(p)$ for which a period p orbit maximizes $\langle F_\gamma \rangle$, good agreement with (3.4) is restored. This is illustrated by the data represented as squares in figure 2. Another important point is that for (4.5) (as for (3.3)) we observe an exponential decrease, as a function of p, of the proportion of phase values γ for which $\langle F_\gamma \rangle$ is maximized on an orbit of period at least p. Thus the result that low period orbits most often are optimal is apparently independent of our choice of F.

The above discussion has been for a one-dimensional map. How do these results carry over into higher dimensionality? To get some indication of the situation we consider two different two-dimensional maps. First we discuss the Kaplan–Yorke map, Yorke (private communication),

$$x_{n+1} = 2x_n \pmod 1, \tag{4.6a}$$

$$y_{n+1} = \lambda y_n + \frac{1}{\pi}\sin(2\pi x_n). \tag{4.6b}$$

The Lyapunov exponents are $\ln 2$ and $\ln \lambda$. Choosing $\lambda = 0.4$ we have an information dimension of $D \approx 1.76$ for the attractor. Results for the optimal period with F chosen to be

$$F_\gamma(x, y) = \cos[2\pi(x + y - \gamma)] \tag{4.7}$$

are shown in the second and third columns of table 2. Also, the scaling of the average

	Kaplan–Yorke Map (4.6)		Hénon Map (4.8)	
p	$f(p)$	$f_{90\%}(p)$	$f(p)$	$f_{90\%}(p)$
1	0.282	0.319	0.427	0.434
2	0	0	0.421	0.424
3	0	0	0	0
4	0.140	0.188	0.0862	0.0857
5	0.223	0.326	0	0
6	0.127	0.139	0.00823	0.0352
7	0.0768	0.0285	0.0415	0.0210
8	0.0466	0	0	0
9	0.0524	0	0	0
10	0.0162	0	0	0
11	0.0169	0	0	0
12	0.00518	0	0.00915	0
13	0.00750	0	0.00531	0
14	0.00274	0	0	0
15	0.00158	0	0	0
16–24	0.00214	0	0.00205	0

TABLE 2. Numerical results for 2D maps

size of the γ interval on which a given period p orbit maximizes $\langle F_\gamma \rangle$ is shown by the triangles in figure 2. These results offer further support for our conjecture.

Next we consider the Hénon map

$$x_{n+1} = a + by_n - x_n^2, \tag{4.8a}$$

$$y_{n+1} = x_n, \tag{4.8b}$$

with the often studied parameter values $a = 1.4$, $b = 0.3$. The periodic orbits of this map were found using the method of Biham & Wenzel (1989), and the function we averaged was

$$F_\gamma(x, y) = \cos[(\pi/2)(x + y - \gamma)]. \tag{4.9}$$

The results are given in the fourth and fifth columns of table 2, and by the crosses in figure 2. Evidently the principle that the optimum is typically achieved by low period orbits, and that near optimum performance can always be achieved by such orbits, continues to hold.

Finally, we note that in all the cases we studied, the Farey tree structure we found in the prototype case of (3.2) and (3.3) is still partially present. In all of the other examples we sometimes observe sudden transitions between γ intervals corresponding to different low periods, but we still find that the high period γ intervals are created by Farey summation (see Hunt & Ott (1996)). The important point is that this implies the appearance of nonperiodic optimizing orbits by the same Farey mechanism as for (3.2) and (3.3), thus supporting our conjecture and indicating the general applicability of the behavior we observed for (3.2) and (3.3).

This research was supported by the U.S. Department of Energy (Offices of Scientific Computing and Energy Research), by the National Science Foundation (Divisions of Mathematical and Physical Sciences), and by the Office of Naval Research. The numerical computations reported in this paper were made possible by a grant from the W. M. Keck Foundation.

REFERENCES

BIHAM, O. & WENZEL, W. 1989 *Phys. Rev. Lett.* **63**, 819; see also GRASSBERGER, P., KANTZ, H. & MOENIG, U. 1989 *J. Phys.* A **22**, 5217.

CHEN, G. & DONG, X. 1993 *Int. J. Bif. and Chaos* **3**, 1363.

DITTO, W. L. & PECORA, L. 1993 *Sci. Am.*, August, 78.

DRESSLER, U. & NITSCHE, G. 1992 *Phys. Rev. Lett.* **68**, 1.

GILLS, Z., IWATA, C., ROY, R., SCHWARTZ, I. B. & TRIANDAF, I. 1992 *Phys. Rev. Lett.* **69**, 3169.

HUNT, B. R. & OTT, E. 1996 *Phys. Rev.* E. To appear.

HUNT, B. R. & YORKE, J. A. 1991 *Trans. Amer. Math. Soc.* **325**, 141.

KAPLAN, J. L. & YORKE, J. A. 1979 In *Functional Differential Equations and Approximation of Fixed Points* (ed. H.-O. Peitgen & H.-O. Walter). Lecture Notes in Mathematics, vol. 730, p. 204. Springer.

OTT, E., GREBOGI, C. & YORKE, J. A. 1990 *Phys. Rev. Lett.* **64**, 1196.

OTT, E., SAUER, T. & YORKE, J. A., EDS. 1994 *Coping with Chaos.* Wiley.

OTT, E. & SPANO, M. L. 1995 *Phys. Today,* May, 34.

PYRAGAS, K. 1993 *Phys. Lett.* A **181**, 203.

ROMEIRAS, F. J., GREBOGI, C., OTT, E. & DAYAWANSA, W. P. 1992 *Physica* D **58**, 165.

SCHWARTZ, I. & TRIANDAF, I. 1992 *Phys. Rev.* A **46**, 7439.

SHINBROT, T., GREBOGI, C., OTT, E., & YORKE, J. A. 1993 *Nature* **363**, 411.

SIGMUND, K. 1972 *Amer. J. Math.* **94**, 31.

SO, P. & OTT, E. 1995 *Phys. Rev.* E **51**, 2955.

YORKE, J. A. Private communication.

Chaotic tracer dynamics in open hydrodynamical flows†

By G. KÁROLYI[1], Á. PÉNTEK[2], T. TÉL[3] AND Z. TOROCZKAI[4]

[1]Research Group for Computational Mechanics of the Hungarian Academy of Sciences, Budapest, Hungary

[2]Institute for Pure and Applied Physical Sciences, University of California, San Diego, La Jolla, USA

[3]Institute for Theoretical Physics, Eötvös University, Budapest, Hungary

[4] Virginia Polytechnic Institute and State University, Blacksburg, USA

We investigate the dynamics of tracer particles in time-dependent open flows. In cases when the time-dependence is restricted to a finite region, we show that the tracer dynamics is typically chaotic but necessarily of transient type. The complex behaviour is then due to an underlying nonattracting chaotic set that is also restricted to a finite domain, and the tracer dynamics corresponds to a kind of chaotic scattering process. Examples are taken from the realm of two-dimensional incompressible flows. The cases of two leapfrogging vortex pairs and of the blinking vortex–sink system illustrate the phenomenon in inviscid fluids, while the von Kármán vortex street problem belongs to the class of viscous flows. Based on these examples, generic features of the scattering tracer dynamics are summarized.

1. Introduction

The advection of particles in hydrodynamical flows is a phenomenon having attracted great recent interest from the side of dynamical system community because these particles can exhibit chaotic motion, see Aref & Balachandar (1986) and Péntek *et al.* (1995b). By particle we mean a light granule of small extension. If it takes on the velocity of the flow very rapidly, i.e. inertial effects are negligible, we call the advection passive, and the particle a passive tracer. Its equation of motion is then

$$\dot{\mathbf{r}} = \mathbf{v}(\mathbf{r}, t), \tag{1.1}$$

where $\mathbf{v}$ represents the velocity field that is assumed to be known. The tracer dynamics is thus governed by a set of ordinary differential equations, like e.g. for driven anharmonic oscillators, whose solution is typically chaotic.

It is a unique feature of chaotic advection in time-dependent *planar incompressible* flows that the fractal structures characterizing chaos in phase space become observable by the naked eye in the form of spatial patterns. In such cases there exists a *stream function* ψ, see Milne-Thomson (1958) and Landau & Lifshitz (1959), whose derivatives can be identified with the velocity components as

$$v_x(x, y, t) = \frac{\partial \psi}{\partial y}, \quad v_y(x, y, t) = -\frac{\partial \psi}{\partial x}, \tag{1.2}$$

and whose level lines provide the streamlines. Note that (1.2) is a consequence of incompressibility because it implies $\nabla \mathbf{v} = 0$. Combining this with (1.1) for a planar flow where $\mathbf{r} = (x, y)$ and $\mathbf{v} = (v_x, v_y)$, one notices that the equations of motion have canonical character with $\psi(x, y, t)$ playing the role of the Hamiltonian and x and y being the

† This paper is dedicated to Professor K. Nagy on the occasion of his 70th birthday.

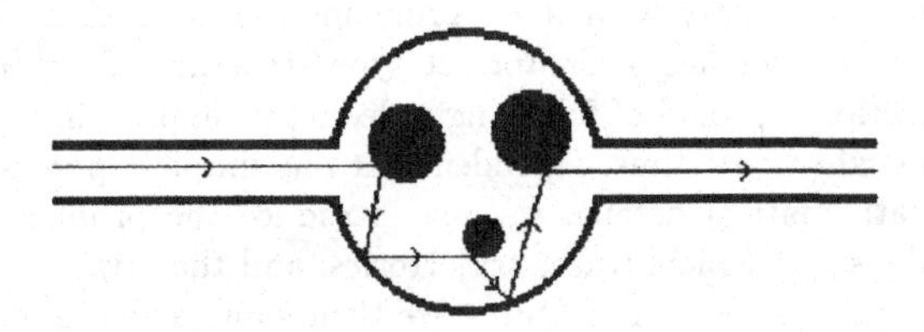

FIGURE 1. Schematic diagram representing the dynamics as chaotic scattering in a system defined by two channels and a scattering region containing billiard scatterers.

canonical coordinates and momenta (or *vice versa*), respectively. Thus, the plane of the flow *coincides* with the particles' phase space. This property makes passive advection in planar incompressible flows especially appealing and a good candidate for an experimental observation of patterns that are hidden otherwise in the abstract phase space. In stationary flows, where ψ is independent of t, problem (1.1), (1.2) is integrable and the particle trajectories coincide with the streamlines. In time-dependent cases, however, particle trajectories and streamlines are different, and the former ones can only be obtained by solving equations (1.1), (1.2).

Here we consider passive advection in *open* flows in cases when complicated tracer movements caused by the time-dependent flow is assumed to be restricted to a *finite* region. This will be called the *mixing region* outside of which the time-dependence of ψ is negligible. It is worth emphasizing that a complicated flow field (turbulence) inside the mixing region is not at all required for complex tracer dynamics and the corresponding fractal patterns. Even simple form of time dependence, e.g. a periodic repetition of the velocity field is sufficient.

For tracers injected into the flow outside of the mixing region, where the flow is still practically stationary, the motion is initially simple and becomes later gradually more complicated as the particle is being advected into the mixing region. The advection dynamics can thus be regarded as a *scattering process*, with the advected particles being 'scattered' on the finite region of nontrivial mixing. The motion in the outflow region is then simple again. Thus, for such processes chaos is necessarily restricted to a finite region both in space and time. We claim that this *transient chaos*, Tél (1990), is the only form of chaos which can appear in the situation studied. Tracer dynamics then corresponds to a kind of *chaotic scattering*, see Smilansky (1992), which is a subfield of nonlinear dynamics with a considerable amount of accumulated knowledge. The interpretation of certain advection phenomena, observable also in experiments, is thus very natural in this framework.

Symbolically, we can indentify the tracer dynamics in the inflow and outflow region with the motion of a point mass in a channel in front of and after a scattering region, respectively. The scattering region characterized either by strongly varying forces or, in billiards, by the presence of scatterers corresponds then to the mixing region in the advection problem (see figure 1). Tracer particles are thus topologically similar to point mass trajectories in a scattering system. We have to bear in mind, however, that the analogy is not complete since the point mass problem's phase space is four-dimensional.

The complicated form of trajectories implies a long time spent in the mixing region. In other words, tracers can be temporarily trapped there. Due to the incompressibility of the flow no attractors can exist, and *almost all* particles escape the mixing region. This *escaping* property is a specific characteristic of chaotic scattering and of the tracer dynamics in open flows of the type we are studying. As a consequence, the underlying chaotic set is a *chaotic saddle* with more pronounced fractal character than chaotic attractors, since the former has a Cantor set type structure along *both* the unstable and the stable direction. Instead of following a deductive approach, in the next section we study examples where the time dependence is the simplest possible one, periodic. Without giving mathematical details, we just introduce the problems and pictorially show the flow fields, some typical tracer trajectories, and the invariant sets obtained in numerical simulations. The generic features are then summarized and discussed in the last, concluding section.

2. Case studies

2.1. *Leapfrogging vortex pairs*

We consider a model of the so-called 'leapfrogging' motion of two smoke rings, see Beigie *et al.* (1994), Van Dyke (1982), Shariff *et al.* (1988), Shariff & Leonard (1992) and Saffman (1992). If the rings have the same sense of rotation and move along the same axis, the rear vortex ring attempts to pass through the front one. The leading ring then widens due to the mutual interaction and slows down. Simultaneously, the other ring shrinks, accelerates and penetrates the first one. This process is then repeated continuously with some period. We studied, Péntek *et al.* (1995a), the two-dimensional analogue of this process: advection in the field of two pairs of ideal point vortices of the same strength and moving along the same symmetry axis (the x-axis) which also exhibit a strictly periodic motion. The equations of motion for the vortices can easily be written down by using the rules of point vortex interactions, Saffman (1992), and can numerically be solved with high accuracy, see Péntek *et al.* (1995a).

Figure 2 exhibits the streamline pattern at two different instants of time, at $t = 0$ and $t = T/2$ in a frame co-moving with the center of mass of the vortex pairs, where T denotes the period of the velocity field. Since the vortex pairs are identical, this period is half of the leapfrogging motion's period. Note the smoothness of the streamlines. The tracer motion in any frozen-in streamline pattern would be simple, it is the temporal variation of ψ that leads to irregular motion.

The stream function $\psi(x, y, t)$ is analytically known, and tracer trajectories can be obtained by solving equations (1.1), (1.2) with this time dependent stream function as an input, see Péntek *et al.* (1995a). Figure 3 contains, in the same co-moving frame, the plot of two complicated, chaotic scattering trajectories of tracers injected into the flow in front of the vortex pairs. The drastical difference in the shape of the trajectories due to a slight change in the initial y coordinates is an example for the sensitive dependence on the initial conditions, which is regarded as a common manifestation of chaos. A comparison with figure 1 shows that the channels towards and away from the scattering region correspond to the regions far away in front of and after the vortex pairs, respectively.

Figure 4 presents the invariant sets associated with the scattering tracer dynamics on a stroboscopic map taken at integer multiples of the period T. The time instant selected is $t = 0 \,(\mathrm{mod}\ T)$. The chaotic saddle (see figure 4(a)) is the union of all unstable bounded trajectories trapped in the mixing region forever. It clearly contains parts that appear to be the direct products of two Cantor sets, e.g. the one around the midpoint between the vortices. Such parts form the *hyperbolic* component of the saddle. There are also rather densely occupied regions forming the *nonhyperbolic* component, situated around curves surrounding the vortices. Note the white regions around the vortices that are not accessible by tracers coming from outside. Thus, in spite of the infinitesimally small extension of the vortices in the Eulerian velocity field, finite vortex 'cores' are formed

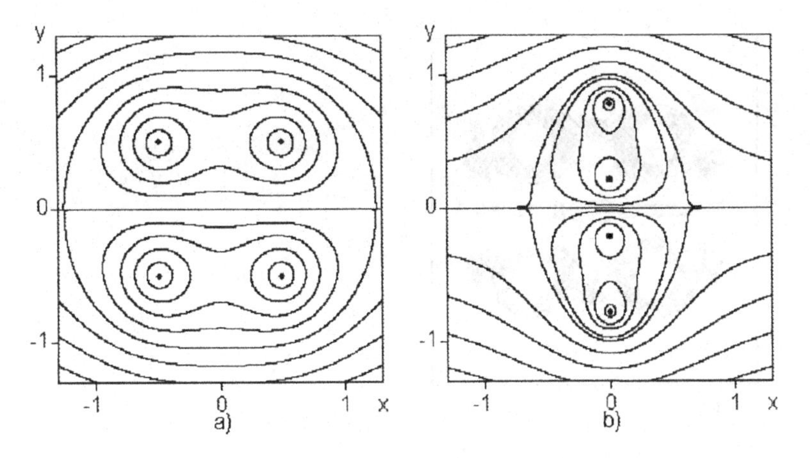

FIGURE 2. Streamlines of the leapfrogging vortex pair problem at (a) $t = 0$ and (b) $t = T/2$. The vortices in the upper (lower) half plane have the same strength, 1 (-1) in dimensionless units. The initial condition $(t = 0)$ is a configuration where the width of both vortex pairs as well as their distance along the x-axis is unity. The vortex centers are denoted by black dots.

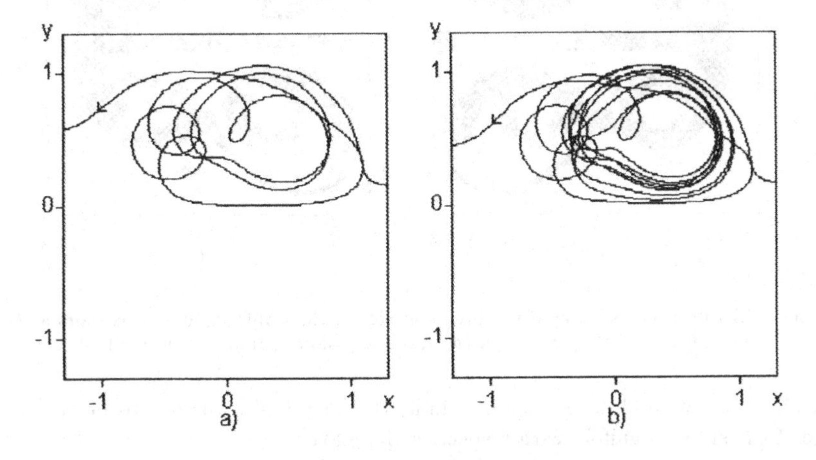

FIGURE 3. Two complicated tracer trajectories with initial y coordinates differing by 10^{-2} in the leapfrogging problem presented in a frame co-moving with the vortex pairs' center of mass.

in the tracer dynamics. The boundaries of the vortex cores are, in the language of dynamical system theory, KAM tori, see Wiggins (1992). The nonhyperbolic component of the saddle surrounds these tori.

Other invariant sets are the stable and unstable manifolds of the chaotic saddle. The stable manifold is the set of initial conditions for tracers that reach the chaotic saddle asymptotically. Since the latter is not an attractor, the stable manifold must be a set of measure zero, i.e. a set whose area is vanishing. It is thus a fractal whose form is given in figure 4(b) at the same instant of time as the saddle of figure 4(a). The unstable manifold (see figure 4(c)) is traced out by trajectories that have approached the saddle with rather high accuracy and left it after staying in its vicinity for a long time. In this

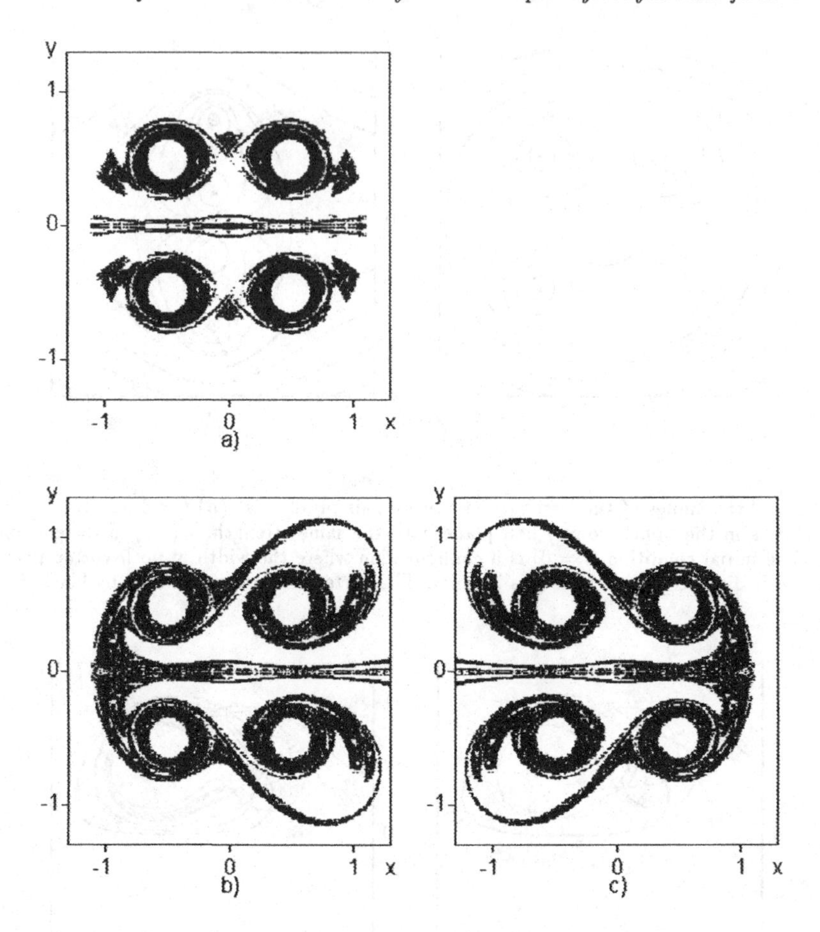

FIGURE 4. Stroboscopic section of the invariant sets of the leapfrogging vortex pairs at $t = 0$ (mod T): (a) Chaotic saddle, (b) stable manifold, (c) unstable manifold.

particular example and at this time instant, this manifold turns out to be the mirror image of the stable manifold with respect to the y axis.

2.2. *The blinking vortex–sink system*

Consider an ideal fluid filling in the infinite plane with a point vortex that is simultaneously sinking. This can be a model of a large bath tub with a sink since a rotational flow is formed around the sink in the course of the outflow. The blinking vortex–sink system introduced originally by Aref *et al.* (1989) is obtained by having two such sinking vortex points some distant apart from each other and being active alternatingly for a duration of $T/2$. This models the outflow from a large bath tub with two sinks that are opened in an alternating manner. In the blinking vortex–sink system the velocity field is periodic with T but in a special way: it is stationary for a half period of $T/2$ and stationary again but of another type for the next half period of $T/2$.

By using the complex function formalism, Milne-Thomson (1958), for describing these stationary flows, one can write down the stream functions explicitly. The streamlines valid for $0 < t \leq T/2$ and $T/2 < t \leq T$ are shown in figures 5(a) and 5(b), respectively.

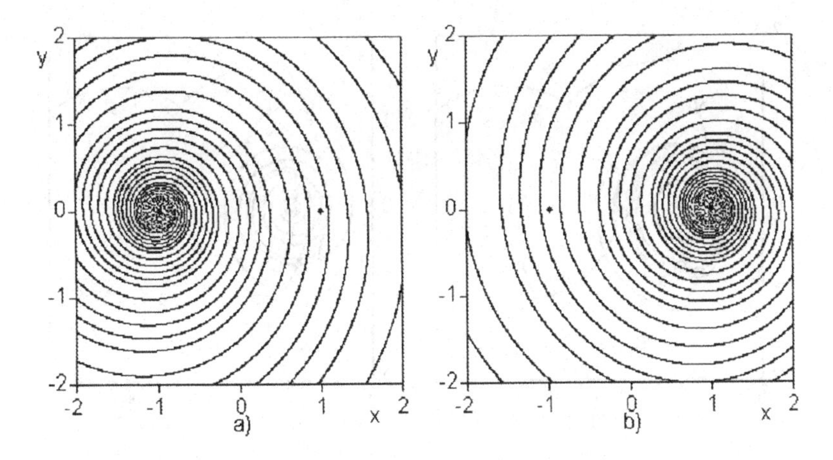

FIGURE 5. Streamlines of the blinking vortex–sink system for (a) $0 < t \leq T/2$ and (b) $T/2 < t \leq T$. The two vortex–sink centers are situated at $x = \pm 1$ in dimensionless units.

Note that sufficiently far away from the origin the differences between the streamlines and the stream functions of the two cases are negligible. Since the time-dependence appears in the form of a jump of ψ at $t = 0 \,(\mathrm{mod}\ T/2)$, the tracer dynamics is analogue to that of a kicked mechanical system. The replacement of a tracer particle can easily be determined within any of the stationary regimes. The comparison of the particles' position right after the first flow field sets in, i.e. at $t = 0^+ \,(\mathrm{mod}\ T)$, leads to a stroboscopic map whose form can be given analytically, see Aref *et al.* (1989). The tracer dynamics is governed in this system by a discrete dynamics rather than by a differential equation and is thus simpler to study. Both sinks are surrounded by a circle of a given radius containing all the points that escape the system within a duration of $T/2$. These two disks correspond to the channel directing particles away from the scattering region of figure 1, while the motion far away from the sinks corresponds to that inside the channel leading towards the scatterers.

The advection problem has two essential dimensionless parameters: the sink strength and the ratio of the vortex and sink strengths. We have carried out a detailed investigation of the tracer dynamics at different values of the parameters, see Károlyi (1995). Two tracer trajectories are shown in figure 6 with long life time before reaching one of the sinks. The breakpoints are due to the sudden jumps between the two different streamline patterns at $t = 0 \,(\mathrm{mod}\ T/2)$.

The invariant sets characterizing this problem are shown in figure 7 on a stroboscopic map taken at time $t = 0^+ \,(\mathrm{mod}\ T)$. The chaotic saddle (see figure 7(a)) seems to be hyperbolic everywhere, i.e. to have a direct product structure. At other ratios of the sink and vortex strengths KAM tori might also appear and in their vicinity the chaotic saddle is nonhyperbolic. The invariant manifolds are given in figures 7(b) and 7(c). The stable manifold (see figure 7(b)) is a fractal curve reaching arbitrarily far away from the vortex–sink centers. Its form is more and more regular when going out to infinity due to the simplicity of the flow in this region. In contrast, the unstable manifold (see figure 7(c)) is a fractal curve bounded to a finite region and connects the chaotic saddle with the right sink closed at $t = 0^- \,(\mathrm{mod}\ T)$, just before taking the stroboscopic map.

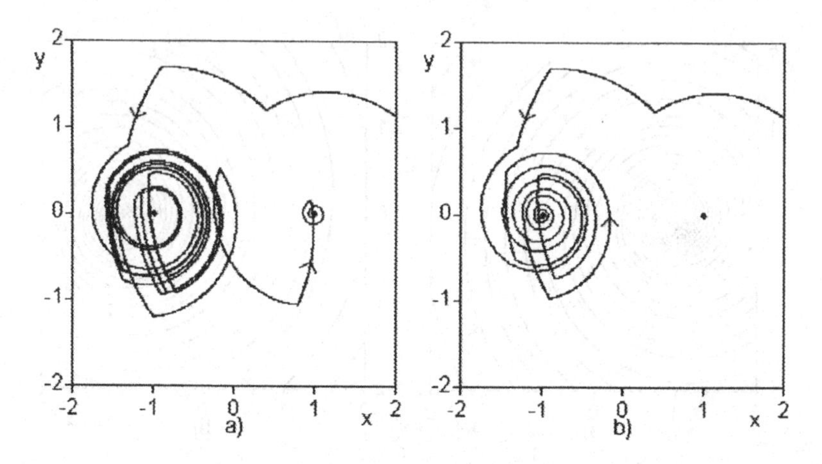

FIGURE 6. Two complicated tracer trajectories of the blinking vortex–sink system. A change of 10^{-2} in the initial coordinates leads to completely different trajectories leaving the system through different sinks. The vortex–sink centers are denoted by black dots.

2.3. *The von Kármán vortex street*

We consider the flow of a viscous fluid around a cylinder with a background velocity pointing along the x-axis. At intermediate background velocities (whose dimensionless measure, the Reynolds number, is on the order of 10^2) no stationary velocity field is stable, instead, a strictly periodic behaviour sets in with period T. Two vortices are created behind the cylinder within each period, one above and the other one below the x-axis. These two vortices are delayed by a time $T/2$. Note that they are now extended vortices with finite velocities even in the vortex centers. The vortices first grow in size, then become detached from the cylinder and start to drift along the channel. This alternating separation of vortices from the upper and lower cylinder surface is called the von Kármán vortex street and is characterized by a strictly periodic velocity field of period T, see Van Dyke (1982) and Beigie *et al.* (1994). After a short length of travel, the vortices are destabilised and destructed due to the viscosity of the fluid. Far away from the cylinder upstream and downstream the flow is, however, practically stationary.

To obtain the velocity distribution one has to solve the two-dimensional inviscid Navier-Stokes equations with no-slip boundary condition along a circle, see Shariff *et al.* (1991) and Jung & Ziemniak (1992). For simplicity we use here an analytic model for the stream function introduced in Jung *et al.* (1993). It was motivated by a direct numerical simulation of the Navier–Stokes flow carried out by Jung & Ziemniak (1992) at Reynolds number 250.

Figures 8(a) and 8(b) show the streamlines of this model at time $t = 0 \pmod T$ and $t = T/4 \pmod T$. The streamlines far away from the cylinder are straight at any instant of time. Thus, the channels towards and away from the scattering region of figure 1 correspond in this case to the upstream and downstream regions, respectively. Tracer trajectories were generated, see Jung *et al.* (1993), Ziemniak *et al.* (1994), Péntek *et al.* (1995b), by solving equations (1.1), (1.2) with the model stream function of Jung *et al.* (1993). Figure 9 exhibits two complicated scattering trajectories that are trapped for a while in the wake of the cylinder.

The invariant sets are shown again on a stroboscopic map taken at integer multiples of the period T in an area surrounding the cylinder. The chaotic saddle (see figure 10(a))

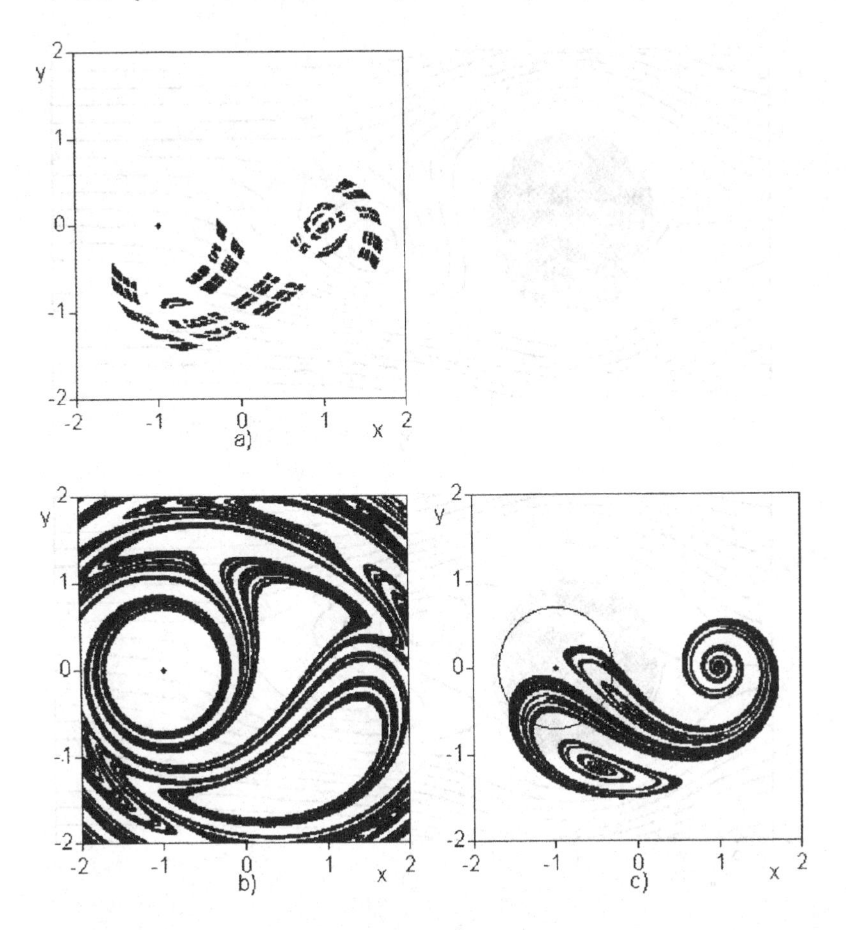

FIGURE 7. Stroboscopic section of the invariant sets of the blinking vortex–sink system at $t = 0^+ \pmod{T}$: (*a*) Chaotic saddle, (*b*) stable manifold, (*c*) unstable manifold. The circle around the left vortex–sink center at $(-1, 0)$ indicates the area from which the tracers leave the system via the left sink during the first half period.

contains now both a hyperbolic and a nonhyperbolic component. The former one is situated away from the cylinder, while the nonhyperbolic component seems to accumulate on the cylinder's surface. This is a nice manifestation of the no-slip boundary condition. Very close the the surface, i.e. in the boundary layer, the velocity must be small and there can exist therefore increasingly many trapped trajectories. In fact, the surface contains an infinity of parabolic orbits, see Jung *et al.* (1993), and it plays a similar role as a KAM surface. This surface is, however, smooth and does not have surrounding cantori. Nevertheless, the effect of both types of tori is similar in collecting the nonhyperbolic component of the chaotic saddle.

The invariant manifolds are exhibited in figures 10(*b*) and 10(*c*). The stable manifold (see figure 10(*b*)) surrounds the cylinder surface and extends to the infinitely far inflow region in a narrow band close to the negative x-axis. In contrast, the unstable manifold (see figure 10(*c*)) touches the cylinder surface along a finite arch only, and extends to the outflow region at infinity in a strongly oscillating way. It is interesting to note, that

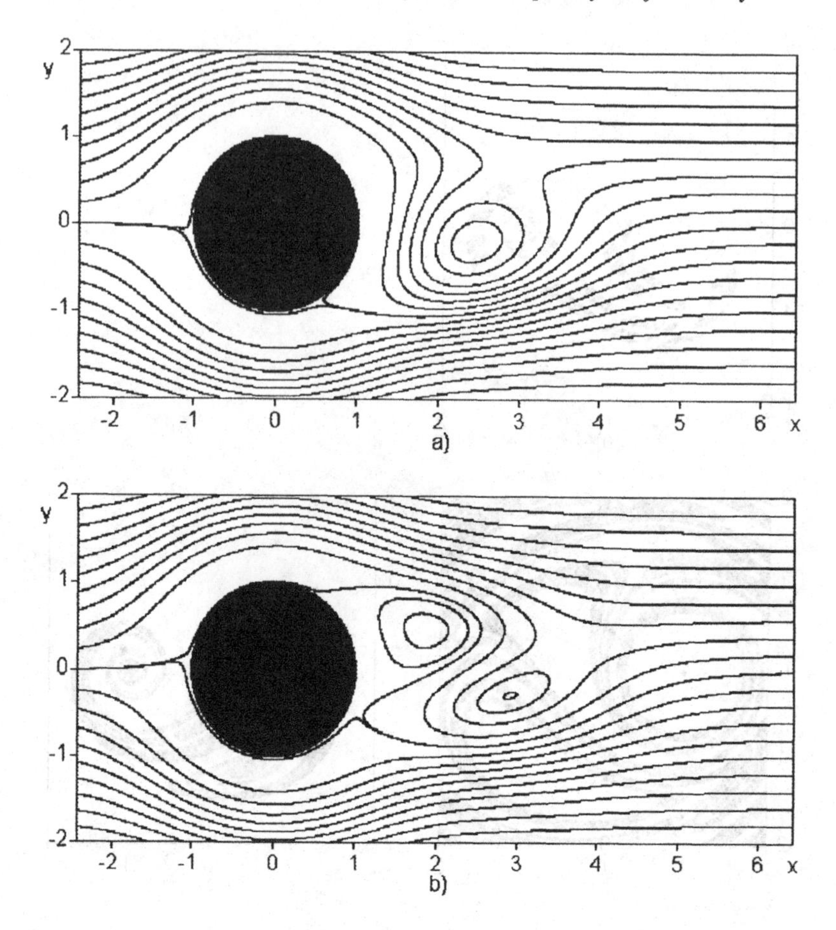

FIGURE 8. Streamlines of the von Kármán vortex street flow at (a) $t = 0$ and (b) $t = T/4$, obtained by using the form and parameters of the model of Jung *et al.* (1993). The radius of the cylinder is 1 in dimensionless units and its center is in the origin. Due to a symmetry of the flow, the streamline pattern at time $t = T/2\,(\mathrm{mod}\ T)$ and $t = 3T/4\,(\mathrm{mod}\ T)$ can be obtained as the mirror images of (a) and (b), respectively.

in this system a small chaotic saddle is formed around $x = 3$ downflow, too, with its own invariant manifolds, see Péntek *et al.* (1995b). This saddle is fully hyperbolic and rather unstable, therefore, its dynamical consequences are not so apparent as the ones discussed above and shown in figure 10(a).

3. Conclusions

Based on the examples, we summarize the most important general features characterizing chaotic passive advection in open flows of the type investigated.

• *The existence of a chaotic saddle* is the key observation in understanding the scattering tracer dynamics. This set, just like a chaotic attractor, is the union of all *bounded orbits*, including periodic ones, never escaping the mixing region. These orbits are all unstable. In contrast to a chaotic attractor, however, the chaotic saddle has practically no

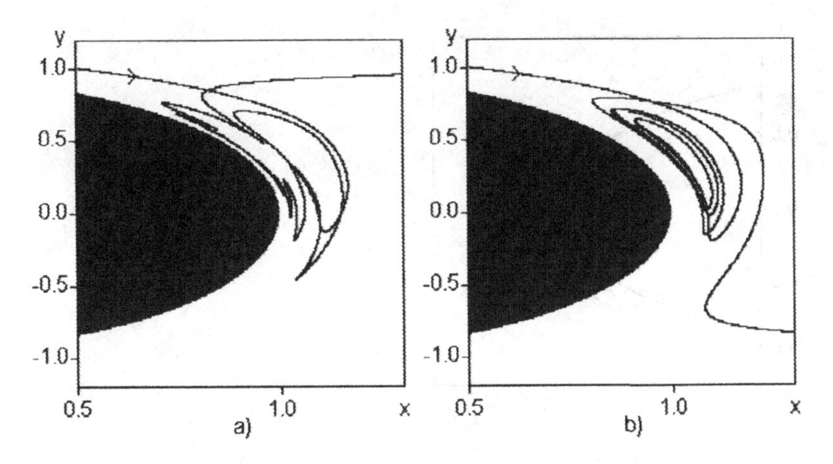

FIGURE 9. Two complicated tracer trajectories in the von Kármán vortex street with initial y coordinates differing by 10^{-2}. The horizontal scale is multiplied by 3 for better visualization.

region of attraction. In other words, the chaotic saddle contains only those orbits which are trapped in the mixing region forever. They are rather exceptional ones and not even their closure covers a finite area of the plane, although they are *infinite* in number. Such orbits form a *fractal subset* of the mixing region.

• *Hyperbolic and nonhyperbolic components* are subsets of the chaotic saddle. The first one contains the strongly unstable trapped orbits with local Lyapunov exponents on the order of unity whose neighbourhood will be left by tracer particles rather rapidly. The nonhyperbolic component is the union of weakly unstable trapped orbits with positive local Lyapunov exponents below a threshold, among which orbits with arbitrarily small positive Lyapunov exponents can also be found. This component lies around KAM tori or other sticky surfaces appearing due to no-slip boundary conditions. The separation of these components is somewhat arbitrary (because the value of the threshold Lyapunov exponent can be freely chosen from the range, say, between 0.1 and 0.01 in dimensionless units) and also geometrically somewhat interwoven. Nevertheless, they are responsible for qualitatively different types of motions (see below) and this is why their distinction is rather useful.

• *The stable manifold* is a complicatedly winding curve along which the chaotic saddle can be reached. Tracers with long lifetime can only be the ones approaching the chaotic saddle close along its stable manifold. Thus the stable manifold can also be considered as the 'basin of attraction' of the saddle. It must have a vanishing area because in a Hamiltonian system no attractor can exist. By means of video techniques this manifold can also be determined in an experiment as suggested by Ziemniak *et al.* (1994) and Péntek *et al.* (1995b). Sprinkle tracers in a domain of the flow, record their trajectories, and keep only those whose lifetime in the mixing region is sufficiently long. By plotting the initial points of these trajectories one obtains a good approximant to the stable manifold.

• *The unstable manifold* of the saddle is the set of points along which particles, after entering a close neighbourhood of the set, leave this neighbourhood. This manifold appears on a stroboscopic map as a rather complicatedly winding curve and extends to the region where particles exit the flow. As a consequence, a droplet of particles injected into the flow in front of the mixing region will, after a long time, trace out the

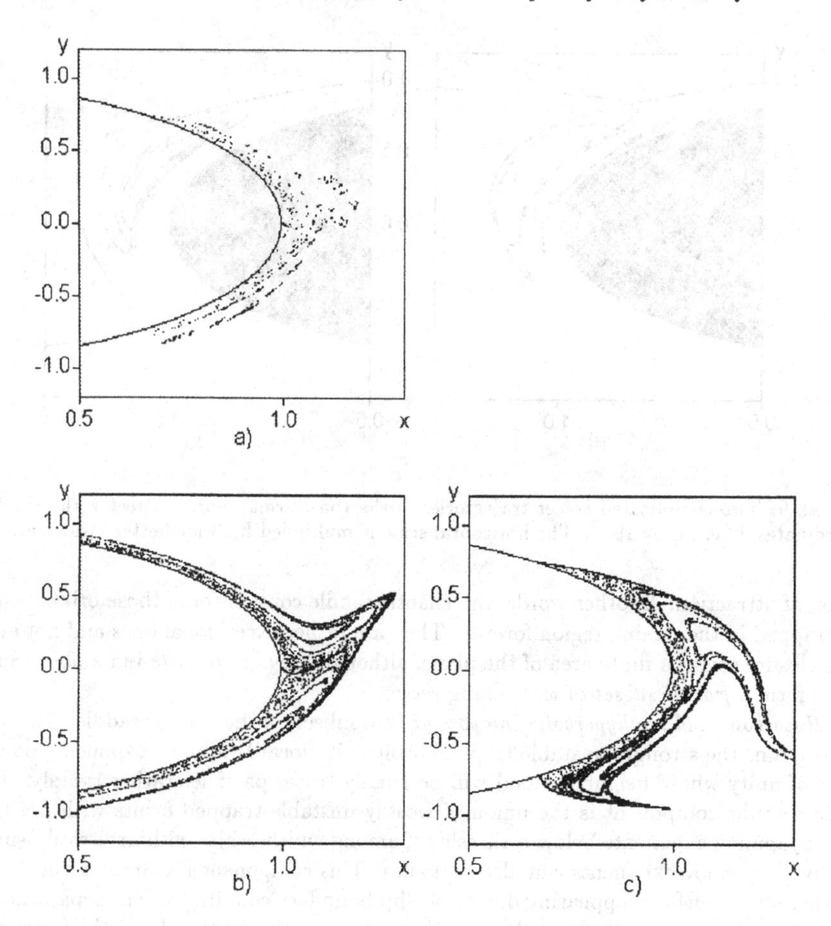

FIGURE 10. Stroboscopic section of the invariant sets in the von Kármán vortex street at $t = 0$ (mod T): (*a*) Chaotic saddle, (*b*) stable manifold, (*c*) unstable manifold. The horizontal scale is multiplied by 3 for better visualization.

unstable manifold, provided the droplet overlaps initially with the stable manifold. This fact makes the unstable manifold a direct physical observable of the passive advection problem, see Ottino (1989) and Beigie *et al.* (1994), and provides the easiest method to plot this manifold experimentally or numerically. (In simulations, the stable manifolds can best be obtained as the unstable ones in the time reversed tracer dynamics. The chaotic saddle is the common part of both invariant manifolds, but there also exist direct powerful methods for its determination, see Nusse & Yorke (1989).)

• *The time delay distribution* is a characteristic of tracer ensembles. Trapped particles have some *time delay* τ relative to the background flow: the longer τ is the more complicated the trajectory becomes. Observe several trajectories (an ensemble) with initial points taken from a closed domain of the flow, or along a straight line. The quantity $P(\tau)d\tau$ is the probability to find a particle with time delay in the interval $(\tau, \tau + d\tau)$. $P(\tau)$ must be a function tending to zero for large times. For intermediate times shorter than some crossover value τ_c it typically decays exponentially,

$$P(\tau) \sim \exp(-\tau/\bar{\tau}). \tag{3.3}$$

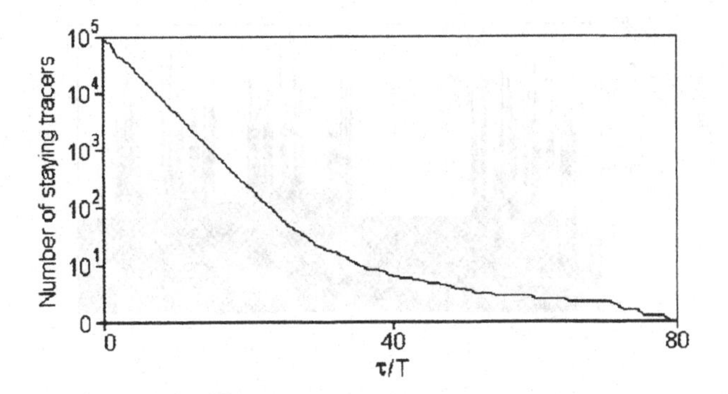

FIGURE 11. A typical time delay distribution $N_0 P(\tau)$ characterizing chaotic tracer dynamics with N_0 as the number of tracer trajectories. For $\tau \ll \tau_c \approx 30T$ and $\tau \gg \tau_c$ equation (3.3) and equation (3.4) holds, respectively. Data are taken from the blinking vortex–sink system with parameter values where KAM tori exist, see Károlyi (1995).

It is only the prefactor, not written out here, that depends on the choice of the initial distribution. For much longer times than the crossover value one observes an algebraic decay

$$P(\tau) \sim \tau^{-\sigma} \tag{3.4}$$

(see figure 11). In periodic flows it is natural to measure τ in units of the velocity field's period, T. We claim that the first behaviour is due to the *hyperbolic* component containing trapped orbits with large local Lyapunov exponents. $\bar{\tau}$ is the average time delay that can then also be considered as the *average lifetime* of chaos on the hyperbolic component and is generally on the order of a few times the flow period. The algebraic decay is a consequence of the presence of neutral trapped orbits, and the exponent σ characterizes the stickiness of the tori. Thus the following qualitative picture emerges concerning the scattering tracer dynamics. Particles starting not exactly on the stable manifold have finite lifetimes in the mixing region. Their motion can be considered as a random walk among the periodic orbits of the chaotic saddle. Those with a not very long lifetime just wander among strictly hyperbolic orbits and have an average lifetime $\bar{\tau}$. The very persistent trajectories must have visited the nonhyperbolic part and their long-time statistics is therefore nonexponential.

• *The time delay function.* A more detailed characterization of the trapping process is based on the observation of individual trajectories. Let us inject tracer particles into the flow along a line and determine the time τ they spend in the mixing region. This defines a function $\tau(y)$ where y denotes the inital position along the line of injection. A unique sign of chaotic tracer scattering is the rather *irregular* appearance of the time delay function as illustrated by figure 12. The intersection of the stable manifold with the line of initial conditions marks points with formally infinite delay times. The different time delays of neighbouring points indicate again the high sensitivity to initial conditions. An irregular scattering function is thus a characteristic of chaotic advection in open flows. The time delay distribution $P(\tau)$ reflects global properties of the scattering process and can easily be derived from the time delay $\tau(y)$ of the trajectories: $P(\tau)\mathrm{d}\tau$ is proportional to the number of tracers whose time delay falls into the interval $(\tau, \tau + \mathrm{d}\tau)$ when integrating over all initial conditions y.

• *Fractal properties of the invariant sets.* First note that the fractal dimension of

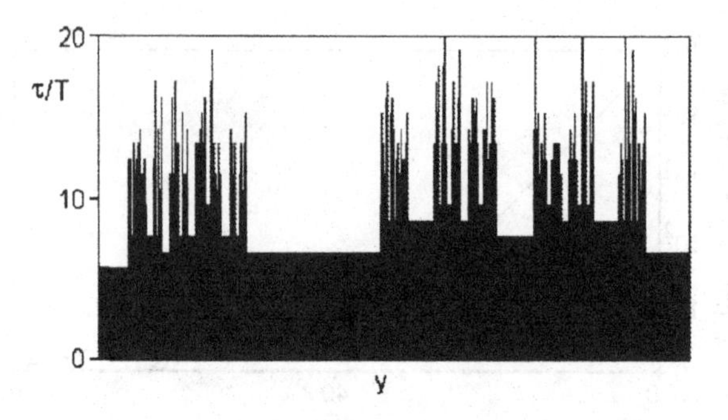

FIGURE 12. A typical irregular time delay function $\tau(y)$ with discrete time delays measured in integer multiples of the flow's period T (data are taken again from the blinking vortex–sink system, see Károlyi (1995)). The large time delay values mark inital conditions with a long trapping in the mixing region. Note the sensitive dependence of τ on y, a criterion for chaotic behaviour.

the Cantor sets whose direct product is the chaotic saddle's hyperbolic component is always a positive number d_0 less than unity. Next, use the fact that the dimension of a direct product of two fractals is the sum of the components' fractal dimensions, see Falconer (1985). Thus, the chaotic saddle's hyperbolic component is of fractal dimension $d_{\text{set}} = 2d_0$. The manifolds emanating from this component are the direct products of a line and a Cantor set, therefore, $d_{\text{manifold}} = 1+d_0$. Note that because of the time reversal symmetry of the canonical equations of motion (1.1), (1.2), the stable and unstable manifolds have common scaling properties and have therefore identical dimensions.

The fractal properties of the nonhyperbolic component are different. In these regions the partial fractal dimension of the chaotic set tends to unity in the limit of extremely fine resolution, see Lau *et al.* (1991). The escape rate and the average Lyapunov exponent are expected to be zero on this component, see Christiansen & Grassberger (1993), in accordance with (3.4). Therefore a fat fractal behaviour becomes evident when increasing the spatial resolution: $d_{\text{set}} \to 2$ and $d_{\text{manifold}} \to 2$. In numerical simulations it might be hard to reach the resolution where the fat fractal property of the manifolds is evident away from the immediate vicinity of the nonhyperbolic component. Thus, we can state that on the practically relevant length scale of 10^{-4} or larger (in dimensionless units) the tracer patterns appear as real fractals with noninteger dimensions.

- *Singularities of the time delay function* sit in points where the line of initial conditions intersect the stable manifold. Thus, with a resolution larger than some critical value ϵ_c, the dimension of singularities is the same value as that of the component Cantor sets: $d_{\text{sing}} = d_0$. On finer scales, one observes a crossover to a fat fractal behaviour and $d_{\text{sing}} \to 1$. The critical resolution ϵ_c, just like the crossover time τ_c, might depend on the position of the line of initial conditions.
- *Fractal tracer boundaries* are easy to generate in open time-periodic flows. One way is to inject dye particles continuously into the flow along a line so that the colours are different in two neighbouring segments, see Péntek *et al.* (1995b), and investigate how the boundary evolves in time. Alternatively, one can also divide the fluid in strips of different colours and study the deformation of the boundaries, see Péntek *et al.* (1995a). After a long time of observation, the boundaries become rather complex containing a fractal part that is nothing but the *unstable* manifold of the chaotic saddle, see Péntek *et al.* (1995b). A different type of boundary is obtained by clouring the inital points according to how

the tracers exit the mixing region. In a problem like the blinking vortex–sink system, the most natural choice is to use different colours for exiting via different sinks, see Aref *et al.* (1989). More generally, one can colour the initial points depending on whether the trajectories cross a preselected line outside of the mixing region in a given segment or not. The fractal part of such boundaries will contain the *stable* manifold of the chaotic saddle, see Péntek *et al.* (1995b). The second type of boundary is an extension of the fractal basin boundary concept, see Grebogi *et al.* (1983), for Hamiltonian systems without any attractors. For both types of boundaries their fractal dimension is the same as that of the manifolds: $d_{\text{boundary}} = d_{\text{manifold}} = 1 + d_0$.

• *Relation between dynamical and fractal properties.* Chaotic advection in open flows is a phenomenon where one clearly sees how the underlying dynamics determines the fractal properties of the chaotic set and its manifolds. Due to a general relation valid for any kind of transient chaos, see Kantz & Grassberger (1985) and Tél (1990), the fractal dimension d_0 of a component Cantor set appears as

$$d_0 \approx 1 - \frac{1}{\bar{\lambda}\bar{\tau}}, \tag{3.5}$$

where $\bar{\lambda}$ is the (positive) average Lyapunov exponent of trajectories spending a long time around the chaotic saddle. This formula says that the deviation of the fractal dimension from unity is proportional to the ratio of two average times characterizing local and global instability. The quantity $1/\bar{\lambda}$ is the average time of the separation of nearby trajectories by a factor of e, while $\bar{\tau}$ is the characteristic time, defined by (3.3), of emptying a given area. Between two sets with the same Lyapunov exponent, the one with larger average chaotic lifetime has the larger dimension.

• *More general flows.* Finally we note that the features summarized here seem to be robust. A weak deviation from the flow's planar character makes the canonical form unvalid for the tracer dynamics, nevertheless, a chaotic saddle will govern the process in the three-dimensional phase space of (1.1). Similarly, if the particles have inertia, the dimension of the phase space doubles due to the appearance of physical momenta, even in the crudest approximation. In the simplest case of a planar flow, the phase space is four dimensional in which a chaotic saddle exists. What we observe as tracer patterns is then related to the projection of the invariant manifolds to the plane of the flow. Furthermore, if the flow is slightly compressible, a large number of periodic attractors are expected to appear in the tracer dynamics. They can thus only have rather tiny basins of attractions. Consequently, the dynamics is dominated by the extended chaotic saddle(s) existing among the attractors producing long average chaotic lifetimes.

We are grateful to Celso Grebogi and Jim Yorke for a most enjoyable and fruitful collaboration. Helpful discussions with them, with J. Kadtke, Z. Kovács, Z. Neufeld, K. G. Szabó and A. Provenzale are acknowledged. This work was partially supported by the Hungarian Science Foundation under grants OTKA T019483, T17493, F17166, and the U.S.–Hungarian Science and Technology Joint Fund under projects JF number 286 and 501.

REFERENCES

AREF, H. & BALACHANDAR, S. 1986 *Phys. Fluids* **29**, 3515.

AREF, H., JONES, S. W., MOFINA, S. & ZAWADSKI, I. 1989 *Physica* D **37**, 423.

BEIGIE, D., LEONARD, A. & WIGGINS, S. 1994 *Chaos Sol. Fract.* **4**, 749.

CHRISTIANSEN, F.& GRASSBERGER, P. 1993 *Phys. Lett.* A **181**, 47.

FALCONER, K. 1985 *The Geometry of Fractal Sets.* Cambridge Univ. Press.

GREBOGI, C., *et al.* 1983 *Phys. Lett.* A **99**, 415; see also 1987 *Physica* D **25**, 347.

JUNG, C., TÉL, T. & ZIEMNIAK, E. 1993 *Chaos* **3**, 555.

JUNG, C. & ZIEMNIAK, E. 1992 *J. Phys.* A **25**, 3929; see also ZIEMNIAK, E. & JUNG, C. 1995 *Phys. Lett.* A **202**, 263.

JUNG, C. & ZIEMNIAK, E. 1994 In *Fractals in the Natural and Applied Sciences* (ed. M. M. Novak). North Holland.

KANTZ, H. & GRASSBERGER, P. 1985 *Physica* D **17**, 75.

KÁROLYI, G. 1995 *Diploma work.* In Hungarian; see also KÁROLYI, G. & TÉL, T. To be published.

LANDAU, L. D. & LIFSHITZ, E. M. 1959 *Fluid Mechanics.* Pergamon.

LAU, Y. T., FINN, J. M. & OTT, E. 1991 *Phys. Rev. Lett.* **66**, 978.

MILNE-THOMSON, L. M. 1958 *Theoretical Aerodynamics.* Macmillan.

NUSSE, H. E. & YORKE, J. 1989 *Physica* D **36**, 137.

OTTINO, J. M. 1989 *The Kinematics of Mixing: Stretching, Chaos and Transport.* Cambridge University Press; see also OTTINO, J. M. 1990 *Ann. Rev. Fluid Mech.***22**, 207.

PÉNTEK, Á., TÉL, T. & TOROCZKAI, Z. 1995a *J. Phys.* A **28**, 2191; see also 1995 *Fractals* **3**, 33.

PÉNTEK, Á., TOROCZKAI, Z., TÉL, T., GREBOGI, C. & YORKE, J. A. 1995b *Phys. Rev.* E **51**, 4076.

SAFFMAN, P. G. 1992 *Vortex Dynamics.* Cambridge Univ. Press; see also AREF, H. 1983 *Ann. Rev. Fluid Mech.* **15**, 345; MELESHKO, V. V., *et al.* 1992 *Phys. Fluids* A **4**, 2779; MELESHKO, V. V. & VAN HEIJST, G. J. F. 1994 *Chaos Sol. Fract.* **4**, 977.

SHARIFF, K., LEONARD, A. 1992 *Ann. Rev. Fluid. Mech.* **24**, 235.

SHARIFF, K., LEONARD, A., ZABUSKY N. J. & FERZIGER, J. H. 1988 *Fluid. Dyn. Res.* **3**, 337.

SHARIFF, K., PULLIAM, T. H. & OTTINO, J. M. 1991 *Lect. Appl. Math.* **28**, 613.

SMILANSKY, U. 1992 In *Chaos and Quantum Physics* (ed. M. J. Giannoni *et al.*). Elsevier; see also JUNG, C. 1992 *Acta Phys. Pol.* **23**, 323; OTT, E. & TÉL, T. 1993 *Chaos* **3**, 417; KOVÁCS, Z. & WIESENFELD, L. 1995 *Phys. Rev.* E **51**, 5476.

TÉL, T. 1990 In *Directions in Chaos* (ed. Hao Bai-Lin), vol. 3, pp. 149–221. World Scientific.

VAN DYKE, M. 1982 *An Album of Fluid Motion.* The Parabolic Press.

WIGGINS, S. 1992 *Chaotic Transport in Dynamical Systems.* Springer.

ZIEMNIAK, E., JUNG, C. & TÉL, T. 1994 *Physica* D **76**, 123.

Homoclinic chaos

By LEONID SHILNIKOV

Research Institute for Applied Mathematics and Cybernetics, Nizhny Novgorod, Russia

We argue that dynamical chaos is always homoclinic. We describe three main classes of strange attractors: hyperbolic, Lorenz type and quasi-attractors. We stress that a complete description of models with quasi-attractors is unrealistic due to uncontrolled bifurcations of structurally unstable Poincaré homoclinic orbits in the Newhouse regions.

1. Introduction

The last decades of XXth century have been marked with great discoveries in nonlinear dynamics of multi-dimensional systems. In a sense, the events that happened can be compared to those in the 1920–1930s, the years of discovery of self-oscillations and creation of a mathematical theory of nonlinear oscillations for planar systems. The present situation is marked with the discovery of a principally new type of motion, chaotic oscillations. In experiments, such oscillations are usually treated as a noise and hence are also called stochastic oscillations. In order to emphasize their dynamical origin, one usually calls them dynamical chaos. Since the discovery of dynamical chaos, many problems of physics, mechanics, ecology, etc., which can be modelled by differential equations, have obtained adequate mathematical description. However, the explanation of some phenomena required the creation and development of new mathematical methods.

Because a central problem to study in nonlinear dynamics are differential equations, we have first to define which objects of a qualitative theory of dynamical systems may claim to be mathematical images of chaotic oscillations. As is now known these are attracting non-wandering sets with unstable trajectories. Such a set is called a strange attractor. The following remark must be made here: a strange attractor cannot be unified universally because it may be a smooth manifold or have a non-trivial structure (from the viewpoint of set topology). In the latter case, one can usually speak of its fractal structure, whose dimension is not an integer. Noticing furthermore that structurally stable strange attractors in concrete applications have not yet been observed, we can imagine how complex mathematical problems of dynamical chaos are. Two principal ones are: the appearance of a strange attractor and its bifurcations. Three-dimensional systems are well understood, but the case of high-dimensional chaos has a lot of 'white spots'. The problem of high-order chaos is crucial, as it is a bridge to understanding the complex dynamics of extended systems. We would also like to mention another important aspect: as all trajectories in a strange attractor are unstable, the description of the motion should be statistical. This means that in order to study chaotic behaviour, one has to use the methods of ergodic theory. This theory started in the 1920–1930s. It was focusing on conservative systems, i.e. those preserving the phase volume. Of special interest for us are geodesic flows on compact manifolds with negative curvature, i.e. flows with unstable trajectories. They are important, because their study stimulated the development of a new branch of mathematics: symbolic dynamics. Ergodic theory and qualitative theory were developing almost independently for a long time. The discovery of strange attractors has changed this situation. Both branches started to interact, thereby forming a powerful mathematical background for dynamical chaos.

We pose the following question: if one admits the possibility of chaos, what sort

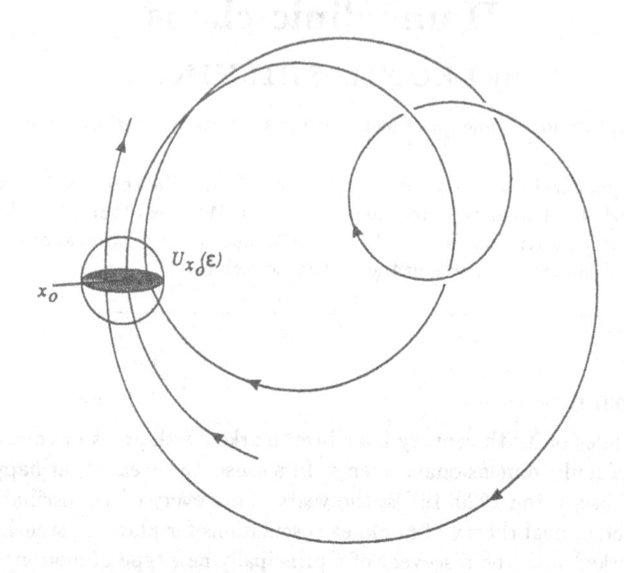

FIGURE 1. An open Poisson stable trajectory passing through the point x_0 intersects the shadowed cross-section inside the ϵ-neighbourhood of the point x_0 infinitely many times.

of trajectories must be associated with it? They are clearly not just the equilibrium states, i.e. periodic or quasi-periodic orbits (if the latter two are Lyapunov stable, they correspond to self-oscillations and modulations, respectively). To answer this question we shall turn to the general theory of dynamical systems of Poincaré and Birkhoff, in which a primary role is played by the *set of central motions* or the *center*. The significance of the center follows from the fact that all trajectories spend most of the time in its neighbourhood. Moreover, for the set of center motions, Poincaré's recurrence theorem holds, and the so-called Poisson stable trajectories are everywhere dense.

2. Poisson stable trajectories

Let a smooth system

$$\dot{x} = X(x), \quad x \in R^n, \tag{2.1}$$

have a trajectory $x(t, x_0)$ passing through the point x_0 at $t = 0$.

DEFINITION 2.1. *The trajectory $x(t, x_0)$ is called a* Poisson stable trajectory *(or* P-trajectory*) if for any $\epsilon > 0$ it comes back to an ϵ-neighbourhood of the point x_0 infinitely many times as $t \to +\infty$ and $t \to -\infty$ (see figure 1).*

We note that this definition is independent of the choice of the point x_0.† Thus if inside the ϵ-neighbourhood we draw a cross-section passing through the point x_0, there

† A weaker property of the return of trajectories can be formulated on the basis of the concept of nonwandering. A point x_0 is called nonwandering if for any neighbourhood U_{x_0}: $T^t U_{x_0} \cap U_{x_0} \neq \emptyset$ for sufficiently large t; here $T^t : R^n \to R^n$ is a flow determined by the vector field (2.1). It is easy to show that the set Ω of all nonwandering points consists of whole trajectories, i.e. it is invariant. Hence, a dynamical system $\{T^t\}$ can be considered on Ω and another nonwandering set Ω_1 can be distinguished in it, and so on. Birkhoff has shown that this process stabilizes at some (transfinite) number of steps β. The obtained set Ω_β is called the center of the dynamical system and is the greatest closed set whose points are all nonwandering

exists a countable sequence

$$\{t_n(\epsilon)\}_{n=-\infty}^{n=+\infty}$$

of the times at which the trajectory $x(t, x_0)$ successively intersects the cross-section. The values

$$\tau_n(\epsilon) = t_n(\epsilon) - t_{n-1}(\epsilon)$$

are called *Poincaré return times*. It is clear that both the equilibrium states and periodic orbits are P-trajectories.

THEOREM 2.1 (BIRKHOFF). *The closure of an open Poisson stable trajectory contains a continuum of open Poisson stable trajectories.*

Two essentially different cases are possible for an open P-trajectory:
(a) the sequence $\{\tau_n(\epsilon)\}$ is bounded by a number $L(\epsilon)$ such that $\lim_{\epsilon\to 0} L(\epsilon) = +\infty$,
(b) the sequence $\{\tau_n(\epsilon)\}$ is unbounded.

In the first case the P_1-trajectory is called *recurrent*. For such a trajectory, all trajectories in its closure are also recurrent, and the closure itself is a minimal set.† A basic property of a recurrent trajectory is that it comes back to the ϵ-neighbourhood of the point x_0 after a time not greater than $L(\epsilon)$. However, in contrast to the periodic or almost-periodic orbits, whose return times are known (the period or almost period), the return time for a recurrent trajectory is not constrained.

In the second case, in the closure of the P_2-trajectory there always are invariant closed subsets which can be either equilibrium states or periodic orbits. Since the P_2-trajectory can approach such subsets arbitrarily closely, Poincaré return times can be arbitrarily large. This means that forecasting is impossible. Therefore, we will consider such P_2-trajectories as chaotic.

The problem of finding a P_2-trajectory is non-trivial. To begin with, let us see what are the properties of a structurally stable system having a P_2-trajectory. First of all, it must be at least three-dimensional, because systems with P_2-trajectories on two-dimensional surfaces are non-rough (structurally unstable).

A rough (structurally stable) system with P_2-trajectory must have a countable set of saddle periodic orbits (see figure 2). If we choose two neighbouring points Q_n and Q_{n+1} and look at them through a magnifying glass, we will see a picture as shown in figure 3. The stable and unstable manifolds of these points intersect each other as shown in figure 3. If we take some iterations of the unstable manifold of the point Q_{n+1}, we discover that the stable and unstable manifolds of the point Q_{n+1} also have an intersection. The intersection point q is called a *homoclinic* one (figure 4).

The concept of a Poincaré homoclinic orbit, i.e. a trajectory biasymptotic to a saddle periodic orbit, is fundamental in the theory of multi-dimensional dynamical systems. Poincaré (1899) was the first who noted that if there is one transverse homoclinic orbit, there are also a countable number of homoclinic orbits of more complex nature (figure 5).

Later, Birkhoff (1935) showed that, in a similar situation for homeomorphisms preserving the square, there is a countable set of saddle periodic orbits. He also formulated the following hypothesis: the set of trajectories near a homoclinic orbit admits a *symbolic description*.

with respect to this set. The closure of the set of all Poisson stable trajectories coincides with the center.

† A set is called *minimal* if it is non-empty, invariant, closed and contains no proper subsets possessing these three properties.

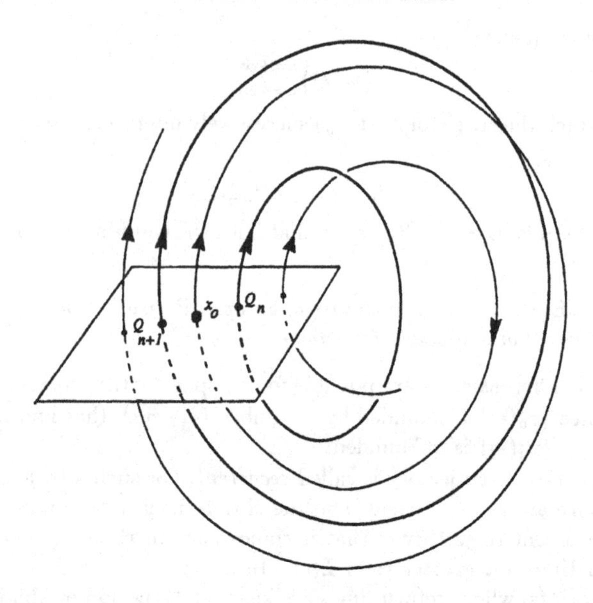

FIGURE 2. The points Q_n and Q_{n+1} are the intersections of two nearby saddle periodic orbits with the cross-section. They are fixed points of the associated Poincaré map along orbits close to the P_2-trajectory.

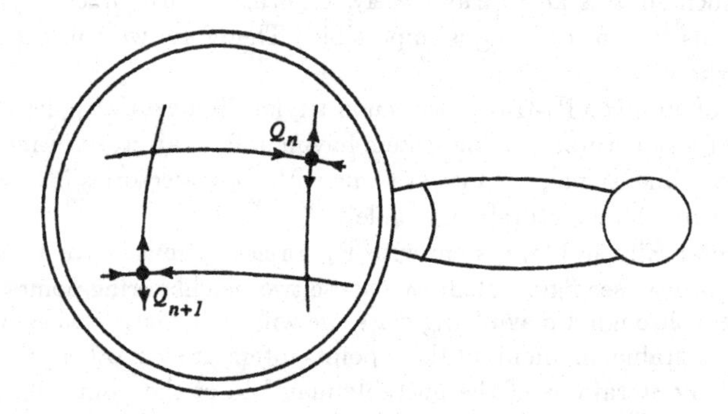

FIGURE 3. The invariant stable and unstable manifolds of the points Q_n and Q_{n+1} intersect each other transversely.

3. The concept of symbolic description

We will show how a symbolic description arises for a trajectory in a neighbourhood of the homoclinic orbit in the case of a flow.

Suppose we have a saddle periodic orbit L with a homoclinic trajectory Γ (figure 6). We surround L and Γ by a neighbourhood U which has the shape of a solid torus to which a handle containing Γ is glued (figure 7). We will code a trajectory lying in U using a symbol 0 if the trajectory makes a complete circuit within the solid torus, and a symbol $\hat{1}$ if it goes along the handle. Thus, the sequence

$$\{0\}_{n=-\infty}^{n=+\infty}$$

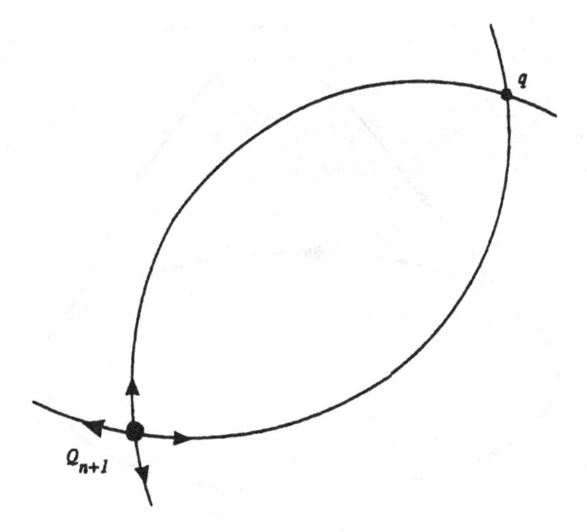

FIGURE 4. A transverse homoclinic orbit.

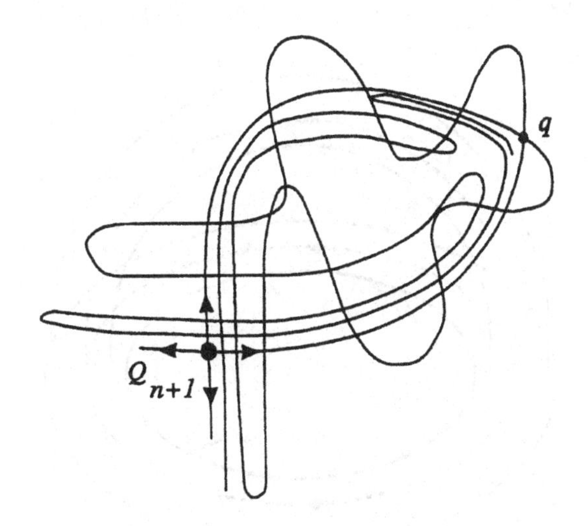

FIGURE 5. Homoclinic tangles.

corresponds to the periodic orbit L. The sequence

$$\{\dots,0,0,\hat{1},0,0,\dots\}$$

corresponds to the homoclinic orbit Γ. The sequence

$$\{j_n\}_{n=-\infty}^{n=+\infty}, \quad j_n = \{0,\hat{1}\}$$

corresponds to an arbitrary trajectory within U. Moreover, each $\hat{1}$ is followed by zeros whose number is not less than $\bar{k}$, where $\bar{k}$ depends on the size of the neighbourhood: the narrower the neighbourhood U, the bigger $\bar{k}$.

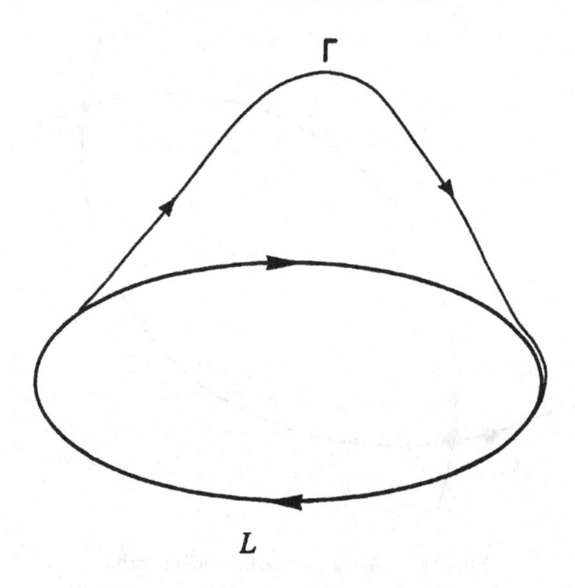

FIGURE 6. A homoclinic curve Γ to the saddle periodic orbit L.

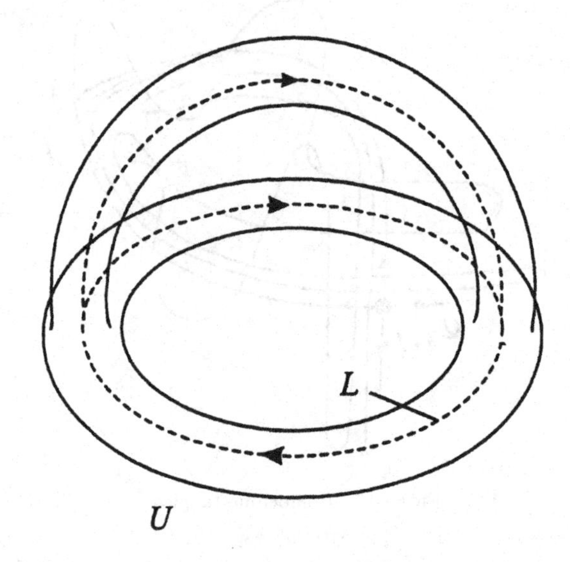

FIGURE 7. The neighbourhood of L and Γ has the shape of a solid torus with a handle.

If we introduce a symbol

$$1 = \left[\hat{1}, \overbrace{0, \ldots, 0}^{\bar{k}}\right],$$

we obtain a new sequence for such a trajectory,

$$\{j_n\}_{n=-\infty}^{n=+\infty}, \quad j_n = \{0, 1\},$$

where j_n can be followed by either 1 or 0. In other words, the set of trajectories lying in

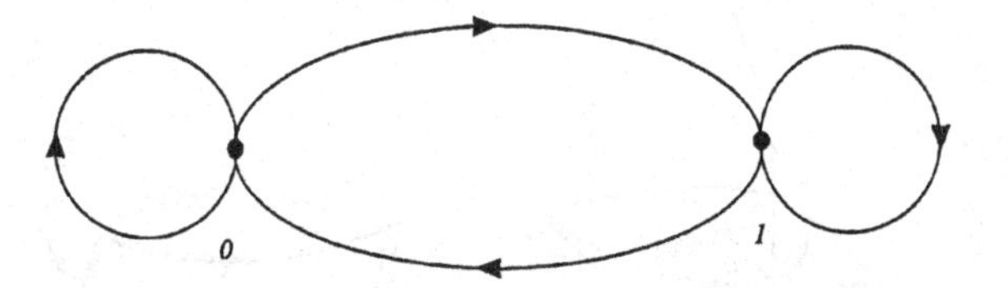

FIGURE 8. The graph of the Bernoulli scheme on two symbols: 0 and 1.

U is in correspondence with the Bernoulli scheme on two symbols (figure 8). Moreover, in this case the inverse statement is also valid. Namely, each trajectory of the original system corresponds to a unique trajectory of the Bernoulli scheme.†

The fact that trajectories can be described by a symbolic language is very important. In the 1920–1930s, Birkhoff, Morse and Hedlund showed that similar symbolic systems always have P-trajectories.

The existence of complex structures near a homoclinic orbit was established by Smale (1983), for example in the famous 'Smale horseshoe'. A complete description of all trajectories in the neighbourhood of homoclinic orbit was given by Shilnikov (1967). He called this problem the Poincaré–Birkhoff problem.

It is also important to note that a complex structure in the neighbourhood of a transverse homoclinic orbit is preserved under small, smooth perturbations. This means that also Poisson stable trajectories are preserved. Hence, it is natural to assume that a universal criterion for chaos in non-linear dynamics is the presence of structurally stable Poincaré homoclinic orbits. In other words, chaos is always *homoclinic*.

We give two examples to show how symbolic dynamics can occur in simple systems.

Example 1. Assume that there is a three-dimensional smooth system

$$\dot{x} = X(x, \mu) \tag{3.2}$$

having, for $\mu = 0$, a non-rough equilibrium state of the saddle–saddle type at the origin. This means that in a small neighbourhood of the origin O, the system can be written in the form

$$\begin{aligned} \dot{x} &= \mu + x^2, \\ \dot{y} &= -y, \\ \dot{z} &= z. \end{aligned}$$

For $\mu < 0$ it has two saddle equilibria O_1 and O_2; O_1 has a two-dimensional stable invariant manifold $W^s_{O_1}$ and a one-dimensional unstable invariant manifold $W^u_{O_1}$, whereas for O_2, $\dim W^s_{O_2} = 1$ and $\dim W^u_{O_2} = 2$ (figure 9).

For $\mu = 0$ the point O is of saddle–saddle type which has a stable manifold W^s_O and unstable manifold W^u_O, homeomorphic to semi-planes (figure 10). With further increase of μ ($\mu > 0$), the equilibrium disappears.

We assume that for $\mu = 0$, W^s_O and W^u_O intersect transversely along the trajectories $\Gamma_1, \ldots, \Gamma_m$, $m \geq 2$.

STATEMENT 3.1 (SHILNIKOV (1969)). *As μ increases through 0, the equilibrium state O disappears and an unstable set Ω is born. Its trajectories are in one-to-one correspondence to the Bernoulli scheme on m symbols. As $\mu \to 0^+$, Ω is tightened to the*

† A more precise description requires the use of a construction called 'suspension under the Bernoulli subshift'.

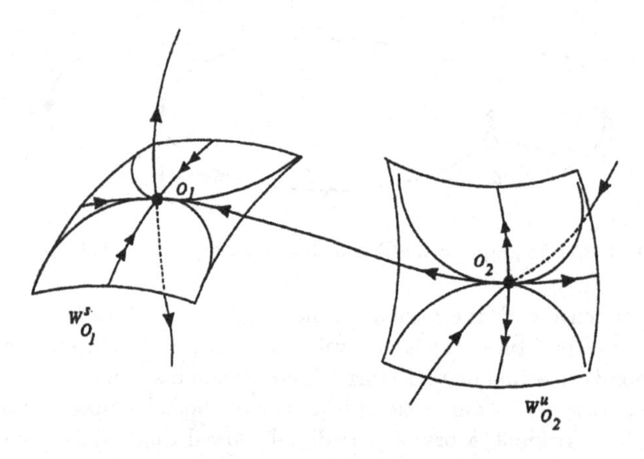

FIGURE 9. Two saddles O_1 and O_2 for $\mu < 0$.

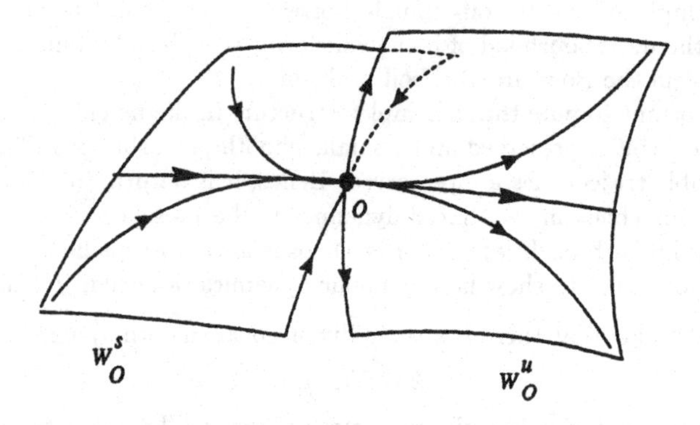

FIGURE 10. A non-rough equilibrium state of saddle–saddle type for $\mu = 0$.

bouquet:

$$\{O \cup \Gamma_1 \cup \ldots \cup \Gamma_m\}.$$

Example 2. We assume that the system (2.1) depends on two parameters, i.e. on $\mu = (\mu_1, \mu_2)$, and for $\mu = 0$, has an equilibrium state O of saddle type, whose characteristic exponents satisfy

(*a*) $\lambda_2 < \lambda_3 < 0 < \lambda_1$,

(*b*) the saddle value $\sigma = \lambda_1 + \lambda_3 > 0$.

Furthermore, there are one-dimensional trajectories (separatrices), Γ_1 and Γ_2, leaving the saddle and returning to the origin as $t \to +\infty$, thereby composing a homoclinic butterfly, see figure 11. We also impose two important general conditions:

(i) The separatrices Γ_1 and Γ_2 return to the saddle along the leading direction corresponding to the characteristic exponent λ_3.

(ii) The separatrix values A_1 and A_2 are non-zero.†

Let the parameters μ_1 and μ_2 be such that μ_1 governs the behaviour of Γ_1 and μ_2 the

† Some details concerning A_1 and A_2 are given in what follows; see also Afraimovich *et al.* (1983) and Shilnikov (1980).

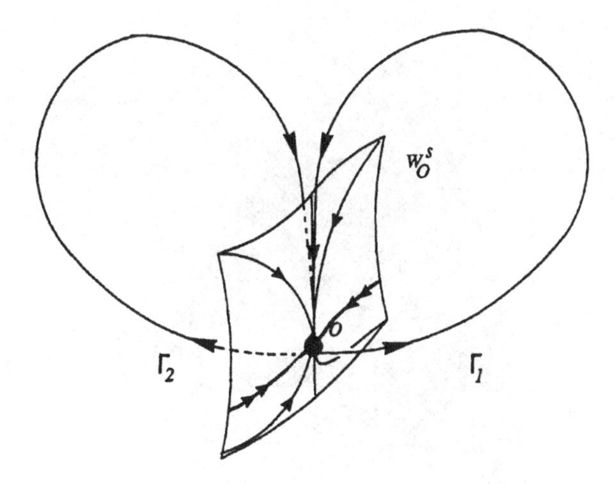

FIGURE 11. Two one-dimensional separatrices Γ_1 and Γ_2 forming a homoclinic butterfly.

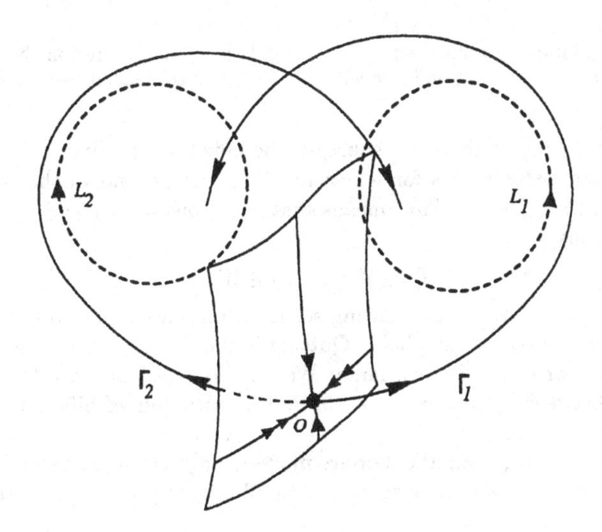

FIGURE 12. On breaking the homoclinic butterfly a saddle periodic orbit L_i is born from each homoclinic loop for $\mu_i > 0$.

behaviour of Γ_2. Moreover, assume that the saddle periodic orbits L_i born from each loop Γ_i (see figure 12) exist for $\mu_i > 0$.

We can now explain the meaning of A_1 and A_2. If $A_i > 0$, the two-dimensional stable manifold $W^s_{L_i}$, and the unstable one $W^u_{L_i}$, are homeomorphic to a cylinder; if $A_i < 0$, they are homeomorphic to a Möbius strip. Figure 13 gives the bifurcation diagram for $0 < A_i < 1$, $i = 1, 2$.

For parameter values from the shaded region there is a non-trivial repeller containing the Poincaré homoclinic orbit. Outside the shaded region one is dealing with either one or two periodic orbits. The curves l_1 and l_2 correspond to parameter values for which homoclinic orbits look like those shown in figure 14.

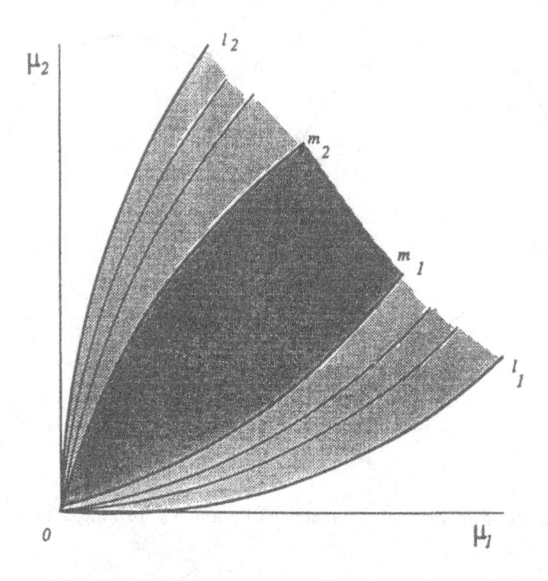

FIGURE 13. The bifurcation diagram for $0 < A_i < 1$. In the darker region the limit set is in the one-to-one correspondence with the Bernoulli scheme on two symbols.

Inside the shaded region there is a subregion bounded by the curves m_1 and m_2. They correspond to parameter values for which the separatrices tend to the saddle periodic orbits, as shown in figure 15. That means that equations for m_1 and m_2 can be found from the conditions

$$\Gamma_1 \in W^s_{L_2}, \quad \Gamma_2 \in W^u_{L_1}.$$

In the subregion in question, the limiting set is also in one-to-one correspondence with the Bernoulli scheme on two symbols. Outside it the description is more complicated: it requires the use of kneading invariants. We also note that between the curves l_1 and m_1, as well as between l_2 and m_2, one can find a continuum of bifurcation curves, but no stable orbits.

In a symmetric case (as in the Lorenz model), only one parameter is needed; for $\mu_1 = \mu_2 > 0$, a set homeomorphic to the Bernoulli scheme is born from the homoclinic butterfly.

As we have noted, all of the above-mentioned sets are unstable. Nevertheless, their role is important because, as parameters are varied further, they can become attractors.

4. Strange attractors

Now we consider the case where a P_2-trajectory belongs to a non-trivial attractor – an attractive limiting set Ω. The fact that this set is transitive means that there is a trajectory which is dense in Ω.

It is worth beginning the discussion of this case with a question posed by Andronov in the 1920s in connection with the mathematical theory of oscillations: can a chaotic P-trajectory be stable? The answer was given by Markov: a *stable* P-trajectory must be almost-periodic.†

† Moreover, we require the P-trajectory to be uniformly Lyapunov stable.

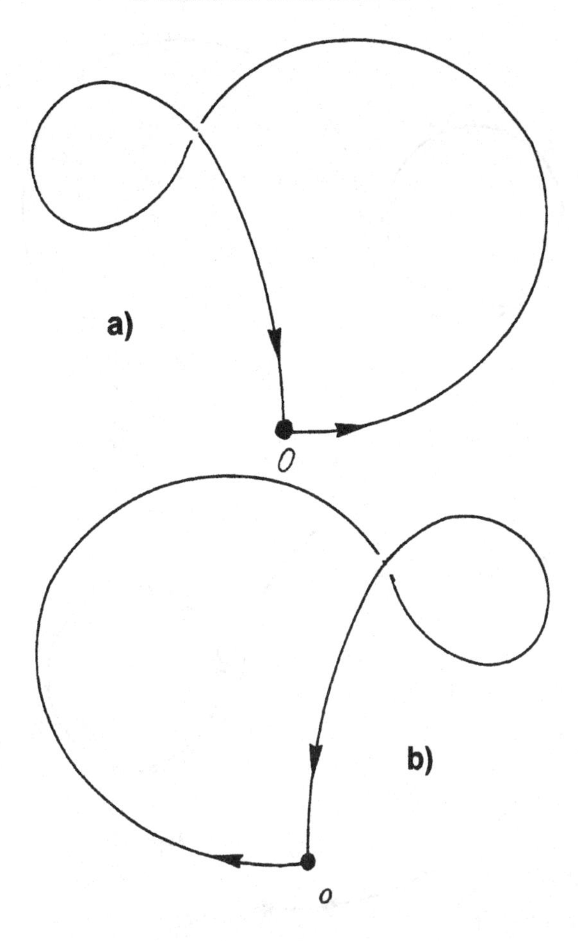

FIGURE 14. The double-circuit homoclinic loops: (*a*) Γ_1 and (*b*) Γ_2.

Thus, the only possible stable limiting regimes are: equilibrium states, periodic and almost-periodic orbits. Probably, this fact is connected with the famous hypothesis of Landau and Hopf on the route to turbulence via a cascade of quasi-periodic regimes:

$$O \to T^1 \to \ldots \to T^n \to T^{n+1} \to \ldots$$

with increasing Rayleigh number.

Hence, we can conclude that a transitive P_2-trajectory is unstable in a non-trivial attractor. But that means that all trajectories in such an attractor are unstable.

Thus, we arrive at a contemporary definition of chaotic attractor: *this is an attractive set of central motions in which all trajectories are unstable.*

It follows from the analysis given in the previous section that strange attractors in nonlinear dynamics, as well as P_2-trajectories, may contain many trajectories of different type. In particular we can mention periodic orbits, homo- and hetero-clinic trajectories and saddle equilibria, if there are any. Although all these trajectories compose a very thin set, in contrast to the set of P_2-trajectories, they deserve a special attention from the point of view of qualitative theory. There are a few reasons for this. First, the bifurcations of periodic orbits, as well as homo- and hetero-clinic cycles are the easiest way on the

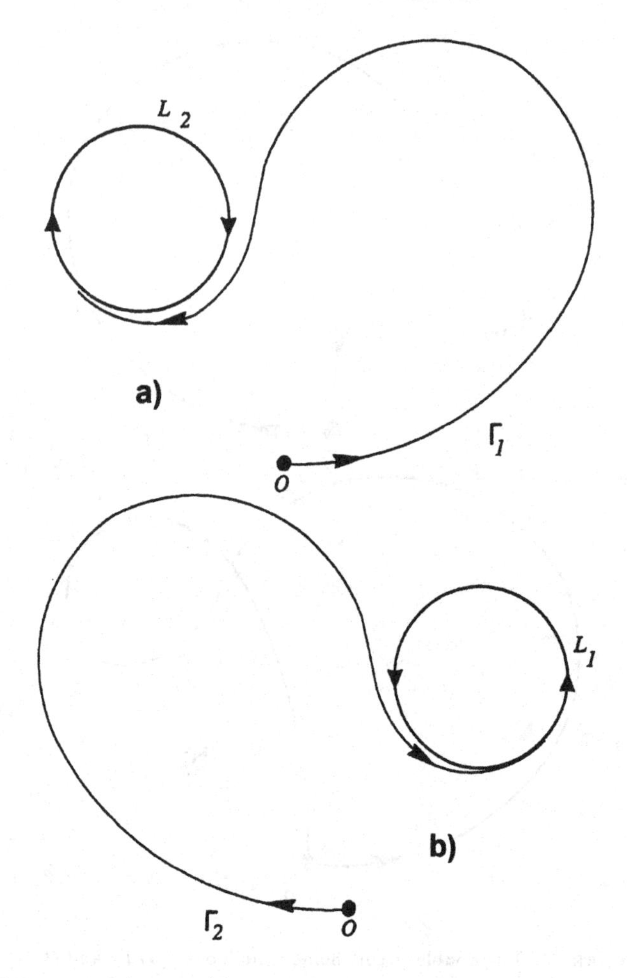

FIGURE 15. The curves: (a) m_1 or (b) m_2 corresponding to situations where the separatrix Γ_1 or Γ_2 tends to the saddle periodic orbit L_2 or L_1.

route to chaos. Second, the principal structural features of a strange attractor can simply be detected by these bifurcations, which is the most comprehensive at its birth. Third, these trajectories are usually responsible for main internal bifurcations of the attractor, up to its disappearance.

Non-trivial attractors in finite-dimensional systems can be divided into three major classes: hyperbolic, Lorenz type and quasi-attractors (an abbreviation for quasi-stochastic attractors).

5. Hyperbolic attractors

Hyperbolic attractors are the limiting sets for which Smale's Axiom A is valid, and hence, they are structurally stable. In such attractors both periodic, homoclinic orbits and P-trajectories are everywhere dense and are of the same saddle type, i.e. the stable (or unstable) manifolds of all trajectories have the same dimension. Examples of these attractors are the Anosov systems, see Anosov (1967), and solenoids of Smale–Williams, see Smale (1967), Williams (1970) and Plykin (1974). These examples were used by Ruelle & Takens (1971) for constructing systems with strange attractors on the three- and four-dimensional tori. Hyperbolic attractors are appropriate objects for ergodic theory. However, in concrete systems, they have not yet been observed. It is possible that one of the reasons for it is the lack of understanding of how hyperbolic structures can arise. However, some progress can be seen. As was established by Shilnikov & Turaev (In preparation), a co-dimension one global bifurcation of a saddle-node periodic orbit of an $(n+1)$-dimensional Morse–Smale system $(n \geq 3)$ can lead to the birth of a hyperbolic attractor, whose intersection with a global cross-section is a Smale–Williams solenoid. The associative geometrical construction which yields the solenoid is valid in three or more dimensions.

Let us consider a solid torus Π in R^3, i.e. $D^2 \times S^1$ where D^2 is a disk and S^1 is a circumference. We now expand Π m times (m is an integer) along the cyclic coordinate and shrink it q times along other coordinates, where $q \leq 1/m$. After that let us embed what we have received into Π, as shown in figure 16, so that this set intersects any section along m disks. We denote this set by Π_1. Now repeat the routine with Π_1 and so on. The set $\Sigma = \subset_{i=1}^{\infty} \Pi_i$ thus obtained is called a solenoid. The solenoid is locally designed as the direct product of an interval and a Cantor set. This construction was first pointed out by Wictorius and Van Danzig. They were trying to build a vector field in Π with an attractor Σ which is a quasi-minimal set of limiting periodic trajectory. Smale noticed that such solenoids may be hyperbolic attractors of diffeomorphisms defined on solid tori. Moreover, similar attractors can be realized as a limit of the inverse spectrum of the expanding cycle map:

$$\bar{\theta} = m\theta \pmod 1.$$

In a number of his papers Williams considered these problems, and even the case of multidimensional solenoids.

The bifurcation described above is very attractive for a few reasons. The point is that if W^{ss} and W^{u} of the cycle have no intersection, and furthermore all trajectories starting from W^{u} come back to the cycle as $t \to +\infty$, then inside T one can construct a Poincaré map which may be written as

$$\bar{x} = f(x, \theta, \mu), \quad \bar{\theta} = t_0(\mu) + m\theta + g(\theta) + h(x, \theta, \mu) \pmod 1,$$

where $x = (x_1, \ldots, x_{n-1})$, θ is a cyclic variable, the functions f and g tend to zero along with their first derivatives as $\mu \to 0$, m is an integer, $g(\theta)$ is a smooth periodic function and $t_0(\mu) \to \infty$ as $\mu \to 0$. The parameter μ governs the shift from the bifurcation surface of saddle-node cycles, so that $t_0(\mu) \sim 1/\sqrt{\mu}$. It is clear that there is a strong convergence along the x-variable. In oder to have an expansion along θ, we must satisfy $|m| \geq 2$. An additional condition which guarantees the Poincaré map to have a solenoid is

$$\max |m + g'(0)| > 1, \quad 0 \leq \theta < 1.$$

In other words, the Poncaré map will have a solenoid for sufficiently small μ, if the one-dimensional map

$$\bar{\theta} = w + m\theta + g(\theta) \pmod 1, \quad 0 \leq w < 1,$$

is expanding. Note that for $n = 2$, $m = 0, \pm 1$. For $n = 3$, m may be an arbitrary integer. Notice also that global bifurcation of a saddle-node torus with a quasiperiodic trajectory on it may imply the appearance of hyperbolic attractors of higher dimension. They may be either the Anosov attractors, which have a smooth structure and hyperbolic

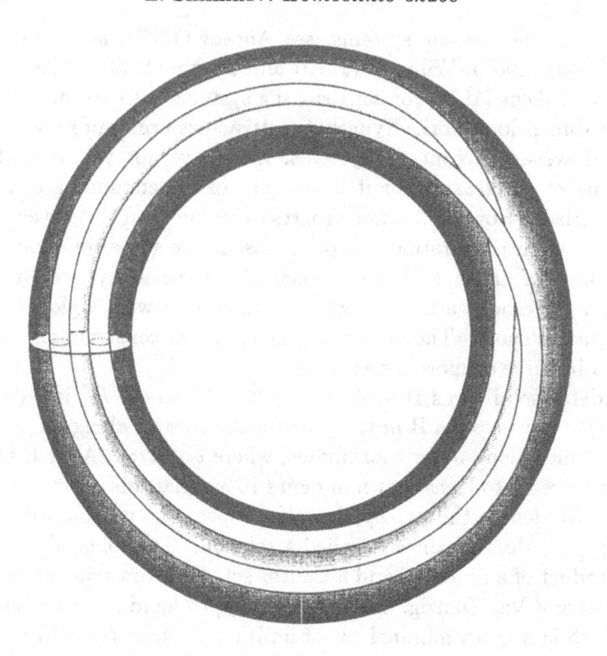

FIGURE 16. A geometrical construction of the solenoid.

behaviour, or attractors which in intersection with a cross-section have dimension greater then one.

6. Lorenz attractors

Lorenz (1961) suggested the model:

$$\begin{aligned} \dot{x} &= -\sigma(x-y), \\ \dot{y} &= rx - y - xz, \\ \dot{z} &= -bz + xy, \end{aligned}$$

which, as numerical simulations showed, contained a strange attractor. Further investigations based on the methods discussed above determined the scenario of the appearance of the attractor and its structure. Similarly to hyperbolic attractors, periodic as well as homoclinic orbits are everywhere dense in the Lorenz attractor. But the Lorenz attractor is structurally unstable. This is due to the embedding of a saddle equilibrium state with a one-dimensional unstable manifold into the attractor. Nevertheless, if the perturbations are small and smooth, stable periodic orbits do not arise.

A verification that the Lorenz model has a strange attractor with such properties was carried out as follows, see Afraimovich *et al.* (1977) and Afraimovich *et al.* (1983): a theoretical model of a two-dimensional discontinuous map T analogous to the Poincaré map for the Lorenz system (figure 17), but more general, was constructed.

The map T was written in the form

$$\begin{aligned} \overline{x} &= f(x,y), \\ \overline{y} &= g(x,y), \end{aligned}$$

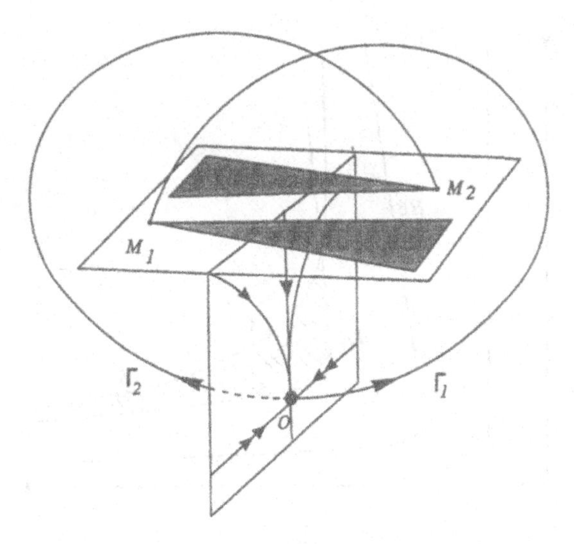

FIGURE 17. First return map for a return plane. Trajectories starting from the plane intersect it within the shaded areas. The points M_1 and M_2 are the first points of intersection with the separatrices and the cross-section.

where the functions f and g satisfy the conditions

$$\left\|\frac{\partial f}{\partial y}\right\| < 1, \quad \left\|\left(\frac{\partial g}{\partial x}\right)^{-1}\right\| < 1,$$

$$\left\|\left(\frac{\partial g}{\partial y}\right)^{-1}\frac{\partial f}{\partial x}\frac{\partial g}{\partial y}\right\| < \left(1 - \left\|\left(\frac{\partial g}{\partial x}\right)^{-1}\right\|\right)\left(1 - \left\|\left(\frac{\partial f}{\partial y}\right\|\right),$$

$$1 - \left\|\left(\frac{\partial g}{\partial x}\right)^{-1}\right\| \cdot \left\|\frac{\partial f}{\partial y}\right\| > 2\sqrt{\left\|\left(\frac{\partial g}{\partial x}\right)^{-1}\right\| \cdot \left\|\frac{\partial g}{\partial y}\right\| \cdot \left\|\left(\frac{\partial g}{\partial x}\right)^{-1}\frac{\partial f}{\partial x}\right\|},$$

where $\|\cdot\| = \sup_{(x,y)} |\cdot|$.

A geometrical meaning of these conditions is that they guarantee a contraction along the y-direction and an expansion along the x-direction, under the action of the map T.

For such a map, the structure of the non-wandering set and that of the attractor, and their bifurcations were studied by Afraimovich *et al.* (1983). The above conditions were checked numerically for the Lorenz model with $b = 8/3$, see Afraimovich *et al.* (1977) and Sinai & Vul (1980). The region of existence of the Lorenz attractor is schematically shown in figure 18. For parameter values to the right of the curve A, the associated Poincaré map looks like that shown in figure 19. This implies the appearance of non-transverse intersections of the stable and unstable manifolds of saddle periodic orbits, and similar intersections of non-rough Poincaré homoclinic curves. At the point C we find the homoclinic connection sketched in figure 20.

In a neighbourhood of the point C, to the right of the curve A, the Lorenz system has stable periodic orbits.

With this theory the bifurcations which give rise to attractors were derived, and the parameter regions corresponding to the existence of symmetric and asymmetric Lorenz type attractors and the Shimizu–Morika attractor, as well as the above conditions, were verified numerically. The criteria leading to conditions for the existence of Lorenz type attractors for arbitrary vector fields are of great interest. Some of these conditions were

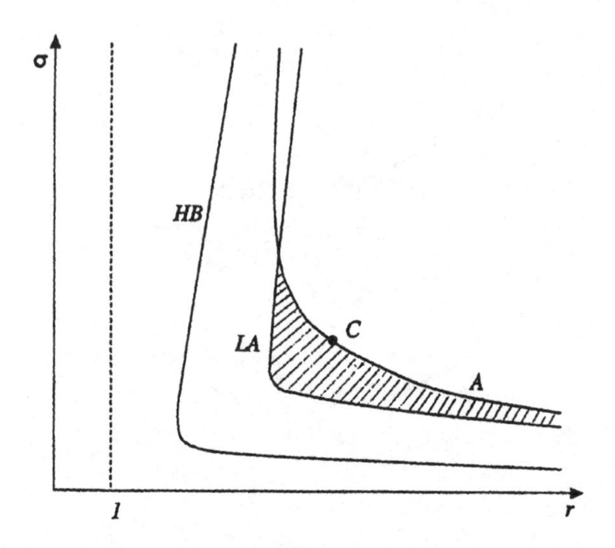

FIGURE 18. The curve labeled HB corresponds to the formation of homoclinic butterfly; on the curve LA the cases sketched in figure 15 occur; on the curve A the separatrix values vanish, Bykov & Shilnikov (1992). The shaded region is the region of existence of a Lorenz attractor.

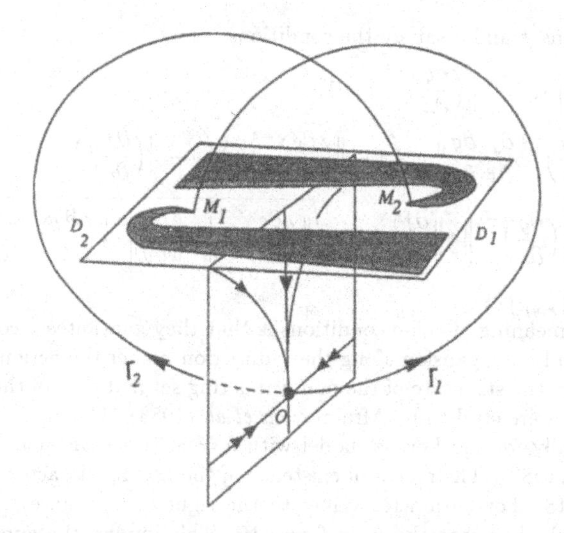

FIGURE 19. The images of the two halves of the return plane under the action of the Poincaré map have developed 'hooks', which are formed as the separatrix value A becomes negative.

obtained by the present author and were first presented at the combined Petrovsky's Seminar and the Moscow Mathematical Society Meeting. Since these conditions turned out to be quite effective and easily checked for certain systems, they will be discussed below.

Consider a finite-number parameter family of vector fields defined by the set of equations

$$\dot{x} = X(x, \mu), \tag{6.3}$$

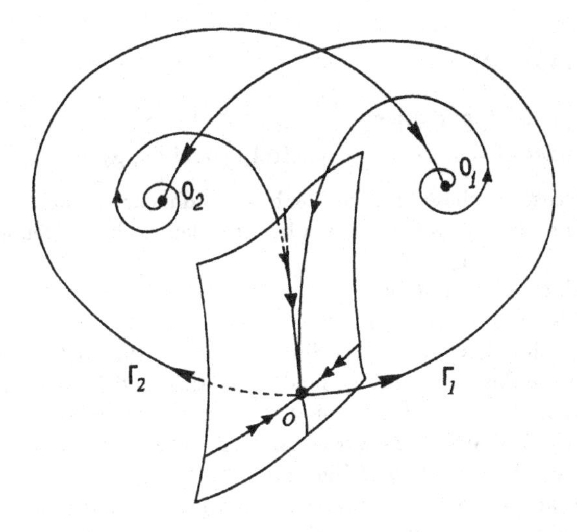

FIGURE 20. The heteroclinic contour including all equilibrium states occurs at the point C of figure 18.

where $x \in R^{n+1}$, $\mu \in R^m$ and $X(x, \mu)$ represents a C^r-smooth function of x and μ.

We assume that the following conditions are satisfied:

(*a*) The system has a saddle-type equilibrium point O(0,0). The eigenvalues of the Jacobian matrix $DX(0,0)$ satisfy

$$\operatorname{Re}\lambda_i < 0,\ i = 1, \dots, n,\ \lambda_0 > 0,\ \lambda_1 \text{ is real and } \operatorname{Re}\lambda_i < \lambda_1,\ i = 2, \dots, n.$$

(*b*) The trajectories Γ_1 and Γ_2, originating from O at $t = 0$, eventually return to O, and are tangent to each other as $t \to \infty$, i.e. $\bar{\Gamma}_1 \cup \bar{\Gamma}_2$ forms a 'figure eight' butterfly.

Then, for $\mu > 0$ there exists an open set V in the parameter space, whose boundary contains the origin, such that the system (6.3) has a Lorenz type attractor in the following cases:

Case 1.

(i) Γ_1 and Γ_2 tending to O are tangential to each other along the dominating direction defined by the eigenvector associated with the eigenvalue λ_1.

(ii) $\frac{1}{2} < \gamma < 1$, $\nu_i > 1$, $\gamma = -\lambda_1\lambda_0^{-1}$, $\nu_i = \operatorname{Re}\lambda_i\lambda_0^{-1}$.

(iii) The separatrix values A_1 and A_2 (defined in Shilnikov (1969)) are equal to zero.

Here we may consider the dimension m of the controlling parameters to be equal to 4; μ_1 and μ_2 control the behaviour of Γ_1 and Γ_2, respectively, and we can choose $\mu_3 = A_1$ and $\mu_4 = A_2$.

Case 2.

(i) $\frac{1}{2} < \gamma < 1$, $\nu_i > 1$.

(ii) λ_2 is real and $\operatorname{Re}\lambda_i < \lambda_2$, $i = 3, \dots, n$.

(iii) Γ_1 and Γ_2 belong to the non-dominating manifold $W_0^{ss} \subset W_0^s$ (i.e. to the eigenspace coresponding to λ_i, $i = 3, \dots, n$) and tending to O are tangential to each other along the dominating direction defined by the eigenvector associated with the eigenvalue λ_1.

Here we may consider $m = 4$; μ_1 and μ_2 control Γ_1 and Γ_2; μ_3 and μ_4 characterize the distances between Γ_1 and Γ_2 and between Γ_1 and W_0^{ss} (at $\mu_1 = \mu_2 = 0$).

Case 3.

(i) $\Gamma_i \not\in W_0^{ss}$, $i = 1, 2$.
(ii) $\gamma = 1$.
(iii) $A_i \neq 0$ and $|A_i| < 2$, $i = 1, 2$.
Here we may consider $m = 3$; μ_1 and μ_2 control Γ_1 and Γ_2; $\mu_3 = \gamma - 1$.

In the symmetric case all these bifurcations have co-dimension two. Note that in each case 1, 2, or 3 it is necessary to study the subclasses defined by the following conditions:
(*a*) orientable: $A_1 > 0,\ A_2 > 0$,
(*b*) semiorientable: $A_1 > 0,\ A_2 < 0$,
(*c*) nonorientable: $A_1 < 0,\ A_2 < 0$.
It was shown in Shilnikov (1986) and Shilnikov (1993) that both subclasses (*a*) and (*c*) are realized in the Shimizu–Morioka model, i.e. this model has both orientable and nonorientable Lorenz type attractors. Using the result of Shilnikov (1969), the birth of lacunas and the death of attractors were explained. The subclass (*a*) was also studied by Rychlic, and the subclass (*c*) by Robinson.

The theory of invariant measures (Sinai–Bowen–Ruelle measure) was applied to Lorenz type attractors, and the results similar to those characterizing hyperbolic attractors were obtained. In this connection Sinai introduced the notion of a stochastic attractor:

DEFINITION 6.1. *By stochastic attractor we mean an invariant closed set A in the phase space of the following properties:*
(*i*) *There exists a neighbourhood U, $A \subset U$, such that if $x \in U$, then*

$$\lim_{t\to+\infty} \text{distance}\,(x(t), A) = 0.$$

(*ii*) *For any initial probability distribution P_0 on A, its shift converges to an invariant distribution P on A as $t \to \infty$, independently of P_0.*
(*iii*) *The probability distribution P is mixing, i.e. the autocorrelation function tends to zero as $t \to \infty$.*

Note that condition (iii) excludes the existence of stable points.

Both hyperbolic and Lorenz type attractors are stochastic attractors and hence can be described in the framework of ergodic theory. Small random perturbations do not influence these attractors, because dynamic stochasticity tends to dominate the white noise.

The problem of the next type of strange attractors (called quasi-attractors) is more complicated, at least for three-dimensional systems, because they are not stochastic. Hence, we believe that the effects due to small perturbations must be included in stochastic analysis of quasi-attractors.

7. Quasi-attractors

The term quasi-attractor (introduced by the present author as an abbreviation for 'quasi-stochastic attractor', see Afraimovich & Shilnikov (1983b)) denotes a limiting set (not necessarily transitive) enclosing periodic orbits of different topological types. This set may be structurally stable (in which case stable periodic orbits coexist with a homoclinic structure) or unstable, the latter case resulting from the presence of homoclinic tangencies, see figure 21. Stable periodic orbits in such a quasi-attractor have large periods and very narrow and tortuous basins. That is why in numerical simulations or in experiments, stable periodic orbits practically cannot be observed, except for a finite number of stability windows; in other respects quasi-attractors resemble deterministic

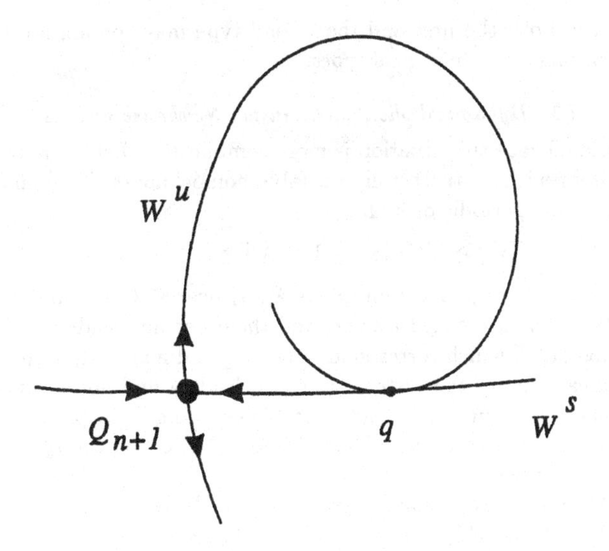

FIGURE 21. Structurally unstable homoclinic orbit.

chaos. Quasi-attractors occur in various systems. For example, they occur in the Lorenz system (to the right of the curve A in figure 18), in the Hénon map, in systems with spiral chaos, in Chua circuits, in connection with the collapse of two-dimensional invariant tori, etc.

The situation for the systems having quasi-attractors is very subtle, since structurally unstable systems may fill out entire regions. In particular, this is due to a phenomenon discovered by Newhouse (1979): any model transverse to a co-dimension-one bifurcation set, corresponding to some homoclinic orbit with quadratic tangency, intersects the regions where systems having structurally unstable homoclinic orbits are dense.

The following C^r-smooth ($r \geq 3$) systems are everywhere dense in the Newhouse regions, see Gonchenko *et al.* (1992) and Gonchenko *et al.* (1993):

(*a*) systems with countable sets of topological invariants;

(*b*) systems having periodic motions with any order of degeneracy for characteristic multipliers equal to 1 and -1;

(*c*) systems with homoclinic tangencies of any order.

For three-dimensional systems with negative divergence, and two-dimensional dissipative diffeomorphisms, the systems with countable sets of stable periodic orbits are everywhere dense in the Newhouse regions, see Newhouse (1979). According to Gonchenko & Shilnikov (1986), Newhouse (1979), Gavrilov & Shilnikov (1973), Gonchenko *et al.* (1992), Gonchenko *et al.* (1993) and Gonchenko *et al.* (1994), this implies a sensitive dependence of the attractor's structure on small variations of the right-hand side of the system.

In order to penetrate the nature of high-dimensional quasi-attractors, one has to study the corresponding high-dimensional homoclinic tangencies. In what follows we give some results which allow us to formulate the following hypothesis: high-dimensional quasi-attractors can be of two different types. The first type is similar to what occurs in three-dimensional systems in the sense that such a quasi-attractor may contain stable periodic orbits and, depending on the dimension of the original system, stable 2D tori and Lorenz-like attractors. The second type is a set without any stable trajectories.

Quasi-attractors of both the first and the second type may contain structurally stable periodic orbits of different topological types.

7.1. *Dynamical phenomena in the Newhouse regions*

In multi-dimensional case the situation is more complicated. Let X be an $(m+n+1)$-dimensional system with a structurally unstable homoclinic orbit Γ, and let λ_i, γ_j be multipliers of a saddle periodic orbit L,

$$|\gamma_n| \geq \ldots \geq |\gamma_1| > 1 > |\lambda_1| \geq \ldots \geq |\lambda_m|.$$

Introduce $\lambda = |\lambda_1|$, $\gamma = |\gamma_1|$. The multipliers λ_i, γ_j nearest to the unit circle (i.e. such that $|\lambda_i| = \lambda$, $|\gamma_j| = \gamma$) are called *leading*, and the other non-leading. The coordinates in a neighbourhood of L which correspond to leading and non-leading eigenspaces of the matrix of the linear part of the system are called leading and non-leading coordinates, respectively. Denote the number of leading stable coordinates by p_s, and the number of leading unstable coordinates by p_u. The pair (p_s, p_u) is called the type of the system. There are four main cases:

$(1,1)$ where λ_1 and γ_1 are real and $\lambda > |\lambda_2|$, $\gamma > |\gamma_2|$;
$(2,1)$ where $\lambda_1 = \overline{\lambda}_2 = \lambda e^{i\varphi}$, γ_1 is real and $\lambda > |\lambda_3|$, $\gamma > |\gamma_2|$;
$(1,2)$ where λ_1 is real, $\gamma_1 = \overline{\gamma}_2 = \gamma e^{i\psi}$, and $\lambda > |\lambda_2|$, $\gamma > |\gamma_3|$;
$(2,2)$ where $\lambda_1 = \overline{\lambda}_2 = \lambda e^{i\varphi}$, $\gamma_1 = \overline{\gamma}_2 = \gamma e^{i\psi}$, and $\lambda > |\lambda_3|$, $\gamma > |\gamma_3|$.

Concerning Γ we suppose that it tends to L tangentially to the leading subspace as $t \to \pm\infty$. Under these conditions it was established in Gonchenko *et al.* (1992) that there are Newhouse regions in a neighbourhood of the given system in the high-dimensional case. In particular, the statements valid for three-dimensional system are also valid for high-dimensional ones. The following proposition shows that the study of dynamics of such system close to $L \cup \Gamma$ reduces to the study of a $(p_s + p_u + 1)$-dimensional system on an invariant manifold.

PROPOSITION 7.1. *Under general conditions, for any system sufficiently close to X, the set of orbits lying entirely in a small neighbourhood of $L \cup \Gamma$ is contained in a $(p_s + p_u + 1)$-dimensional invariant C^1-manifold $\mathcal{M}_C$ tangential to leading directions everywhere on L. A contraction (expansion) more strong than that on $\mathcal{M}_C$ takes place along non-leading stable (unstable) directions.*

Introduce the value D equal to the absolute value of the product of all leading multipliers, $D = \lambda^{p_s}\gamma^{p_u}$.

PROPOSITION 7.2. *If $D < 1$, then systems with infinitely many periodic orbits stable on $\mathcal{M}_C$ are dense in the Newhouse regions. If $D > 1$, then (in a small neighbourhood of $L \cup \Gamma$) neither X nor the systems close to it have orbits stable on $\mathcal{M}_C$.*

PROPOSITION 7.3. *Let $D < 1$ and let some of the leading multipliers of L be complex. Divide such systems into two groups:*

(a) systems of type $(2,1)$ at $\lambda\gamma > 1$, systems of type $(1,2)$, and systems of type $(2,2)$ at $\lambda^2\gamma < 1$;

(b) systems of type $(2,2)$ at $\lambda^2\gamma > 1$.

Then, there are dense systems in the Newhouse regions, having infinitely many periodic orbits with two (in the case (a)) or three (in the case (b)) multipliers equal to unity in absolute value.

Note that a two-dimensional invariant torus can be born from the orbit with two unit multipliers, whereas from the triply degenerate orbits strange attractors (actually,

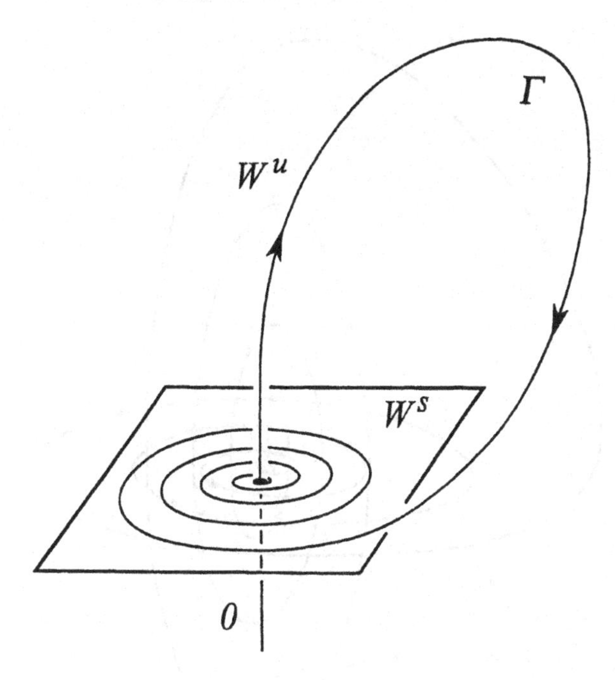

FIGURE 22. A homoclinic loop at to a saddle-focus characterized by a two-dimensional stable manifold W^s ($\rho < 0$) and a one-dimensional unstable manifold W^u ($\lambda > 0$).

attractors on the center manifold) can be born . Thus, if there are two -1 and one $+1$, then a Lorenz-like attractor can arise, see Shilnikov *et al.* (1993); it was established there that a truncated normal form for such bifurcation is the Shimizu–Morioka model:

$$\begin{aligned}\dot{x} &= y,\\ \dot{y} &= x - \lambda y - xz,\\ \dot{z} &= -\alpha z + x^2,\end{aligned}$$

for which the presence of a region in the parameter space where the Lorenz attractor exists was proved in Shilnikov (1986) and Shilnikov (1993).

7.2. *Dynamical phenomena in systems with homoclinic loop to a saddle-focus*

Systems with homoclinic loop to a saddle-focus are well known to be principal subjects of both spiral and screw-like attractors. We present a number of recent results which emphasize a very sophisticated nature of spiral attractors.

Consider a C^r-smooth ($r \geq 4$) dynamical system and suppose that the following conditions are satisfied:

(*a*) The system has an isolated equilibrium point O which is a saddle focus. We distinguish three types of saddle foci according to the location of characteristic roots:

(1) $\lambda_{1,2} = -\gamma \pm i\omega$, $\gamma > 0$, $\omega \neq 0$, $\lambda_3 = \lambda > 0$; $\sigma_1 = \lambda - \gamma > 0$;
(2) $\lambda_1 = -\gamma < 0$, $\lambda_{2,3} = \lambda \pm i\alpha$, $\lambda > 0$, $\alpha \neq 0$; $\sigma_1 = \lambda - \gamma < 0$;
(3) $\lambda_{1,2} = -\gamma \pm i\omega$, $\lambda_{3,4} = \lambda \pm i\alpha$, $\alpha\omega \neq 0$; $\lambda, \gamma > 0$; $\sigma_1 = \lambda - \gamma \neq 0$;

(the remaining roots are supposed to lie to the left of the line $\mathrm{Re}\, e(\cdot) = -\gamma$).

(*b*) The system has an orbit Γ homoclinic to O.

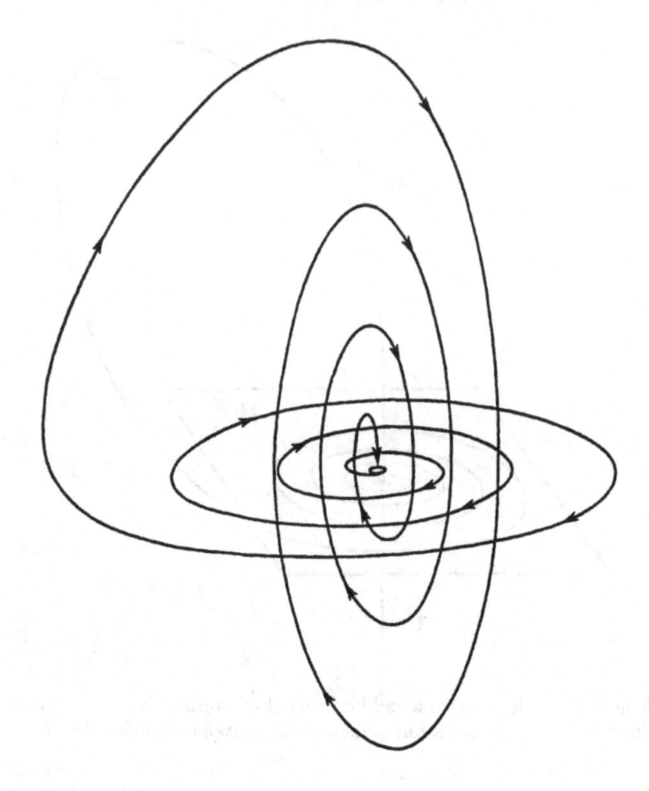

FIGURE 23. Typical homoclinic orbit of a saddle-focus of type $(2+, 2)$ with a two-dimensional unstable manifold and a two-dimensional stable one.

(*c*) As $t \to +\infty$, the curve Γ tends to O tangentially to the leading stable eigenspace (which is one-dimensional for type (2) and two-dimensional for types (1) and (3)).

(*d*) The separatrix value Δ is different from zero. This means that the contact of the stable and unstable manifolds at Γ is, in a sense, simple.

Thus, depending on the type of saddle focus, we distinguish three cases.

In the first case (1), the system has a one-dimensional unstable manifold and a two-dimensional leading plane in the stable manifold. Typical representatives are three-dimensional systems (see figure 22); for them, conditions (*c*) and (*d*) are automatically satisfied. In this case the type of the saddle focus will be denoted by $(2+, 1)$, where the $+$ sign indicates that it is possible to have any number of 'non-leading' stable coordinates.

In the second case (2), the system has a two-dimensional unstable manifold and a one-dimensional leading direction in the stable manifold. This class also includes three-dimensional systems which automatically satisfy conditions (*c*) and (*d*). It is clear that when $n = 3$, the second case reduces to the first upon changing the time direction. However, this is not the case for $n > 3$, because of the existence of non-leading coordinates. In general, the type of such a saddle-focus will be denoted by $(1+, 2)$.

Finally, in the third case (3), the system has a two-dimensional unstable manifold and a two-dimensional leading plane in the stable manifold. Typical representatives are four-dimensional systems with two pairs of complex conjugate roots (see figure 23 which is schematic). We will denote the type of such a saddle focus by $(2+, 2)$.

Systems with the properties mentioned above form a co-dimension one submani-

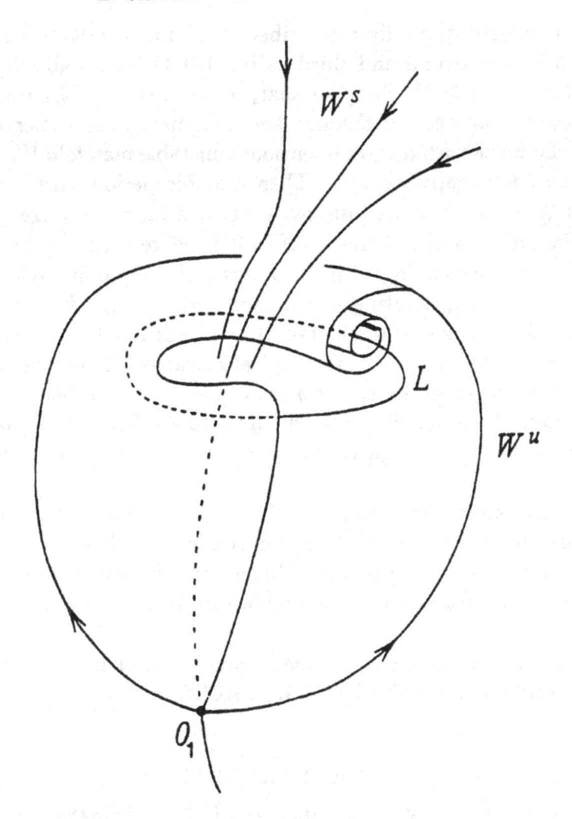

FIGURE 24. An attractive whirlpol forming a basin of a spiral attractor.

fold $\mathcal{B}^1$ of the space of dynamical systems in the C^r-topology. It is well known, see Shilnikov (1970), that for such systems there are infinitely many hyperbolic sets in any neighbourhood of the loop (note that this is not the case if we change the sign of the saddle value σ_1). We define the second saddle value σ_2 as follows:

$$\begin{aligned} \sigma_2 &= \lambda - 2\gamma && \text{for the first case (1),} \\ \sigma_2 &= 2\lambda - \gamma && \text{for the second case (2),} \\ \sigma_2 &= 2(\lambda - \gamma) && \text{for the third case (3).} \end{aligned}$$

The condition $\sigma_2 < 0$ determines an open subset $\mathcal{B}^1_{\mathrm{s}}$ of $\mathcal{B}^1$.

PROPOSITION 7.4. *Systems with a countable number of stable periodic orbits are dense in* $\mathcal{B}^1_{\mathrm{s}}$.

Suppose now that $\sigma_2 > 0$ in all three cases. This condition determines an open subset $\mathcal{B}^1_{\mathrm{u}}$ of $\mathcal{B}^1$. We consider a sufficiently small neighbourhood $U_{\mathrm{O}\Gamma}$ of the loop $\mathrm{O}\cup\Gamma$ of some system in $\mathcal{B}^1_{\mathrm{u}}$.

PROPOSITION 7.5. *Given any system in* $\mathcal{B}^1_{\mathrm{u}}$, *there is a sufficiently small neighbourhood* $U_{\mathcal{B}}$ *in the space of dynamical systems such that no systems in* $U_{\mathcal{B}}$ *have stable periodic orbits in* $U_{\mathrm{O}\Gamma}$.

PROPOSITION 7.6. *Systems with structurally unstable periodic orbits are dense in* $\mathcal{B}^1$.

The following construction, first described in Shilnikov (1983), clarifies the meaning of the theorem in the second and third cases. Let O be a stable equilibrium point in n-dimensional space ($n \geq 3$). Suppose that, under a change of parameters, the pair of complex conjugate roots passes through the imaginary axis. After loss of stability, O becomes a saddle-focus with a two-dimensional unstable manifold W^u_O. Suppose that the Lyapunov value l is negative, $l < 0$. Then a stable periodic orbit L, which will be a boundary of W^u_O, arises from O. Suppose that with further change of parameters, that multiplier of the cycle L which corresponds to W^u_O merges with another one and converges to the real axis, while remaining with the others inside the unit circle. The manifold W^u_O will now wind onto L, generating a whirlpool (figure 24). Furthermore, W^u_O can be tangent to W^s_O along a homoclinic curve. In the whirlpool there will be an attracting set of a complicated nature, a so-called spiral attractor. It will be a quasi-attractor if $\sigma_2 < 0$. But if $\sigma_2 > 0$, proposition 7.3 guarantees that the behaviour of orbits has a hyperchaotic character in a sufficiently small neighbourhood of the loop O$\cup\Gamma$, although the divergence of the field is negative in the second case for $n > 3$, and in the third case for $n > 4$.

The transition to spiral chaos along the lines of the above scenario is usually preceded by either a cascade of periodic doubling bifurcations (if three-dimensional volumes are contracting) or by a breakdown of a two-dimensional invariant torus arising from L (see Shilnikov (1991) and Afraimovich & Shilnikov (1983a)).

This research was supported by Russian Foundation of Basic Research and the EC–Russia Collaborative Project ESPRIT P9282-ACTS.

REFERENCES

AFRAIMOVICH, V. S., BYKOV, V. V. & SHILNIKOV, L. P. 1977 On the appearance and structure of Lorenz attractor. *DAN SSSR* **234**, 336–339.

AFRAIMOVICH, V. S., BYKOV, V. V. & SHILNIKOV, L. P. 1983 On the structurally unstable attracting limit sets of Lorenz attractor type. *Tran. Moscow Math. Soc.* **2**, 153–215.

AFRAIMOVICH, V. S., SHILNIKOV, L. P. 1983a Invariant two-dimensional tori, their breakdown and stochasticity. In *Methods of Qualitative Theory of Differential Equations.* Gorky Univ. Press. English translation: *Amer. Math. Soc. Trans.* 2, **149**, 201–212.

AFRAIMOVICH, V. S. & SHILNIKOV, L. P. 1983b Strange attractors and quasi-attractors. In *Nonlinear Dynamics and Turbulence* (ed. G. I. Barenblatt, G. Iooss & D. D. Joseph), pp. 1–28. Pitman.

ANOSOV, D. 1967 Geodesic flows on compact Riemannian manifolds of negative curvature. *Trudy Math. Inst. im. V.A. Steclova* **90**. In Russian.

BIRKHOFF, G. D. 1935 Nouvelles recherche sur les systèmes dynamiques. *Memorie Pont Acad. Sci. Novi Lyncaei* **1**, 85–216.

BYKOV, V. V. & SHILNIKOV, A. L. 1992 On boundaries of the region of existence of the Lorenz attractor. *Selecta Math. Sovietica* 4, **11**, 375–382.

GONCHENKO, S.V. & SHILNIKOV, L. P. 1986 On dynamical systems with structurally unstable homoclinic curves. *Sov. Math. Dokl.* 234–238.

GONCHENKO, S. V., TURAEV, D. V. & SHILNIKOV, L. P. 1992 On models with a structurally unstable homoclinic Poincaré curve. *Sov. Math. Dokl.* 2, **44**, 422–426.

GONCHENKO, S. V., TURAEV, D. V. & SHILNIKOV, L. P. 1993 On models with non-rough Poincaré homoclinic curves. *Physica* D **62**, 1–14.

GONCHENKO, S. V., TURAEV, D. V. & SHILNIKOV, L. P. 1994 Dynamical phenomena in the systems with a non-rough Poincaré homoclinic curve. *Chaos.* In press.

LORENZ, E. N. 1961 Deterministic non-periodic flow. *J. Atmos. Sci.* **20**, 130–141.

NEWHOUSE, S. E. 1979 The abundance of wild hyperbolic sets and non-smooth sets for diffeomorphism. *Publ. Math. IHES* **50**, 101–151.

PLYKIN, R. 1974 Sources and sinks of A-diffeomorphism of surfaces. *Math. Sbornik* **94**, 243–264.

POINCARÉ, H. 1899 *Les Méthodes Nouvelles de la Mècanique Céleste,* vol. 3.

RUELLE, D. & TAKENS, F. 1971 On the nature of turbulence. *Comm. in Math. Phys.* **20**, 167–192.

SHILNIKOV, L. P. 1967 On a Poincaré–Birkhoff problem. *Math. USSR Sbornik* **3**, 353–371.

SHILNIKOV, L. P. 1969 On a new type of bifurcation of multi-dimensional dynamical systems. *Soviet Math. Dokl.* **10**, 1368–1371.

SHILNIKOV, L. P. 1970 A contribution to the problem of the structure of an extended neighbourhood of a rough equilibrium state of saddle-focus type. *Math. USSR Sb.* **10**, 91–102.

SHILNIKOV, L. P. 1980 Theory of bifurcations and the Lorenz model. Appendix to Russian edition of *The Hopf Bifurcation and its Applications* (ed. J. E. Marsden. & M. McCracken). Mir.

SHILNIKOV, L. P. 1983 Bifurcation theory and turbulence. In *Nonlinear and Turbulent Processes in Physics,* vol. 3, pp. 1627–1635. Harwood Academic Publishers.

SHILNIKOV, A. L. 1986 Bifurcations and chaos in the Shimizu system. In *Methods of Qualitative Theory of Diff. Equations* pp. 180–193. Gorky State Univ. In Russian.; see also 1991 *Trans. Selecta Math. Sov.* **10**, 105–117.

SHILNIKOV, L. P. 1991 The theory of bifurcations and turbulence. *Selecta Math. Sovietica* 10, **1**, 43–53.

SHILNIKOV, A. L. 1993 On bifurcations of the Lorenz attractor in the Morioka–Shimizu model. *Physica* D 1/4, **62**, 338–346.

SHILNIKOV, L. & TURAEV, D. In preparation.

SHILNIKOV, A. L. & SHILNIKOV, L. P. 1991 On the nonsymmetric Lorenz model. *Bifurcation and Chaos* 4, **1**, 773–776.

SHILNIKOV, A. L., SHILNIKOV, L. P. & TURAEV, D. V. 1993 Normal forms and Lorenz attractors. *Bifurcation and Chaos* 5.

SINAI, JA. G. & VUL, E. B. 1980 Hyperbolicity conditions for the Lorenz model. *Physica* D **2**, 3–7.

SMALE, S. 1963 Diffeomorphisms with many periodic points. In *Diff. and Comb. Topology* (ed. E. Cairus), pp. 63–86. Princeton Univ. Press.

SMALE, S. 1967 Differentiable dynamical systems. *Bull. Amer. Math. Soc.* **73**, 747–817.

WILLIAMS, R. 1970 Classification of one-dimensional attractors. *Proc. Symp. in Pure Math.*, pp. 361–393.

See also

GAVRILOV N. K. & SHILNIKOV, L. P. 1973 On three-dimensional systems close to systems with a structurally unstable homoclinic curve. *Math. USSR Sb.* 1, **17**, 446–485; see also GAVRILOV, N. K. & SHILNIKOV, L. P. 1973 On three-dimensional systems close to systems with a structurally unstable homoclinic curve. *Math. USSR Sb.* 2, **19**, 1–37.

GUCKENHEIMER, J. & WILLIAMS 1979 Structural stability of the attractor in a Lorenz model. *Inst. Hautes Etudes Sci. Publ. Math.* **50**, 59–72.

OVSYANNIKOV, I. M. & SHILNIKOV, L. P. 1987 On systems with saddle-focus homoclinic curve. *Mathem. Sbornik* 2, **58**, 557–574.

OVSYANNIKOV, I. M. & SHILNIKOV, L. P. 1992 Systems with homoclinic curve of the multidimensional saddle-focus and spiral chaos. *Mathem. Sbornik* 2, **73**, 415–443.

Hydrodynamics of relativistic probability flows

By IWO BIALYNICKI-BIRULA

Center for Theoretical Physics PAN, Warsaw, Poland
and
Abteilung für Quantenphysik, Universität Ulm, Ulm, Germany

A hydrodynamic form of relativistic wave equations is derived. The formulation is quite general and can be applied to any set of first order wave equations. The most interesting examples are the Weyl equation and the Maxwell equations. The set of hydrodynamic variables includes the density field, the velocity field and the momentum field. The reduction in the number of independent variables requires a quantization condition that relates the curl of the momentum field to a vector built from the remaining fields.

---

1. Introduction

The hydrodynamic formulation of wave mechanics is almost as old as the Schrödinger equation, see Madelung (1926). This approach has been extended by Takabayasi (1952), Takabayasi (1953), Takabayasi (1955), Takabayasi (1983a) and others: Bohm & Schiller (1955), Bohm *et al.* (1955) and Janossy & Ziegler-Naray (1965), to the nonrelativistic wave equation for spinning particles (Pauli equation). A review of the hydrodynamic formulation of nonrelativistic wave mechanics including a large number of references can be found in Ghosh & Deb (1982) (for a textbook treatment, see Bialynicki-Birula *et al.* (1992)). A hydrodynamic formulation of relativistic wave mechanics of the Dirac particle was given by Takabayasi (1956). His formulation, however, cannot be applied to the massless case treated in the present work.

In the present paper I shall present a general derivation that can be applied to any set of first order equations and yields a hydrodynamic form of these equations. The application of this method to Maxwell equations offers a new look at the dynamics of photons. The hydrodynamic formulation of wave mechanics enables one to visualize the flow of probability by providing us with an intuitive picture of this flow described in terms of the familiar variables of classical hydrodynamics. Such a visualization is of particular interest for the Maxwell theory owing to the unquestionable dominance of electromagnetic phenomena in many areas of physics. There were many attempts to provide a 'mechanical picture' of electromagnetism and the hydrodynamic formulation may also play a role in this category. The main purpose of my study, however, is to exhibit an intricate relationship between the linear (the Weyl equation and the Maxwell equations) and nonlinear (hydrodynamic-like equations) theories. The Maxwell equations and their hydrodynamic version offer a very good platform to discuss this relationship. On the one hand there is a very well understood theory describing the propagation of the electromagnetic field governed by linear field equations while on the other hand there is a nonlinear, hydrodynamic-like theory describing the transport of various physical quantities (energy, momentum, etc.).

In the hydrodynamic formulation of nonrelativistic wave mechanics the flow of probability is described by the hydrodynamic variables: the probability density ρ and the velocity of the probability flow $\vec{v}$. For spinning particles we need additional variables describing the internal degrees of freedom. In general, the number of hydrodynamic variables exceeds the number of components of the wave function. In order to make the hydrodynamic formulation completely equivalent to the original one, one must impose a condition on the initial values of the hydrodynamic variables. The condition restricting the number of the degrees of freedom is related to the Bohr–Sommerfeld quantization condition of quantum theory.

2. General approach to the hydrodynamic formulation

In the simplest case of a nonrelativistic spinless particle, described by the Schrödinger equation, the hydrodynamic variables and the wave function are related through the formulas

$$\rho = |\psi|^2, \quad m\vec{v} = \vec{p} = \nabla S, \tag{2.1}$$

where S is the phase of the wave function (R is assumed real)

$$\psi = R\exp(\mathrm{i}S/\hbar). \tag{2.2}$$

The equations of motion for the hydrodynamic variables ρ and $\vec{v}$ can be obtained by the repeated use of the Schrödinger equation. For the force-free case they have the form

$$\partial_t\rho + (\vec{v}\cdot\nabla)\rho = -(\partial_k v_k)\rho, \tag{2.3}$$

$$\partial_t v_i + (\vec{v}\cdot\nabla)v_i = \frac{\hbar^2}{4m^2\rho}\partial_k(\rho\,\partial_i\partial_k \ln\rho). \tag{2.4}$$

The requirement that the wave function be single-valued imposes the following restriction – the quantization condition – on the velocity field

$$\oint_C \mathrm{d}\vec{l}\cdot\vec{v} = \frac{2\pi\hbar}{m}n, \tag{2.5}$$

where C is an arbitrary closed contour. By the Stokes theorem, the quantization condition can also be expressed in terms of a surface integral representing the vorticity flux,

$$\int_S \mathrm{d}\vec{S}\cdot(\nabla\times\vec{v}) = \frac{2\pi\hbar}{m}n, \tag{2.6}$$

where S is any surface spanned by the closed contour C. Thus, the quantization condition states that the motion of the probability fluid is irrotational almost everywhere except possibly at a discrete set of vortex lines whose strength is quantized in units of $2\pi\hbar/m$. This condition reduces the number of independent components to two: the density ρ and the longitudinal part of the velocity vector $\vec{v}$. The transverse part of the momentum vector is fixed.

The starting point of my analysis will be a general relativistic wave equation, obeyed by an N-component wave function ϕ_a, of the form

$$\partial_t\phi = -\alpha^k\partial_k\phi - \mathrm{i}\beta\phi, \tag{2.7}$$

where α^k and β are $N\times N$ matrices. Every set of linear equations with higher derivatives can be reduced to this form by introducing additional components of the wave function. For example, for the Klein–Gordon equation,

$$\left[\Box + (mc/\hbar)^2\right]\psi = 0, \tag{2.8}$$

one can define

$$\chi = \frac{\hbar}{mc^2}\partial_t\psi, \tag{2.9}$$

$$\chi_k = \frac{\hbar}{mc}\partial_k\psi, \tag{2.10}$$

to rewrite (2.8) as

$$\partial_t\psi = (mc^2/\hbar)\chi, \tag{2.11}$$
$$\partial_t\chi = c\nabla\cdot\vec{\chi} - (mc^2/\hbar)\psi, \tag{2.12}$$
$$\partial_t\vec{\chi} = c\nabla\chi. \tag{2.13}$$

This set of equations is clearly of the form (2.7). In the present case one also needs an auxiliary condition

$$\nabla\psi = (mc/\hbar)\vec{\chi} \tag{2.14}$$

imposed on the initial data and preserved in time. Similar auxiliary conditions will be present in other cases.

In what follows I assume that all four matrices α_k and β are Hermitian. This is required for the conservation of probability (the Hamiltonian must be Hermitian) and is, of course, satisfied in all cases of interest. I shall derive a hydrodynamic form of the general wave equation (2.7) in four steps.

In the first step, I define the following hydrodynamic variables (summation convention is used throughout):

- a scalar density of the 'probabilistic fluid' ρ,

$$\rho = \phi_a^*\phi_a, \tag{2.15}$$

- a Hermitian matrix w_{ab} of trace 1,

$$\rho w_{ab} = \phi_a\phi_b^*, \tag{2.16}$$

- and the momentum vector $\vec{u}$,

$$\rho\vec{u} = \frac{1}{2\mathrm{i}}\phi_a^*\overleftrightarrow{\nabla}\phi_a. \tag{2.17}$$

The matrix w carries all the information about the direction of the complex vector ϕ in the N-dimensional space. The length of this vector is determined by ρ. In order to complete the reconstruction one only needs the phase of ϕ. In complete analogy with the treatment of the Schrödinger equation, one may obtain the information about the phase from the longitudinal part of the momentum vector $\vec{u}$. I shall keep referring to this vector as momentum even though its dimension – inverse of length – is that of a wave vector. In order to obtain the correct dimension I would have to multiply $\vec{u}$ by the Planck constant but such an operation would only introduce $\hbar$ into various formulas. Thus, my basic hydrodynamic variables in the general case will comprise the density ρ, $2N-2$ independent components of w and the longitudinal part of $\vec{u}$.

In the second step, with the help of (2.7), I derive the following two evolution equations

$$\partial_t(\rho w) = -(\vec{\alpha}\cdot\nabla)(\rho w) + \rho(\vec{\alpha}\cdot\vec{z}^\dagger - \vec{z}^\dagger\cdot\vec{\alpha}) - \mathrm{i}\rho\,[\beta, w]\,, \tag{2.18}$$
$$\partial_t(\rho z_k) = -\vec{\alpha}\cdot\nabla(\rho z_k) + \rho(\vec{\alpha}\cdot z_k\vec{z}^\dagger - z_k\vec{z}^\dagger\cdot\vec{\alpha}) - \mathrm{i}\rho\,[\beta, z_k]\,, \tag{2.19}$$

where the two auxiliary vector matrices $\vec{z}$ and $\vec{z}^\dagger$ are defined through the formulas

$$\rho\vec{z}_{ab} = (\nabla\phi_a)\phi_b^*, \quad \rho\vec{z}^\dagger_{ab} = \phi_a\nabla\phi_b^*, \tag{2.20}$$

and the matrix multiplication is understood whenever applicable. The use of the vectors $\vec{z}$ and $\vec{z}^\dagger$ greatly simplifies the formulas.

In the third step I derive the following formulas that relate the matrices $\vec{z}$ and $\vec{z}^\dagger$ to the basic hydrodynamic variables

$$\vec{z} = \left(\frac{1}{2\rho}\nabla\rho + \mathrm{i}\vec{u} + \nabla w\right)w, \quad \vec{z}^\dagger = w\left(\frac{1}{2\rho}\nabla\rho - \mathrm{i}\vec{u} + \nabla w\right). \tag{2.21}$$

These relations can be verified by substituting the definitions of ρ, $\vec{u}$ and w on the right hand side and performing all the differentiations. I shall not carry out the substitution of (2.21) into (2.18) and (2.19) in the general case because it does not lead to anything simple or transparent. I shall only do it in the special cases discussed in the forthcoming sections.

Finally, in the fourth step, I derive the quantization condition that fixes the transverse part of $\vec{u}$. This condition can be expressed in terms of w by using again $\vec{z}$ and $\vec{z}^{\dagger}$ in the intermediate steps. From the definition of $\vec{u}$ one obtains

$$\nabla \times \rho\vec{u} = \mathrm{i}\rho \mathrm{Tr}\,(\vec{z} \times \vec{z}^{\dagger}). \tag{2.22}$$

With the help of (2.21) one arrives at a very simple formula

$$\nabla \times \vec{u} = \mathrm{Tr}\,(w\nabla w \times \nabla w). \tag{2.23}$$

This relation holds everywhere except on quantized vortex lines, where $\nabla \times \vec{u}$ has a surface delta-function singularity. Therefore, the quantization condition in the general case has the form

$$\int_S \mathrm{d}\vec{S} \cdot \left[\nabla \times \vec{u} - \mathrm{Tr}\,(w\nabla w \times \nabla w)\right] = 2\pi n. \tag{2.24}$$

In the following sections, this general formalism will be applied to the Weyl equation and to the Maxwell equations.

3. Hydrodynamic form of the Weyl equation

The wave equation describing the time evolution of the wave function for a massless, spin one-half particle (the Weyl equation) has the form

$$\partial_t \phi = -c\,\vec{\sigma}\cdot\nabla\phi, \tag{3.25}$$

where ϕ, has two complex components. The hydrodynamic variables for neutrino are defined as follows

$$\rho = \phi_a^* \phi_a, \quad \rho\vec{v} = c\,\phi_a^* \vec{\sigma}_{ab} \phi_b, \quad \rho\vec{u} = \frac{1}{2\mathrm{i}} \phi_a^* \overleftrightarrow{\nabla} \phi_a. \tag{3.26}$$

The variable $\vec{v}$ appearing in the second equation has a fixed length c and can be interpreted as the particle velocity. For spinning particles the velocity $\vec{v}$ and momentum $\vec{u}$ are distinct – in general, they have different directions.

In the present case the matrix w_{ab} has three independent components that can be expressed by the components of the velocity vector $\vec{v}$

$$w_{ab} = \tfrac{1}{2}(\delta_{ab} + \vec{\sigma}_{ab}\cdot\vec{v}/c). \tag{3.27}$$

This leads to the following formula for $\vec{z}_{ab}$

$$\vec{z}_{ab} = \tfrac{1}{4}\left[(\nabla \ln\rho + 2\mathrm{i}\vec{u})(1 + \vec{\sigma}\cdot\vec{v}/c) + (\nabla v_k - \mathrm{i}\varepsilon_{ijk} v_i \nabla v_j)\sigma_k\right]_{ab}, \tag{3.28}$$

and its Hermitian conjugate for $\vec{z}^{\dagger}$. With their help I obtain the following set of evolution equations for the hydrodynamic variables, see Bialynicki-Birula (1995),

$$\partial_t \rho + (\vec{v}\cdot\nabla)\rho = -(\nabla\cdot\vec{v})\rho, \tag{3.29}$$

$$\partial_t \vec{v} + (\vec{v}\cdot\nabla)\vec{v} = \vec{v} \times (\vec{v} \times \nabla\ln\rho - 2c\vec{u} - 2\nabla \times \vec{v}), \tag{3.30}$$

$$\partial_t \vec{u} + (\vec{v}\cdot\nabla)\vec{u} = \frac{1}{2c\rho}\partial_i(\rho\,\varepsilon_{ikl} v_k \nabla v_l). \tag{3.31}$$

These equations have a characteristic hydrodynamic form with substantial derivatives of the variables appearing on the left hand side, as in the nonrelativistic case.

In order to restrict the solutions of the hydrodynamic equations (3.29)–(3.31) to only those that are obtained from a two-component wave function, I impose constraints on the initial conditions. The first constraint is purely algebraic; it determines the length of the velocity vector, $|\vec{v}| = c$. The second constraint – the quantization condition – is obtained from the general formula (2.24). In terms of the hydrodynamic variables, this condition reads

$$\int_S d\vec{S}\cdot\left[\nabla\times\vec{u} - \frac{1}{4c^3}\varepsilon_{ijk}\, v_i\,(\nabla v_j)\times(\nabla v_k)\right] = 2\pi n. \tag{3.32}$$

Thus, the time evolution of the neutrino wave function is described in the hydrodynamic formulation as a flow of probability with velocity $\vec{v}$. The velocity vector is precessing around the vector $\vec{v}\times\nabla\ln\rho - 2(c\vec{u}+\nabla\times\vec{v})$.

4. Hydrodynamic form of the Maxwell equations

In order to find the hydrodynamic form of the Maxwell equations I rewrite them in the form (2.7) in order to apply my general formalism. This is done with the use of the photon wave function that was described in detail in Bialynicki-Birula (1994) and Bialynicki-Birula (1996). The photon wave function $\vec{F}$ is the complex three-vector constructed from the fields $\vec{D}$ and $\vec{B}$ as follows

$$\vec{F}(\vec{r},t) = \frac{\vec{D}(\vec{r},t)}{\sqrt{2\varepsilon}} + i\frac{\vec{B}(\vec{r},t)}{\sqrt{2\mu}}. \tag{4.33}$$

The Maxwell equations written in terms of $\vec{F}$ have the form

$$\partial_t\vec{F}(\vec{r},t) = -ic\nabla\times\vec{F}(\vec{r},t), \tag{4.34}$$

$$\nabla\cdot\vec{F}(\vec{r},t) = 0. \tag{4.35}$$

The important physical quantities can be expressed in terms of $\vec{F}$ as bilinear combinations:

- Energy density

$$\mathcal{E} = \vec{F}^*\cdot\vec{F} = \frac{\vec{D}^2}{2\varepsilon} + \frac{\vec{B}^2}{2\mu}, \tag{4.36}$$

- Energy flux (Poynting vector)

$$\vec{P} = \frac{c}{2i}\vec{F}^*\times\vec{F} = c^2\vec{D}\times\vec{B}, \tag{4.37}$$

- Maxwell stress tensor

$$T_{ij} = F_i^*F_j + F_j^*F_i - \delta_{ij}\vec{F}^*\cdot\vec{F} = \frac{D_iD_j}{\varepsilon} + \frac{B_iB_j}{\mu} - \delta_{ij}\left(\frac{\vec{D}^2}{2\varepsilon} + \frac{\vec{B}^2}{2\mu}\right). \tag{4.38}$$

The well known continuity equations for the energy density and the momentum density

$$\partial_t\mathcal{E} = -\nabla\cdot\vec{P}, \quad \partial_tP_i = c^2\partial_jT_{ij}, \tag{4.39}$$

may serve as hints how to introduce convenient hydrodynamic variables. Five hydrodynamic variables will be derived from the energy density, the energy flux and the stress tensor. The sixth variable determines the phase of $\vec{F}$ and is defined as in my general treatment.

$$\rho = \mathcal{E} = \frac{\vec{D}^2}{2\varepsilon} + \frac{\vec{B}^2}{2\mu}, \quad \rho\vec{v} = \vec{P} = c^2\vec{D}\times\vec{B}, \tag{4.40}$$

$$\rho t_{ij} = cT_{ij} + \delta_{ij}c\rho = \frac{cD_iD_j}{\varepsilon} + \frac{cB_iB_j}{\mu}, \tag{4.41}$$

$$\rho\vec{u} = \frac{1}{2i}F_i^* \overleftrightarrow{\nabla} F_i = \frac{c}{2}(D_i\nabla B_i - B_i\nabla D_i). \tag{4.42}$$

Only one component of t_{ij} is free to choose. The remaining five are fixed by the conditions (summation convention)

$$t_{ii} = 2c, \quad v_i t_{ik} = 0, \quad t_{ij}t_{ij} = 4c^2 - 2\vec{v}^2. \tag{4.43}$$

The momentum vector $\vec{u}$ is defined in analogy to wave mechanics but there is one difference; it transforms as a pseudovector under space-reflections. This is due to the fact that $\vec{F}$ becomes complex conjugate under reflections since $\vec{B}$ is a pseudovector.

The curl of $\vec{u}$ is subject to the quantization condition that can be obtained from the general formula (2.24),

$$\int d\vec{S}\cdot\left[\nabla\times\vec{u} - \frac{1}{8c^3}\varepsilon_{ijk}(v_i\nabla v_j\times\nabla v_k + v_i\nabla t_{jl}\times\nabla t_{kl} - 2t_{il}\nabla t_{jl}\times\nabla v_k)\right] = 2\pi n. \tag{4.44}$$

In addition to the quantization condition, the initial values of the hydrodynamic variables must satisfy two equations that guarantee that the fields $\vec{D}$ and $\vec{B}$ are divergenceless

$$\tfrac{1}{2}\partial_k(\rho t_{ik}) + \rho\varepsilon_{ijk}v_j u_k + \frac{\rho}{4c}(t_{jk}\partial_k t_{ij} - t_{ij}\partial_k t_{jk} + v_k\partial_k v_i - v_i\partial_k v_k) = 0, \tag{4.45}$$

$$\tfrac{1}{2}\partial_k(\rho\varepsilon_{ikl}v_l) + \rho t_{ik}u_k + \frac{\rho}{4c}\left[\varepsilon_{jkl}(t_{il}\partial_k v_j - v_j\partial_k t_{il}) + \varepsilon_{ijl}(t_{kl}\partial_k v_j - v_j\partial_k t_{kl})\right] = 0. \tag{4.46}$$

Having completed all preparatory steps, I may now write the evolution equations in terms of the hydrodynamic variables

$$\partial_t\rho = -\nabla\cdot(\rho\vec{v}), \tag{4.47}$$

$$\partial_t(\rho v_i) = \partial_j(-c^2\rho\delta_{ij} + c\rho t_{ij}), \tag{4.48}$$

$$\begin{aligned}\partial_t(\rho t_{ij}) = {}& \delta_{ij}\rho v_k\partial_l t_{kl} - \delta_{ij}cv_k\partial_k\rho + \frac{c}{2}(v_i\partial_j + v_j\partial_i)\rho \\ & + c\rho\varepsilon_{ikl}u_k t_{lj} + c\rho\varepsilon_{jkl}u_k t_{li} + \rho(v_k\partial_k t_{ij} - t_{ij}\partial_k v_k) \\ & + \frac{\rho}{2}(t_{ik}\partial_k v_j + t_{jk}\partial_k v_i - v_i\partial_k t_{kj} - v_j\partial_k t_{ki} - v_k\partial_i t_{kj} - v_k\partial_j t_{ki}),\end{aligned} \tag{4.49}$$

$$\partial_t(\rho u_i) = \partial_j\left[-\rho u_i v_j + \frac{\rho}{4c}\varepsilon_{jkl}(t_{km}\partial_i t_{ml} + v_k\partial_i v_l)\right]. \tag{4.50}$$

These equations can also be written in a form clearly exhibiting their hydrodynamic structure, with all substantial derivatives on the left

$$\partial_t\rho + (\vec{v}\cdot\nabla)\rho = -\rho(\nabla\cdot\vec{v}), \tag{4.51}$$

$$\partial_t v_i + (\vec{v}\cdot\nabla)v_i = \frac{1}{\rho}\partial_j(-c^2\rho\delta_{ij} + \rho v_i v_j + \rho t_{ij}), \tag{4.52}$$

$$\begin{aligned}\partial_t t_{ij} + (\vec{v}\cdot\nabla)t_{ij} = {}& \frac{1}{\rho}\left[t_{ij}v_k\partial_k\rho - c\delta_{ij}v_k\partial_k\rho + \frac{c}{2}(v_i\partial_j + v_j\partial_i)\rho\right] \\ & + \delta_{ij}v_k\partial_l t_{kl} + 2v_k\partial_k t_{ij} + c\varepsilon_{ikl}u_k t_{lj} + c\varepsilon_{jkl}u_k t_{li} \\ & + (v_k\partial_k t_{ij} - t_{ij}\partial_k v_k) + \tfrac{1}{2}(t_{ik}\partial_k v_j + t_{jk}\partial_k v_i) \\ & - \tfrac{1}{2}(v_i\partial_k t_{kj} + v_j\partial_k t_{ki} + v_k\partial_i t_{kj} + v_k\partial_j t_{ki}),\end{aligned} \tag{4.53}$$

$$\partial_t u_i + (\vec{v}\cdot\nabla)u_i = \frac{1}{4c\rho}\partial_j\left[\rho\varepsilon_{jkl}(t_{km}\partial_i t_{ml} + v_k\partial_i v_l)\right]. \tag{4.54}$$

The hydrodynamic form of Maxwell equations has a dual interpretation. On the one hand, one may use purely classical notions of the energy flow and momentum flow of the classical electromagnetic field. On the other hand, one may use the quantum notion of the flow of the probability density and the additional variables characterizing the probabilistic fluid. I prefer the second interpretation because of the intimate connection with the hydrodynamic formulation of the Schrödinger equation. In particular the variable $\vec{u}$ with its quantization condition is deeply rooted in quantum theory, its role in the classical interpretation being purely formal.

5. Conclusions

The main conclusion of this study is that the flow of probability associated with the Weyl equation and the Maxwell equations can be described in purely hydrodynamic terms. The probabilistic fluid moves with the speed $\vec{v}$ and is endowed with additional degrees of freedom – the longitudinal part of a vector field $\vec{u}$ and an additional variable for the Maxwell theory. As compared with relativistic dynamics of a perfect fluid, the flow of the probability fluid corresponding to relativistic wave equations is fairly complex. In the face of these complications, are there any advantages of using the hydrodynamic description? It certainly offers a totally different look at the wave function and its time evolution. The new form involves only observable quantities bilinear in the wave function. Moreover, the hydrodynamic description clearly separates the local dynamical laws and a nonlocal, global, quantization condition. This important property has been utilized in the past to derive the quantization condition for the magnetic charge, see Bialynicki-Birula & Bialynicka-Birula (1971), and to clarify the interpretation of the Aharonov–Bohm effect, see Takabayasi (1983b). Whether it will lead to some new insights into the quantum properties of massless particles, remains an open question.

Finally, I would like to mention a possible connection of the hydrodynamic formulation of wave mechanics of massless particles to the string theory. In the hydrodynamic formulation an essential role is assigned to quantized vortex lines. Such vortex lines are very similar to relativistic strings. They move with the speed of light, but in contradistinction to free strings they interact, just like vortex lines do in ordinary fluid dynamics. It might be possible to extract some interesting relativistic dynamics of the vortex lines – the strings – from the hydrodynamic equations of the probabilistic fluid.

I would like to thank Wolfgang Schleich for his warm hospitality at the University of Ulm and the Alexander von Humboldt Foundation for an award that made my visit in Ulm possible.

REFERENCES

BIALYNICKI-BIRULA, I. 1994 *Acta Phys. Polon.* A **86**, 97.

BIALYNICKI-BIRULA, I. 1995 *Acta Phys. Polon.* A **26**, 1201.

BIALYNICKI-BIRULA, I. 1996 in *Proceedings of Coherence and Quantum Optics VII* (ed. J. H. Eberly, L. Mandel & E. Wolf).

BIALYNICKI-BIRULA, I. & BIALYNICKA-BIRULA, Z. 1971 *Phys. Rev.* D **3**, 2410.

BIALYNICKI-BIRULA, I., CIEPLAK, M. & KAMINSKI, J. 1992 *Theory of Quanta*, p. 105. Oxford University Press.

BOHM, D. & SCHILLER, R. 1955 *Suppl. Nuovo Cimento* **1**, 67.

BOHM, D., SCHILLER, R. & TIOMNO, J. 1955 *Suppl. Nuovo Cimento* **1**, 48.

GHOSH, S. K. & DEB, B. M. 1982 *Physics Reports* **92**, 1.

JANOSSY, L. & ZIEGLER-NARAY, M. 1965 *Acta Phys. Hung.* **20**, 23.

MADELUNG, E. 1926 *Z. Phys.* **40**, 322.

TAKABAYASI, T. 1952 *Prog. Theor. Phys.* **8**, 143.

TAKABAYASI, T. 1953 *Prog. Theor. Phys.* **9**, 187.

TAKABAYASI, T. 1955 *Prog. Theor. Phys.* **14**, 283.

TAKABAYASI, T. 1956 *Nuovo Cimento* **3**, 233.

TAKABAYASI, T. 1983a *Prog. Theor. Phys.* **70**, 1.

TAKABAYASI, T. 1983b *Prog. Theor. Phys.* **69**, 1323.

Waves in ionic reaction–diffusion–migration systems

By P. HASAL, V. NEVORAL, I. SCHREIBER, H. ŠEVČÍKOVÁ, D. ŠNITA AND M. MAREK

Department of Chemical Engineering, Prague Institute of Chemical Technology, Prague, Czech Republic

Two examples of ionic reaction–diffusion–migration systems supporting oscillations and travelling waves are discussed: Ca^{2+} signalling in cells, and the Belousov–Zhabotinskii reaction. Two models of Ca^{2+} kinetics (referred to as the CICR and the ICC models) are chosen from the literature, excitability conditions are adjusted and several spatially extended systems with added transport phenomena are defined and studied: (i) An array of diffusionally coupled sites with CICR kinetics is periodically stimulated at one end and signal propagation through the array is studied with the use of bifurcation analysis and continuation techniques and properties of resonance regimes are examined. (ii) Effects of an electric field on propagating calcium wave patterns in a spatially 1D system are studied for the ICC model and a method of controlling a chaotic spatiotemporal pattern by an external electric field, reducing it to a periodic patern, is described. (iii) An electric field applied to a spatially 2D system with CICR kinetics leads to the development of crescent structures followed by the creation of spirals. Finally, we experimentally examine wave patterns under an electric field in a two-dimensional Belousov–Zhabotinskii reaction medium. We find qualitatively the same scenario of the evolution of crescent patterns into spirals as that predicted from modelling calcium waves.

1. Introduction

Excitability is a property shared by some chemical systems and a large number of biological systems. In general, a system is excitable if it undergoes large excursions away and then back to a steady state, when perturbed by a suprathreshold stimulus. In an excitable medium, suprathreshold stimuli are propagated as pulse- or front-waves by a mechanism combining local excitatory kinetics with transport (e.g. by diffusion or ionic migration), see Holden *et al.* (1990). The Belousov–Zhabotinskii reaction, i.e. oxidation of malonic acid by bromate with $Ce^{4+/3+}$ or $Fe^{3+/2+}$ catalysts, has been used most often in studies of excitability in both lumped parameter (continuous flow stirred cell) or distributed systems, see Field & Burger (1990). Typical examples of oscillatory and excitable phenomena in biology are Ca^{2+} oscillations and waves in cells. Intracellular Ca^{2+} release is connected with many receptor-induced cell signals, controlling processes ranging from secretion and heart rate frequency control to transcription and cell division, see Berridge & Dupont (1994) and Berridge (1995). The introduction of digital Ca^{2+} imaging techniques in the last ten years has visualized spatiotemporal patterns of Ca^{2+}, ranging from various types of waves (propagating fronts, circular, spiral and complex waves) to synchronized oscillations, see e.g. Lechleiter & Clapham (1992) and Kasai *et al.* (1993). The Ca^{2+} waves in biological systems are mostly treated within the framework of excitable media with ionic transport, see Lechleiter & Clapham (1992). These waves can often be either elicited by electric stimulation or influenced by an electric field, see e.g. Madergaard (1994).

Several models of nonlinear kinetics of Ca^{2+} ions, describing oscillations in lumped parameter systems, were presented in recent years. Among those extensively studied were the models of Meyer & Stryer (1988) and Goldbeter *et al.* (1990). Models of

Ca^{2+} waves based on these kinetics were also proposed, see Berridge & Dupont (1994), Meyer & Stryer (1991) and Dupont & Goldbeter (1992). However, until now the analysis of both Ca^{2+} kinetic relations and Ca^{2+} waves was mostly qualitative. Quantitative, mostly numerical methods of analysis of oscillatory and excitable systems are now available, see Kubíček & Marek (1983), Marek & Schreiber (1995), Marek & Schreiber (1993) and Schreiber & Marek (1995). These methods were applied to chemical systems, where the nonlinear kinetic models and transport relations are relatively well developed. Since for chemical systems, quantitative experimental data obtained under well defined conditions are available, a comparison between experimental and modelling studies is possible. In this paper we illustrate the main ideas of application of numerical methods of nonlinear analysis by two examples: the analysis of Ca^{2+} kinetics and waves. Experimental data on the Ca^{2+} spatiotemporal patterns are now increasingly becoming available, and so we believe that a proper interpretation by comparison with modelling results can help advance the understanding of signal propagation in biological systems.

We present two models of Ca^{2+} kinetics, and demonstrate nonlinear dynamical methods of analysis of stimulated single and coupled cell systems. Then we discuss balance relations for ionic reaction–diffusion–migration systems in an electric field and apply them to the propagation of waves of Ca^{2+} ions. We present concrete results on the propagation of Ca^{2+} waves in 1D for the Meyer–Stryer model, and discuss the Ca^{2+} chaotic spatiotemporal patterns, and possibilities of their control. Modelling of the effects of an electric field on propagating Ca^{2+} patterns in spatially 2D systems, leading to the evolution of crescent structures and spirals, are also discussed. Such an evolution of complex spatiotemporal patterns in Ca^{2+} models is finally compared with similar patterns experimentally studied by means of digital image processing techniques in the BZ reaction system. The role of individual ionic components, their reactions and transport will be stressed. In conclusion, we briefly speculate on the problems of ionic signal propagation in an electric field, and on the role of nonlinear dynamics research in the studies of waves in biological systems.

2. Modelling of excitations and wave patterns in cytosolic calcium

Oscillations in cytosolic calcium ions are widespread in various types of cells. They arise either spontaneously or in response to an extracellular signal. Temporal oscillations can be either sinusoidal or relaxational. They lead to sharp spikes above a baseline, which are related to excitability – a suprathreshold stimulation elicits a spike from an originally quiescent state. In the presence of some transport, such as diffusion or ionic migration, the spike may then spread through cytosol as a wave. In this sense, the cytosol can act as an excitable medium, through which waves can propagate. Such waves can be repeatedly initiated at a particular spot. The wave spreading depends crucially on the autocatalytic nature of Ca^{2+} kinetics. Different models of the kinetics differently emphasize the role of various species involved in the activatory and regulatory feedback loops. The key species in all these models are the inositol 1,4,5-triphosphate (IP_3), the cytosolic Ca^{2+} (Ca_i) and the calcium ions sequestered in an intracellular store (Ca_s). We have chosen two prominent models: the Calcium-Induced Calcium Release (CICR) model, Dupont & Goldbeter (1993), and the IP_3–Ca_i Cross-Coupling (ICC) model, Meyer & Stryer (1991). From a mechanistic viewpoint, in the CICR model Ca_i alone has a positive self-feedback loop, whereas in the ICC model the positive feedback loop also involves IP_3 in addition to Ca_i; the species Ca_s plays a regulatory role. That is, a spike occurs whenever a suprathreshold stimulus, such as an increased level of IP_3 or Ca_i, causes the control maintained by the sequestered calcium to be unable to hold back

Param.	Value		Param.	Value		Param.	Value
k_f	1.0	min^{-1}	K_2	0.50	μM	n	2.0
k	10.0	min^{-1}	V_{M3}	50.00	$\mu\mathrm{M\,min}^{-1}$	m	2.0
v_1	1.7	$\mu\mathrm{M\,min}^{-1}$	K_R	1.00	μM	p	4.0
V_{M2}	325.0	$\mu\mathrm{M\,min}^{-1}$	K_A	0.45	μM		

TABLE 1. Parameter values for the CICR model

a rapid production of autocatalytic species. By this process, the sequestered calcium is released into cytosol, and eventually its level drops and causes the autocatalysis to stop. The cycle is completed by removal of the autocatalytic species. In particular, Ca_i is pumped back to the internal store and, in addition, IP_3 is removed by a decay process in the ICC model.

2.1. *Chemical kinetics of intracellular calcium*

The CICR model

Here we use a one-pool variant of CICR kinetics, see Dupont & Goldbeter (1993), which assumes that the internal Ca_s pool is sensitive to both cytosolic calcium and IP_3, the latter being an external parameter represented in the model by a saturation parameter β. In a gradientless (lumped parameter) environment, the dynamics is governed by the mass balance equations

$$\frac{\mathrm{d}x}{\mathrm{d}t} = f(x,y) = V_{\mathrm{in}} - V_2 + V_3 + k_f y - kx, \tag{2.1}$$

$$\frac{\mathrm{d}y}{\mathrm{d}t} = g(x,y) = V_2 - V_3 - k_f y, \tag{2.2}$$

(2.3)

where

$$V_{\mathrm{in}} = V_0 + V_1\beta, \tag{2.4}$$

$$V_2 = V_{M2}\frac{x^n}{(K_2^n + x^n)}, \tag{2.5}$$

$$V_3 = \beta V_{M3}\frac{y^m}{(K_R^m + y^m)}\frac{x^p}{(K_A^p + x^p)}, \tag{2.6}$$

and x is the concentration of Ca_i, y is the concentration of Ca_s, V_{in} is the total constant entry of Ca^{2+} into the cytosol, consisting of V_0 (the constant influx) and $V_1\beta$ (the IP_3-stimulated influx from extracellular medium), V_2 is the rate of pumping into the internal store, V_3 is the rate of release from the store, k_f is the coefficient of passive efflux from the store, k is the rate of passive efflux from the cytosol. The terms V_2 and V_3 follow Hill kinetics. In all calculations, we use the parameter values given in table 1. Since we are interested in excitable conditions, we treat V_0 and β as adjustable parameters to set up the appropriate dynamics. We find two different kinds of excitability in this system. One is characterized by a nodal steady state that is perturbed by *adding* x to get an excitatory response, and the other steady state is a focus and undergoes excitation by *removing* x. We have chosen the corresponding two pairs (V_0, β) to be $(1.7, 0.25)$ and $(5.25, 0.126)$.

Param.	Value		Param.	Value		Param.	Value	
A	20.0	s^{-1}	K_1	0.50	μM	E_v	1.000	$\mu\mathrm{M}^{-4}\,\mathrm{s}^{-1}$
B	40.0	$\mu\mathrm{M\,s}^{-1}$	K_2	0.15	μM	F_v	0.020	s^{-1}
C	1.1	$\mu\mathrm{M\,s}^{-1}$	K_3	1.00	μM	R	0.025	
D	2.0	s^{-1}	L	0.01	s^{-1}			

TABLE 2. Parameter values for the ICC model

The ICC model

The inositol 1,4,5-trisphosphate–calcium crosscoupling model of calcium spiking in living cells, see Meyer & Stryer (1991), has, in addition to x and y (the concentration of $\mathrm{Ca_i}$ and $\mathrm{Ca_s}$, respectively) two other variables: u (the concentration of $\mathrm{IP_3}$) and v (the fraction of receptors through which the sequestered calcium is released into cytosol). In a gradientless environment, the evolution equations read

$$\frac{\mathrm{d}x}{\mathrm{d}t} = vJ_{\mathrm{channel}} - J_{\mathrm{pump}}, \tag{2.7}$$

$$\frac{\mathrm{d}y}{\mathrm{d}t} = J_{\mathrm{pump}} - vJ_{\mathrm{channel}}, \tag{2.8}$$

$$\frac{\mathrm{d}u}{\mathrm{d}t} = k_{\mathrm{PLC}} - Du, \tag{2.9}$$

$$\frac{\mathrm{d}v}{\mathrm{d}t} = E_v x^4 v - F_v(1-v). \tag{2.10}$$

where

$$J_{\mathrm{channel}} = \left[\frac{Au^4}{(u+K_1)^4} + L\right] y, \tag{2.11}$$

$$J_{\mathrm{pump}} = \frac{Bx^2}{x^2+K_2^2}\,. \tag{2.12}$$

$$k_{\mathrm{PLC}} = C\left[1 - \frac{K_3}{(x+K_3)(1+R)}\right]. \tag{2.13}$$

Unlike the CICR model, $\mathrm{Ca_i}$ is neither supplied from, nor removed to, the extracellular environment. On the other hand, there is an additional activatory variable u (produced at the rate k_{PLC} and removed by a first order decay process) and a major regulatory role is played by the channel activation variable v rather than by $\mathrm{Ca_s}$ alone; J_{channel} is the rate of $\mathrm{Ca_s}$ release, J_{pump} is the rate of $\mathrm{Ca_i}$ sequestration and k_{PLC} is the rate of production of $\mathrm{IP_3}$ catalyzed by the enzyme phospholipase C.

The values of the kinetic parameters (see Meyer & Stryer (1991)) are summarized in table 2. They were chosen so that the system is excitable by adding x (or, equivalently, by adding u, since both species are autocatalytic). The excitability by removal of x (or u) was not found with this model.

2.2. *Spreading of excitation in coupled cell systems*

In this section we assume a discrete linear N array of sites at which the dynamics follows the CICR kinetics, where adjacent sites are mutually coupled via discrete diffusion terms. A site can have two interpretations, see Berridge & Dupont (1994); either it is a particular internal calcium store in a cell, or the array describes a chain of cells. The first case corresponds to an excitation appearing at a particular site due to kinetics being propagated along a one-dimensional array of calcium stores within a long and narrow cell, via diffusion. Thus, following a perturbation, a wave of excitation may spread through one cell. The second case assumes a gradientless environment within each cell and describes a wave propagation through an array of such cells. The evolution equations for the concentrations x_i, y_i at the ith site are $(i = 1, \ldots, N)$

$$\frac{dx_i}{dt} = V_{in} - V_2 + V_3 + k_f y - kx + d_x(x_{i+1} + x_{i-1} - 2x_i), \tag{2.14}$$

$$\frac{dy_i}{dt} = V_2 - V_3 - k_f y + d_y(y_{i+1} + y_{i-1} - 2y_i), \tag{2.15}$$

with boundary conditions

$$(x_0, y_0) = (x_1, y_1), \quad (x_{N+1}, y_{N+1}) = (x_N, y_N).$$

Since Ca_s is restricted to the internal store, $d_y = 0$. We assume that the first unit is repeatedly perturbed by a rapid addition (or removal, depending on the type of excitability) of Ca_i with an amplitude $\delta x_0 = A$ and period T. This mimics spontaneous spiking at a marginal site in the simplest way. We vary A, T and d_x, and follow the dynamics of the system. We find that a vast majority of choices of the parameters leads asymptotically to periodic responses, although we do find quasiperiodic and chaotic responses in limited regions of parameter space. This feature is typical of systems with two vastly different time scales. To accurately find both the stable and unstable periodic regimes, we use a continuation method, see Kubíček & Marek (1983) and Marek & Schreiber (1995), specially adapted for systems forced by periodic pulses, see Schreiber *et al.* (1988). The periodic regimes must have a period qT, where q is an integer. If, on the ith site, there are p_i excitations within the period qT, then we call $\nu_i = p_i/q$ the *firing number* for the ith cell. Two ways of presenting the results are used. One is based on delineating regions of distinct dynamics in the forcing-period–forcing-amplitude plane. Another one is to vary T at a fixed value of A, and follow the response amplitude δx_i of the autocatalytic variable x_i in each of the cells. Since the dynamics is of an 'all or none' nature, all stable periodic regimes have either a very small or a near-to-maximal response amplitude, and can be unambiguously assigned a firing number. We find that smooth transitions from small to large amplitude response regimes (and therefore discrete jumps in the firing number) do exist, but the connecting periodic orbits with intermediate amplitudes are all unstable, unless the input forcing period is much smaller than the characteristic time of the excitation event.

Perturbation of a focus

Consider first the case when the unperturbed dynamics possesses a stable focal steady state, and an excitation is elicited by removing Ca_i (i.e. by lowering x). Figure 1 (a–c) shows δx versus T for regimes with $q = 1, 2, 3, 4$ in one unit perturbed at a small, medium and large level of the forcing amplitude A. As pointed out, only regimes with a near-to-maximum and near-to-minimum response amplitude are stable. Hence figure 1(a) shows that small forcing amplitude leads to no excitation, except for a rather narrow region, from $T \approx 2.5$ to $T \approx 3.0$, where stable 1:2 and 1:1 resonances occur. Outside this region, 0:1 resonance dominates. Intermediate amplitudes imply complex repeating structures, see figure 1(b). Period 2, 3, and 4 regimes are shown only at the leftmost lobe of the period 1 branch, but analogous branches occur at every lobe. This complexity is a consequence of the focal nature of the steady state of the unperturbed system, since the phase point winding toward the focus can be shifted by the pulse either above or below the excitability threshold. This occurs only when A is close to the distance (measured along the x-coordinate) of the focus from the threshold set. It implies that intervals of stable 0:1 and 1:1 resonances alternate. The hierarchy of p:q, $q > 1$, resonances is more involved, and will be the subject of further studies; those few shown in the figure have all firing numbers 1:q. The response to a large forcing amplitude is shown in figure 1(*c*). There are no repeated structures, and plateaus of 1:4, 1:3, 1:2 and 1:1 resonances are now broad. Also, the plateaus fit together tightly, so that there is very little space for observing higher order resonances or nonperiodic behaviour.

An overall bifurcation structure is conveniently represented by a diagram in figure 2, where some bifurcation curves in the A–T parameter plane for regimes with $q = 1$ and $q = 2$ are plotted. These curves correspond to three generic codimension-one bifurcations that can occur: the limiting (or turning) point, the period doubling and the torus (or secondary Hopf's) bifurcation. Parts of these curves that correspond to stability boundaries delimitate regions with different firing numbers, as indicated in the figure. Some conclusions that can be drawn from figure 2 are:

(*a*) torus bifurcations occur at high forcing frequencies, and quasiperiodic dynamics alternating with higher order resonances exist nearby,

(*b*) 1:q resonance plateaus dominate,

(*c*) a very complex bifurcation structure, related to the autonomous focus, exists at the transition from 0:q to p:q with $p > 0$,

(*d*) the bifurcation structure possesses all essential features of resonance horns, known from forced oscillators, see McGehee & Peckham (1994), although the overall arrangement is different, and there are extra features that will be discussed in detail elsewhere.

Next we examine the response amplitudes of $q = 1$ regimes in two and three coupled units, to see how the excitation spreads across the array. This is shown in figure 3. In two cells (figure 3(*a*)) there are three major stable resonances characterized by the following pairs (ν_1, ν_2) of firing numbers: (0:1, 0:1), (1:1, 0:1) and (1:1, 1:1). Thus we have an excitation failure, a propagation failure and a completed propagation. The succession of these regimes repeats itself as T is increased due to the autonomous focus. We have seen this regularity in our earlier experiments with the Belousov–Zhabotinskii reaction, as well as in the corresponding mathematical models, see Kosek & Marek (1993) and Kosek *et al.* (1994). However, our present calculations suggest that such a regular behaviour is only present in rather small ranges of forcing amplitude, forcing period and transport coefficient. Three coupled units are represented in figure 3(*b*), which shows (0:1, 0:1, 0:1) and (1:1, 0:1, 0:1) stable regimes only, that is, at the same value of the transport coefficient as in two units, we have an excitation failure and a propagation failure only. A complete spectrum of propagation patterns can be established by raising d_x.

Perturbation of a node

In this case an excitation can be elicited by adding Ca_i. The bifurcation diagram for a single unit (figure 4), analogous to that given in figure 2 for perturbed focus, shows all bifurcation curves for $q = 1$, and those for $q = 2, 3, 4, 5$ that delimitate regions of stable resonances. Except for the complex structure due to focus, the basic features of both diagrams are the same. There is a 'quasiperiodicity plus higher order resonances' region, enclosed by torus bifurcation curves at high forcing frequencies, and the major low resonances are packed tightly so that nonperiodic dynamics are restricted to a set of negligible measure. The transition zone from 0:q to p:q with $p > 0$ lacks the bifurcational complexity of its focal counterpart.

The response amplitudes in one, two and three coupled units are shown in figure 5. Figure 5(*a*) is to be compared with figure 1(*c*). There is no possibility for repeated patterns on period one branch in this case, comparable to figure 1(*b*). However, the period two, three and four branches are similar to those in figure 1(*c*). They are even more closely packed, leaving virtually no gaps for other dynamics. Figure 5(*b*) shows response

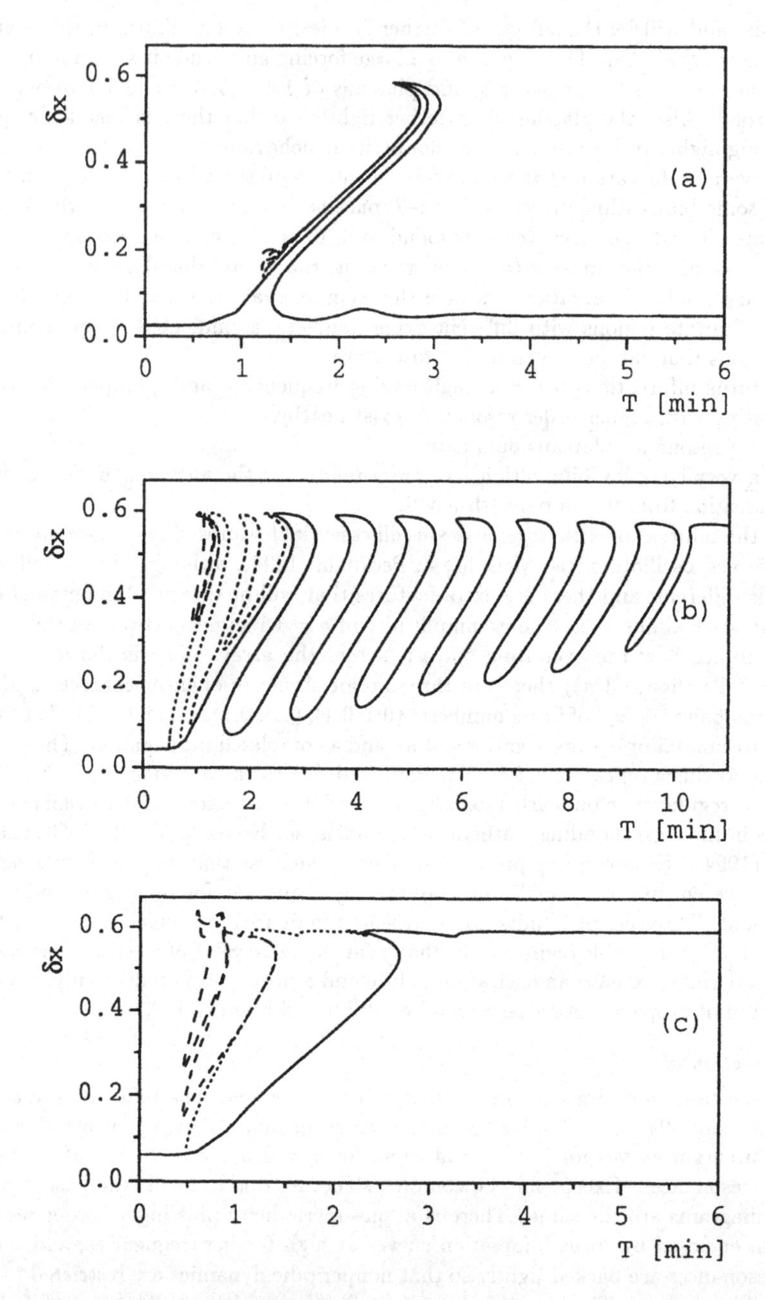

FIGURE 1. The response amplitude δx for periodic regimes with $q = 1$ (solid line), $q = 2$ (dotted line), $q = 3$ (dashed line) and $q = 4$ (long dashed line) in a single unit, versus forcing period T: (a) $A = 0.025$ μM, (b) $A = 0.04236$ μM, (c) $A = 0.06$ μM.

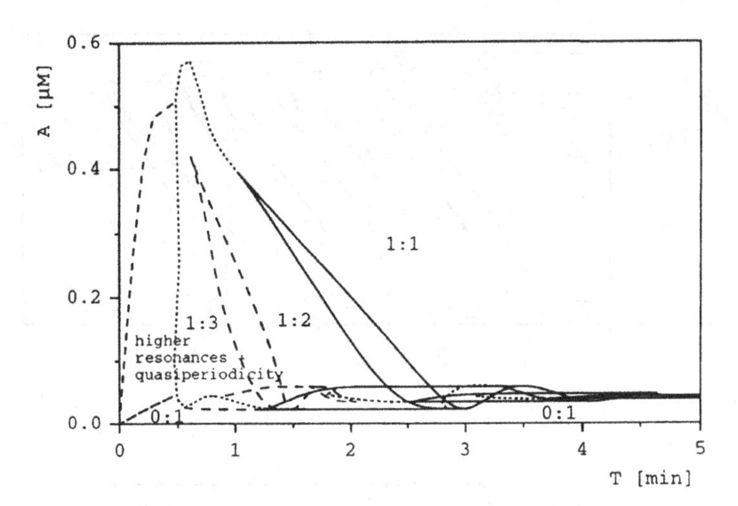

FIGURE 2. Bifurcation diagram in the A–T parameter plane, the choice of fixed parameters corresponding to perturbed focus; the limiting point on 1-periodic regime (solid line), period doubling on 1-periodic regime (dotted line), torus bifurcation on 1-periodic regime (dashed line), limiting point on 2-periodic regime (solid line).

amplitudes for period one in two coupled cells. As one increases T, the propagation pattern ranges from failed excitation to failed propagation, and to completed propagation. The same sequence also exists in three coupled units, as shown in figure 5(c). However, the ranges of T where a particular propagation pattern exists are very uneven. There is a broad range of an excitation in the first unit that fails to propagate, and a broad range corresponding to propagation through all three units. In between is a narrow interval (around $T \approx 8.25$) where an excitation in the first cell propagates to the second, but fails to propagate to the third. The width of this region cannot be blown up by adjusting the free parameters A and d_x. Thus the system appears to prefer certain propagation patterns and suppress others.

The problem of excitations becoming extinct somewhere along their way through the array, a propagation failure phenomenon, is of crucial importance for signal transmission, see Lauffenburger & Lindermann (1993). It is also related to the problem of travelling waves in continuous excitable media. Signal transmission via excitations, either propagating or failing to propagate, provides a possible basis for frequency coding of information carried through the array. This coding will depend on both the properties of the input signal (forcing amplitude and frequency) and those of the unperturbed excitable kinetics (oscillatory or nonoscillatory approach to the steady state). A continuation technique seems to be an efficient tool for getting insight into this problem. The problem of travelling waves in continuous excitable media applied to calcium dynamics is studied in the following sections.

2.3. *General reaction–diffusion–migration model for continuous media*

Ionic reaction–diffusion–migration systems are ubiquitous in living organisms. For example, calcium ions act as intra- and intercellular second messengers (signals) that control the functions of many types of living cells and their assemblies. Calcium is often released in the form of oscillations and waves, due to the interaction of reactions and trans-

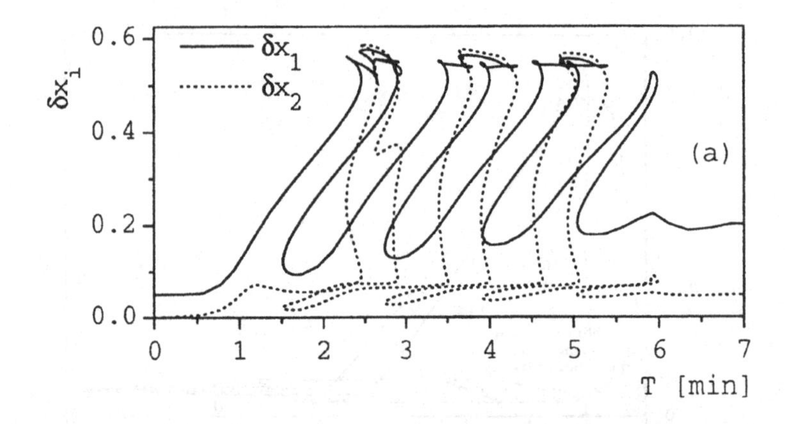

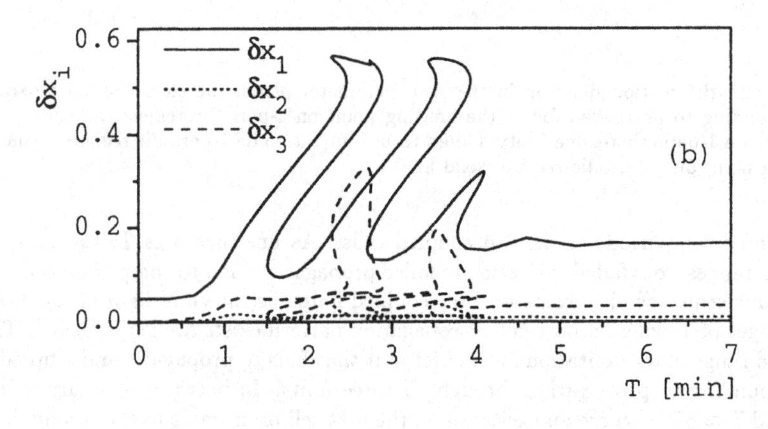

FIGURE 3. The response amplitude δx_i for periodic regimes with $q = 1$ in two and three coupled units, versus forcing period T; $A = 0.05$ μM, $d_x = 0.9$ min^{-1}; (*a*) two cells, (*b*) three cells.

port, see Berridge (1995), Lechleiter & Clapham (1992), Dupont & Goldbeter (1992), Av-Ron *et al.* (1991), Smith & Saldana (1992), Clapham (1995) and Ghosh & Greenberg (1995). The intra- and intercellular propagation of calcium waves can be affected by gradients of electric field. For example, in neuronal-glial networks, calcium signalling is intimately connected with local electric field gradients formed due to spatiotemporal patterns of ionic currents, see Cooper (1995). It has been proposed that the frequency of generated spatiotemporal Ca^{2+} patterns and ionic currents in living systems be used for the control of signalling the behaviour of various cell and tissue functions. Several reaction-diffusion models of spatiotemporal patterns for calcium waves have recently been proposed, see Berridge (1995), Lechleiter & Clapham (1992), Meyer & Stryer (1988), Mayer & Stryer (1991), Dupont & Goldbeter (1992), Dupont & Goldbeter (1993), Smith & Saldana (1992), but their validity is still being questioned. None of the proposed models takes into account the specifically ionic character of components, or effects of migration in an electric field. Propagating calcium waves can become chaotic in a certain range of model parameters. The application of an external electric field further complicates the behaviour of such spatiotemporal patterns. The coexistence of propagat-

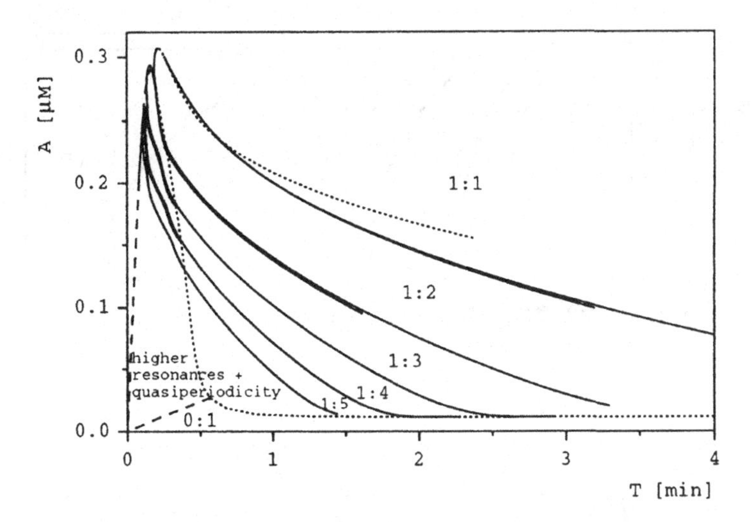

FIGURE 4. Bifurcation diagram in the A–T parameter plane, the choice of fixed parameters corresponding to perturbed node; the limiting point on 1-periodic regime (solid line), period doubling on 1-periodic regime (dotted line), torus bifurcation on 1-periodic regime (dashed line), limiting point on 2-periodic regime (solid line).

ing and non-propagating calcium waves was observed experimentally in living cells, see Boitano *et al.* (1992). Furthermore, localized spatiotemporal patterns ('calcium sparks') are known to be associated with the propagation of calcium waves, see Cheng *et al.* (1993).

General model

We begin with the formulation of evolution equations. The local mass balance equation for each of the N components in the reacting mixture can be written

$$\frac{\partial c_i}{\partial t} = R_i - \nabla \cdot \vec{J}_i, \quad i = 1, \dots, N, \tag{2.16}$$

where the subscript i refers to ith component of the mixture; $c_i \equiv c_i(\vec{l}, t)$ is the concentration, $\vec{l}$ represents the spatial coordinates, t is time, $R_i \equiv R_i(c_1, \dots, c_N)$ is the source term due to chemical reactions and $\vec{J}_i \equiv \vec{J}_i(\vec{l}, t)$ is molar flux intensity

$$\vec{J}_i = -D_i \nabla c_i + D_i z_i c_i \vec{E} F/(R_g T). \tag{2.17}$$

Here D_i is the diffusion coefficient, z_i is the charge number, $\vec{E} \equiv \vec{E}(\vec{l}, t)$ is the electric field, F is the Faraday constant, R_g is the universal gas constant and T the absolute temperature.

The electric field can be written as the gradient of an electric potential $\varphi \equiv \varphi(\vec{l}, t)$:

$$\vec{E} = -\nabla \varphi. \tag{2.18}$$

Gauss' law of electrostatics

$$\varepsilon \nabla \cdot \vec{E} = q \tag{2.19}$$

determines the divergence of the electric field, in a system with constant permittivity ε, in terms of the charge density

$$q \equiv q(\vec{l}, t) = F \sum_{i=1}^{N} z_i c_i. \tag{2.20}$$

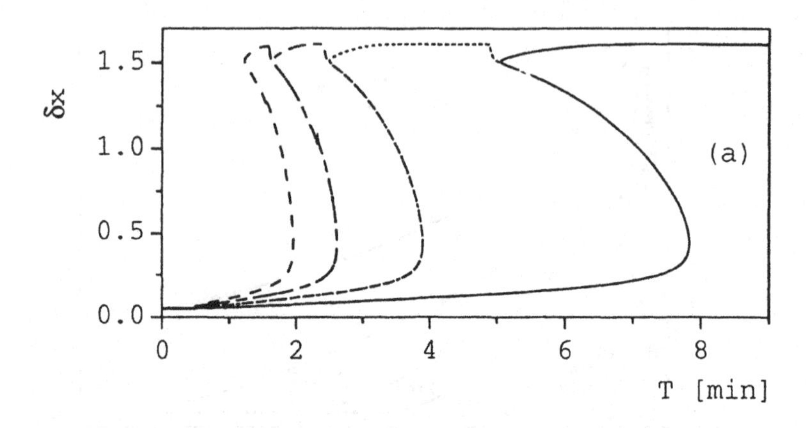

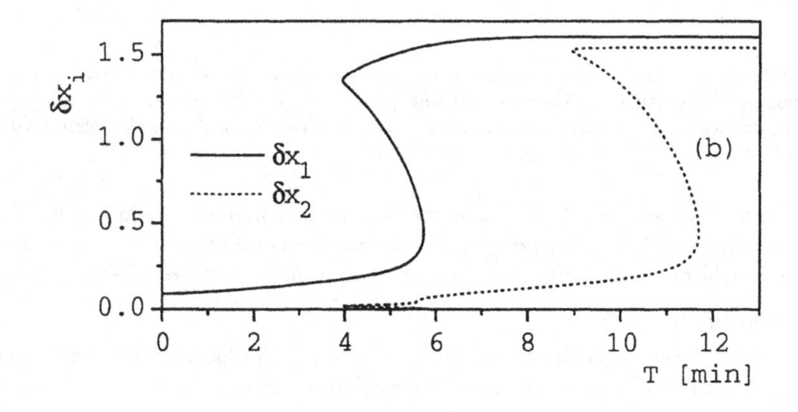

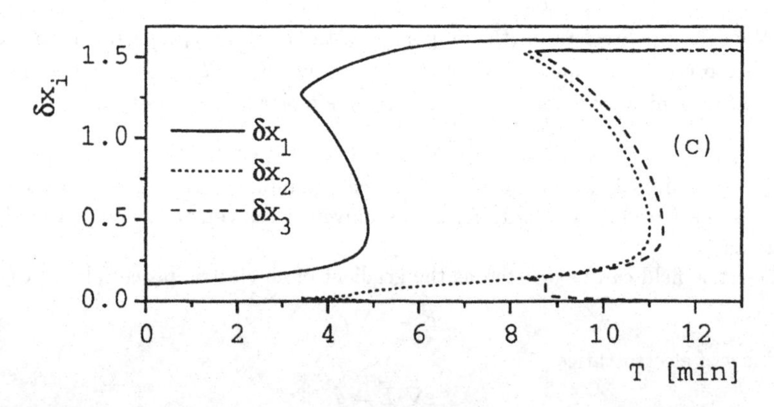

FIGURE 5. The response amplitude δx_i for periodic regimes in one, two and three coupled units, versus forcing period T; (a) one cell, $q = 1$ (solid line), $q = 2$ (dotted line), $q = 3$ (dashed line) and $q = 4$ (long dashed line); $A = 0.05$ μM; (b) two cells, $q = 1$, $A = 0.09$ μM, $d_x = 0.1$ min^{-1}; (c) three cells, $q = 1$, $A = 0.105$ μM, $d_x = 0.11$ min^{-1}.

A combination of equations (2.18) and (2.19) gives Poisson's equation:

$$\nabla^2 \varphi = -\frac{q}{\varepsilon} = -\frac{F}{\varepsilon} \sum_{i=1}^{N} z_i c_i. \tag{2.21}$$

The electric current intensity $\vec{I} \equiv \vec{I}(\vec{l}, t)$ in an electrolyte is given by

$$\vec{I} = F \sum_{i=1}^{N} z_i \vec{J}_i. \tag{2.22}$$

The charge conservation law, to be fulfilled in the chemical reactions, leads to

$$\sum_{i=1}^{N} z_i R_i = 0. \tag{2.23}$$

The local charge balance equation (linear combination of equations (2.16)) has the form:

$$\frac{\partial q}{\partial t} = -\nabla \cdot \vec{I}. \tag{2.24}$$

Simplified model

Let us assume local electroneutrality, constant conductivity and an *a priori* known divergenceless electric field, see e.g. Šnita & Marek (1994). Local mass balance equations for components in spatially one-dimensional systems then take the form:

$$\frac{\partial c_i}{\partial t} = D_i \frac{\partial^2 c_i}{\partial l^2} - D_i z_i \frac{\partial c_i}{\partial l} \frac{EF}{R_g T} + R_i(c_1, \ldots, c_N), \quad i = 1, \ldots, N, \tag{2.25}$$

where l is a spatial coordinate and E is an electric field; in a spatially two-dimensional system they can be written as:

$$\frac{\partial c_i}{\partial t} = D_i \left(\frac{\partial^2 c_i}{\partial l_1^2} + \frac{\partial^2 c_i}{\partial l_2^2} \right) - D_i z_i \left(\frac{\partial c_i}{\partial l_1} E_1 + \frac{\partial c_i}{\partial l_2} E_2 \right) \frac{F}{R_g T} + R_i, \quad i = 1, \ldots, N, \tag{2.26}$$

where l_1, l_2 are spatial coordinates and E_1, E_2 are components of the electric field.

2.4. *Waves in a 1D medium with ICC kinetics*

In a one-dimensional diffusion medium, based on an ionic extension of the ICC model of calcium waves due to Meyer & Stryer (1991), not only regular but also spatiotemporal chaotic waves can arise.

The ICC kinetics is assumed to take place within a one-dimensional reaction–diffusion–electromigration medium, and the value of the electric field intensity E throughout the system is considered constant, due to a high ionic strength of the cytosol. The cytosolic Ca^{2+} ions (variable x) and cytosolic IP_3 ions (variable u) are regarded as mobile components. The Ca^{2+} ions sequestered in endoplasmic reticulum (ER) stores (variable y) are assumed to be immobile, similarly to ER membrane ionic channels (the variable v characterizing the fraction of opened channels). The model equations (component mass balance equations) then read:

$$\frac{\partial x}{\partial t} = \left[\frac{Au^4}{(u+K_1)^4} + L \right] yv - \frac{Bx^2}{x^2 + K_2^2} + D_x \frac{\partial^2 x}{\partial l^2} + \frac{z_x F D_x}{R_g T} E \frac{\partial x}{\partial l}, \tag{2.27}$$

$$\frac{\partial y}{\partial t} = \frac{Bx^2}{x^2 + K_2^2} - \left[\frac{Au^4}{(u+K_1)^4} + L \right] yv, \tag{2.28}$$

$$\frac{\partial u}{\partial t} = C \left[1 - \frac{K_3}{(x+K_3)(1+R)} \right] - Du + D_u \frac{\partial^2 u}{\partial l^2} + \frac{z_u F D_u}{R_g T} E \frac{\partial u}{\partial l}, \tag{2.29}$$

$$\frac{\partial v}{\partial t} = E_v x^4 v - F_v(1 - v). \tag{2.30}$$

We recall that F is the Faraday constant, R_g is the universal gas constant, T is the absolute temperature, t is time and l denotes a spatial coordinate. The diffusion coefficients and charge numbers of respective species are $D_x = 6\ \mu\text{m}^2\,\text{s}^{-1}$, $D_u = 200\ \mu\text{m}^2\,\text{s}^{-1}$, $z_x = 2$ and $z_u = -6$. The values of other model parameters (see Meyer & Stryer (1991)) are summarized in table 2.

The total length of the system was $L = 150\ \mu$m. Spatial derivatives in equations (2.27) and (2.29) were replaced by finite differences (the total number of mesh points was $N_{\text{mp}} = 257$). The resulting set of $2N_{\text{mp}}$ ODEs, together with N_{mp} equations (2.28) and N_{mp} equations (2.30) was solved using the LSODE package described in Hindmarsh (1983) and Petzold (1983). Neumann boundary conditions were imposed. Steady state concentrations (at $R = 0.025$) of all components were used as initial conditions. A small positive perturbation of the variable u was applied in the central grid point at $t = 0$. These initial conditions yield a pair of excitable pulse calcium waves moving in opposite directions towards system boundaries. Perturbations applied at the boundary grid points yield a pair of waves propagating towards the center of the system.

When there is no external electric field, the wave pairs propagate with constant velocity, except for small regions very close to system boundaries (see figures 7(*a*) and 7(*e*)). When an external electric field is applied, the wave propagating in the direction of the electric field is accelerated, and that propagating in the opposite direction is decelerated (see figures 7(*b*) and 7(*f*)). The velocity of the accelerated wave is a linear function of the electric field intensity E (see figure 6). The decelerated waves, however, can exist only up to a relatively low intensity of the electric field (at parameter values taken from table 2, the limiting value of E is approximately equal to $-3.5\ \text{V}\,\text{cm}^{-1}$). At higher values of E, the decelerated wave is destroyed (see figures 7(*d*) and 7(*h*)). Splitting of the decelerated wave is observed at E values close to the limiting value (see figures 7(*c*) and 7(*g*)). These phenomena can be associated with the influence of the imposed electric field on the diffusion and migration of IP_3 ions (variable u): the electrophoretic mobility of the IP_3 ion is high due to its high diffusion coefficient, and its charge is negative ($z_u = -6$). Therefore the calcium waves triggering level of IP_3 concentration rapidly propagates towards the anode, and the calcium wave is accelerated. In the opposite direction, the diffusional propagation of IP_3 ions is strongly limited by ionic migration. Hence, at sufficiently high electric field intensity, the electromigration of IP_3 ions prevails, and they begin to be transported back into the refractory region of the reaction–diffusion medium. As a result, the Ca^{2+} wave vanishes, see figures 7(*d*) and 7(*h*). The IP_3 pulses propagating within the system are very broad as compared to the width of Ca^{2+} pulses. At a proper electric field intensity (just below the threshold for the wave destruction) the diffusional transport only slightly exceeds the migrational transport of IP_3 ions, and the IP_3 pulses split and cause the secondary Ca^{2+} waves to appear at the tail of the primary wave, moving both in the direction coinciding and opposite to the electric field, see figures 7(*c*) and 7(*g*). Secondary waves, however, can propagate only for a limited period of time, until the IP_3 level falls below the Ca^{2+} waves triggering threshold, due to the continuing diffusional IP_3 pulse propagation. The wave phenomena shown in figures 7 and 6 clearly demonstrate possible effects of the external electric field on the propagation of calcium waves in tissue cells. The Ca^{2+} waves based signal transmission between cells can be e.g. violated or completely damaged by an external electric field or, on the contrary, enhanced by a speeding up of the waves.

The parameter R in (2.29) denotes fractional activation of the cell-surface receptors. It determines the calcium spiking frequency in a gradientless environment, and also the

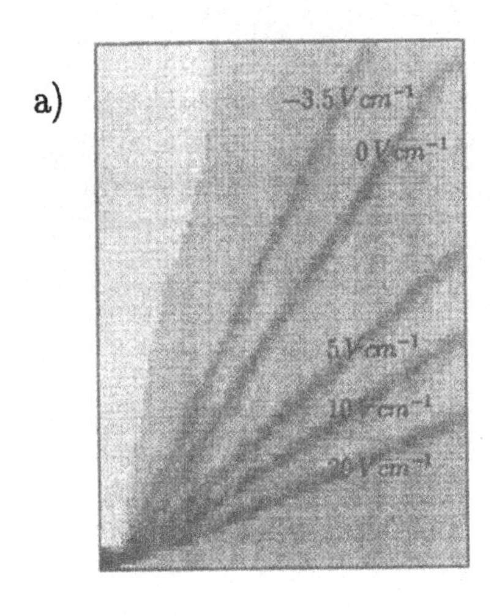

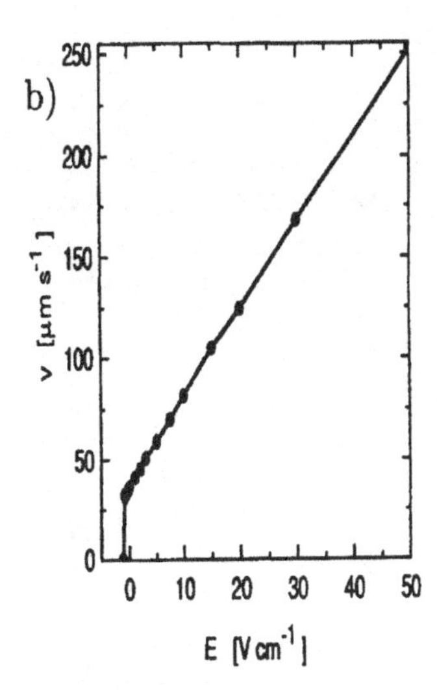

FIGURE 6. Velocity of cytosolic calcium waves in the ICC model in a spatially one-dimensional system: (*a*) overlay of the space–time traces of pulse waves at different electric field intensity (the range of the space and time coordinates is the same as in figure 7). Dark/light areas correspond to high/low cytosolic Ca^{2+} concentration; (*b*) wave velocity as a function of electric field intensity. See table 2 for parameter values.

frequency of calcium bulk oscillations in a spatially distributed system. Excitable calcium waves can be triggered in the system by a local perturbation of a spatially homogeneous stationary state, but only at low values of R. At $R = 0.0256$, the system undergoes Hopf's bifurcation, and long-period non-homogeneous bulk oscillations are excited there. The period of bulk oscillations decreases as one increases R. Non-homogeneous bulk oscillations at $R = 0.475$, resulting from the interaction betwen oscillatory and excitatory regimes, are shown in figure 8(*a*). Chaotic spatiotemporal patterns can be evoked in the system at $R > 0.475$, e.g. by a local perturbation of the variable u (see figure 8(*b*)). For calcium waves in a chaotic state, neither the velocity nor the amplitude is constant, and the waves split and annihilate chaotically. The inter- and intracellular signalling can therefore be corrupted at medium levels of cell-surface receptor activation. The oscillatory and/or chaotic patterns disappear at $R = 0.585$, where the stable spatially homogeneous steady state is again established via Hopf's bifurcation.

The analysis of chaotic spatiotemporal data shown in figure 8(*b*) indicates high dimensionality of the chaotic attractor. Based on the application of the Karhunen–Loève decomposition, see Aubry *et al.* (1991) and Hasal *et al.* (1994), the embedding dimension of the attractor was estimated as equal to 8. Three positive Lyapunov exponents were found for the attractor reconstructed from the data shown in figure 8(*b*). The presence of three positive Lyapunov exponents was confirmed by repeating evaluation with increasing embedding dimension. The chaotic pattern shown in figure 8(*b*) can therefore be called hyperchaotic. Hyperchaos can be connected with intensive mixing in the phase space. Its biological significance was the subject of recent speculations. Chaotic propagation of calcium waves within the tissue cells is clearly undesirable from the point of view of intra- and intercellular calcium signalling. Therefore the possibility of stabilizing the chaotic calcium wave motion to any of the unstable periodic orbits embedded within the chaotic attractor was investigated. Sev-

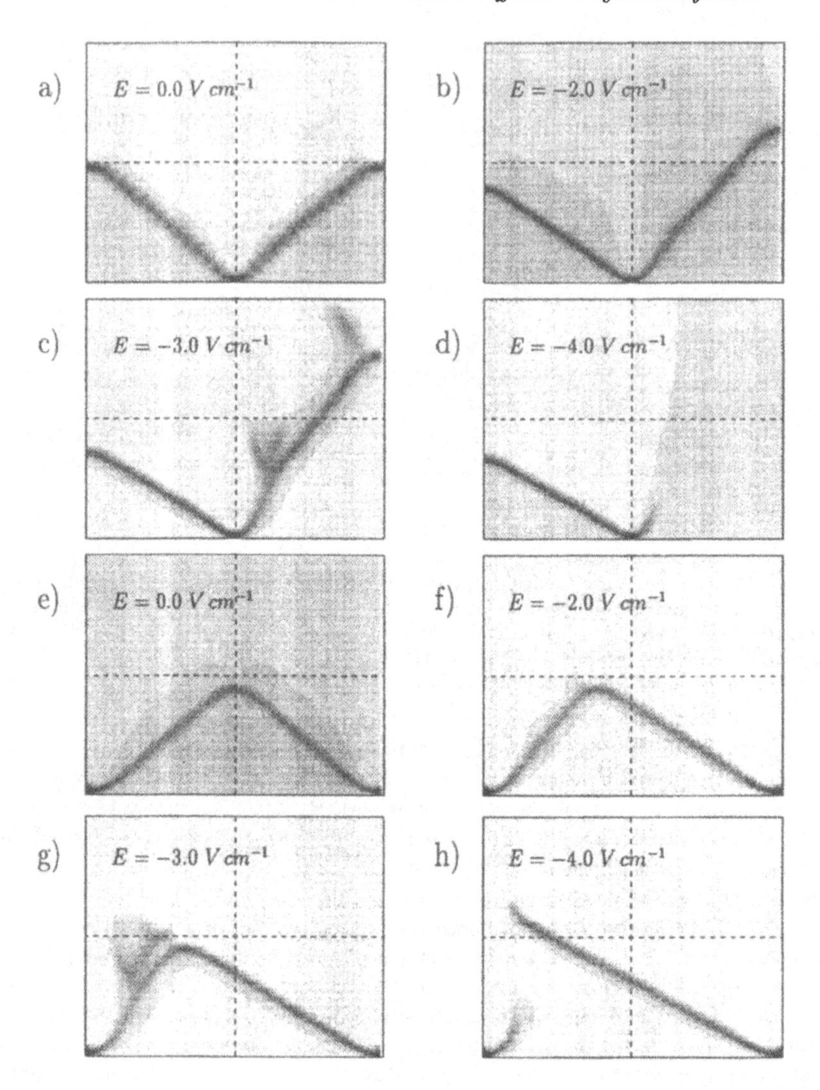

FIGURE 7. Cytosolic calcium waves in the ICC model in a spatially one-dimensional system: (*a*) an unaffected pair of pulse waves triggered in the center by IP_3 concentration pulse, (*b*) wave acceleration and deceleration in a weak electric field, (*c*) wave splitting and reversal in a medium strength electric field, (*d*) wave destruction in a strong electric field, (*e*)–(*h*) a pair of counter-propagating pulse waves at the same electric field intensities as in figures (*a*)–(*d*). Dark/light areas correspond to high/low cytosolic Ca^{2+} concentration. The range of space coordinate (horizontal axis) corresponds to total system length $L = 150\ \mu m$; time (vertical axis) is running in the interval 0–4 s.

eral methods of chaos control in spatially distributed systems were described, see e.g. Gang & Kaifen (1993), Qin *et al.* (1994), Gay & Ray (1995), Chakravarti *et al.* (1995) and Lourenço (1995). The system dynamics can also be controlled by adjusting the value of a suitable global system parameter. The model equations (2.27) – (2.30) involve just one such parameter, i.e. the external electric field intensity E. Stabilization

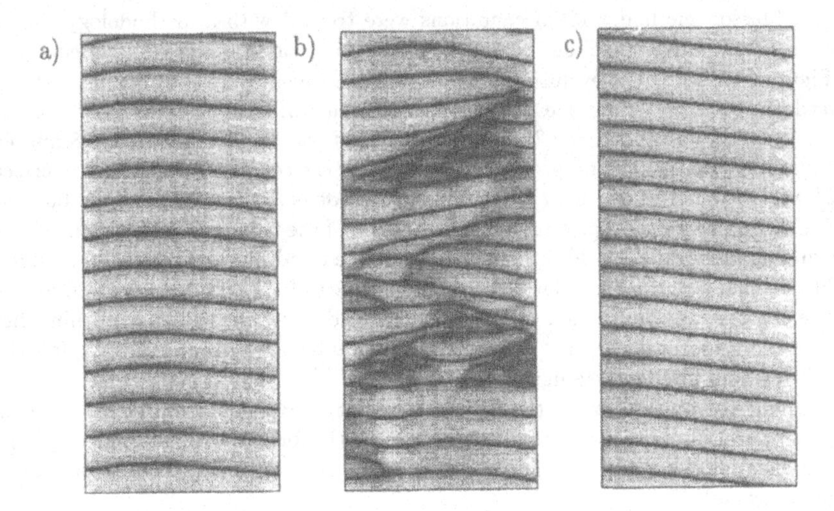

FIGURE 8. Spatiotemporal patterns of cytosolic calcium in the ICC model in a spatially one-dimensional system: (*a*) non-homogeneous bulk oscillations ($R = 0.475$, $E = 0.0\ \mathrm{V\,cm^{-1}}$), (*b*) chaotic waves ($R = 0.500$, $E = 0.0\ \mathrm{V\,cm^{-1}}$), (*c*) stabilized chaotic waves ($R = 0.500$, $E = 1.0\ \mathrm{V\,cm^{-1}}$). For other parameter values, see table 2. Dark/light areas corespond to high/low cytosolic Ca^{2+} concentration. The range of space coordinate (horizontal axis) corresponds to $L = 150\ \mu\mathrm{m}$; time (vertical axis) is running downward in the interval 400–450 s.

of the chaotic waves of figure 8(*b*) to periodic ones is attained simply by application of the d.c. electric field of intensity $E = +1.0\ \mathrm{V\,cm^{-1}}$ to system boundaries. This electric field shifted the system dynamics to travelling calcium pulse waves originating at the left system boundary, i.e. $l = 0\ \mu\mathrm{m}$ (where the system trajectory resembles bulk oscillatory regime) and moving to the opposite boundary, see figure 8(*c*). Stabilization of wave motion can be assigned to an overwhelming effect of the imposed electric field on the transport of IP_3 ions. The direction of wave motion can be simply reversed by reversing the electric field. Control of the chaotic waves is poor if $|E| < 1.0\ \mathrm{V\,cm^{-1}}$, as in that case the electromigration contribution to the net transport of ionic reaction components is too small. On the other hand, further increase of the electric field intensity only leads to acceleration of stabilized waves, without any change in the frequency of their generation. Hence the application of a relatively low external electric field to the cell offers the possibility of a simple and efficient stabilization of the calcium signalling process against chaotic signal propagation.

2.5. *Waves in 1D and 2D media with CICR kinetics*

Equations (2.25) (or (2.26)) were solved with the use of CICR kinetics:

$$N = 2, \quad c_1 = x, \quad c_2 = y, \tag{2.31}$$

$$R_1 = f(x, y), \quad R_2 = g(x, y), \quad D_1 = 360\ \mu\mathrm{m}^2\,\mathrm{min}^{-1}, \quad D_2 = 0, \tag{2.32}$$

see equations (2.1)–(2.6) and table 1.

Spatial derivatives in the resulting differential equations (ordinary or partial) were replaced by finite differences, where the total number of mesh points was $N_{\mathrm{mp}} = 800$ in 1D and $N_{\mathrm{mp}} = 400 \times 400$ in 2D. The resulting set of ODEs was solved by standard Runge–

Kutta–Merson method. Initial conditions were treated with a methodology similar to that used in the ICC medium. Neumann's boundary conditions were imposed.

Figure 9 presents the evolution of a Ca^{2+} pulse wave in the two-dimensional system after local perturbation at the center, i.e. an increase of local cytosolic Ca^{2+} concentration ($V_0 = 1.7\ \mu$M and $\beta = 0.25$). No electric field was applied until $t = 0.8$ min and a circular wave developed and propagated away from the center of the system. The electric field was switched on at time $t = 0.8$ min (the anode is located at the upper boundary). The Ca^{2+} pulse wave propagating in the direction of the field is accelerated and the pulse moving opposite to the field direction is decelerated and finally annihilated at sufficient field strength (see the annihilation of the upper part of the circular wave and formation of the crescent wave in the figure 9). When the field is switched off ($t = 1.2$ min) the free tips of the crescent wave start to wind into the center and form a pair of spirals which finally start to interact (see figure 9).

Without the application of the field, the circular wave would continue to propagate. The temporary application of the field has caused the formation of a localized 'geometric pacemaker' as the pair of winding spirals emits circular waves continuously into the neighbourhood.

The evolution of Ca^{2+} waves in the spatialy one-dimensional system after a perturbation (i.e. local decrease of cytosolic Ca^{2+} concentration in the left part of the system) is shown in figure 10 ($V_0 = 5.25\ \mu\mathrm{M\,min}^{-1}$ and $\beta = 0.126$). The perturbation ceases to elicit any response, if there is no electric field (see the middle column in figure 10). If an electric field of sufficient intensity is applied, the perturbation elicits a wave packet propagating in the direction of the field, while the number of waves in the packet continuously increases (see the left hand column in figure 10). If the field is switched off, the wave packet ceases to exist (see the right hand column in figure 10). However, if the field is switched on, the 'remnants' of ceased structures act as perturbations, and the wave packet is reconstructed and develops further (see the middle column in figure 10).

A homogeneous state is stable and non-excitable if there is no electric field. However, in the presence of an electric field this state becomes unstable, and the smallest perturbation (realizable numerically) causes the formation of the wave packet. The same phenomenon for the spatially two-dimensional system is presented in figure 11. The number of structures formed in the direction of the field increases. In the direction perpendicular to the field one can observe diffusion and subsequent spreading of complex spatiotemporal patterns.

3. Experiments with crescent and spiral waves in the Belousov–Zhabotinskii (BZ) reaction

A shallow layer of the BZ reacting medium is often used for an experimental investigation of excitable pulse waves (spirals or circles in spatially two-dimensional systems, and plane waves in one-dimensional systems) and their interaction with an oriented migration flow of a reaction species, evoked by an imposed d.c. electric field. Depending on the initial concentrations of substrates, one can observe either oxidation waves (i.e. a narrow region consisting of an oxidized form of a catalyst, propagating through a reduced medium, see Zaikin & Zhabotinskii (1970), Winfree (1972), Field & Noyes (1974)) or reduction waves (i.e. a zone of a reduced catalyst propagating through an oxidized medium, see Smoes (1980), Marek *et al.* (1994)). Electric field interaction with oxidation waves has been studied for more than ten years, see Feeney *et al.* (1981), Ševčíková & Marek (1983), Agladze & De Kepper (1992), Steinbock *et al.* (1992) and

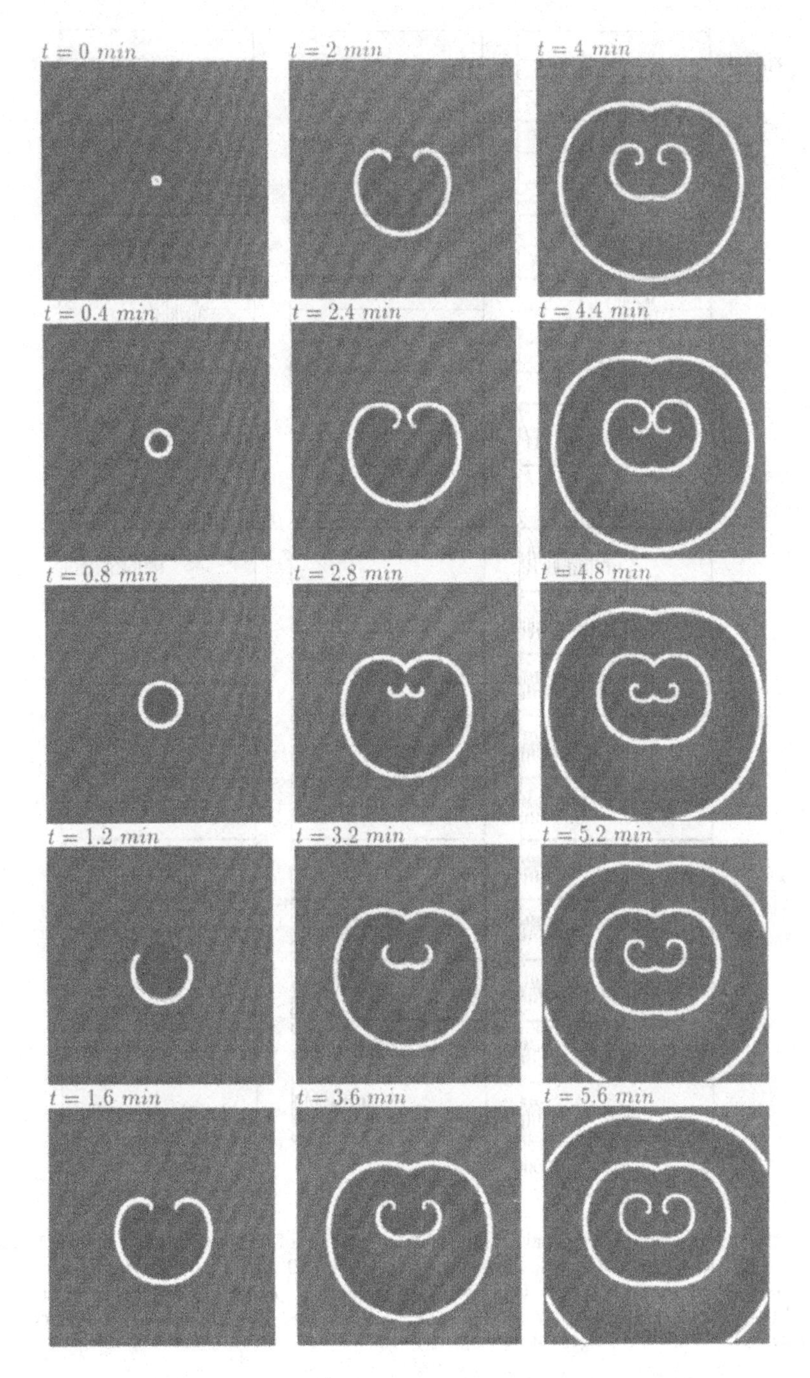

FIGURE 9. 2D dynamics of the CICR model on the square 2000×2000 μm^2; $\beta = 0.25$, $V_0 = 1.7\ \mu M\,min^{-1}$, $E_1 = 0\ V\,cm^{-1}$, $E_2(t < 0.8\,min) = 0\ V\,cm^{-1}$, $E_2(0.8\,min < t < 1.2\,min) = -5\ V\,cm^{-1}$ (anode at the upper boundary), $E_2(1.2\,min < t) = 0\ V\,cm^{-1}$; dark/light areas correspond to low/high cytosolic Ca^{2+} concentration.

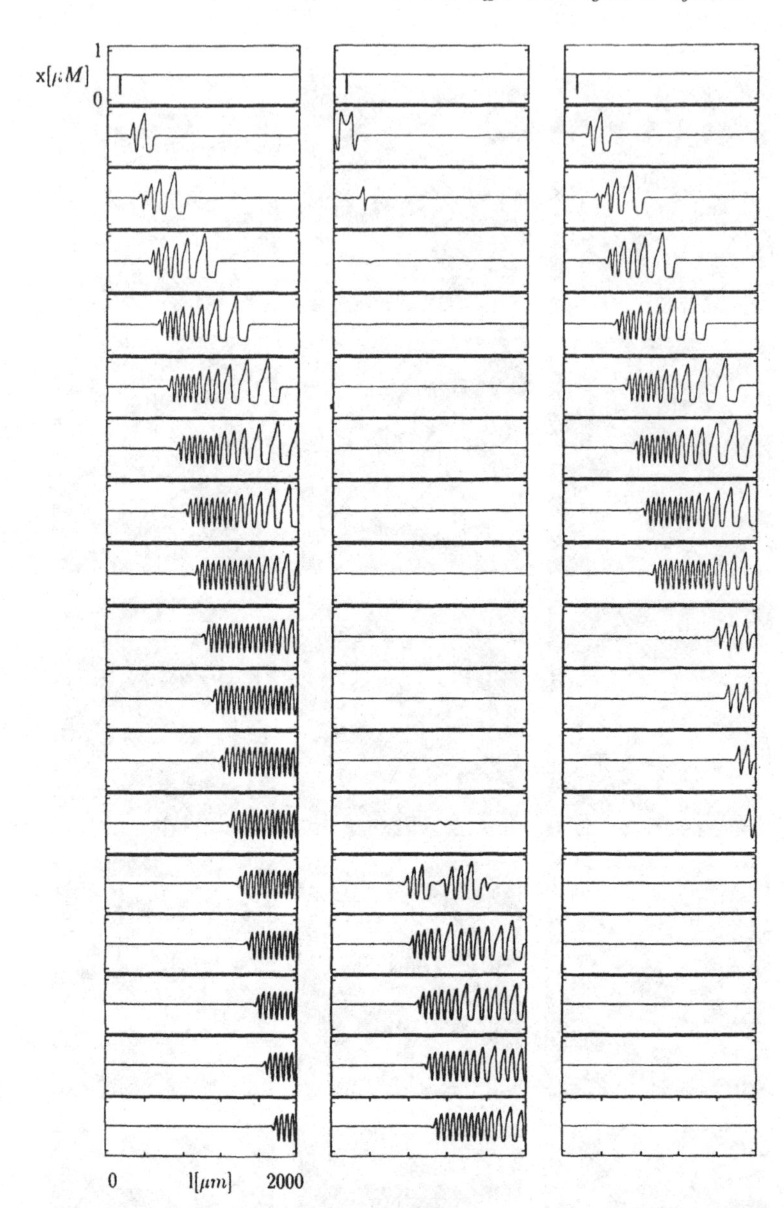

FIGURE 10. Evolution of spatial concentration profile of cytosolic form of Ca^{2+} ions; time (vertical axis) increases downwards from $t = 0$ to $t = 6.8$ min; time difference between two profiles is 0.4 min; left column: $E = 5\,\mathrm{V\,cm^{-1}}$; middle column: $E(t \leq 3.2\,\mathrm{min}) = 0\,\mathrm{V\,cm^{-1}}$, $E(t > 3.2\,\mathrm{min}) = 5\,\mathrm{V\,cm^{-1}}$; right column: $E(t \leq 3.2\,\mathrm{min}) = 5\,\mathrm{V\,cm^{-1}}$, $E(t > 3.2\,\mathrm{min}) = 0\,\mathrm{V\,cm^{-1}}$.

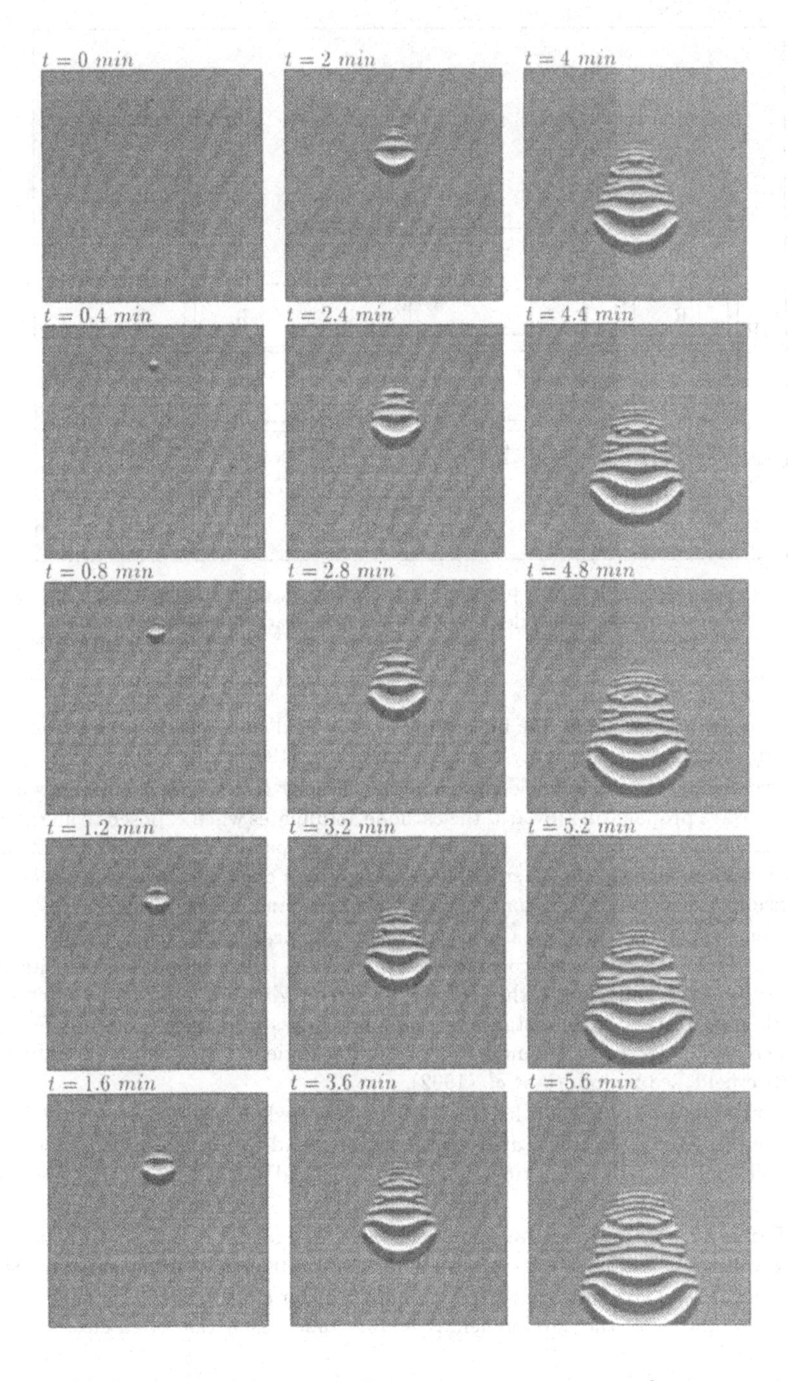

FIGURE 11. 2D dynamics of the CICR model on the square 2000×2000 μm^2; $V_0 = 4.8\ \mu M\ min^{-1}$, $\beta = 0.19$, $E_1 = 0\ V\,cm^{-1}$, $E_2 = -5\ V\,cm^{-1}$ (anode at the upper boundary); dark/light areas correspond to low/high cytosolic Ca^{2+} concentration.

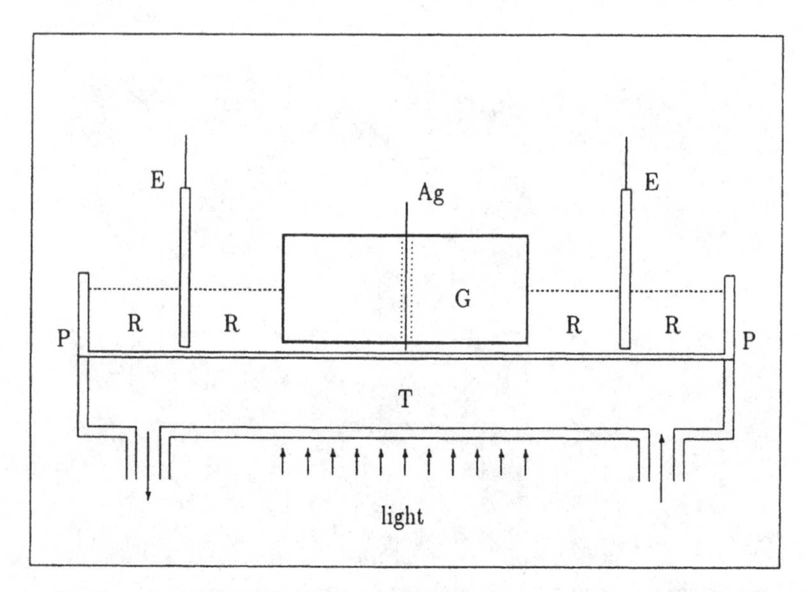

FIGURE 12. Experimental setup: P – Petri dish, T – termostated double bottom, G – glass seal, E – electrodes, Ag – Ag wire, R – reaction mixture.

Muñuzuri *et al.* (1994), while the first results on the field's influence on reduction waves have been published only recently, see Kaštánek *et al.* (1995).

First experiments with a single circular wave, Feeney *et al.* (1981), have shown that part of the wave propagating towards the cathode annihilates when exposed to an external d.c. electric field. A formed crescent wave further propagates and preserves its shape until the field is turned off. Then the free ends of the crescent wave start to curl in, giving rise to two vortices that continuously emit spiralling waves.

Later on, application of a d.c. electric field to spiral wave structures was found to cause a drift of the spiral core toward the anode, and, simultaneously, its shift in a direction perpendicular to the electric field, see Agladze & De Kepper (1992) and Steinbock *et al.* (1992). The distance of two spiral cores in a spiral pair was found to either increase or decrease, depending on the mutual orientation of spiral chirality and the electric field, see Steinbock *et al.* (1992).

A drift and shift of spiral cores has also been observed when an alternating electric field was applied to the system. The drift direction was found to depend on both the initial phase angle between the spiral tip velocity and the field vector, and on the frequency, see Muñuzuri *et al.* (1994).

The formation and propagation of spiral waves in excitable cardiac tissue is being intensively studied, because of its proposed connection to heart fibrillation and sudden cardiac death, see Davidenko *et al.* (1992) and Starobin *et al.* (1994). Detailed mechanistic understanding of the development of spirals under the action of an electric field is of particular importance.

Quantitative data on propagation velocity in an imposed d.c. electric field have been obtained for the plane BZ waves propagating along a spatially one-dimensional system (capillary tubular reactor), see Ševčíková & Marek (1983). The propagation velocity increases when the wave moves toward the anode, and decreases when it moves toward the

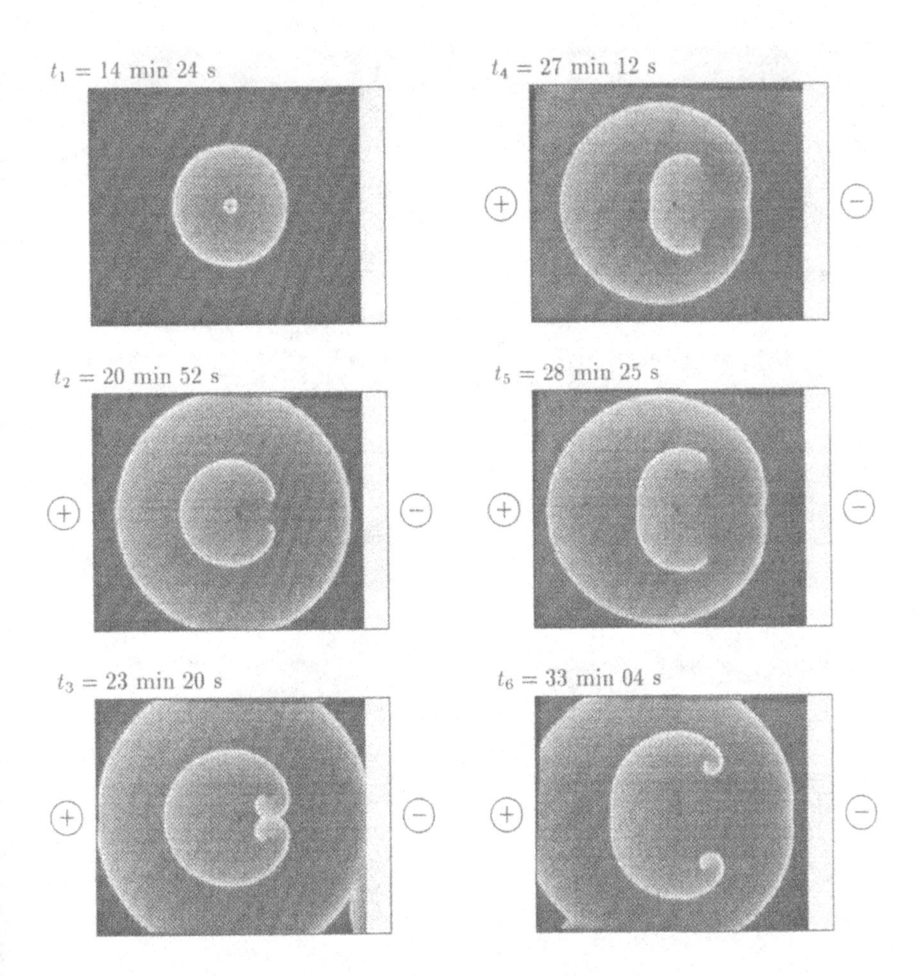

FIGURE 13. Second wave break-up and spiral formation in the imposed electric field of intensity $E = 3.0\ \mathrm{V\,cm^{-1}}$. Light/dark areas correspond to oxidized/reduced form of the catalyst. The electric field was turned on at time $t = 10$ min 28 s. The monitored region is $3.3 \times 4.0\ \mathrm{cm}^2$.

cathode. The dependence on field intensity is nonlinear, and when a 'velocity-decreasing' field of supercritical magnitude is applied, the wave ceases to exist.

One-dimensional studies with oxidation waves have shown, in addition to quantitative effects on the velocity of propagation and wave stability, also qualitative changes of spatiotemporal structures. These can be wave reversal and splitting, see Ševčíková & Marek (1983) and Ševčíková *et al.* (1992). Wave reversal has been observed at electric field intensities only slightly lower than the supercritical. In the course of wave splitting, which occurs at the range of even lower field intensities, new waves propagating towards the anode emerge from the back of the wave propagating towards the cathode. The life time of this temporal wave source (and hence the number of emitted waves) depends on the magnitude of the electrical field gradient.

Recently, the splitting off of a single wave from an almost planar one has also been

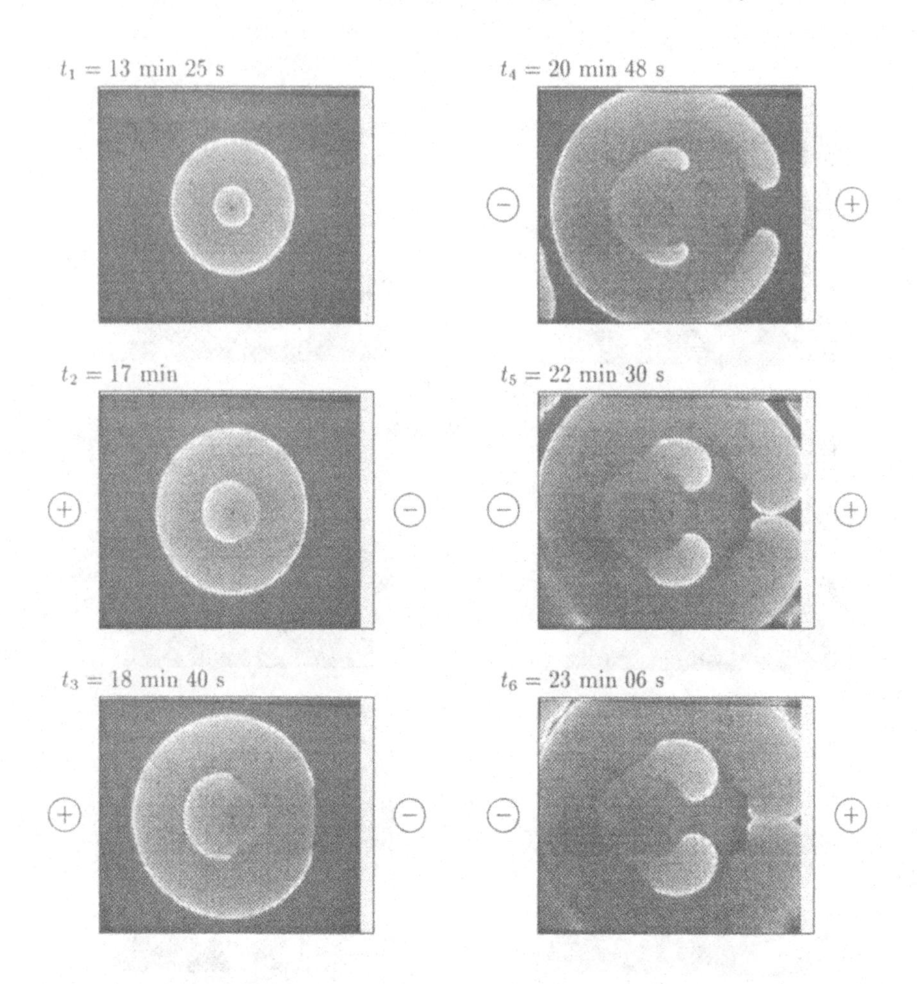

FIGURE 14. Successive break-up of both waves in the pattern, stabilization of free ends, and spiral wave formation after the field direction has been reversed; $E = 6.5\ \mathrm{V\,cm^{-1}}$. The electric field was turned on approximately at time $t = 16$ min 30 s and reversed approximately at time $t = 29$ min 30 s. The monitored region is $3.3 \times 4.0\ \mathrm{cm^2}$.

observed in a spatially two-dimensional system, see Ševčíková *et al.* (1996). Distinct effects of an imposed d.c. electric field on individual waves in a periodic wave pattern, leading to the development of disordered spatiotemporal structures via the process of wave segmentation and rejoining, have been found, see Ševčíková *et al.* (1996).

The explanation of the field effects on the BZ wave structures is based on a commonly adopted mechanism of BZ wave propagation. It accounts for the interaction of excitable kinetics with diffusive transport of reacting species in spatial gradients, see Field & Noyes (1972). The BZ wave propagation is controlled by the concentration of bromide ions in a small region in front of the wave, see Field & Noyes (1974) and Field & Noyes (1972). This concentration changes both due to mutual reactions of bromide with bromous acid and other reactants, and by molecular diffusion (in a field-free case). It can also change due to the inflow from or the outflow into the surrounding medium by migration (when an electric field is applied). From the concentration profile of bromide ions along the wave, see Field & Noyes (1972), one can conclude that the bromide is supplied when the wave moves towards the cathode; it is depleted when the wave moves towards the anode, see Ševčíková & Marek (1983). Bromide depletion leads to wave acceleration, and bromide supply to wave deceleration.

The migration supply of bromide ions increases when increasing field intensity, and implies the following sequence of phenomena: (i) propagation deceleration, (ii) wave splitting, (iii) wave reversal and (iv) wave annihilation. This sequence was observed in a spatially one-dimensional system, where the propagation velocity is paralell to the electric field, see Ševčíková & Marek (1983).

When a circular wave propagates in an electric field of constant direction, the effective bromide flow paralell to the normal velocity vector changes along the wave arc, and thus the applied field affects different parts of the wave in different ways.

Here we demonstrate characteristic experimental observations of electric field effects on a pattern of two concentric circular waves that account for local wave front annihilation, stabilization of free ends of crescent waves and the formation of spiral centers.

The experiments were performed with the modified BZ reaction mixture, consisting of 0.205 M $HBrO_3$, 0.007 M KBr, 0.05 M Malonic Acid and a 0.002 M ferroin catalyst, bounded in agar gel (0.4 %) in order to avoid possible induced convective flows.

The experimental setup is based on a specifically adapted Petri dish (see figure 12). The dish is filled with rubber-like material (Lukopren) so that a rectangular channel (5×12 cm^2) is formed. Along the longer sides of the channel, two stripes of teflon (0.6 mm high, 12 cm long) are fixed on the dish bottom. The channel is filled with the reaction mixture and than a specially shaped glass seal ($4.7 \times 4.7 \times 3.0$ cm^3) is placed on the teflon stripes. Thus a thin layer (depth 0.6 mm) of the reaction medium is formed under the glass seal. The rest of the mixture is squeezed out by the glass seal, and two layers 1.4 cm high are created on both sides of the seal. After solidification of the gel, an Ag wire is introduced through a tiny hole (diameter 0.3 mm) in the middle of the glass seal into the thin layer of the medium. When two waves are initiated at the Ag wire, the wire is removed. Plated stainless steel electrodes are immersed into the mixture on both sides of the glass seal, and an electric field of chosen intensity is applied. Thermostated water (15° C) flows through the double bottom of the Petri dish, ensuring constant temperature in the course of the experiment.

A horizontally placed Petri dish is lighted from below, and the course of the spatiotemporal pattern development is monitored and recorded on video tape, using a CCD camera. The data are then processed by the hardware and software for computerized image processing.

The evolution of a circular pattern of two waves in a field of intensity $E = 3.5$ V cm^{-1} is shown in figure 13. After the field is applied to the system, a small part of the second wave annihilates (at time t_2) and the free ends of an arising crescent wave start to curl into spirals. The first wave is not broken by the field, and at time t_3 slowly leaves the monitored region. A new continuous wave front is formed after the arms of both spirals collide (time t_3). A new crescent wave breaks away and the free ends keep spiralling (times t_4 through t_6). Comparing the positions of spiral cores at times t_3 and t_6, one can see that the spiral cores are driven away from each other in a direction perpendicular to the electric field. This observation is in agreement with earlier results given in Agladze & De Kepper (1992) and Steinbock *et al.* (1992). On the other hand, no shift of spiral cores toward the anode was observed in this experiment.

The effect of a higher field intensity ($E = 6.5$ V cm^{-1}) on the two-wave pattern created without the imposed electric field (t_1) is illustrated in figure 14. After the field is turned on, the full right half of the second wave circle is broken (t_2). The first wave is broken a little later (t_3) and only a small part of its right arc annihilates. When the field is reversed (between times t_3 and t_4) almost the full left half of the second wave crescent annihilates, and at the same time the free ends of the original crescent start to grow towards the anode and curl in (t_4). The first wave is not broken by the reversed field (at least not when it stays in the monitored region). The free ends of the first wave crescent form spirals in the reversed field (from t_4 to t_6). Free ends do not curl into spirals for the field pointing toward the cathode. Thus a stabilization of free ends by the electric field of a suitable strength and orientation is observed.

When we compare the above experimental observations with the modelling of the development of Ca^{2+} spiral waves described in the previous section, we can claim that spiral formation and evolution in both systems obviously have common generic features. The well defined chemical system used here allows for a detailed interpretationon of the basis of the reaction mechanism and the migration of individual species.

4. Conclusions

The above plethora of propagating excitations and waves in relatively simple chemical and biological excitable systems, without and with an electric field, illustrate a ubiquitous presence of nonlinear phenomena in chemical and biological systems. This is so in particular with information transfer. An efficient analysis of both spatially discrete models (coupled cells) and spatially continuous models (1D and 2D) requires the use of bifurcation theory combined with appropriate numerical techniques (continuation methods). The experimental patterns are obtained by methods of digital image processing. A juxtaposition of controlled experiments with the results of mathematical analysis and modelling, allows for detailed studies of the evolution mechanisms for spatiotemporal patterns. One can also propose methods for their control. Generalizations and generic results are possible. However, we have to keep in mind that each case is specific and dependent on many constraints, primarily on the kinetics which is the major contributor to specific nonlinearities present in the system. The methods of interpretation of data corresponding to experimentally measured complex spatiotemporal patterns, now becoming increasingly available, will hopefully help make progress in understanding the nature of information coding, transmission and decoding in living matter.

The authors wish to thank the Grant Agency of the Czech Republic for financial support of this project (grants 104/93/0272 and 201/94/1111).

REFERENCES

Agladze, K. I. & De Kepper, P. 1992 Influence of electric field on rotating spiral waves in the Belousov–Zhabotinsky reaction. *J. Phys. Chem.* **96**, 5239–5242.

Aubry, N., Guyonnet, R. & Lima, R. 1991 Spatiotemporal analysis of complex signals: theory and applications. *J. Stat. Phys.* **64**, 683.

Av-Ron, E., Parnas, H. & Segel, L. A. 1991 A minimal biophysical model for an excitable and oscillatory neuron. *Biol. Cybern.* **65**, 487–500.

Berridge, M. J. & Dupont, G. 1994 Spatial and temporal signalling by calcium. *Curr. Opinion Cell Biol.* **6**, 267–274.

Berridge, M. J. 1995 Calcium signalling and cell proliferation. *BioEssays* **17**, 491–500.

Boitano, S., Dirksen, E. R. & Sanderson, M. J. 1992 Intercellular propagation of calcium waves mediated by inositol trisphosphate. *Science* **258**, 292–295.

CHAKRAVARTI, S., MAREK, M. & RAY, W. H. 1995 Reaction–diffusion system with Brusselator kinetics: control of a quasi-periodic route to chaos. *Phys. Rev.* E **52**, 2407.

CHENG, H., LEDERER, W. J. & CANELL, M. B. 1993 Calcium sparks: elementary events underlying excitation–contraction coupling in heart muscle. *Science* **262**, 740–747.

CLAPHAM, D. E. 1995 Replenishing the stores. *Nature* **375**, 634–635.

COOPER, M. S. 1995 Intercellular signaling in neuronal–glial networks. *BioSystems* **34**, 65–85.

DAVIDENKO, J. M., PERTSOV, A. V., SALOMONSZ, R., BAXTER, W. & JALIFE, J. 1992 Stationary and drifting spiral waves of excitation in isolated cardiac muscle. *Nature* **355**, 349–351.

DUPONT, G. & GOLDBETER, A. 1992 Oscillations and waves of cytosolic calcium: insights from theoretical models. *BioEssays* **14**, 485–493.

DUPONT, G. & GOLDBETER, A. 1993 One-pool model for Ca^{2+} oscillations involving Ca^{2+} and inositol 1,4,5-trisphosphate as co-agonists for Ca^{2+} release. *Cell Calcium* **14**, 311–322.

FEENEY, R., SCHMIDT, S. & ORTOLEVA, P. 1981 Experiments on electric field – BZ chemical wave interactions: annihilation and the crescent wave. *Physica* D **2**, 536–544.

FIELD, R. J. & BURGER, M., EDS. 1990 *Oscillations and Travelling Waves in Chemical Systems*. Wiley.

FIELD, R. J. & NOYES, R. M. 1972 Explanation of spatial band propagation in the Belousov reaction. *Nature* **237**, 390–392.

FIELD, R. J. & NOYES, R. M. 1974 Oscillations in chemical systems. Quantitative explanation of band migration in the Belousov–Zhabotinskii reaction. *J. Am. Chem. Soc.* **96**, 2001.

HOLDEN, A. V., MARKUS, M. & OTHMER, M. G., EDS. 1990 *Nonlinear Wave Processes in Excitable Media*. Plenum Press.

GANG, H. & KAIFEN, H. 1993 Controlling chaos in systems described by partial differential equations. *Phys. Rev. Lett.* **71**, 3794–3797.

GAY, D. H. & RAY, W. H. 1995 Identification and control of distributed parameter systems by means of the singular value decomposition. *Chem. Eng. Sci.* **50**, 1519–1539.

GHOSH, A. & GREENBERG, M. E. 1995 Calcium signaling in neurons: molecular mechanisms and cellular consequences. *Science* **268**, 239–247.

GOLDBETER, A., DUPONT, G. & BERRIDGE, M. J. 1990 Minimal model for signal-induced Ca^{2+} oscillations and for their frequency encoding through protein phosphorylation. *Proc. Nat. Acad. Sci. USA* **87**, 1461–1465.

HASAL, P., MÜNSTER, A. F. & MAREK, M. 1994 Spatiotemporal chaos in an electric current driven ionic reaction–diffusion system. *Chaos* **4**, 531–546.

HINDMARSH, A. C. 1983 In *Scientific Computing* (ed. R. S. Stepleman *et al.*), p. 55. North-Holland.

KASAI, H., LI, Y. X. & MIYASHITA, Y. 1993 Subcellular distribution of Ca^{2+} release channels underlying Ca^{2+} waves and oscillations in exocrine pancreas. *Cell* **74**, 669–677.

KAŠTÁNEK, P., KOSEK, J., ŠNITA, D., SCHREIBER, I. & MAREK, M. 1995 Reduction waves in the BZ reaction: circles, spirals and effects of electric field. *Physica* D **84**, 79–94.

KOSEK, J. & MAREK, M. 1993 Coupled excitable cells. *J. Phys. Chem.* **97**, 120–127.

KOSEK, J., SCHREIBER, I. & MAREK, M. 1994 Phase mappings from diffusion-coupled excitable chemical systems. *Phil. Trans. R. Soc. Lond.* A **347**, 643–660.

KUBÍČEK, M. & MAREK, M. 1983 *Computational Methods in Bifurcation Theory and Dissipative Structures*. Springer.

LAUFFENBURGER, D. A. & LINDERMANN, J. J. 1993 *Receptors. Models for Binding, Trafficking and Signalling*. Oxford University Press.

LECHLEITER, J. L. & CLAPHAM, D. E. 1992 Molecular mechanisms of intracellular calcium excitability in X. laevis oocytes. *Cell* **69**, 283–294.

LOURENÇO, C., HOUGARDY, M. & BABLOYANTZ, A. 1995 Control of low-dimensional spatiotemporal chaos in Fourier space. *Phys. Rev.* E **52**, 1528–1532.

MAREK, M. & SCHREIBER, I. 1993 Chaos in forced and coupled chemical oscillators and excitators. In *Chaos in Chemistry and Biochemistry* (ed. R. J. Field & L. Györgyi). World Scientific.

MAREK, M., KAŠTÁNEK, P. & MÜLLER, S. C. 1994 Ring-shaped waves of inhibition in the Belousov–Zhabotinsky reaction. *J. Phys. Chem.* **98**, 7452–7454.

MAREK, M. & SCHREIBER, I. 1995 *Chaotic Behaviour of Deterministic Dissipative Systems.* Cambridge University Press.

MCGEHEE, R. P. & PECKHAM, B. M. 1994 Resonance surface for forced oscillators. *Experimental Mathematics* **3**, 221–244.

MEDERGAARD, M. 1994 Direct signalling from astrocytes to neurones in cultures of mammalian brain cells. *Science* **263**, 1768–1771.

MEYER, T. & STRYER, L. 1988 Molecular model for receptor-stimulated calcium spiking. *Proc. Nat. Acad. Sci. USA* **85**, 5051–5055.

MEYER, T. & STRYER, L. 1991 Calcium spiking. *Ann. Rev. Biophys. Biophys. Chem.* **20**, 153–174.

MUÑUZURI, A. P., GÓMEZ-GESTEIRA, M. PÉREZ-MUÑUZURI, V., KRINSKY, V. I. & PÉREZ-VILLAR, V. 1994 Parametric resonance of a vortex in an active medium *Phys. Rev.* E **50**, 4258–4261.

PETZOLD, L. R. 1983 *SIAM J. Sci. Stat. Comput.* **4**, 136.

QIN, F., WOLF, E. E. & CHANG, H.-C. 1994 Controlling spatiotemporal patterns on a catalytic wafer. *Phys. Rev. Lett.* **72**, 1459–1462.

SCHREIBER, I., DOLNIK, M., CHOC, P. & MAREK, M. 1988 Resonance behaviour in two-parameter families of periodically forced oscillators. *Phys. Lett.* A **128**, 66–70.

SCHREIBER, I. & MAREK, M. 1995 Dynamics of oscillatory chemical systems. In *Modelling the Dynamics of Biological Systems* (ed. E. Mosekilde & O. G. Mouritsen). Springer.

SMITH, D. A. & SALDANA, R. 1992 Model of the Ca^{2+} oscillator for shuttle streaming in physarum polycephalum. *Biophys. J.* **61**, 368–380.

SMOES, M. L. 1980 Chemical waves in the oscillatory Zhabotinskii system. A transition from temporal to spatiotemporal organization. In *Dynamics of Synergetic Systems* (ed. H. Haken), p. 80. Springer.

ŠNITA, D. & MAREK, M. 1994 Transport and reaction in ionic chemical systems. *Physica* D **75**, 521–540.

ŠEVČÍKOVÁ, H. & MAREK, M. 1983 Chemical waves in electric field. *Physica* D **9**, 140–156.

ŠEVČÍKOVÁ, H., MAREK, M. & MÜLLER, S. C. 1992 The reversal and splitting of waves in an excitable medium caused by an electric field. *Science* **257**, 951–954.

ŠEVČÍKOVÁ, H., KOSEK, J. & MAREK, M. 1996 Splitting of 2D waves of excitation in d.c. electric field. *J. Phys. Chem.* **100**, 1666–1675.

STAROBIN, J., ZILBERTER, Y. I. & STARMER, C. F. 1994 Vulnerability in one-dimensional media. *Physica* D **70**, 321–341.

STEINBOCK, O., SCHÜTZE, J. & MÜLLER, S. C. 1992 Electric-field-induced drift and deformation of spiral waves in an excitable medium. *Phys. Rev. Lett.* **68**, 248–251.

WINFREE, A. T. 1972 Spiral waves of chemical activity. *Science* **175**, 634.

ZAIKIN, A. N. & ZHABOTINSKII, A. M. 1970 Concentration wave propagation in two-dimensional liquid-phase self-oscillating system. *Nature* **225**, 535.

Anomalous scaling in turbulence: a field theoretical approach

By VICTOR LVOV AND ITAMAR PROCACCIA

Department of Chemical Physics, The Weizmann Institute of Science, Rehovot, Israel

We present a short qualitative overview of some recent progress in understanding the universal statistics of fully developed turbulence. It is explained that field theoretical techniques shed light on the breakdown of dimensional analysis and on the mechanisms for the appearance of the inner (dissipative) and the outer (integral) scales of turbulence as renormalization scales. In general, correlation functions exhibit anomalous scaling with a dependence on both renormalization lengths.

1. Introduction

Modern concepts about high *Re* number turbulence started to evolve with Richardson's insightful contributions, Richardson (1922), which contained the famous 'poem' that paraphrased J. Swift: 'Big whirls have little whirls that feed on their velocity, and little whirls have lesser whirls and so on to viscosity – in the molecular sense'. In this way Richardson conveyed an image of the creation of turbulence by large scale forcing, setting up a cascade of energy transfers to smaller and smaller scales by the nonlinearities of fluid motion, until the energy dissipates at small scales by viscosity, turning into heat. This picture led in time to innumerable 'cascade models' that tried to capture the statistical physics of turbulence by assuming something or other about the cascade process. Indeed, no one in their right mind is interested in the full solution of the turbulent velocity field at all points in space-time. The interest is in the statistical properties of the turbulent flow. Moreover, the statistics of the velocity field itself is too heavily dependent on the particular boundary conditions of the flow. Richardson understood that universal properties may be found in the statistics of velocity *differences* $\delta\mathbf{u}(\mathbf{r}_1,\mathbf{r}_2) \equiv \mathbf{u}(\mathbf{r}_2)-\mathbf{u}(\mathbf{r}_1)$ across a separation $\mathbf{R} = \mathbf{r}_2-\mathbf{r}_1$. In taking such a difference we subtract the non-universal large scale motions (known as the 'wind' in atmospheric flows). In experiments (see e.g. Monin & Yaglom (1973), Anselmet *et al.* (1984), Sreenivasan & Kailasnath (1993), Benzi *et al.* (1993), Praskovskii & Oncley (1994) and Frisch (1995)) it is common to consider one-dimensional cuts of the velocity field, $\delta u_\ell(R) \equiv \delta\mathbf{u}(\mathbf{r}_1,\mathbf{r}_2)\cdot\mathbf{R}/R$. The interest is in the probability distribution function of $\delta u_\ell(R)$ and its moments. These moments are known as the 'structure functions' $S_n(R) \equiv \langle \delta u_\ell(R)^n\rangle$ where $\langle\cdot\rangle$ stands for a suitably defined ensemble average. For Gaussian statistics, the whole distribution function is governed by the second moment $S_2(R)$. There is no information to be gained from higher order moments. In contrast, hydrodynamic experiments indicate that turbulent statistics is extremely non-Gaussian, and the higher order moments contain important new information about the distribution functions.

Possibly the most ingenious attempt to understand the statistics of turbulence is due to Kolmogorov (1941) who proposed the idea of universality (turning the study of small scale turbulence from mechanics to fundamental physics) based on the notion of the 'inertial range'. The idea is that for very large values of *Re* there is a wide separation between the 'scale of energy input' L and the typical 'viscous dissipation scale' η at which viscous friction becomes important and dissipates the energy into heat. In the stationary situation, when the statistical characteristics of the turbulent flow are time independent, the rate of energy input at large scales (L) is balanced by the rate of energy dissipation at small scales (η), and must also be the same as the flux of energy from larger to smaller scales (denoted $\bar{\epsilon}$), as it is measured at any scale R in the so-called 'inertial' interval $\eta \ll R \ll L$. Kolmogorov proposed that the only relevant parameter in the inertial interval is $\bar{\epsilon}$, and that L and η are irrelevant for the statistical characteristics of motions on the scale of R. This assumption means that R is the only available length for the development of dimensional analysis. In addition we have the dimensional parameters $\bar{\epsilon}$ and the mass density of the fluid ρ. From these three parameters we can form combinations $\rho^x \bar{\epsilon}^y R^z$ such that, with a proper choice of the exponents x, y, z, we form any desired dimensionality. This leads to detailed predictions about the statistical physics of turbulence. For example, to predict $S_n(R)$ we note that the only combination of $\bar{\epsilon}$ and R that gives the right dimension for S_n is $(\bar{\epsilon}R)^{n/3}$. In particular for $n = 2$ this is the famous Kolmogorov '2/3' law which in Fourier representation is also known as the '$-5/3$' law. The idea that one extracts universal properties by focusing on statistical quantities can also be applied to the correlations of gradients of the velocity field. An important example is the rate at which energy is dissipated into heat due to viscous damping $\epsilon(\mathbf{r}, t)$. This rate is roughly equal to $\nu|\nabla \mathbf{u}(\mathbf{r}, t)|^2$. One is interested in the fluctuations of the energy dissipation $\epsilon(\mathbf{r}, t)$ about their mean $\bar{\epsilon}$, $\hat{\epsilon}(\mathbf{r}, t) = \epsilon(\mathbf{r}, t) - \bar{\epsilon}$, and how these fluctuations are correlated in space. The answer is given by the often-studied correlation function $K_{\epsilon\epsilon}(R) = \langle \hat{\epsilon}(\mathbf{r} + \mathbf{R}, t)\hat{\epsilon}(\mathbf{r}, t)\rangle$. If the fluctuations at different points were uncorrelated, this function would vanish for all $R \neq 0$. Within the Kolmogorov theory one estimates $K_{\epsilon\epsilon}(R) \simeq \nu^2 \bar{\epsilon}^{4/3} R^{-8/3}$, which means that the correlation decays as a power, like $1/R^{8/3}$.

Experimental measurements show that Kolmogorov was remarkably close to the truth. The major aspect of his predictions, i.e. that the statistical quantities depend on the length scale R as power laws is corroborated by experiments. On the other hand, the predicted exponents seem not to be exactly realized. For example, the experimental correlation $K_{\epsilon\epsilon}(R)$ decays according to a power law, $K_{\epsilon\epsilon}(R) \sim R^{-\mu}$ for $\eta \ll R \ll L$, with μ having a numerical value of $0.2 - 0.3$ instead of $8/3$, see Sreenivasan & Kailasnath (1993). The structure functions also behave as power laws, $S_n(R) \simeq R^{\zeta_n}$, but the numerical values of ζ_n deviate progressively from $n/3$ when n increases, see Anselmet *et al.* (1984) and Benzi *et al.* (1993). Something fundamental seems to be missing. The uninitiated reader might think that the numerical value of this or that exponent is not a fundamental issue. However, one needs to understand that the Kolmogorov theory exhausts the dimensions of the statistical quantities under the assumption that $\bar{\epsilon}$ is the only relevant parameter. Therefore, a deviation in the numerical value of an exponent from the prediction of dimensional analysis requires the appearance of another dimensional parameter. Of course, there exist two dimensional parameters, i.e. L and η, which may turn out to be relevant. Indeed, experiments indicate, that for the statistical quantities mentioned above the energy-input scale, L is indeed relevant, and appears as a normalization scale for the deviations from Kolmogorov's predictions: $S_n(R) \simeq (\bar{\epsilon}R)^{n/3} (L/R)^{\delta_n}$ where $\delta_n = n/3 - \zeta_n$. Such a form of scaling which deviates from the predictions of dimensional analysis is referred to as 'anomalous scaling'. The realization that the experimental results for the structure functions were consistent with L rather than η as the normalization scale developed over a long time. This involved a large number of experiments; recently the accuracy of determination of the exponents increased appreciably as a result of a clever method of data analysis by Benzi *et al.* (1993). Similarly, a careful demonstration of the appearance of L in the dissipation correlation was achieved by Sreenivasan & Kailasnath (1993). A direct analysis of scaling exponents ζ_n and μ in a high Reynolds number flow was presented by Praskovskii & Oncley (1994) leading to the same conclusions.

2. Turbulence as a field theory

Theoretical studies of the universal small scale structure of turbulence can be classified broadly into two main classes. Firstly, there is a large body of phenomenological models that, by attempting to achieve agreement with experiments, reached important insights on the nature of the cascade or the statistics of the turbulent fields, see Frisch (1995). In particular influential ideas appeared, following Mandelbrot (1974), about the fractal geometry of highly turbulent fields which allow scaling properties that are sufficiently complicated to include non-Kolmogorov scaling. Parisi & Frisch (1985) showed that by introducing multifractals one can accomodate the nonlinear dependence of ζ_n and n. However, these models are not derived directly from the equations of fluid mechanics; one is always left with uncertainties about the validity or relevance of such models. The second class of approaches is based on the equations of fluid mechanics. Typically one acknowledges the fact that fluid mechanics is a (classical) field theory, and resorts to field theoretical methods in order to compute statistical quantities. In spite of nearly 50 years of continuous effort in this direction, the analytic derivation of the scaling laws for $K_{\epsilon\epsilon}(R)$ and $S_n(R)$ from the Navier–Stokes equations, and the calculation of the numerical value of the scaling exponents μ and ζ_n, have been among the most elusive goals of theoretical research. Why did it turn out to be so difficult?

To understand the difficulties, we need to elaborate a little on the nature of the field theoretical approach. Suppose that we want to calculate the average response of a turbulent fluid at some point $\mathbf{r}_0$ to forcing at point $\mathbf{r}_1$. The field theoretical approach allows us to consider this response as an infinite sum of the following processes. Firstly, there is the direct response at point $\mathbf{r}_0$ due to the forcing at $\mathbf{r}_1$. This response is instantaneous if we assume that the fluid is incompressible (and therefore the speed of sound is infinite). Then there is the process of forcing at $\mathbf{r}_1$ with a response at an intermediate point $\mathbf{r}_2$, which then acts as a forcing for the response at $\mathbf{r}_0$. This intermediate process can take time, and we need to integrate over all possible positions of point $\mathbf{r}_2$ and all times. This is the second-order term in perturbation theory. Then we can force at $\mathbf{r}_1$, the response at $\mathbf{r}_2$ acting as a forcing for $\mathbf{r}_3$, and the response at $\mathbf{r}_3$ forces a response at $\mathbf{r}_0$. We need to integrate over all possible intermediate positions $\mathbf{r}_2$ and $\mathbf{r}_3$ and all intermediate times. This is the third-order term in perturbation theory, and so on. The actual response is the infinite sum of all these contributions. In applying this field theoretical method one encounters three main difficulties:

(*a*) The theory has no small parameter. The usual procedure is to develop the theory perturbatively around the linear part of the equation of motion. In other words, the zeroth order solution of the Navier–Stokes equations is obtained by discarding the terms which are quadratic in the velocity field. The expansion parameter is then obtained from the ratio of the quadratic to the linear terms; this ratio is of the order of the Reynolds number *Re* which was defined above. Since we are interested in $Re \gg 1$, naive perturbation expansions are badly divergent. In other words, the contribution of the various processes described above increases as $(Re)^n$ with the number n of intermediate points in space-time.

(*b*) The theory exhibits two types of nonlinear interactions. Both are hidden in the nonlinear term $\mathbf{u}\cdot\nabla\mathbf{u}$ in the Navier–Stokes equations. The larger of the two is known to any person who watched how a small floating object is entrained in the eddies of a river and swept along a complicated path with the turbulent flow. In a similar way, any small scale fluctuation is swept along by all the larger eddies. Physically this sweeping couples any given scale of motion to all the larger scales. Unfortunately the largest scales contain most of the energy of the flow; these large scale motions are what is experienced as gusts of wind in the atmosphere or an ocean swell. In the perturbation theory for $S_n(R)$ one has the consequences of the sweeping effect from all scales larger than R, with the main contribution coming from the largest, most intensive gusts on the scale of L. As a result these contributions diverge when $L \to \infty$. In the theoretical jargon this is known as 'infrared divergences'. Such divergences are common in other field theories, the best known example being quantum electrodynamics. In that theory the divergences are of similar strength in higher order terms in the series, and they can be removed by introducing finite constants to the theory, like the charge and the mass of the electron. In the hydrodynamic theory the divergences become stronger with the order of the contribution, and to eliminate them in this manner, one needs an infinite number of constants. In the jargon such a theory is called 'not renormalizable'. However, sweeping is just a kinematic effect that does not lead to energy redistribution between scales, and one may hope that if the effect of sweeping is taken care of in a consistent fashion, a renormalizable theory might emerge. This redistribution of energy results from the second type of interaction, that stems from the shear and torsion effects that are sizable only if they couple fluid motions of comparable scales. The second type of nonlinearity is smaller in size but crucial in consequence, and it may certainly lead to a scale-invariant theory.

(*c*) Nonlocality of interaction in $\mathbf{r}$ space. One recognizes that the gradient of the pressure is dimensionally the same as $(\mathbf{u}\cdot\nabla)\mathbf{u}$, and the fluctuations in the pressure are quadratic in the fluctuations of the velocity. This means that the pressure term is also nonlinear in the velocity. However, the pressure at any given point is determined by the velocity field everywhere. Theoretically one sees this effect by taking the divergence of the Navier–Stokes equations. This leads to the equation $\nabla^2 p = \nabla\cdot[(\mathbf{u}\cdot\nabla)\mathbf{u}]$. The inversion of the Laplacian operator involves an integral over all space. Physically this stems from the fact that in the incompressible limit of the Navier–Stokes equations, the sound speed is infinite and velocity fluctuations at all distant points are instantaneously coupled.

Indeed, these difficulties seemed to complicate the application of field theoretical methods to such a degree that a wide-spread feeling appeared to the effect that it is impossible to gain valuable insight into the universal properties of turbulence along these lines, even though they proved so fruitful in other field theories. The present authors (as well as other researchers starting with Kraichnan (1965) and recently Migdal (1994), Polyakov (1993), Eyink (1993), etc.) think differently, and in the rest of this paper we will explain why.

The first task of a successful theory of turbulence is to overcome the existence of the interwoven nonlinear effects that were explained in *difficulty* (*b*). This is not achieved by directly applying a formal field-theoretical tool to the Navier–Stokes equations. It does not matter whether one uses standard field theoretical perturbation theory, see Wyld (1961), path integral formulation or renormalization group, see Yakhot & Orszag (1986), ϵ-expansion or large N-limit, see Mou & Weichman (1995), or one's choice formal method. One needs to take care of the particular nature of hydrodynamic turbulence as embodied in *difficulty* (*b*) *first*, and *then* proceed using formal tools.

The removal of the effects of sweeping is based on Richardson's remark that universality in turbulence is expected for the statistics of velocity *differences across a length scale R* rather than for the statistics of the velocity field itself. The velocity fields are dominated by the large scale motions that are not universal, since they are produced directly by the agent that forces the flow. This forcing agent differs in different flow realizations (atmosphere, wind tunnels, channel flow etc.). Richardson's insight was developed by Kraichnan (1965), who attempted to cast the field theoretical approach in terms of Lagrangian paths, meaning a description of the fluid flow which follows the paths of every individual fluid particle. Such a description automatically removes the large scale contributions. Kraichnan's approach was fundamentally correct, and gave rise to important and influential insights in the description of turbulence, but did not provide a convenient technical way to consider all orders of perturbation theory. The theory does not provide transparent rules for how to consider arbitrarily high terms in perturbation theory. Only low order truncations were considered.

A way to overcome *difficulty* (*b*) was suggested by Belinicher & Lvov (1987). They introduced a novel transformation that allowed on the one hand for the elimination of the sweeping that leads to infrared divergences, and on the other hand for the development of simple rules for writing down arbitrary order terms in the perturbation theory for statistical quantities. An essential idea in this transformation is the use of a coordinate frame in which velocities are measured relative to the velocity of *one* chosen fluid particle. The use of this transformation allowed for the examination of the structure functions of velocity differences $S_n(R)$ to all orders in perturbation theory. Of course, *difficulty* (*a*) remains; the perturbation series still diverges rapidly for large values of Re, but now standard field theoretical methods can be used to reformulate the perturbation expansion such that the viscosity is changed by an effective 'eddie viscosity'. A theoretical tool that achieves this exchange is known in quantum field theory as the Dyson line resummation, see Lvov & Procaccia (1995a). The result of this procedure is that the effective expansion parameter is no longer Re, but an expansion parameter of the order of unity. Of course, such a perturbation series may still diverge as a whole. Notwithstanding this it is crucial to examine first the order-by-order properties of series of this type.

Such an examination leads to a major surprise: every term in this perturbation theory remains finite when the energy-input scale $L \to \infty$ and the viscous-dissipation scale $\eta \to 0$, see Lvov & Procaccia (1995a). The meaning of this is that the perturbative theory for S_n does not indicate the existence of any typical length-scale. Such a length is needed in order to represent deviations in the scaling exponents from the predictions of Kolmogorov's dimensional analysis in which both scales L and η are assumed irrelevant. In other areas of theoretical physics in which anomalous scaling has been found, it is usual for the perturbative series to indicate this phenomenon. In many cases this is seen in the appearance of logarithmic divergences that must be tamed by truncating the integrals at some renormalization length. Hydrodynamic turbulence seems different at this point. The nonlinear Belinicher–Lvov transformation changes the underlying linear theory such that the resulting perturbative scheme for the structure functions is of finite order by order, see Belinicher & Lvov (1987) and Lvov & Procaccia (1995a). The physical meaning of this result is that, much as can be seen from this perturbative series, the main effects on the statistical quantities for velocity differences across a scale R come from activities of scales comparable to R. This is the perturbative justification of the Richardson–Kolmogorov cascade picture in which widely separated scales do not interact.

Consequently, the main question still remains: how does a renormalization scale appear in the statistical theory of turbulence?

It turns out that there are two different mechanisms that furnish a renormalization scale, and that finally *both* L *and* η appear in the theory. The viscous scale η appears via a rather standard mechanism that can be seen in perturbation theory as logarithmic divergences, but in order to see it one needs to consider the statistics of gradient fields rather than the velocity differences themselves, see Lvov and Lebedev (1995) and

Lvov & Procaccia (1995b). For example, considering the perturbative series for $K_{\epsilon\epsilon}(R)$, which is the correlation function of the rate of energy dissipation $\nu|\nabla\mathbf{u}|^2$, leads immediately to the discovery of logarithmic ultraviolet divergences in every order of the perturbation theory. These divergences are controlled by an ultraviolet cutoff scale which is identified as the viscous-dissipation scale η, acting here as the renormalization scale. The summation of the infinite series results in a factor $(R/\eta)^{2\Delta}$ with some anomalous exponent Δ which is, generally speaking, of the order of unity. The appearance of such a factor means that the actual correlation of two R-separated dissipation fields is much larger, when R is much larger than η, than the naive prediction of dimensional analysis. The physical explanation of this renormalization, see Lvov & Lebedev (1995) and Lvov & Procaccia (1995c), is the effect of the multi-step interaction of two R-separated small eddies of scale η with a large eddy of scale R via an infinite set of eddies of intermediate scales. The net result on the scaling exponent is that the exponent μ changes from 8/3 as expected in the Kolmogorov theory to $8/3 - 2\Delta$.

At this point it is important to understand what is the numerical value of the anomalous exponent Δ. In Lvov & Procaccia (1995b) there was found an exact sum that forces a relation between the numerical value of Δ and the numerical value of the exponent ζ_2 of $S_2(R)$, $\Delta = 2 - \zeta_2$. Such a relation between different exponents is known in the jargon as a 'scaling relation' or a 'bridge relation'. Physically this relation is a consequence of the existence of a universal nonequilibrium stationary state that supports an energy flux from large to small scales, see Lvov & Procaccia (1995b) and Lvov & Procaccia (1996a). The scaling relation for Δ has far-reaching implications for the theory of the structure functions. It was explained that with this value of Δ, the series for the structure functions $S_n(R)$ diverge when the energy-input scale L approaches ∞ as powers of L, like $(L/R)^{\delta_n}$. The anomalous exponents δ_n are the deviations of the exponents of $S_n(R)$ from their Kolmogorov value. This is a very delicate and important point, and we therefore expand on it. Think about the series representation of $S_n(R)$ in terms of lower order quantities, and imagine that one has succeeded to resum it into an operator equation for $S_n(R)$. Typically such a resummed equation may look like $[1 - \hat{O}]S_n(R) = \text{r.h.s.}$, where $\hat{O}$ is some integro-differential operator which is not small compared to unity. If we expand this equation in powers of $\hat{O}$ around the r.h.s., we regain the infinite perturbative series that we started with. However, we now realize that the equation also possesses homogeneous solutions, solutions of $[1 - \hat{O}]S_n(R) = 0$, which are inherently nonperturbative since they can no longer be expanded around the r.h.s. These homogeneous solutions may be much larger than the inhomogeneous perturbative solutions. Of course, homogeneous solutions must be matched with the boundary conditions at $R = L$, and this is the way that the energy input scale L appears in the theory. This is particularly important when the homogeneous solution diverges in size when $L \to \infty$, as is indeed the case for the problem at hand.

The next step in the theoretical development is to understand how to compute the anomalous exponents δ_n. The divergence of the perturbation theory for $S_n(R)$ with $L \to \infty$ forces us to seek a nonperturbative handle on the theory. One finds this in the idea that there always exists a global balance between energy input and dissipation, which may be turned into a nonperturbative constraint on each n-th order structure function, see Lvov & Procaccia (1996a). Using the Navier–Stokes equations one derives the set of equations of motion

$$\frac{\partial S_n(R,t)}{\partial t} + D_n(R,t) = \nu J_n(R,t), \tag{2.1}$$

where D_n and J_n stem from the nonlinear and viscous terms in the Navier–Stokes equations, respectively. To understand the physical meaning of this equation, note that $S_2(R)$ is precisely the mean kinetic energy of motions of size R. The term $D_2(R)$ whose meaning is the rate of energy flux through the scale R is known exactly: $D_2(R) = \mathrm{d}S_3(R)/\mathrm{d}R$. The term $\nu J_2(R)$ is precisely the rate of energy dissipation due to viscous effects. The higher order equation for $n > 2$ is a direct generalization of this to higher order moments. In the stationary state, the time derivative vanishes and one has the balance equation $D_n(R) = \nu J_n(R)$. For $n = 2$ it reflects the balance between energy flux and energy dissipation. The evaluation of $D_n(R)$ for $n > 2$ requires dealing with the *difficulty* (*c*) of the nonlocality of the interaction, but it does not pose conceptual difficulties. It was shown, see Lvov & Procaccia (1996a), that $D_n(R)$ is of the order of $\mathrm{d}S_{n+1}/\mathrm{d}R$. On the other hand, the evaluation of $J_n(R)$ raises a number of very interesting issues whose resolution lies at the heart of the universal scaling properties of turbulence. Presently not all of these issues have been resolved, and we briefly mention here some ground on which progress has been made by the present authors.

From the derivation of (2.1), one finds that $J_n(R)$ consists of a correlation of $\nabla^2 u$ with $n-2$ velocity differences across a scale $R = |\mathbf{r}_1 - \mathbf{r}_2|$: $\langle \nabla^2 u(\mathbf{r}_1)[\delta u_\ell(\mathbf{r}_1, \mathbf{r}_2)]^{n-2} \rangle$. The question is how to evaluate such a quantity in terms of the usual structure functions $S_n(R)$. Recall that a gradient of a field is the difference in the field values at two points divided by the separation when the latter goes to zero. In going to zero one necessarily crosses the dissipative scale. To understand what happens in this process, one needs first to introduce many-point correlation functions of a product of n velocity differences:

$$F_n(\mathbf{r}_0|\mathbf{r}_1, \ldots, \mathbf{r}_n) \equiv \langle \delta u_\ell(\mathbf{r}_0, \mathbf{r}_1) \cdots \delta u_\ell(\mathbf{r}_0, \mathbf{r}_n) \rangle \,. \tag{2.2}$$

Next we need to formulate rules for the evaluation of such correlation functions of velocity differences when some of the coordinates get very close to each other. For example, a gradient $\partial/\partial r_\alpha$ can be formed from the limit $\mathbf{r}_1 \to \mathbf{r}_0$ when we divide by $r_{1,\alpha} - r_{0,\alpha}$. These rules are known in the theoretical jargon as 'fusion rules'. The fusion rules for hydrodynamic turbulence were presented in Lvov & Procaccia (1996b). They show that when p coordinates in F_n are separated by a small distance r, and the remaining $n-p$ coordinates are separated by a large distance R, then the scaling dependence on r is like that of $S_p(r)$, i.e. r^{ζ_p}. This is true until r crosses the dissipative scale. Assuming that below the viscous-dissipation scale η, derivatives exist and the fields are smooth, one can estimate gradients at the end of the smooth range by dividing differences across η by η. The question is, what is the appropriate cross-over scale to smooth behaviour? Is there just one cross-over scale η, or is there a multiplicity of such scales, depending on the function one is studying? For example, when does the above n-point correlator become differentiable as a function of r when p of its coordinates approach $\mathbf{r}_0$? Is that typical scale the same as the one exhibited by $S_p(r)$ itself, or does it depend on p and n and on the remaining distances of the remaining $n-p$ coordinates that are still far away from $\mathbf{r}_0$?

The answer is that there is a multiplicity of cross-over scales. For the n-point correlator discussed above we denote the dissipative scale as $\eta(p, n, R)$, and it depends on each of its arguments, see Lvov & Procaccia (1996b) and Lvov & Procaccia (to be published). In particular it depends on the inertial range variables R and this dependence must be known when one attempts to determine the scaling exponents ζ_n of the structure functions. In brief, this line of thought leads to a set of non-trivial scaling relations. For example we confirm the phenomenologically conjectured 'bridge relation', see Frisch (1995), $\mu = 2 - \zeta_6$ (in close agreement with the experimental values) and predict that although the ζ_n are not Kolmogorov, they are nevertheless linear in n for large n.

3. Summary

It appears that there are four conceptual steps in the construction of a theory of the universal anomalous statistics of turbulence on the basis of the Navier–Stokes equations. First one needs to take care of the sweeping interactions that mask the scale invariant theory, see Belinicher & Lvov (1987) and Lvov & Procaccia (1995a). After doing so, the perturbation expansion converges order by order, and the Kolmogorov scaling of the velocity structure functions is found as a perturbative solution. Secondly, one understands the appearance of the viscous-dissipation scale η as the natural normalization scale in the theory of the correlation functions of the gradient fields, see Lvov and Lebedev (1995) and Lvov & Procaccia (1995b). This step is similar to critical phenomena and it leads to a similarly rich theory of anomalous behaviour of the gradient fields. Only the tip of the iceberg was considered above. In fact, when one considers the correlations of tensor fields which are constructed from $\partial u_\alpha / \partial r_\beta$ (rather than the scalar field ϵ) one finds that every field with a different transformation property under the rotation of the coordinates has its own independent scaling exponent which is the analog of Δ above, see Lvov *et al.* (1996). The third step is the understanding of the divergence of the diagrammatic series for the structure functions as a whole, see Lvov & Procaccia (1996a). This sheds light on the emergence of the energy-input scale L as a normalization length in the theory of turbulence. This means that the basic Kolmogorov assertion, that there is no typical scale in the expressions for statistical quantities involving correlations across a scale R when $\eta \ll R \ll L$ is doubly wrong. In general both lengths appear in dimensionless combinations and change the exponents from the predictions of dimensional analysis. Examples of correlation functions in which both normalization scales L and η appear simultaneously were given explicitly, see Lvov *et al.* (1996). Last but not least is the formulation of the fusion rules and the exposition of the multiplicity of the dissipative scales which should eventually result in a satisfactory description of all the scaling properties, see Lvov & Procaccia (to be published).

The road ahead is not fully charted, but it seems that some of the conceptual difficulties have been surmounted. We believe that the crucial building blocks of the theory are now available, and they begin to delineate the structure of the theory. We hope that the remaining four years of this century will suffice to achieve a proper understanding of the anomalous scaling exponents in turbulence. Considerable work, however, is still needed in order to fully clarify many aspects of the problem, and most of them are as exciting and important as the scaling properties. There are universal aspects that go beyond exponents, such as distribution functions and the eddy viscosity, and there are important non-universal aspects like the role of inhomogeneities, the effect of boundaries and so on. Progress on these issues will bring the theory closer to the concern of engineers. The marriage of physics and engineering will be the challenge of the 21st century.

We profitted from discussions and collaboration with Victor Belinicher, Adrienne Fairhall, Gregory Falkovich, Uriel Frisch, Omri Gat, Robert Kraichnan, Vladimir Lebedev and Evgeni Podivilov. This work was supported in part by the US–Israel Bi-National Science Foundation, the German–Israeli Foundation and the Naftali and Anna Backenroth-Bronicki Fund for Research in Chaos and Complexity.

REFERENCES

Anselmet, F., Gagne, Y., Hopfinger, E. J. & Antonia, R. A. 1984 *J. Fluid Mech.* **140**, 63.

Belinicher, V. I. & Lvov, V. S. 1987 *Sov. Phys. JETP* **66**, 303.

BENZI, R., CILIBERTO, S., TRIPICCIONE, R., BAUDET, C., MASSAIOLI, F. & SUCCI, S. 1993 *Phys. Rev.* E **48**, R29.

EYINK, G. L. 1993 *Phys. Lett.* A **172**, 355.

FRISCH, U. 1995 *Turbulence: The Legacy of A. N. Kolmogorov.* Cambridge University Press.

KOLMOGOROV, A. N. 1941 *Dokl. Acad. Nauk SSSR* **30**, 9.

KRAICHNAN, R. H. 1965 *Phys. Fluids* **8**, 575.

LVOV, V. S. & LEBEDEV, V. 1995 *Europhys. Lett.* **29**, 681.

LVOV, V. S., PODIVILOV, E. & PROCACCIA, I. 1996 *Phys. Rev. Lett.* **76**, 3963.

LVOV, V. S. & PROCACCIA, I. 1994 In *Fluctuating Geometries in Statistical Mechanics and Field Theory* (ed. F. David, P. Ginsparg & J. Zinn-Justin). 1995 Les Houches session LXII. Elsevier.

LVOV, V. S. & PROCACCIA, I. 1995a *Phys. Rev.* E **52**, 3840.

LVOV, V. S. & PROCACCIA, I. 1995b *Phys. Rev.* E **52**, 3858.

LVOV, V. S. & PROCACCIA, I. 1995c *Phys. Rev. Lett.* **74**, 2690.

LVOV, V. S. & PROCACCIA, I. 1996a Exact resummation in the theory of hydrodynamic turbulence: III. Scenarios for multiscaling and intermittency. *Phys. Rev.* E. **53**, 3468.

LVOV, V. S. & PROCACCIA, I. Exact resummation in the theory of hydrodynamic turbulence: IV. Fusion rules and the multiplicity of viscous scales. *Phys. Rev.* E. To be published.

LVOV, V. S. & PROCACCIA, I. 1996b Fusion rules in turbulent systems with flux equilibrium. *Phys. Rev. Lett.* **76**, 2898.

MANDELBROT, B. B. 1974 *J. Fluid Mech.* **62**, 331.

MIGDAL, A. A. 1994 *Int. J. of Mod. Phys.* A **9**, 1197.

MONIN, A. S. & YAGLOM, A. M. 1973 *Statistical Fluid Mechanics: Mechanics of Turbulence*, vol. 2. MIT Press.

MOU, C.-Y. & WEICHMAN, P. B. 1995 *Phys. Rev.* E **52**, 3738.

PARISI, G. & FRISCH, F. 1985 In *Turbulence and Predictability in Geophyiscal Fluid Dynamics* (ed. M. Ghil, R. Benzi & G. Parisi). North Holland.

POLYAKOV, A. M. 1993 *Nucl. Phys.* B **396**, 367.

PRASKOVSKII, A. & ONCLEY, S. 1994 *Phys. Fluids* **6**, 2886.

RICHARDSON, L. F. 1922 *Weather Prediction by Numerical Process.* Cambridge.

SREENIVASAN, K. R. & KAILASNATH, P. 1993 *Phys. Fluids* A **5**, 512.

WYLD, H. W. 1961 *Ann. Phys.* **14**, 143.

YAKHOT, V. & ORSZAG, S. A. 1986 *Phys. Rev. Lett.* **57**, 1722.

Abelian sandpile cellular automata

By MÁRIA MARKOŠOVÁ

Institute of Measurement, Slovak Academy of Sciences, Bratislava, Slovak Republic

Analytical methods developed to study the self organized critical state in abelian sandpile cellular automata are reviewed. For calculations concerning models with nearest neighbour topplings, properties of the Toeplitz matrices are emoployed.

---

1. Introduction

The sandpile, as a paradigm representing a spatially extended system, was first introduced by Bak *et al.* (1988). Their paper immediately provoked wide scientific activity. The authors claimed that their sandpile cellular automaton, the model they created to demonstrate the dynamics of avalanches on a pile of sand, evolves spontaneously, and without any fine tuning of system parameters, to a stationary stable state, with long range spatial and temporal correlations. This state was called a self organized critical (SOC) state. It was speculated that this special type of cellular automaton can help us understand the underlying mechanism of spontaneous creation of scale invariant space and time structures, manifesting themselves as fractal structures and flicker noise time signals.

It is known from the theory of equilibrium systems, that long range correlations are established when the order parameter is fine tuned to its critical value. Therefore it seems natural to suppose that scale invariant structures in nature are also consequences of some critical state, but this state should arise spontaneously without any fine tuning of parameters, and this difference is of crucial importance. Moreover, such a critical state must be rather insensitive to perturbations. The model suggested in Bak *et al.* (1988) fulfills all the above mentioned demands.

The sandpile-like cellular automaton is a model of a spatially extended nonequilibrium dynamical system with short range interactions, consisting of many almost identical elements ('sand grains'). It is defined on a discrete lattice by the local dynamical rules and local critical stability condition, and works in two dynamical modes: input process and relaxation. The input process mimics the slow natural driving forces. Sand grains are added at randomly chosen sites on the lattice, until somewhere the local stability criterion is violated (e.g. the local slope of a real sandpile exceeds the critical slope). Then the quick relaxation (avalanche) starts from that point. In the avalanche some number of particles roll to neighbouring sites and possibly destabilize them. The process continues, until all sites become subcritical. Then the input mode is applied again. Boundary conditions are open, for example, and the particles can leave the system at the boundary. Thanks to the interplay between input, relaxation and boundary conditions, the dynamics, on average, conserves the number of particles in the system, and thus such automata are called conservative sandpiles.

In the SOC state, there is no characteristic length or time scale. This leads to the power law behaviuour of the characteristic quantities, such as avalanche size distribution $g(s)$ and avalanche cluster lifetime distribution $g(\tau)$. Hence the main indication of the SOC state are the power laws

$$g(s) \sim s^{-\beta} \tag{1.1}$$

and

$$g(\tau) \sim \tau^{-\delta}. \tag{1.2}$$

The lifetime of the cluster of size s scales with s as

$$\tau_s \sim s^x. \tag{1.3}$$

In (1.1) and (1.3) s and τ_s denote the size of the avalanche and the avalanche lifetime, respectively.

It was argued in Bak *et al.* (1988), that scaling like (1.2) and (1.3) leads to the $\frac{1}{f}$ power spectrum of sliding particles, with nontrivial ($\alpha \neq 2$) exponent. These $\frac{1}{f}$ time signals are very common in Nature, but their origin has not yet been explained. Bak *et al.* (1988) speculated that systems with an SOC state could explain this fact. The flicker noise time signals and fractal structures are fingerprints of an SOC state in time and space, respectively. However, it was shown in Jensen *et al.* (1989) and Kertész & Kiss (1990) that these expectations were too optimistic. Kertész & Kiss (1990) have shown that the occurrence of flicker noise depends on the exponents β and x, and the inenequality $2x + \beta > 3$ has to be fulfilled in order to get a nontrivial α exponent. Nevertheless, the concept of an SOC state is still very appealing and widely studied. Numerical and analytical studies of nonconservative models, given in Feder & Feder (1991), Christensen & Olami (1992), Christensen *et al.* (1992), Jánosi & Kertész (1993), Grassberger (1994) and Markošová (1995b) gave a new impulse for the theory of self organized criticality.

Properties of the SOC state were first studied numerically. In this lecture I would like to present a review of the analytical calculations. I focus on the algebraic aspects of one special family of cellular automata, known as abelian sandpiles.

2. Abelian models

In what follows I describe a class of models called abelian sandpiles. In spite of their name, in fact they do not model the dynamical behaviuour of real sand avalanches, but rather suggest how a quantity (stress, energy etc.) spreads through a lattice. Unless otherwise stated, all models considered are defined on a two dimensional square lattice (the generalization to more dimensions is possible and usually straightforward), and the local critical condition is given as the local critical sand column height (critical height sandpile). The dynamical rules define two types of behaviuour:

Input process:

A sandgrain is added at a randomly chosen site every time unit. This increases the height of sand column at the site in question by one,

$$h_i \to h_i + 1, \tag{2.4a}$$
$$h_i \leq h_{ic},$$
$$h_{ic} = \Delta_{ii}, \quad i = 1, 2, 3, \ldots, N.$$

In (2.4a) h_{ic} denotes the critical height at ith site and N is the number of sites in the lattice. The lattice positions are labelled from upper left to lower right corner, see figure 1.

If the local critical height is accidentally exceeded by adding a grain to the previous configuration, the input of particles is stopped, and the abrupt relaxation process, the avalanche, starts.

Avalanche:

$$h_j \to h_j - \Delta_{ij}, \tag{2.4b}$$

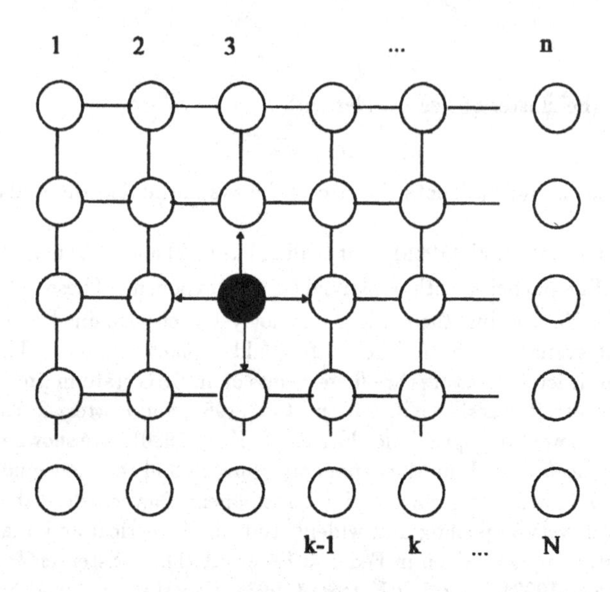

FIGURE 1. Abelian model with nearest neighbour topplings on a square lattice. A black circle denotes the supercritical site.

$$h_i > h_{ic}.$$

We assume that Δ_{ii} particles leave the ith site, being added to other sites on the lattice, where $\mathbf{\Delta}$ represents the integer *toppling matrix* satisfying the conditions:

$$\begin{aligned} &\Delta_{ii} > 0 \quad \text{for all } i = 1, 2, 3, \ldots, N, \\ &\Delta_{ij} \leq 0 \quad \text{for all } i \neq j, \\ &\sum_{j=1}^{N} \Delta_{ij} \geq 0. \end{aligned} \tag{2.4c}$$

We also assume that boundaries are open, so that the particles leave the system at the boundaries. In the SOC state the automaton is conservative.

The best known example of these models is the automaton with nearest neighbour topplings of Bak *et al.* (1988) (BTW pile). Here $h_{ic} = h_c = 4$ and $\mathbf{\Delta}$ in the two-dimensional case is a block matrix

$$\mathbf{\Delta} = \mathbf{T}_2^n = \begin{bmatrix} \mathbf{T}_1^n & -\mathbf{I} & \mathbf{0} & \mathbf{0} & \cdots & \mathbf{0} \\ -\mathbf{I} & \mathbf{T}_1^n & -\mathbf{I} & \mathbf{0} & \cdots & \mathbf{0} \\ \mathbf{0} & -\mathbf{I} & \mathbf{T}_1^n & -\mathbf{I} & \ddots & \mathbf{0} \\ \vdots & \ddots & \ddots & \ddots & \ddots & \vdots \\ \mathbf{0} & \mathbf{0} & \cdots & -\mathbf{I} & \mathbf{T}_1^n & -\mathbf{I} \\ \mathbf{0} & \mathbf{0} & \cdots & \mathbf{0} & -\mathbf{I} & \mathbf{T}_1^n \end{bmatrix}, \tag{2.5a}$$

where

$$\mathbf{T}_1^n \equiv \mathbf{T}_1^n(4) = \begin{bmatrix} 4 & -1 & 0 & 0 & \dots & 0 \\ -1 & 4 & -1 & 0 & \dots & 0 \\ 0 & -1 & 4 & -1 & \ddots & 0 \\ \vdots & \ddots & \ddots & \ddots & \ddots & \vdots \\ 0 & 0 & \dots & -1 & 4 & -1 \\ 0 & 0 & \dots & 0 & -1 & 4 \end{bmatrix} \tag{2.5b}$$

and **I** is the unit matrix.

As can be seen from (2.4) and (2.5), in an avalanche dynamical mode four particles are distributed from the overcritical site to the nearest neighbour positions in each toppling. Nearest neighbours thus gain one more particle, see figure 1. Matrices, such as $\mathbf{T}_1^n$ (2.5*b*) and $\mathbf{T}_2^n$ (2.5*a*) are known in mathematics as Toeplitz matrices of first and second generation, respectively. The first generation Toeplitz matrix is a tridiagonal matrix of order n, see Chau & Cheng (1993),

$$\mathbf{T}_1^n(a) = \begin{bmatrix} a & -1 & 0 & 0 & \dots & 0 \\ -1 & a & -1 & 0 & \dots & 0 \\ 0 & -1 & a & -1 & \ddots & 0 \\ \vdots & \ddots & \ddots & \ddots & \ddots & \vdots \\ 0 & 0 & \dots & -1 & a & -1 \\ 0 & 0 & \dots & 0 & -1 & a \end{bmatrix}, \tag{2.6}$$

where $a \geq 2$.

The eigenvalues and eigenvectors of (2.6) are given by ($j, k = 1, 2, \dots, n$):

$$\lambda_k = a - 2\cos\left(\frac{k\pi}{n+1}\right), \tag{2.7a}$$

$$\psi_j = \sqrt{\frac{2}{n+1}}\sin\left(\frac{jk\pi}{n+1}\right). \tag{2.7b}$$

The mth generation Toeplitz matrix $\mathbf{T}_m^n$ is a block matrix consisting of $n \times n$ blocks:

$$\mathbf{T}_m^n = \begin{bmatrix} \mathbf{T}_{m-1}^n & -\mathbf{I} & \mathbf{0} & \mathbf{0} & \dots & \mathbf{0} \\ -\mathbf{I} & \mathbf{T}_{m-1}^n & -\mathbf{I} & \mathbf{0} & \dots & \mathbf{0} \\ \mathbf{0} & -\mathbf{I} & \mathbf{T}_{m-1}^n & -\mathbf{I} & \ddots & \mathbf{0} \\ \vdots & \ddots & \ddots & \ddots & \ddots & \vdots \\ \mathbf{0} & \mathbf{0} & \dots & -\mathbf{I} & \mathbf{T}_{m-1}^n & -\mathbf{I} \\ \mathbf{0} & \mathbf{0} & \dots & \mathbf{0} & -\mathbf{I} & \mathbf{T}_{m-1}^n \end{bmatrix}. \tag{2.8}$$

In order to develope the method by which one can find the eigenvalues of the higher generation Toeplitz matrices, we define the mth generation matrix $\mathbf{H}_m$:

$$\mathbf{H}_m = \mathbf{T}_m^n - \mathbf{A}_m, \tag{2.9}$$

where $\mathbf{A}_m$ is a diagonal matrix having the same order as $\mathbf{T}_m^n$ with a on the diagonal. The first generation matrix $\mathbf{H}_1$ is diagonalized by the matrix $\mathbf{Q}_1$, which has elements

$$Q_1(r,s) = \sqrt{\frac{2}{n+1}}\sin\left(\frac{rs\pi}{n+1}\right). \tag{2.10}$$

Making use of $\mathbf{H}_1$ and $\mathbf{Q}_1$, one can easily create the **H**-matrices and also the Toeplitz matrices of higher generation, and find their eigenvalues. For example

$$\mathbf{H}_2 = \mathbf{H}_1 \otimes \mathbf{I} + \mathbf{I} \otimes \mathbf{H}_1, \tag{2.11}$$

$$\mathbf{Q}_2 = \mathbf{Q}_1 \otimes \mathbf{Q}_1,$$

and

$$\mathbf{H}_3 = \mathbf{H}_1 \otimes \mathbf{I} \otimes \mathbf{I} + \mathbf{I} \otimes \mathbf{H}_1 \otimes \mathbf{I} + \mathbf{I} \otimes \mathbf{I} \otimes \mathbf{H}_1, \tag{2.12}$$
$$\mathbf{Q}_3 = \mathbf{Q}_1 \otimes \mathbf{Q}_1 \otimes \mathbf{Q}_1,$$

where $\otimes$ denotes tensor multiplication. The matrices of the mth generation $\mathbf{H}_m$ and $\mathbf{Q}_m$ can be obtained similarly. To calculate $\mathbf{H}_m$ we simply sum m tensor multiplications in which we permute $\mathbf{H}_1$ and $m-1$ unit matrices. Diagonalizing $\mathbf{H}_m$ and taking into account (2.9), it is easy to find the eigenvalues of $\mathbf{H}_m$ and $\mathbf{T}^n_m$. For example, the eigenvalues of $\mathbf{T}^n_2$ are $(j, k = 1, 2, \ldots, n)$:

$$\Lambda(j,k) = a - 2\cos\left(\frac{j\pi}{n+1}\right) - 2\cos\left(\frac{k\pi}{n+1}\right). \tag{2.13}$$

Let us now turn to the sandpiles. The main task in studies of the BTW model (and other sandpile models) is to characterize the SOC state as completely as possible. The characteristic properties, such as the power laws (1.1)–(1.3), single site height distributions, two-point correlations etc., were first studied numerically, see Bak *et al.* (1988), Kadanoff *et al.* (1989) and Grassberger (1990). A significant step in the analytical calculations was done by Dhar and coworkers. Their analytical results were published in a series of inspiring papers, see Dhar & Ramaswamy (1989), Dhar (1990), Dhar & Majumdar (1990) and Majumdar & Dhar (1991).

In a two-dimensional model (2.4), Dhar (1990) defined N operators a_i, acting in a space of stable configurations, such that $a_i C = C^*$; C^* is a stable configuration obtained from C by adding a particle at site i and letting the system relax. Two operators a_i and a_j commute

$$[a_i, a_j] = 0, \tag{2.14}$$

because the situation on a pile will be the same if one puts a grain first at the ith position (with subsequent relaxation) and then at the jth position, or if one reverses the order. In this operation the sth position at the lattice gains $|\Delta_{si} + \Delta_{sj}|$ grains, which is symmetric in i and j. This commutative property gives its name to the abelian sandpiles. There are two possible types of stable configuration:

(*a*) Reccurent configuration, defined by

$$a_i^{m_i} C = C.$$

(*b*) Transient configurations, i.e. all nonreccurent configurations.

It follows from (2.14) that if C is reccurent, so is C^*. This simply means that if the system once gets to the space of reccurent configurations, it cannot leave it. Hence, as was shown by Dhar (1990), in the SOC state, only reccurent configurations are of nonzero probability. Moreover, this probability is the same for all configurations.

The most important analytical results of Dhar (1990) show that the number of distinct reccurent configurations is

$$N_R = \det \mathbf{\Delta}, \tag{2.15}$$

and the two-point correlation function, i.e. the probability that the site j will topple if we add a particle at the site i, is

$$G_{ij} = [\mathbf{\Delta}^{-1}]_{ij}. \tag{2.16}$$

There are no other limitations imposed on the toppling matrix Δ, except for (2.4*c*). In the special case of nearest neighbour topplings on a two-dimensional square lattice, $\mathbf{\Delta}$ is a Toeplitz matrix of second generation. Therefore we can easily calculate the number of reccurent configurations with the help of (2.13), see Chau & Cheng (1993) ($N = n \times n$),

$$N_R = \prod_{l=1}^{n} \prod_{k=1}^{n} \left[4 - 2\cos\left(\frac{l\pi}{n+1}\right) - 2\cos\left(\frac{k\pi}{n+1}\right) \right]$$
$$= 4^N \prod_{l=1}^{n} \prod_{k=1}^{n} \left[\sin^2\left(\frac{l\pi}{2(n+1)}\right) + \sin^2\left(\frac{k\pi}{2(n+1)}\right) \right]. \tag{2.17}$$

In the thermodynamic limit, $n \to \infty$, we assume that $N_R \approx \xi^N$, and from equation (2.17) we get

$$\ln\frac{\xi}{4} = \frac{4}{\pi^2} \int_0^{\frac{\pi}{2}} \int_0^{\frac{\pi}{2}} \ln[\sin^2(x) + \sin^2(y)]\, dx dy. \tag{2.18}$$

Evaluating the integrals one obtains $N_R \approx 3.212^N$, see Chau & Cheng (1993).

Let us now imagine the situation in which, in the input process, grains are not added accidentaly to all sites, but only to one site, say, the central one. The automaton is then fully deterministic and in the SOC state the configurations behave periodically. Thus, besides the power law distributions (1.1), the SOC state is also characterized by the period, which depends on the number of lattice sites, see Wiesenfeld *et al.* (1990), Markošová & Markoš (1992) and Markošová (1995a). The analytical calculation of this period is based on the properties of the Toeplitz matrices, see Markošová & Markoš (1992) and Markošovă (1995a).

In a two-dimensional deterministic sandpile with nearest neighbour topplings we define the spilling number $x_i(\tau)$, as the number of topplings of the site i ($i = 1, 2, \ldots, N$) during the time interval τ. The change of the column height at i during τ is due to the following processes: relaxation of the site i, which decreases the number of particles at i, and relaxation of the nearest neighbours and input, which, on the contrary, increase the number of grains at i:

$$h_i(t) \to h_i(t+\tau) = h_i(t) + \sum_{\text{nn}} x_{\text{nn}}(\tau) - 4x_i(\tau) + \tau\delta_{ii^*}, \tag{2.19}$$

where i^* is the index of the input site and nn denotes the nearest neighbours. If the pile is in the SOC state, the configuration $C = \{h_i\}$ repeats itself after T time units and thus

$$h_i(t) = h_i(t+T). \tag{2.20}$$

From (2.19) and (2.20) we obtain

$$\sum_{\text{nn}} x_{\text{nn}} - 4x_i = -T\delta_{ii^*}, \tag{2.21}$$

where $x_i = x_i(T)$, and x_i and T are both integers.

Our aim is to solve equation (2.21) in integers. For this purpose we rewrite it in matrix form

$$\mathbf{\Delta}\vec{x}_2 = \vec{t}_2, \tag{2.22}$$

where $\mathbf{\Delta} = \mathbf{T}_2^n = \mathbf{H}_2 + \mathbf{A}_2$ is a Toeplitz matrix of second generation (2.5a), and $\mathbf{A}_2$ is a diagonal matrix having 4 on the diagonal; $\vec{t}_2$ is a *time vector* with all components equal to zero, except for the i^*th one, which is equal to $-T$. The solution of (2.22) can be expressed as follows

$$\vec{x}_2 = \mathbf{G}_2\vec{t}_2, \tag{2.23}$$

where $\mathbf{G}_2 = \mathbf{\Delta}^{-1}$ (which is the two-point correlation matrix (2.16)), and

$$\mathbf{G}_2 = \mathbf{Q}_2\mathbf{\Lambda}^{-1}\mathbf{Q}_2^{-1}, \tag{2.24}$$

where

$$\mathbf{\Lambda} = \mathbf{Q}_2^{-1}\mathbf{\Delta}\mathbf{Q}_2, \tag{2.25}$$

which is diagonal. Using (2.23)–(2.25) along with (2.11) one gets

$$x_k = T\frac{P_k}{Q_k} = TG_2(k), \tag{2.26}$$

where

$$G_2(k) = \frac{P_k}{Q_k} = \frac{2}{(n+1)^2}\sum_{u,v=0}^{\frac{n+1}{2}} \frac{(-1)^{u+v}2\sin\left(\frac{(2u+1)i_1\pi}{n+1}\right)\sin\left(\frac{(2v+1)i_2\pi}{n+1}\right)}{4-2\cos\left(\frac{(2u+1)\pi}{n+1}\right)-2\cos\left(\frac{(2v+1)\pi}{n+1}\right)}, \tag{2.27}$$

$$k = i_2 + (i_1 - 1)n,$$

and i_1, i_2 are the coordinates of the kth site on the square lattice. Evaluating P_k and Q_k for all k, in order to keep the solution x_k integer, we choose a 'period' T_k equal to Q_k. The solution x_k is therefore also an integer and is equal to P_k. Doing this for all sites, we find a set of integer T_k and x_k. The period T is then found as the smallest T_k divisible by all denominators Q_k. Numerical analysis shows that $T \sim \exp(0.11n^2)$ and $T \sim \exp(0.04n^3)$ for the two- and three-dimensional cases, respectively.

If we want to calculate the configuration period of a deterministic automaton with two opposite opened and two opposite periodic boundaries, the matrix

$$\mathbf{H}_1^p = \begin{bmatrix} 0 & -1 & 0 & 0 & \dots & -1 \\ -1 & 0 & -1 & 0 & \dots & 0 \\ -0 & -1 & 0 & -1 & \ddots & 0 \\ \vdots & \ddots & \ddots & \ddots & \ddots & \vdots \\ 0 & 0 & \dots & -1 & 0 & -1 \\ -1 & 0 & \dots & 0 & -1 & 0 \end{bmatrix} \tag{2.28}$$

enters the scene. The eigenvalues of $\mathbf{H}_1^p$ are given by $(k = 1, 2, \dots, n)$

$$\lambda_k^p = 2\cos\left(\frac{2k\pi}{n}\right), \tag{2.29}$$

and $\mathbf{H}_1^p$ is diagonalizable by the matrix $\mathbf{Q}_1^p$ with elements

$$Q_1^p(r,s) = \sqrt{\frac{1}{n}}\exp\left(\frac{\mathrm{i}2rs\pi}{n}\right), \tag{2.30}$$

where i is the imaginary unit. Second generation matrices $\mathbf{H}_2^p$ and $\mathbf{Q}_2^p$ can be expressed in terms of $\mathbf{H}_1^p$, $\mathbf{H}_1$ and $\mathbf{Q}_1^p$, $\mathbf{Q}_1$:

$$\begin{aligned} \mathbf{H}_2^p &= \mathbf{H}_1^p \otimes \mathbf{I} + \mathbf{I} \otimes \mathbf{H}_1, \\ \mathbf{Q}_2^p &= \mathbf{Q}_1^p \otimes \mathbf{Q}_1. \end{aligned} \tag{2.31}$$

Doing the same as in the case of open boundaries we get the periods. Numerical analysis shows that for the two-dimensional lattice $T \sim \exp(0.316n^2)$.

It is useful now to look more precisely at the correlation matrix $\mathbf{G}_2$, see (2.24),

$$G_2(i_1,i_2) = \sum_{p,q=1}^{n}\psi_p(i_1)\psi_q(i_2)\frac{1}{4-2\cos\left(\dfrac{p\pi}{n+1}\right)-2\cos\left(\dfrac{q\pi}{n+1}\right)}\psi_p(i_1^*)\psi_q(i_2^*), \tag{2.32}$$

where i_1^* and i_2^* are the coordinates of the site where the grain is added. The vectors ψ are given by (2.7*b*).

In the denominators of the fractions in the sum (2.32) we find the eigenvalues (2.13) of the second generation Toeplitz matrix . One of these eigenvalues equals $4 - 4\cos\frac{\pi}{n+1}$. In the thermodynamic limit, $n \to \infty$, this eigenvalue will tend to zero, and therefore some elements of the correlation matrix will be rather large. As was argued by Chau & Cheng (1993), this fact might be responsible for the long range correlations in the SOC state. If the sandpile is nonconservative, some grains are dissipated to the environment in each toppling. In the case of the nearest neighbour toppling, this will be reflected in the toppling matrix having $a > 4$, and also in the denominator of the two point correlation function, where now instead of 4 we have a constant $a > 4$. Hence from (2.32) it follows that the denominator can never tend to zero, and the elements of the two-point correlation function will never tend to infinity. Therefore exponentially decreasing correlations are assumed, see Chau & Cheng (1993). The nonconservative properties of the model yield typical length and time scales of the system.

We supplemented these considerations by analytical studies of the nonconservative model with nearest neighbour topplings on the Bethe lattice, see Markošová (1995b). A similar but conservative automaton was studied by Dhar & Majumdar (1990). They succeeded in a complete analytical description of the SOC state by single site height distributions, as well as pair height distributions and avalanche size distribution. They showed that the avalanche size distribution obeys a power law

$$g(s) \sim s^{-\frac{3}{2}} \tag{2.33}$$

for large s.

We used the method of Dhar and Majumdar and calculated analytically the same quantities for a nonconservative cellular automaton. Our nonconservative sandpile is defined by the following dynamical rules:

Input process:

$$\begin{aligned} &h_i \to h_i + 1, \\ &h_i < h_c, \\ &h_c = 1 + z + \tilde{n}. \end{aligned} \tag{2.34a}$$

Avalanche:

$$\begin{aligned} &h_i \to h_i - (z + \tilde{n}) = 1, \\ &h_{nn} \to h_{nn} + 1. \end{aligned} \tag{2.34b}$$

That means that in each toppling z grains are distributed to the nearest neighbours nn, $\tilde{n}$ grains dissipate to the environment, and one grain rests at the original site. Boundaries are left open. If $\tilde{n} = 0$, one has the conservative model of Dhar and Majumdar and all analytical results should be the same as those in Dhar & Majumdar (1990). Under the dynamical rules, the model (2.34) evolves to the stationary stable state in which $1, 2, \ldots, h_{cc} (= h_c - 1)$ grains can be found at the inner lattice site. If N is a number of lattice sites, h_{cc}^N distinct column height configurations are possible, but not all of them are allowed (see figure 2). The allowed configurations are of the two types:

(a) strongly allowed,

(b) weakly allowed.

Let C be an allowed configuration at the Bethe tree T rooted on the vertex α. We connect α with another site β containing only one grain (see figure 3). If the new configuration $\tilde{C}$ is still allowed, C will be called strongly allowed. If $\tilde{C}$ is forbidden, the old configuration C will be called weakly allowed.

If one removes the site α from T, the tree T breaks into two subtrees T_1 and T_2, see figure 3. If $N_\mathrm{w}(\mathrm{T}, h_\alpha)$ and $N_\mathrm{s}(\mathrm{T}, h_\alpha)$ denote weakly and strongly allowed configurations,

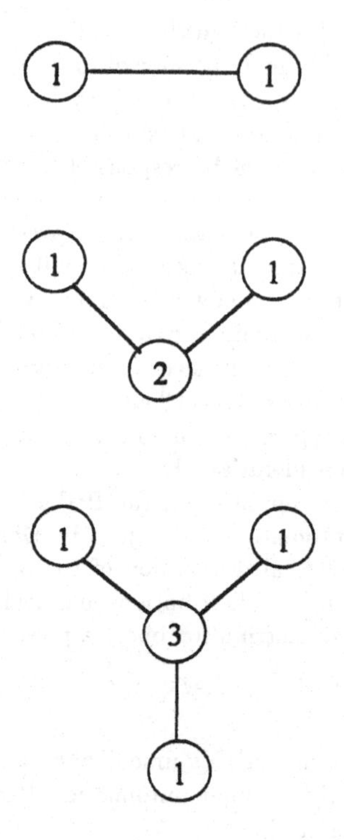

FIGURE 2. Three examples of forbidden configurations on the Bethe lattice. The sites with only one grain have just toppled. Therefore the previous configuration would have to contain some zero particle sites, which is impossible.

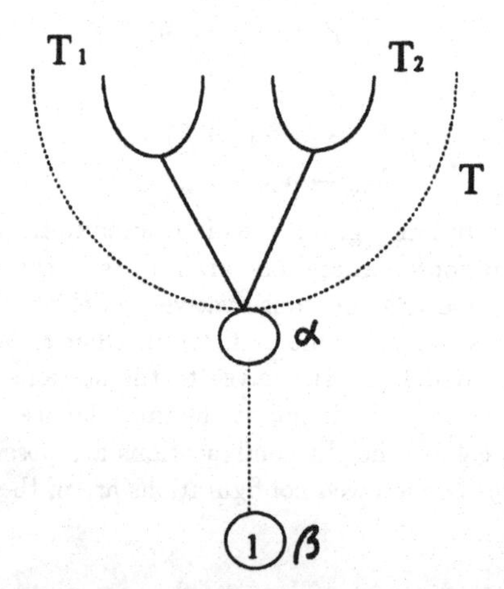

FIGURE 3. To the definition of the strongly and weakly allowed configurations.

provided that the column height at α is h_α, the total number of weakly and strongly allowed configurations on T can be expressed as follows

$$N_{\mathrm{w}}(\mathrm{T}) = \sum_{h_\alpha=1}^{h_{\mathrm{cc}}} N_{\mathrm{w}}(\mathrm{T}, h_\alpha), \tag{2.35}$$
$$N_{\mathrm{s}}(\mathrm{T}) = \sum_{h_\alpha=1}^{h_{\mathrm{cc}}} N_{\mathrm{s}}(\mathrm{T}, h_\alpha),$$

where

$$N_{\mathrm{w}}(\mathrm{T},1) = N_{\mathrm{s}}(\mathrm{T}_1)N_{\mathrm{s}}(\mathrm{T}_2), \tag{2.36a}$$
$$N_{\mathrm{s}}(\mathrm{T},1) = 0,$$

$$N_{\mathrm{w}}(\mathrm{T},2) = N_{\mathrm{s}}(\mathrm{T}_1)N_{\mathrm{w}}(\mathrm{T}_2) + N_{\mathrm{w}}(\mathrm{T}_1)N_{\mathrm{s}}(\mathrm{T}_2), \tag{2.36b}$$
$$N_{\mathrm{s}}(\mathrm{T},2) = N_{\mathrm{s}}(\mathrm{T}_1)N_{\mathrm{s}}(\mathrm{T}_2),$$

$$N_{\mathrm{w}}(\mathrm{T},3) = N_{\mathrm{w}}(\mathrm{T}_1)N_{\mathrm{w}}(\mathrm{T}_2), \tag{2.36c}$$
$$N_{\mathrm{s}}(\mathrm{T},3) = N_{\mathrm{s}}(\mathrm{T}_1)N_{\mathrm{s}}(\mathrm{T}_2) + N_{\mathrm{w}}(\mathrm{T}_1)N_{\mathrm{s}}(\mathrm{T}_2) + N_{\mathrm{s}}(\mathrm{T}_1)N_{\mathrm{w}}(\mathrm{T}_2),$$

$$N_{\mathrm{w}}(\mathrm{T}, h_\alpha = 4, \ldots, h_{\mathrm{cc}}) = 0, \tag{2.36d}$$
$$\begin{aligned} N_{\mathrm{s}}(\mathrm{T}, h_\alpha = 4, \ldots, h_{\mathrm{cc}}) &= N_{\mathrm{s}}(\mathrm{T}_1)N_{\mathrm{s}}(\mathrm{T}_2) + N_{\mathrm{w}}(\mathrm{T}_1)N_{\mathrm{s}}(\mathrm{T}_2) \\ &\quad + N_{\mathrm{s}}(\mathrm{T}_1)N_{\mathrm{w}}(\mathrm{T}_2) + N_{\mathrm{w}}(\mathrm{T}_1)N_{\mathrm{w}}(\mathrm{T}_2). \end{aligned}$$

In the limit $N \to \infty$ (N being the number of lattice sites), T_1 and T_2 can be considered the equivalent trees of the previous generation. If T is a tree of mth generation, the recurrence relation for $x^{(m)} = N_{\mathrm{w}}(\mathrm{T}^{(m)})/N_{\mathrm{s}}(\mathrm{T}^{(m)})$ can be derived from (2.35) and (2.36):

$$x^{(m+1)} = \frac{x^{(m)}+1}{\tilde{n}(x^{(m)}+1)+2}. \tag{2.37a}$$

Equation (2.37a) has one positive, physically relevant, fixed point

$$x^* = \frac{\sqrt{(\tilde{n}+1)^2 + 4\tilde{n}} - (\tilde{n}+1)}{2\tilde{n}}. \tag{2.37b}$$

Hence for a large lattice, $x^{(m)}$ can be replaced by x^* in calculations.

Now we shall calculate the single site height distributions. Any position O, which is far from the boundary, is connected to the three trees T_1, T_2 and T_3 (see figure 4). The number of allowed configurations, if the column height at O is h_{O}, is given by

$$N(\mathrm{T}, h_{\mathrm{O}} = 1) = \prod_{\omega=1}^{3} N_{\mathrm{s}}(\mathrm{T}_\omega), \tag{2.38a}$$

$$N(\mathrm{T}, h_{\mathrm{O}} = 2) = \prod_{\omega=1}^{3} N_{\mathrm{s}}(\mathrm{T}_\omega)\Big[1 + \sum_{j=1}^{3} x_j\Big], \tag{2.38b}$$

$$N(\mathrm{T}, h_{\mathrm{O}} = 3) = \prod_{\omega=1}^{3} N_{\mathrm{s}}(\mathrm{T}_\omega)\Big[1 + \sum_{j=1}^{3} x_j + \sum_{j<k} x_j x_k\Big], \tag{2.38c}$$

$$N(\mathrm{T}, h_{\mathrm{O}} = 4, \ldots, h_{\mathrm{cc}}) = \prod_{\omega=1}^{3} N_{\mathrm{s}}(\mathrm{T}_\omega)\Big[1 + \sum_{j=1}^{3} x_j + \sum_{j<k} x_j x_k + x_1 x_2 x_3\Big], \tag{2.38d}$$

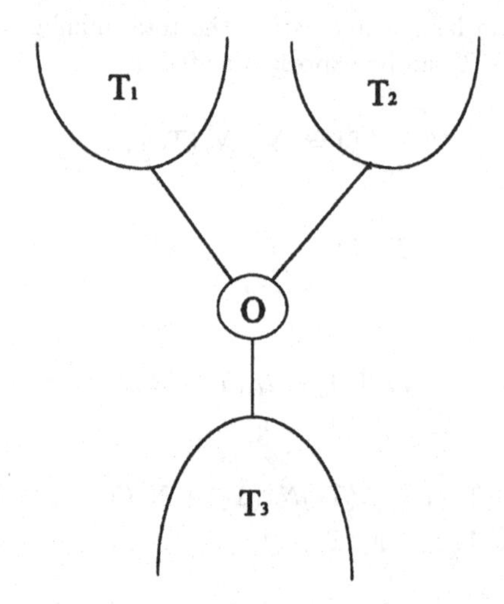

FIGURE 4. For a calculation of single site probabilities. The site O is far from the boundary.

$\tilde{n}$	$P(h_O = 1)$	$P(h_O = 2)$	$P(h_O = 3)$	$P(h_O = 4)$	$P(h_O = 5)$	$P(h_O = 6)$
0	$\frac{1}{12}$	$\frac{4}{12}$	$\frac{7}{12}$			
1	$\frac{1}{2(3+\sqrt{2}}$	$\frac{11\sqrt{2}-12}{14}$	$\frac{7-3\sqrt{2}}{2(3+\sqrt{2}}$	$\frac{3\sqrt{2}-2}{7}$		
2	$\frac{1}{5+\sqrt{17}}$	$\frac{3\sqrt{17}-5}{4(5+\sqrt{17})}$	$\frac{29-3\sqrt{17}}{8(5+\sqrt{17})}$	$\frac{13+5\sqrt{17}}{16(5+\sqrt{17})}$	$\frac{13+5\sqrt{17}}{16(5+\sqrt{17})}$	
3	$\frac{9}{46+16\sqrt{7}}$	$\frac{9(\sqrt{7}-1)}{46+16\sqrt{7}}$	$\frac{24-3\sqrt{7}}{46+16\sqrt{7}}$	$\frac{22+10\sqrt{7}}{3(46+16\sqrt{7})}$	$\frac{22+10\sqrt{7}}{3(46+16\sqrt{7})}$	$\frac{22+10\sqrt{7}}{3(46+16\sqrt{7})}$

TABLE 1. Single site probabilities for the conservative ($\tilde{n} = 0$) and all three nonconservative versions of the model (2.34*a*) and (2.34*b*); $\tilde{n}$ denotes the number of dissipated particles

where

$$x_j = \frac{N_w(\mathrm{T}_j)}{N_s(\mathrm{T}_j)}.$$

As O is situated deep inside the lattice, $x_j = x^*$ in (2.38), and the equation for the single site height distributions reads

$$P(h_O, \tilde{n}) = \frac{N(\mathrm{T}, h_O)}{N_{\mathrm{total}}}, \tag{2.39}$$

where N_{total} is the sum of all allowed configurations,

$$N_{\mathrm{total}} = \sum_{i=1}^{h_{cc}} N(\mathrm{T}, h_O = i).$$

Putting $\tilde{n} = 0$, the single site height distributions of the conservative model of Dhar & Majumdar (1990) are found. Numerical values of the single site probabilities are collected in table 1. Three versions of the model (2.38) were studied. They differ in $\tilde{n}$, i.e. in the number of dissipated particles in each toppling.

Playing the game with the strongly and weakly allowed configurations, and using similar but a little more sophisticated arguments, it is possible to calculate analytically the two-point distribution functions and the avalanche size distributions, see Markošová

(1995b) and Dhar & Majumdar (1990). For the nonconservative cellular automaton, the result for the avalanche size distribution is

$$g(s) \sim \exp\left(-\frac{s}{s_0}\right) s^{-\frac{3}{2}}. \tag{2.40}$$

In the conservative limit, $\tilde{n} = 0$, (2.40) reduces to the conservative power law distribution (2.33), see Markošová (1995b). The dissipation incorporates a typical $\tilde{n}$ dependent length scale s_0 into the distribution, see Markošová (1995b) and Chau & Cheng (1993). This indicates that the nonconservative, randomly driven automata defined by (2.38*a*) cannot reach the SOC state.

An analytical calculation of the single site and pair probabilities, as well as of avalanche size distributions on a square lattice is more difficult. A significant step was made by Majumdar & Dhar (1991). Let us consider the model (2.4) with $\mathbf{\Delta}$ being the second generation Toeplitz matrix. To show whether the given column height configuration on the square lattice is allowed, i.e. whether it has a nonzero probability in the SOC state, the 'burning algorithm' was created, see Dhar (1990): One deletes from the configuration (burns) that site whose height strictly exceeds the number of its unburnt neighbours. If the configuration is allowed, no site remains unburnt after a few repetitions of the burning procedure. The burning algorithm follows from Dhar's definition of a forbidden configuration, see Dhar (1990): The configuration F is forbidden, if for all sites j in F the following equation holds:

$$h_j \leq \sum_{i \neq j} (-\Delta_{ij}).$$

Site i is also from F.

As an example, we shall calculate the probability that the column height $h_{\mathrm{O}} = 1$ at site O inside the lattice. The burning algoritm shows that two adjacent sites cannot have one grain. Therefore four neighbours of O (N, E, S and W) may only contain 2, 3 or 4 grains. From the allowed configuration C at the lattice L with $h_{\mathrm{O}} = 1$ we create a new configuration C' by reducing the N, S and W column heights by one. Then we define a new abelian model on the lattice L$'$; L$'$ is obtained from L by deleting bonds ON, OS and OW (see figure 5). Then we decrease the maximum height allowed at the two sites of the bond by 1, so that the model is again conservative. For this new sandpile automaton the toppling matrix $\mathbf{\Delta}'$ is given by

$$\mathbf{\Delta}' = \mathbf{\Delta} + \mathbf{B}, \tag{2.41a}$$

where the only nonzero elements of $\mathbf{B}$ are

$$\begin{aligned} &B(\mathrm{N},\mathrm{N}) = B(\mathrm{W},\mathrm{W}) = B(\mathrm{S},\mathrm{S}) = -1, \quad B(\mathrm{O},\mathrm{O}) = -3, \\ &B(\mathrm{O},\mathrm{N}) = B(\mathrm{N},\mathrm{O}) = B(\mathrm{S},\mathrm{O}) = B(\mathrm{O},\mathrm{S}) = B(\mathrm{W},\mathrm{O}) = B(\mathrm{O},\mathrm{W}) = 1. \end{aligned} \tag{2.41b}$$

All sites of C are burnt under the burning procedure in a certain sequence. It is easy to show that the sites in C' on L$'$ are burnt in the same sequence. That means that C' is allowed under $\mathbf{\Delta}'$ only if C is allowed under $\mathbf{\Delta}$. There exists a one to one correspondence between the allowed configurations having one particle at O and the allowed configurations on L$'$. The number of allowed configurations with $h_{\mathrm{O}} = 1$ is equal to the number of allowed configurations under the toppling rules defined by $\mathbf{\Delta}'$, which is det $\mathbf{\Delta}'$. Therefore

$$P(h_{\mathrm{O}} = 1) = \frac{\det \mathbf{\Delta}'}{\det \mathbf{\Delta}} = \det(\mathbf{I} + \mathbf{G}_2\mathbf{B}) = \frac{2}{\pi}(1 - \frac{2}{\pi}) = 0.073636, \tag{2.42}$$

where $\mathbf{G}_2 = \mathbf{\Delta}^{-1}$. Using similar arguments, it is possible to calculate the probabilities

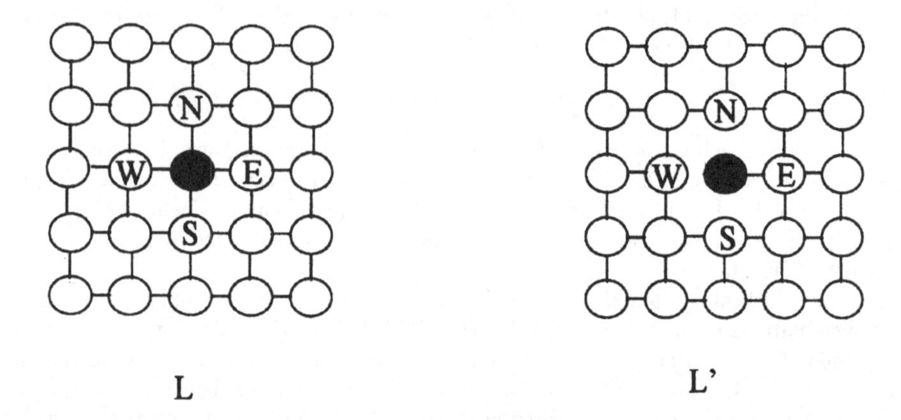

FIGURE 5. Abelian models with nearest neighbour topplings defined on lattices L and L'. On the lattice L' the maximum height on the N, S and W sites are three grains, the maximum height on the black site is one grain.

of some special subconfigurations consisting of more sites, and to establish inequalities between single site probabilities. The results presented in Majumdar & Dhar (1991) show that

$$P(1) \leq P(2) \leq P(3) \leq P(4),$$

and $P(2) \geq 0.131438$. The last result is consistent with the numerical value $P(2) \approx 0.174$ found by Manna (1990).

3. Concluding remarks

I have presented a short review of the analytical methods developed to investigate abelian sandpiles. The analytical results enable us to better understand the physical and mathematical aspects of the SOC state in abelian models. But, on the other hand, they pose further questions, for example those about the role of conservativeness in self organized criticality.

It was shown numerically by Feder & Feder (1991) and Christensen & Olami (1992a), that models describing the dynamics of the Earth crust evolve, after some transients, to the SOC state with power law distributions of avalanche sizes and lifetimes. These models were further studied by Jánosi & Kertész (1993) and Grassberger (1994).

Apart from sandpile cellular automata and the above mentioned continuously driven models, the continuous analogues of sandpiles were also studied, see Grinstein (1991). These conservative nonequilibrium dynamical systems driven by uncorrelated noise are described by the Langevine equation. In a simpler case it has the form

$$\frac{\partial h(\vec{x},t)}{\partial t} = -\Gamma f_{\vec{x}}(h) + \eta(\vec{x},t), \tag{3.43}$$

where η is a gaussian noise variable,

$$\langle \eta(\vec{x},t)\eta(\vec{x}',t')\rangle = D\,\delta(\vec{x}-\vec{x}')\,\delta(t-t').$$

It was shown analytically, see Grinstein (1991), that only if the deterministic part is conservative, will the space correlations be generically algebraic and fall off more slowly then $1/x^d$ in d dimensions. The role of conservativeness in the SOC state is an important and widely studied problem. As we have mentioned above, see Markošová (1995b), randomly driven nonconservative cellular automata can never reach the SOC state. But the conditions in which the scale invariances arise in continuously driven systems, are not yet completely clear. Nor is the relation between the systems described by the Langevine equations and sandpile cellular automata.

REFERENCES

BAK, P., TANG, C. & WIESENFELD, K. 1988 *Phys. Rev.* A **38**, 364.
CHAU, H.F. & CHENG, K.S. 1993 *Phys. Rev.* E **47**, 2394.
CHRISTENSEN, K. & OLAMI, Z. 1992 *Phys. Rev.* A **46**, 1829.
CHRISTENSEN, K., OLAMI, Z. & BAK, P. 1992 *Phys. Rev. Lett.* **68**, 2417.
DHAR, D. 1990 *Phys. Rev. Lett.* **64**, 1613.
DHAR, D. & MAJUMDAR, S. N. 1989 *J. Phys. A: Math. Gen.* **23**, 4333.
DHAR, D. & RAMASWAMY, R. 1989 *Phys. Rev. Lett.* **63**, 1659.
FEDER, H.J. & FEDER, J. 1991 *Phys. Rev. Lett.* **66**, 2669.
GRASSBERGER, P. & MANNA, S. 1990 *J. Phys. France* **51**, 1077.
GRASSBERGER, P. 1994 *Phys. Rev.* E **49**, 2436.
GRINSTEIN, G. 1991 *J. Appl. Phys.* **69**, 5441.
JÁNOSI, J.M. & KERTÉSZ, J. 1993 *Physica* A **200**, 179.
JENSEN, H.J., CHRISTENSEN, K. & FOGEDBY, H.C. 1989 *Phys. Rev.* B **40**, 7425.
KADANOFF, L.P., NAGEL, S.A., WU, L. & ZHOU, S. 1989 *Phys. Rev.* A **39**, 6524.
KERTÉSZ, J. & KISS, L.B. 1990 *J. Phys. A: Math. Gen.* **23**, L433.
MAJUMDAR, S.N. & DHAR, D. 1991 TIFR Report TH/91-3.
MANNA, S.S 1990 *J. Stat. Phys.* **59**, 509.
MARKOŠOVÁ, M. 1995a *Physica* D **80**, 41.
MARKOŠOVÁ, M. 1995b *J. Phys. A: Math. Gen.* **28**, 6903.
MARKOŠOVÁ, M. & MARKOŠ, P. 1992 *Phys. Rev.* A **46**, 3531.
WIESENFELD, K., THEILER, J. & MC NAMARA, B. 1990 *Phys. Rev. Lett.* **65**, 949.

Transport in an incompletely chaotic magnetic field

By FLORIN SPINEANU

Institute of Atomic Physics, Magurele Bucharest, Romania

We discuss test particle transport in magnetic configurations consisting of a strong confining field perturbed by a stochastic transversal component. Detailed analytical calculations performed for the nonchaotic regime prove that the range of validity of the model is related to the condition of convergence of a series of high order correlations of the stochastic perturbations. This condition singles out a nondimensional parameter, the same as that found previously by the two-point theories of the chaotic regimes.

The effect of coherent structures on the diffusive motion is examined using simple analytical and numerical models.

We perform an analytical investigation of the physical processes which are neglected by the widely used Corrsin factorization and show that trapping effects are the most important component.

1. Introduction

As a dynamical system, the magnetic field in a plasma confining device or in astrophysics can have a broad range of analytical behaviours, from integrable structures to erratic wandering in space. The source of this complexity is primarily the high nonlinearity which is inherent in ideal magnetohydrodynamics. In addition, the presence of small dissipation (resistivity) induces reconnections of the magnetic lines with generation of even more complex structures like magnetic islands. These are surrounded by regions where the field lines exhibit high sensitivity to initial conditions, i.e. exponential instability, which is the signature of chaos. As in many other cases, although the system is purely deterministic, a statistical description can be appropriate and the field is described as a stochastic variable. For a confining device like the tokamak, the (simplified) configuration of the field consists of a strong unidirectional component and much smaller transversal perturbations which we assume to be stochastic variables with given statistical properties. We focus on the effect which these small perturbations have on the motion of charged particles since this is one of the causes of the anomalous particle and heat transport observed in experiments, see Balescu (1988). We briefly review some basic results, see Krommes *et al.* (1983).

The simplest case consists of a noncollisional plasma with particles performing ballistic motion along the magnetic lines. It is a well known fact that the magnetic field lines in a statistical ensemble of realizations of the transversal perturbations behave diffusively. The magnetic diffusion is $D_m = \beta^2 \lambda_\parallel$ and yields a diffusion of particles: $D = D_m v_{th}$. Here β is the amplitude of the magnetic perturbations normalized to the equilibrium value, $\lambda_\parallel$ is the correlation length along the direction of the field at equilibrium and v_{th} is the thermal particle velocity. This diffusion is very large (due to the large value of v_{th}) which clearly shows that magnetic stochasticity is a very efficient transport mechanism. We now perform a slight modification of the above picture, allowing for parallel collisions of the particles along the lines. The collisions replace the ballistic motion of the particles by a diffusion along each line. It is easy to show by heuristic arguments (see Krommes *et al.* (1983); below we give an analytical derivation) that the combination of this diffusion and of the magnetic field line diffusion leads to a subdiffusive behaviour of the particles in transversal planes. This may appear as a rather surprising suppression of the effectiveness of the magnetic stochasticity induced transport and is a first manifestation of a sort of trapping effect.

The subdiffusive situation is very 'fragile' and any amount of perpendicular collisionality restores diffusion. The transverse collisions provide transport independent of the magnetic stochasticity. However, their most important effect arises from a coupling with the parallel motion through the Lagrangian argument of the fluctuating magnetic field. Two different manifestations exist. First, the transverse collisions induce displacements of the particles between different lines, thus allowing a wider sampling of the statistical ensemble of the turbulent lines. If the transversal jumps (given by velocity $\eta_\perp$) are correlated with the transversal deviations of the lines ($b\eta_\parallel$), they can give large contributions to the mean square displacement. We show analytically that opposite correlations are possible as well, and partial cancellations occur. As a consequence, for specified statistics of the field and of the collisions, these higher order correlations require a large time to build up substantial contributions. This is simply related to the fact that sequences of highly correlated displacements are rare in the probabilistic sense. The analysis of these effects requires extensive calculations using certain technical methods to perform averages (the functional method), see Spineanu & Vlad (1995). The result consists in expressing the particle mean square displacement as an infinite series of terms, each representing such a high order correlation. The condition of convergence singles out a certain parameter (s, see (2.23)) and defines the region of validity of this approach.

A different manifestation of $\eta_\perp$ appears in the two-point approach to the transport in a stochastic magnetic field, see Rochester & Rosenbluth (1978) and Wang *et al.* (1995). It is assumed that the field is chaotic, i.e. two initially close lines separate exponentially along the parallel direction. Since the magnetic flux is conserved, a bundle of lines map a regular spatial region of the section (e.g. a circle) into a highly irregular figure, with long thin branches along the local exponentiation directions. The collisions induce jumps across these branches such that the particle can easily decorrelate from the initial bundle of lines. This is a purely chaotic regime and it is an interesting observation that the same parameter s is identified as governing the limits of validity of this model. Naturally the range of s here is complimentary to that found in the first approach. We note, however, that only in the first (nonchaotic regime) approach is the meaning of s clearly identified as the ratio of two successive cumulants.

Since in real plasmas, coherent structures appear and interact with transport, we shall examine two aspects: a test particle performing large excursions due to correlated jumps on quasi-regular magnetic structures and, alternatively, the trapping effect of the magnetic islands or intermittent vortex motions.

Finally, we will discuss the quasi-integrable motions associated with the trapping with the aim of proving that they must be identified as effects neglected by the Corrsin approximation.

2. Test particle diffusion in a collisional plasma and a stochastic magnetic field

2.1. *The mean square displacement*

The main objective of this part is to provide a basis for a quantitative treatment of the problem of the test particle in a collisional plasma immersed in a stochastic magnetic field, see Spineanu & Vlad (1995). We start by defining, in a simple geometry given by a strong equilibrium magnetic field B_{0z}, three independent stochastic processes: a transversal magnetic field $b_{x,y}$ and the parallel and perpendicular collisional fluctuations of the particle velocity $\eta_{\parallel}, \eta_{\perp x,y}$. The spectrum of the magnetic field components b_{α} is $(\alpha = x, y)$

$$\mathcal{B}(k_x, k_y, k_{\parallel}) \equiv \left\langle b_{\alpha}(\vec{k}) b_{\beta}(-\vec{k}) \right\rangle = \delta_{\alpha\beta} \beta^2 \frac{\lambda_{\parallel} \lambda_{\perp}^2}{(2\pi)^{3/2}} \exp\left(-\tfrac{1}{2} (k_{\parallel}^2 \lambda_{\parallel}^2 + k_{\perp}^2 \lambda_{\perp}^2) \right).$$

The parameters $\lambda_{\perp}, \lambda_{\parallel}$ are the effective spatial extensions of the transversal and parallel correlations. The mixed Lagrangian correlations of the magnetic field vanish: $\langle b_{\alpha}[\vec{r}(t)] b_{\beta}[\vec{r}(t')] \rangle = 0$ for $\alpha \neq \beta$. We construct the processes $\eta_{\parallel}$ and $\eta_{\perp x,y}$ by solving Langevin equations with white noise, as will be shown later.

We shall restrict considerations to the realistic situations pertaining to confining devices. The spectrum of transverse wavelengths is limited by the smallest width of a filamentation channel, i.e. there is a maximum value $k_{\perp \max}$ ('ultraviolet cutoff') in wavenumber space. We also introduce a limitation in the 'observation time' for all time dependent processes which we consider. These reasonable *a priori* restrictions define what we shall call the 'physically meaningful' domain.

The main steps of the calculation are the following: starting from the particle equation of motion (Langevin equation) we write the mean square displacement (MSD) and then use the Fourier representation of the magnetic field and the Corrsin factorization to 'expand' the Lagrangian nonlinearity. Averaging over the magnetic fluctuations yields an expression depending nonlinearly on $\eta_{\perp}$. Then, although $\eta_{\perp}$ itself is Gaussian, averaging over $\eta_{\perp}$ introduces an infinite number of cumulants. Each cumulant is a series of powers of β^2, from which we retain the zero and first order terms. This is consistent with our objective of calculating the MSD through β^4-accuracy. Since the dependence on $\eta_{\parallel}$ is also nonlinear, we again obtain an infinite number of cumulants from the $\eta_{\parallel}$ averaging. Both $\eta_{\perp}$ and $\eta_{\parallel}$ averaging procedures rely heavily on the functional method and are performed in two steps. Initially, the first four cumulants of each average are calculated explicitly. At this level, only the terms through order β^2 are retained in the expressions of the cumulants arising from the $\eta_{\parallel}$ averaging, for consistency with the previous β^2 truncation. In the second step, the expressions of the higher order cumulants are derived. At the outset, the series generated by the $\eta_{\perp}$ averaging ('$\eta_{\perp}$ series') appears in every term of the infinite sum resulting from the $\eta_{\parallel}$ averaging. The reason to proceed in two steps is so as to find a useful truncation of the series of cumulants together with its range of validity.

The equations of motion are:

$$\begin{aligned} \dot{x}(t) &= b_x(\vec{r}(t)) \eta_{\parallel}(t) + \eta_{\perp x}(t) \equiv \xi_x, \\ \dot{y}(t) &= b_y(\vec{r}(t)) \eta_{\parallel}(t) + \eta_{\perp y}(t) \equiv \xi_y, \\ \dot{z}(t) &= \eta_{\parallel}(t) \equiv \xi_z. \end{aligned} \tag{2.1}$$

In order to calculate the mean square displacement (MSD) in the transverse direction, we formally integrate the equations of motion and perform the average. For the x-coordinate:

$$\langle \Delta x^2(t) \rangle = \int_0^t \mathrm{d}t_1 \int_0^t \mathrm{d}t_2 \, \langle \xi_x(t_1) \xi_x(t_2) \rangle \,. \tag{2.2}$$

The integrand is the Lagrangian correlation of the random transversal displacements:

$$\langle \xi_x(t_1) \xi_x(t_2) \rangle = \langle b_x(\vec{r}(t_1)) b_x(\vec{r}(t_2)) \eta_{\parallel}(t_1) \eta_{\parallel}(t_2) \rangle + \langle \eta_{\perp x}(t_1) \eta_{\perp x}(t_2) \rangle = T_1 + \chi_{\perp} t. \tag{2.3}$$

The anomalous perpendicular diffusion comes entirely from T_1. Fourier transforming and replacing the difference $\vec{r}(t_2) - \vec{r}(t_1)$ in the exponent by the formal solution of the equations of motion, we get

$$T_1 = \int d\vec{k}\, dk_\parallel\, \mathcal{B}(\vec{k}_\perp, k_\parallel) \Big\langle \Big\langle \exp\left(ik_\parallel \int_{t_1}^{t_2} dt\, \eta_\parallel(t) - \tfrac{1}{2} k_x^2 \int_{t_1}^{t_2} dt \int_{t_1}^{t_2} dt' \, \langle b_x b_x \rangle \, \eta_\parallel(t) \eta_\parallel(t') \right)$$
$$\times \exp\left(-\tfrac{1}{2} k_y^2 \int_{t_1}^{t_2} dt \int_{t_1}^{t_2} dt' \, \langle b_y b_y \rangle \, \eta_\parallel(t) \eta_\parallel(t') \right)$$
$$\times \exp\left(ik_x \int_{t_1}^{t_2} dt \eta_{\perp x}(t) + ik_y \int_{t_1}^{t_2} dt\, \eta_{\perp y}(t) \right) \Big\rangle_\perp \eta_\parallel(t_1) \eta_\parallel(t_2) \Big\rangle_\parallel . \tag{2.4}$$

2.2. *Averaging over collisional fluctuations*

Both $\eta_\perp$ and $\eta_\parallel$ are fluctuating particle velocities which are defined by the equation

$$\dot{\eta} = -\nu \eta + \alpha(t), \tag{2.5}$$

where

$$\langle \alpha(t) \rangle = 0 \quad \text{and} \quad \langle \alpha(t) \alpha(t') \rangle = 2D\delta(t - t').$$

We perform the averages over $\eta_\parallel$ and $\eta_\perp$ using the functional method, see Spineanu & Vlad (1994). The generating functional of the correlations is

$$Z_{\bar{J}} \equiv \int \mathcal{D}[\eta(t)] \mathcal{D}[k(t)] \exp\Big(i \int_0^T dt \big[-k(t)$$
$$\times \left(\dot{\eta}(t) + \nu \eta(t) \right) + ik^2(t) D + J_1(t) \eta(t) + J_2(t) k(t) \big] \Big). \tag{2.6}$$

The path which extremizes the action is obtained by solving the Euler–Lagrange equations. We obtain (summation over $k, l = 1, 2$):

$$Z_{\bar{J}} = \exp\left(\frac{i}{2} \int_0^T dt \int_0^T dt' \, J_k(t) \Delta_{kl}(t, t') J_l(t') \right). \tag{2.7}$$

Here the propagator which will be used in the following calculation is

$$\Delta_{11}(t, t') = \frac{iD}{\nu} \left[e^{\nu(t-t')} \Theta(t' - t) + e^{-\nu(t-t')} \Theta(t - t') \right], \tag{2.8}$$

where Θ is the Heaviside function. We can now obtain averages of products of $\eta(t)$, with various time arguments, by taking functional differentiation of $Z_{\bar{J}}$ with respect to the appropriate current $J_i(t)$. The simplest application is

$$\langle \eta_\parallel(t_1) \eta_\parallel(t_2) \rangle = \frac{\delta}{i\delta J_1(t_1)} \frac{\delta}{i\delta J_1(t_2)} Z_{\bar{J}} \Big|_{\bar{J}=0} = \tfrac{1}{2} \chi_\parallel \nu \, e^{-\nu(t_2 - t_1)}. \tag{2.9}$$

Applying this method, the result of averaging T_1 over $\eta_\perp$, retaining up to the fourth cumulant is

$$T_1 = \int d\vec{k}\, dk_\parallel\, \mathcal{B}(\vec{k}_\perp, k_\parallel) e^{-k_\perp^2 g_\perp(t_1, t_2)}$$
$$\times \left\langle \exp\left(ik_\parallel \int_{t_1}^{t_2} dt\, \eta_\parallel(t) - \int_{t_1}^{t_2} dt \int_{t_1}^{t_2} dt' \, \eta_\parallel(t) \eta_\parallel(t') \Gamma(t, t') \right) \eta_\parallel(t_1) \eta_\parallel(t_2) \right\rangle_\parallel , \tag{2.10}$$

where:

$$\Gamma(t, t') \equiv k_\perp^2 \Lambda(t, t') + k_\perp^4 \Omega(t_1, t_2, t, t'), \tag{2.11}$$

$$\Lambda(t, t') = \tfrac{1}{2} \beta^2 \left(1 + \frac{2g_\parallel(t, t')}{\lambda_\parallel^2} \right)^{-1/2} \left(1 + \frac{2g_\perp(t, t')}{\lambda_\perp^2} \right)^{-1}, \tag{2.12}$$

$$\Omega(t_1, t_2, t, t') = \beta^2 \frac{\chi_\perp^2 \nu^2}{16 \lambda_\perp^2} \left(1 + \frac{2g_\parallel(t, t')}{\lambda_\parallel^2} \right)^{-1/2} \left(1 + \frac{2g_\perp(t, t')}{\lambda_\perp^2} \right)^{-2} v(t_1, t_2, t, t'). \tag{2.13}$$

Here

$$g_\alpha = \frac{\chi_\alpha}{2\nu}\left(\tau + \mathrm{e}^{-\tau} - 1\right), \quad \alpha = \parallel, \perp$$

and v is a multiple integral over the propagator Δ_{11}. In order to perform the averaging over $\eta_\parallel$, we first calculate

$$\left\langle \exp\left(\mathrm{i}k_\parallel \int_{t_1}^{t_2} \mathrm{d}t\, \eta_\parallel(t) - \int_{t_1}^{t_2} \mathrm{d}t \int_{t_1}^{t_2} \mathrm{d}t'\, \eta_\parallel(t)\eta_\parallel(t')\Gamma(t,t') \right) \right\rangle_{\vec{J}}. \tag{2.14}$$

The subscript $\vec{J}$ means that the functional action is modified as usual by a perturbing current $\vec{J}$, in order to obtain averages by functional differentiation with respect to it. The following formula is used to calculate this average (F and Δ are in general operators):

$$\exp\left(\tfrac{1}{2}\frac{\delta}{\delta J}F\frac{\delta}{\delta J}\right)\exp(\tfrac{1}{2}J\Delta J) = \exp\left(\tfrac{1}{2}J\left(\Delta^{-1} - F\right)^{-1}J\right)\exp(-Tr\ln(1 - F\Delta)). \tag{2.15}$$

To the same order β^2 in the exponent, the result is

$$T_1 = \int \mathrm{d}\vec{k}_\perp\, \mathrm{d}k_\parallel\, \mathcal{B}(\vec{k}_\perp, k_\parallel)\exp\Big(-k_\perp^2(g_\perp + f_1 + k_\perp^2\varphi_1 + \cdots) - k_\parallel^2(g_\parallel + k_\perp^2 f_2 + k_\perp^4\varphi_2 + \cdots)\Big)$$
$$\times\Big[h_0 - k_\perp^2 h_2 - k_\perp^4 h_3 - \cdots - k_\parallel^2(h_1 + k_\perp^2 h_4 + k_\perp^4 h_5 + \cdots)\Big]. \tag{2.16}$$

The functions f_k, φ_k and h_k are integrals over products of the functions Λ and Ω with the propagator Δ. Their expressions are given in Spineanu & Vlad (1995).

2.3. *Higher order correlations*

The series appearing in the expression of T_1 with increasing powers of $k_\perp^2$ raises the problem of convergence. This can only be examined by calculating the higher order cumulants in the averages over $\eta_\perp$ and $\eta_\parallel$. Skipping numerous intermediate steps, we find that the parallel part of the magnetic spectrum is modified in the following way:

$$-\tfrac{1}{2}k_\parallel^2\lambda_\parallel^2 \rightarrow -\tfrac{1}{2}k_\parallel^2\lambda_\parallel^2\,E, \tag{2.17}$$

where

$$E = 1 + \frac{2g_\parallel}{\lambda_\parallel^2} - \left(k_\perp^2\lambda_\perp^2\right)\frac{2f_2}{\lambda_\perp^2\lambda_\parallel^2} - \left(k_\perp^2\lambda_\perp^2\right)^2\frac{2(\varphi_2 - f_4)}{\lambda_\perp^4\lambda_\parallel^2} - \sum_{\ell=2}^{\infty}\left(k_\perp^2\lambda_\perp^2\right)^{\ell+1}$$
$$\times\left[\frac{1}{\lambda_\parallel^2\lambda_\perp^{4\ell+2}}\frac{\beta^2}{2}(\chi_\parallel\tau_c)^2(\chi_\perp\tau_c)^{2\ell}\right]\left[\tilde{\varphi}_{2\ell+1} + (-1)^{\ell+1}\left(\beta^2\frac{\chi_\parallel\lambda_\perp^2}{\chi_\perp^2\tau_c}\right)^\ell \tilde{f}_{\ell+1}^{(2)}\right]. \tag{2.18}$$

The functions $\tilde{\varphi}_{2\ell+1}$ and $\tilde{f}_{\ell+1}^{(2)}$ consist of multiple integrals over the propagator Δ_{11}. Using the large-τ behavior of the functions $\tilde{\varphi}_{2\ell+1}$ and $\tilde{f}_\ell^{(2)}$ the last expression in brackets becomes

$$\tilde{\varphi}_{2\ell+1}(\tau) + (-1)^{\ell+1}\left(\beta^2\frac{\chi_\parallel\lambda_\perp^2}{\chi_\perp^2\tau_c}\right)^\ell \tilde{f}_{\ell+1}^{(2)}(\tau) \simeq \frac{4}{4\ell^2 - 1}\frac{\lambda_\parallel\lambda_\perp^{2\ell+2}}{(\chi_\parallel\tau_c)^{1/2}(\chi_\perp\tau_c)^{\ell+1}}$$
$$\times\left[\tau^{\ell+1/2} + (-1)^{\ell+1}\lambda_\parallel^\ell\lambda_\perp^{\ell-1}\frac{(\chi_\parallel\tau_c)^{\ell/2}}{(\chi_\perp\tau_c)^{(3\ell-1)/2}}\tau\right]. \tag{2.19}$$

In the range of parameters which we have adopted as relevant (i.e. in the confinement devices), the second term in brackets is dominant for finite τ. At orders $\ell \gg 1$, the limitation of this range in terms of the maximum 'time of observation' reads

$$\tau < \tau_{\max} \ll \lambda_\parallel\lambda_\perp\frac{(\chi_\parallel\tau_c)^{1/2}}{(\chi_\perp\tau_c)^{3/2}}. \tag{2.20}$$

Assuming that this condition is satisfied, we evaluate the magnitude of the ℓth order term:

$$\left(k_\perp^2 \lambda_\perp^2\right)^{\ell+1} \frac{\chi_\parallel \tau_c}{\lambda_\parallel^2} \left(\beta^2 \frac{\lambda_\parallel \sqrt{\chi_\parallel \tau_c}}{\lambda_\perp \sqrt{\chi_\perp \tau_c}}\right)^{\ell+1} \tau .$$

It reaches its maximum value at the limits of the physical domain. In this case the requirement to be negligible relative to the first term in E, yields

$$x_{\max} \tau_{\max}^{1/(\ell+1)} \ll \frac{1}{\beta^2} \frac{\lambda_\perp \sqrt{\chi_\perp}}{\lambda_\parallel \sqrt{\chi_\parallel}} \left(\frac{\lambda_\parallel^2}{\chi_\parallel \tau_c}\right)^{1/(\ell+1)}, \tag{2.21}$$

where $x_{\max} \equiv \left(k_{\perp \max}\right)^2 \lambda_\perp^2$. In this case, the maximal value of the ratio of two consecutive terms in E, of order $\ell+1$ and ℓ, respectively, is

$$x_{\max}\, s \ll 1, \tag{2.22}$$

where we have introduced the notation:

$$s \equiv \beta^2 \frac{\lambda_\parallel \sqrt{\chi_\parallel}}{\lambda_\perp \sqrt{\chi_\perp}} . \tag{2.23}$$

This parameter is thus of major importance in restricting the range of validity of our approach. Equation (2.22) defines the domain in which the test particle approach (i.e. one-point analysis) can describe the transport. This corresponds to the non-chaotic regimes, since beyond this range it is necessary to consider the exponentiation of the trajectories with a two-point analysis.

We now provide two illustration of the usefulness of (2.16) where various truncations are possible. When the magnetic field depends only on z (the case usually called 'linear'), we can take the limit $\lambda_\perp \to \infty$ and obtain

$$\begin{aligned} T_1 &= \beta^2 \left(1 + \frac{2 g_\parallel}{\lambda_\parallel^2}\right)^{-3/2} \left[h_0 \left(1 + \frac{2 g_\parallel}{\lambda_\parallel^2}\right) - \frac{h_1}{\lambda_\parallel^2}\right] \\ &= \beta^2 \left(\frac{\chi_\parallel}{2\tau_c}\right) \left[\mathrm{e}^{-\tau} \left(1 + \frac{\chi_\parallel \tau_c}{\lambda_\parallel^2} \left(\tau + \mathrm{e}^{-\tau} - 1\right)\right)^{-1/2} \right. \\ &\quad \left. - \frac{\chi_\parallel^2}{4 \lambda_\parallel^2 \tau_c^2} \left(1 - \mathrm{e}^{-\tau}\right)^2 \left(1 + \frac{\chi_\parallel \tau_c}{\lambda_\parallel^2} \left(\tau + \mathrm{e}^{-\tau} - 1\right)\right)^{-3/2}\right]. \end{aligned} \tag{2.24}$$

When integrated over time in (2.2), this formula leads to an increase of the particle MSD as $t^{1/2}$, the subdiffusive behavior discussed previously. We shall calculate the explicit form of the dispersion $\left\langle \Delta x_\perp^2 \right\rangle$ in the case when only the first order effect of the transverse collisions is retained. Then (2.16) gives

$$\left\langle \Delta x_\perp^2 \right\rangle = \nu^{-2} \int_0^{\nu t} \mathrm{d}\tau \, (\nu t - \tau) \beta^2 \left[\frac{\mathrm{e}^{-\tau} \chi_\parallel \nu / 2}{(1 + a\tau)^{1/2}} - \frac{\chi_\parallel^2 \left(1 - \mathrm{e}^{-\tau}\right)^2 / (4 \lambda_\parallel^2)}{(1 + a\tau)^{3/2}}\right] \frac{1}{1 + c\tau}, \tag{2.25}$$

where t is dimensional (sec), $a = \chi_\parallel / (\nu \lambda_\parallel^2)$ and $c = \chi_\perp / (\nu \lambda_\perp^2) + \beta^2 \chi_\parallel / (2 \nu \lambda_\perp^2)$. Returning to physical parameters, (2.25) gives the diffusion in the Kadomtsev–Pogutse regime, see Kadomtsev & Pogutse (1979):

$$\left\langle \Delta x_\perp^2 \right\rangle = \frac{\pi}{4} \beta^2 \frac{\lambda_\parallel}{\lambda_\perp} \sqrt{\chi_\parallel \chi_\perp}\, t. \tag{2.26}$$

2.4. *The asymptotic regime*

The role of the parameter s appears more clearly in the large-time asymptotic regime. We return to the higher order correlations which are explicitly shown by $\eta_\parallel$ averaging, in (2.14):

$$\exp\Big(-\tfrac{1}{2}k_{\parallel}^{2}\lambda_{\parallel}^{2}a\Big[\tilde{f}_{0}^{(2)}(\tau)-(\omega x)\tilde{f}_{1}^{(2)}(\tau)+(\omega x)^{2}\tilde{f}_{2}^{(2)}(\tau)-\cdots\Big]\Big)$$
$$\times\exp\Big(-(\omega x)\tilde{f}_{1}^{(1)}(\tau)+\tfrac{1}{2}(\omega x)^{2}\tilde{f}_{2}^{(1)}(\tau)-\tfrac{1}{3}(\omega x)^{3}\tilde{f}_{3}^{(1)}(\tau)+\cdots\Big),$$

where $\omega\equiv\beta^{2}\chi_{\parallel}\tau_{c}/\lambda_{\perp}^{2}$. Replacing the asymptotic form of these functions we obtain

$$\exp\Big(-\tfrac{1}{2}k_{\parallel}^{2}\lambda_{\parallel}^{2}a\tau\Big[1-(\omega xI_{\Lambda})+(\omega xI_{\Lambda})^{2}-\cdots\Big]\Big)$$
$$\times\exp\Big(-\omega x\tau\,(1-I_{\Lambda})-\tau\Big[(\omega xI_{\Lambda})-\tfrac{1}{2}(\omega xI_{\Lambda})^{2}+\tfrac{1}{3}(\omega xI_{\Lambda})^{3}-\cdots\Big]\Big),$$

where $I_{\Lambda}=\pi\nu\lambda_{\perp}\lambda_{\parallel}/(\chi_{\perp}\chi_{\parallel})^{1/2}$ (compatible with the truncation of $\tilde{\Gamma}$). We note the relation: $\omega I_{\Lambda}=s$. The domain of parameters specified before lies inside the common circle of convergence of the above series and we can write, in the asymptotic-τ range:

$$\exp\Big(-\tfrac{1}{2}k_{\parallel}^{2}\lambda_{\parallel}^{2}a\tau\frac{1}{1+sx}\Big)\exp\Big(-\omega x\tau\,(1-I_{\Lambda})-\tau\ln(1+sx)\Big). \tag{2.27}$$

In this regime, where convergence is ensured, we use truncations which allow us to retain the exact (i.e. not asymptotic) form for the functions. It is clear that for a large domain in parameter space, where $sx>1$, the resummation is questionable since we are beyond the radius of convergence. We can only say that, from the compact expressions in (2.27), valid in the large-τ regime, we can obtain the above series through formal Taylor expansion. We note, however, that much higher values of s are required by the two-point theory of the chaotic regimes. Thus our approach can still be valid in the intermediate range.

3. Diffusion in the presence of coherent structures

3.1. *Magnetic stochasticity and the regime of criticality*

It is difficult to perform experimental measurements of the magnetic field inside a hot plasma like that of a tokamak. Only indirect data can be obtained in the form of fluctuations which are recorded at the boundary of the plasma, from which one must infer the amplitude and the phase of the magnetic fluctuations inside. However, it is known that the fast particles arising from additional heating are transported mainly by these fluctuations. Simple arguments show that a small amount of random magnetic fluctuation is sufficient to determine a large rate of transport, which can easily overcome that given by the electrostatic instabilities, see Vlad *et al.* (1996) and Misguich *et al.* (1995). The excitation of tearing modes is a fundamental mechanism for continuous generation of random perturbations of the equilibrium magnetic structure.

The presence of small scale magnetic field stochasticity is a reasonable assumption for high-β plasmas. It should probably be represented as a continuous process of growth, saturation and decay of small magnetic islands due to random filamentation of the current taking place in a finite volume. Chaotic behavior, often assumed in theoretical models, arises from a random superposition of islands, leading to exponential separation of two initially close magnetic lines. The fact that this largely increases the transport (as compared to electrostatic instabilities) can be qualitatively explained by the occurrence in the dynamics of transport of the particle parallel velocity which is a very large parameter.

We start from these known facts to examine the problem of the ubiquity of magnetic stochasticity, as it is usually assumed. If the presence of magnetic stochasticity renders the transport very large, one can expect that the reservoir of free energy of the instability which is at the origin of magnetic fluctuations will be emptied at a fast rate and the fluctuations quenched. This is similar to the situation of an electrostatic instability having a highly efficient induced transport which locally flattens the gradient so disappears. This picture has recently been investigated in relation with the spontaneous evolution of some open driven systems toward a critical-like state. This property, called self-organized criticality, implies that the system places itself in a stable state consisting of fluctuations around marginal stability, any excursion into unstable states being penalized by a fast loss of free energy (with transport) and fall onto a deeper stable state. From this the combined effect of sources and diffusion drives the system back towards marginal stability. The magnetic stochasticity in a finite shell volume inside the confining region of the tokamak seems to provide the necessary condition for a self-organization of the plasma at criticality. It is then more reasonable to assume that the magnetic field is, on the average, slightly under the stochasticity threshold and that it becomes stochastic at random and for a very short time. In other words, the stochasticity is an intermittent phenomenon. For transport, this implies that most of the time it is slowly diffusive with very short bursts of large increases of the diffusion coefficient during the stochasticity events. Most of the transport takes place in these intermittent events (isolated bursts).

According to this picture, when the system fluctuates far below marginal stability (i.e. the free energy available for the magnetic instability is small), the chains of magnetic islands are barely touching. A particle can jump between these islands and perform large excursions but can also be trapped within an island. We are interested to find the characteristics of the transport in the presence of such quasi-regular structures like the islands. In an analytical approach we shall use the functional method presented above in the determination of some probability distributions related to random walks. Some more realistic situations are more difficult to handle analytically and so we shall employ numerical simulations of the diffusion in the presence of trapping.

3.2. *Quasiregular magnetic surfaces*

We start with the simplest configuration consisting of a finite region where neighbouring magnetic surfaces are perturbed by chains of small amplitude, coherent magnetic islands. Two neighbouring chains can touch, and in the region of contact reconnections are possible due to the finite resistivity. This can be represented as a jump of a magnetic line to a neighbouring chain of islands, where it can remain for some time before the occurrence of another jump. These events appear randomly in time, and at each time step the line is displaced with the width of the island, Δ. One can describe this process by a continuum time random walk with equal spatial steps $\pm\Delta$ on a lattice, or alternatively as a random walk with Poisson distribution of the times of the jumps. We employ a functional formalism and introduce the probability density functional of a trajectory which obeys the Langevin equation $\dot{x}(t) = \eta(t)$:

$$\mathcal{P}\left[x(t), k(t)\right] = \mathcal{N} \exp\left(-\mathrm{i} \int_0^T \mathrm{d}t\, k(t)\dot{x}(t)\right) \left\langle \exp\left(\mathrm{i} \int_0^T \mathrm{d}t\, k(t)\eta(t)\right)\right\rangle. \tag{3.28}$$

The jumps (represented by the noise $\eta(t)$) are independent, take values $\pm\Delta$ at random and occur at moments τ_n which are distributed on the time axis according to the Poisson law ($\eta(t)$ is the 'white Poisson noise') with density τ_m^{-1}. The averaging is performed using the generalized cluster expansion. The probability of a trajectory can be written:

$$\mathcal{P}\left[x(t)\right] = \mathcal{N} \int \mathcal{D}\left[k(t)\right] \exp\left(-\mathrm{i} \int_0^T \mathrm{d}t\, k(t)\dot{x}(t) + \frac{1}{\tau_m} \int_0^T \mathrm{d}t\, \left[\widetilde{\varphi}(k(t)) - 1\right]\right), \tag{3.29}$$

where $\widetilde{\varphi}$ is the Fourier transform of the distribution of η : $\widetilde{\varphi}(q) = \cos(q\Delta)$. As before, we introduce a generating functional

$$\mathcal{Z}_J \equiv \int \mathcal{D}[x(t)]\,\mathcal{D}[k(t)] \exp\left(\mathrm{i}\int_0^T \mathrm{d}t\, \mathcal{L}_J[x(t),k(t)]\right), \tag{3.30}$$

where

$$\mathcal{L}_J[x(t),k(t)] = -k(t)\dot{x}(t) - \frac{\mathrm{i}}{\tau_m}\left[\cos\left(k(t)\Delta\right) - 1\right] + J(t)x(t). \tag{3.31}$$

After performing the functional integrations we obtain

$$\mathcal{Z}_J = \exp\left(\frac{1}{\tau_m}\int_0^T \mathrm{d}t\left[\cos\left(\Delta\int_t^T \mathrm{d}t'\,J(t')\right) - 1\right]\right). \tag{3.32}$$

Using this generating functional, we calculate the probability P that a particle which starts at the origin can be found in $x = s\Delta$ at $t = t_0$ (s is an integer). It is convenient to measure the time in units of τ_m introducing $q \equiv t_0/\tau_m$:

$$P = \frac{\Delta}{2\pi}\mathrm{e}^{-q}\int_0^{2\pi} \mathrm{d}\lambda\, \mathrm{e}^{\mathrm{i}\lambda s\Delta + q\cos(q\Delta)} = \mathrm{e}^{-q} I_s(q). \tag{3.33}$$

Since this is a result for a random walk with time step τ_m and space step Δ, the diffusion can be larger as compared with that induced by the electrostatic instabilities if $\lambda_\perp < \Delta$ and/or $\gamma^{-1} > \tau_m$ ($\lambda_\perp$ and γ are respectively the wavelength and the growth rate of the instability). However, only a fraction of particles are involved in this lattice-type diffusion, while other particles are actually trapped by the islands. It is more and more likely that the transport must be considered as a result of largely different contributions: a small population can drive almost all transport while the bulk of particles can give a small contribution. It is then worthwhile to examine an even more efficient diffusion, related to the above example but with finite correlation between the amplitudes of the jumps, see Maruyama & Shibata (1988). It is assumed for simplicity that the jumps are performed in a periodic sequence with basic timestep τ_0. The noise is: $\eta(t) = \sum_{n=1}^{\infty} \eta_n \delta(t - n\tau_0)$, with $\eta_n = \pm\Delta$ and exponentially decaying correlation: $\langle \eta(t)\eta(t')\rangle = \Delta^2 \exp\left(-\gamma\,|t-t'|\right)$. Using the functional formalism we calculate the probability P to find a particle in x at t:

$$P = (2\pi)^{-1} b^{-\frac{1}{2}} \mathrm{e}^{-a}\left\{\sum_{m=0}^{\infty} \frac{a^{2m}}{(2m)!}\left(1 + \frac{a}{2m+1}\right) H_m(y) + D(a,y) + E(a,y)\right\}, \tag{3.34}$$

where

$$H_m(y) = \sum_{l=0}^{m} \binom{m}{l} \frac{(-1)^l}{2l+1}\left[{}_1F_1(2l+1, 2l+2; \mathrm{i}y) + {}_1F_1(2l+1, 2l+2; -\mathrm{i}y)\right] \tag{3.35}$$

and $a = \gamma t/2$, $b = 4\Delta^2/\gamma^2$, $y = x/b^{\frac{1}{2}}$. We introduced the following notations:

$$D(a,y) \equiv 2\int_0^{\infty} \mathrm{d}u \frac{u}{\sqrt{1+u^2}} \cos(au)\cos\left(y\sqrt{1+u^2}\right), \tag{3.36}$$

$$E(a,y) \equiv 2\int_0^{\infty} \mathrm{d}u \frac{1}{\sqrt{1+u^2}} \sin(au)\cos\left(y\sqrt{1+u^2}\right). \tag{3.37}$$

This formula must be calculated numerically and exhibits large values of the probability for a certain range of parameters. As we have noted, this approach must be complemented with an evaluation of the number of particles which participate in these particular motions. The same problem arises for the class of problems related to the trapping effect of the magnetic islands or of the vortex motions .

3.3. *Trapping effect of the coherent structures*

For certain plasma regimes, the magnetic stochasticity cannot occupy a finite volume. Instead, chains of magnetic islands are easily excited around the magnetic surfaces which correspond to rational values of the safety factor q. The transport is determined by the electrostatic instabilities which yield a diffusive motion. For some of the diffusive particles, the chains of magnetic islands act as traps with finite time of residence. To analyze the effect of these barriers, we examine a simple one-dimensional diffusive motion in the presence of m spatially periodic traps separated by a distance b. We assume that the spatial step r of the particle in the electrostatic turbulence is given by the probability distribution $p(r) = A\exp(-ar)$, with A a normalization constant. The problem is equivalent to a non-nearest neighbor random walk with traps, see Rubin & Weiss (1982), and the probability P_k that a particle starting from the origin is trapped in the island at $x = kb\ (1 \leq k \leq m)$ is

$$P_k = \left[\sinh\left((m-k+1)\Gamma\right) - \sinh\left((m-k)\Gamma\right)\right] / \sinh\left(m\Gamma\right), \tag{3.38}$$

where $\sinh\left(\Gamma/2\right) = \left(b/\rho\right)^{1/2}\sinh\left(a\rho/2\right)$ and ρ is a unit of length. We take $b = \left(ndq/dr\right)^{-1}$, where n is the toroidal number of the magnetic perturbation, and choose a such that the average step size is $\lambda_\perp$, the characteristic wavelength in the electric spectrum: $a = \rho^{-1}\ln\left(\lambda_\perp/\left(\lambda_\perp - \rho\right)\right)$. For $\lambda_\perp \gg \left(ndq/dr\right)^{-1}$, the probability becomes $P_k \simeq 1/m$, independent of k and rather small, which means that the trapping is not important at short times. In the opposite limit, we have $P_k = 1 - \exp(-\Gamma)$ for $\Gamma \to \infty$, which means that the particle is almost surely trapped in the first chain of islands. Since every particle is trapped and spends a certain time moving along the lines inside a magnetic island, the diffusion coefficient will be reduced due to this delay.

We now consider a two-dimensional case. A particle performs diffusive motion in the presence of traps distributed randomly on the plane. We assume that once the particle arrives at a trap, it is absorbed and remains there for a certain time before being released to continue its diffusive motion. We consider that the duration of residence in a trap is a constant, Δ. To completely specify the problem we need the diffusion coefficient of the motion between the traps and the law of distribution of the positions of the trapping centres in the plane. We shall refer to the problem in this setting as the 'Eulerian problem'. Alternatively, we adopt an approach in which the moments τ_i when the particle encounters a trap obey a distribution law expressed as a function of the 'proper time' of the particle (the 'Lagrangian problem'). Assuming that the distribution of τ_i is the Poisson law, we shall calculate the MSD of a particle which performs diffusive motion interrupted by trapping events of fixed duration Δ. The Langevin equation is

$$\dot{x}(t) = \eta_{\mathrm{tr}}(t) = \begin{cases} \eta(t) & \text{for } t \notin \bigcup_i(\tau_i, \tau_i + \Delta) \\ 0 & \text{for } t \in \bigcup_i(\tau_i, \tau_i + \Delta) \end{cases}$$

and $\langle \eta(t)\eta(t')\rangle = 2D\delta(t-t')$. The problem is solved using the functional formalism. After performing averaging over the noise, the generating functional is

$$\mathcal{Z}_J = \int \mathcal{D}[x(t)]\,\mathcal{D}[k(t)]\exp\Big(\mathrm{i}\int \mathrm{d}t\left[-k\dot{x} + Jx + \mathrm{i}Dk^2 + \mathrm{i}\nu\left(1 - \exp\left(\Delta Dk^2\right)\right)\right]\Big),$$

with ν the average frequency of the trapping events. Performing the functional integration we obtain

$$\mathcal{Z}_J = \exp\Big(-\int_0^T \mathrm{d}t\Big(D\big[\int_t^T \mathrm{d}t' J(t')\big]^2 + \nu\Big\{1 - \exp\Big(\Delta D\big[\int_t^T \mathrm{d}t' J(t')\big]^2\Big)\Big\}\Big)\Big).$$

The MSD is obtained as in (2.9) and we have

$$\left\langle x^2(t)\right\rangle = 2D\left(1 - \nu\Delta\right)t.$$

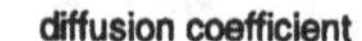

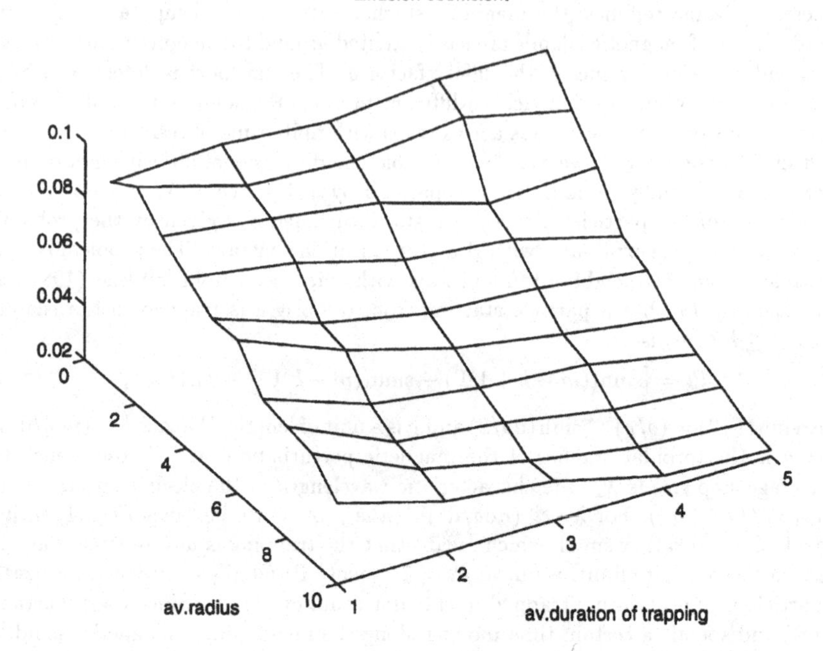

FIGURE 1. Dependence of the diffusion coefficient on the spatial extension of the trapping regions and on the duration of trapping.

The decrease of the diffusion coefficient can be important for large residence time and for a large number of trappers. The regime of very dense trapping must be examined without certain approximations which are included in the above treatment.

For illustration of the effective reduction in the diffusion coefficient when traps are present, we present some numerical simulations. In the first case, 500 particles perform 10^5 steps of a Gaussian diffusion process in a plane with a fixed distribution of trapping centres obeying the Poisson law. The radii of the circles where trapping occurs and the durations of trapping are chosen at random and are then multiplied with constant factors in several runs. The diffusion coefficient obtained by linear regression exhibits the expected decrease but also shows fluctuations (see figure 1). The second case corresponds with several realizations of the trapping structure: the positions are distributed in each realization such that the density of trappers fluctuates, obeying the Poisson law. The radii and the residence times are chosen at random. Figure 2 presents the averaged dispersions of the positions against time in 10 realizations. Note the large fluctuations which an average over the realizations would ignore. This confirms that the presence of trapping centres induces significant changes in transport which can appear as fluctuations of the diffusion coefficient in some regimes.

4. Beyond the Corrsin approximation is trapping

We have shown in the previous section that the existence of coherent structures in a stochastic field has a strong influence on the observed effective diffusion coefficient due to particle trapping in such structures. The process of trapping is actually also present

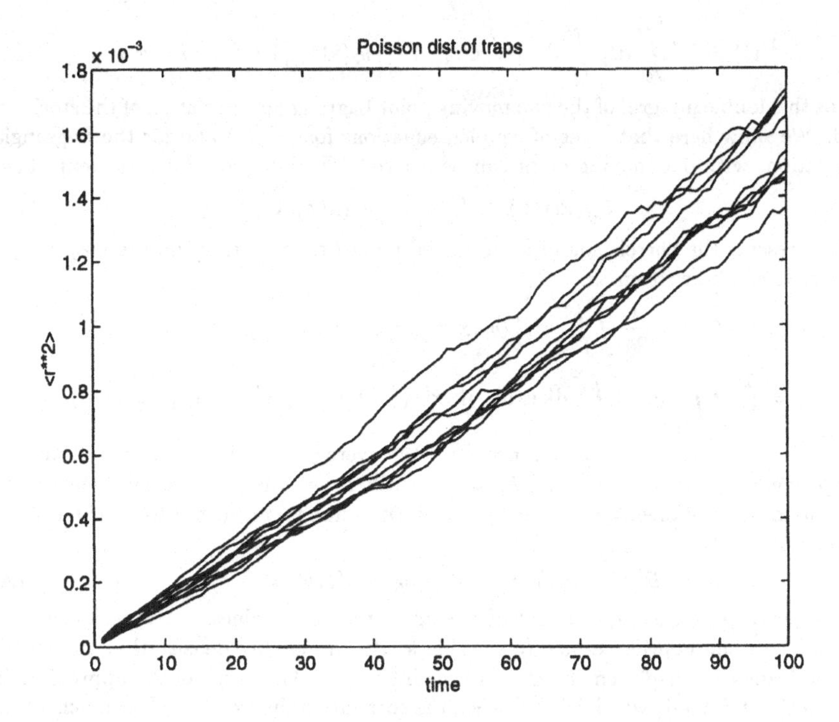

FIGURE 2. Mean square displacement for diffusing particles in 10 realizations of the random distribution of the trapping centres. The fluctuation in the diffusion coefficient is larger than 20%.

in a purely stochastic field when the correlation time of the field is large compared to the average time during which the particles traverse a correlation length. Numerical calculation show that in such conditions, particle trajectories become very complicated having parts which are confined to small regions for long time intervals and other parts which represent long jumps. Excepting a qualitative estimate based on an analogy with the percolation problem in stochastic landscapes (see Isichenko (1991)), the results concerning particle evolution in this strongly nonlinear limit rely on numerical calculations. We have recently developed a statistical approach which seems to describe the influence of these self-consistent trappings on the effective diffusion coefficient.

We consider the Langevin equations (2.1) for the guiding center trajectories, but in the collisionless limit ($\eta_{\parallel} = V = \text{const}$, $\eta_{\perp} = 0$), since we are interested in evaluating the influence of the Lagrangian nonlinearity:

$$\frac{\mathrm{d}\mathbf{x}}{\mathrm{d}t} = \mathbf{b}(\mathbf{x}, z)V, \tag{4.39}$$

where $\mathbf{x} = (x, y)$. The divergence free stochastic magnetic field $\mathbf{b}(\mathbf{x}, z)$ can be represented by the vector potential $\Psi(\mathbf{x}, z)\mathbf{e}_z$, ($\mathbf{b}(\mathbf{x}, z) = \nabla \times \Psi(\mathbf{x}, z)\mathbf{e}_z$), see Wang *et al.* (1995). The motion along the unperturbed magnetic field (along z) is simply $z = Vt$. The mean square displacement of the particles is determined from the formal solution of equation (4.39) as

$$\langle x_i^2(t)\rangle = V^2\int_0^t \mathrm{d}t_1 \int_0^t \mathrm{d}t_2 \left\langle b_i\left(\mathbf{x}(t_1), Vt_1\right) b_i\left(\mathbf{x}(t_2), Vt_2\right)\right\rangle, \quad i = x, y, \tag{4.40}$$

i.e. as the double integral of the two moving point Lagrangian correlation of the stochastic field. We show here that a set of coupled equations for $\langle x_i^2(t)\rangle$ and for the Lagrangian correlations with one moving point can be derived. The later correlation is defined as

$$L_{ij}(\mathbf{x}, t; \tau) \equiv \left\langle b_i\left(\mathbf{x}, Vt\right) b_j\left(\mathbf{x}(\tau), V\tau\right)\right\rangle \tag{4.41}$$

and represents the correlation of b_i in a fixed point $(\mathbf{x}, Vt)$ with b_j on the trajectory at time τ. It can be written as

$$\begin{aligned} L_{ij}(\mathbf{x}, t; \tau) &= \frac{1}{2\pi}\iint_{-\infty}^{\infty} \mathrm{d}\mathbf{x}' \left\langle b_i\left(\mathbf{x}, Vt\right) b_j\left(\mathbf{x}', V\tau\right) \delta\left(\mathbf{x}' - \mathbf{x}(\tau)\right)\right\rangle \\ &= \frac{1}{2\pi}\iint_{-\infty}^{\infty} \mathrm{d}\mathbf{x}' \iint_{-\infty}^{\infty} \mathrm{d}\mathbf{k} \exp(-\mathrm{i}\mathbf{k}\cdot\mathbf{x}') \left\langle b_i\left(\mathbf{x}, Vt\right) b_j\left(\mathbf{x}', V\tau\right) \exp(\mathrm{i}\mathbf{k}\cdot\mathbf{x}(\tau))\right\rangle . \end{aligned} \tag{4.42}$$

In the usual treatments of this problem, the average in (4.42) is factorized into two independent averages: one over $b_i b_j$ and the other over the exponential. The average over the field components which appear at fixed points is considered to be the Eulerian correlation

$$\mathcal{B}_{ij}(\mathbf{x} - \mathbf{x}', V(t - \tau)) \equiv \left\langle b_i\left(\mathbf{x}, Vt\right) b_j\left(\mathbf{x}', V\tau\right)\right\rangle . \tag{4.43}$$

The average over the exponential of the trajectory determines, after performing the integral over the Fourier conjugate variable $\mathbf{k}$, the probability density that the particle motion leads to a displacement $\mathbf{x}'$ in a time interval τ. This is a Corrsin approximation (see McComb (1990)) which is valid when the correlation time $\tau_c = \lambda_\parallel/V$ is much smaller than the transit time of the particles over a correlation length, $\tau_{tr} = \lambda_\perp/(\beta V)$, i.e. for $\alpha \equiv \tau_c/\tau_{tr} = \beta\lambda_\parallel/\lambda_\perp \ll 1$ (quasilinear limit). Extrapolated to large values of α, the Corrsin approximation leads to a diffusion coefficient proportional to α (Bohm limit), see Wang *et al.* (1995). This is not a correct result since when $\alpha \to \infty$, the particles are moving in a frozen field along closed lines representing constant lines of the potential Ψ. Consequently, the diffusion coefficient has to be zero as $\alpha \to \infty$.

The average in (4.42) can be calculated without factorization as follows. First, the average of $\exp\left[\mathrm{i}\mathbf{k}\cdot\mathbf{x}(\tau) + \mathrm{i}J_1 b_i(\mathbf{x}, Vt) + \mathrm{i}J_2 b_j(\mathbf{x}', V\tau)\right]$ is determined using the second cumulant approximation. Then, both sides of the resulting equation are differentiated with respect to J_1 and J_2, and finally the two parameters are eliminated, $J_1 = J_2 = 0$. The calculation sequence is straightforwardly continued by integrating over k_x, k_y. The condition of isotropy in the (x, y) plane is used, $\langle x^2(t)\rangle = \langle y^2(t)\rangle$, $\langle x(t)y(t)\rangle = 0$. The following set of coupled equations for the Lagrangian correlations is obtained:

$$\begin{aligned} L_{ij}(\mathbf{x}, t; \tau) = &\iint_{-\infty}^{\infty} \mathrm{d}\mathbf{x}' \, \mathcal{B}_{ij}(\mathbf{x} - \mathbf{x}', V(t - \tau))\, \gamma(\mathbf{x}', \tau) \\ &+ D_{im}(\mathbf{x}, t; \tau) \iint_{-\infty}^{\infty} \mathrm{d}\mathbf{x}' D_{jn}(\mathbf{x}', \tau; \tau) \frac{\partial^2 \gamma(\mathbf{x}', \tau)}{\partial x'_m \partial x'_n}, \end{aligned} \tag{4.44}$$

where $i, j, m, n = x, y$. The functions $D_{ij}(\mathbf{x}, t; \tau)$ are defined as:

$$D_{ij}(\mathbf{x}, t; \tau) \equiv \int_0^\tau \mathrm{d}\tau_1 L_{ij}(\mathbf{x}, t; \tau_1) = \left\langle b_i(\mathbf{x}, Vt) x_j(\tau)\right\rangle . \tag{4.45}$$

Here $\gamma(\mathbf{x}, \tau)$ is the probability density of reaching a distance $\mathbf{x}$ in a time interval τ:

$$\gamma(\mathbf{x}, \tau) = \frac{1}{2\pi \langle x^2(\tau)\rangle} \exp\left(-\frac{x^2 + y^2}{2\langle x^2(\tau)\rangle}\right). \tag{4.46}$$

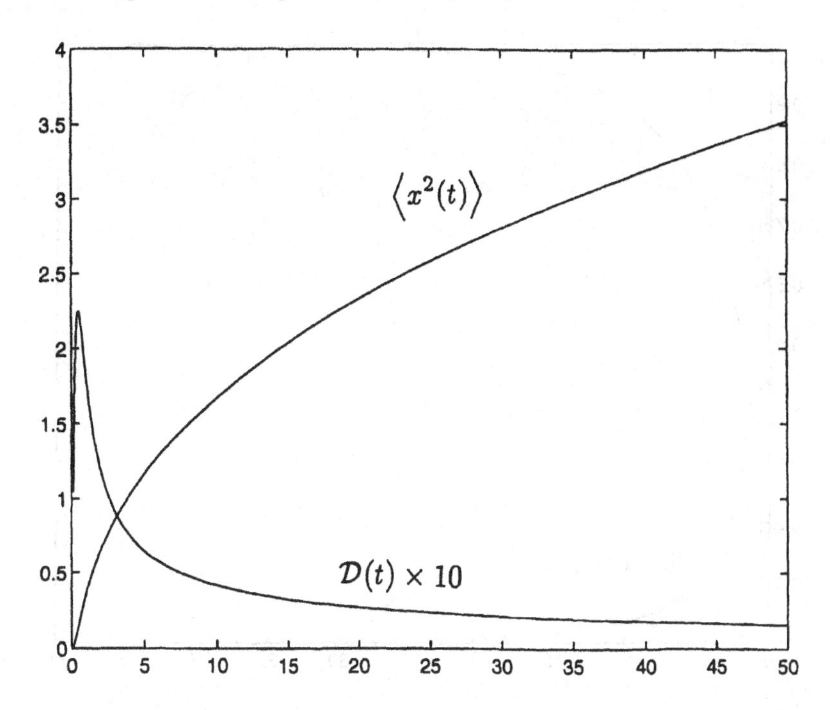

FIGURE 3. Mean square displacement and the running diffusion coefficient obtained from (4.47)–(4.49) and (4.51) in the highly nonlinear limit $\alpha \to \infty$.

Equations (4.44)–(4.46) determine the Lagrangian correlations of the stochastic field components. The first term on the r.h.s. of (4.44) is identical with the result of Corrsin factorization. It is determined by the Eulerian correlation of the field components and represents a source term. The second term is a self-effect of the Lagrangian correlations. This term is non-local and has a non-Markovian character. It is negligible at small times τ when D_{ij} are small, but can become important at large τ. In the quasilinear limit $\alpha \ll 1$ which corresponds to large $\lambda_\perp$, i.e. to a weak dependence of the Eulerian correlation on $\mathbf{x}$, the dependence on $\mathbf{x}$ of $L_{ij}(\mathbf{x},t;\tau)$ and $D_{ij}(\mathbf{x},t;\tau)$ is also weak and the integral in the second term of (4.44) can be approximated by

$$D_{ij}(\mathbf{0},\tau;\tau)\iint_{-\infty}^{\infty}\mathrm{d}\mathbf{x}'\frac{\partial^2\gamma(\mathbf{x}',\tau)}{\partial x'_m\partial x'_n}=0.$$

Thus, equation (4.44) reduces to the Corrsin result in the quasilinear limit.

The set of equations for the Lagrangian correlations can be significantly simplified using the symmetries of the Eulerian correlation function $\mathcal{B}_{ij}$. They can be derived (see Wang *et al.* (1995)) from the correlation of the potential $\Psi(\mathbf{x},z)$. Due to the isotropy condition, the later has to be a function of (x^2+y^2) and consequently $\mathcal{B}_{ii}$ is even in x and y while $\mathcal{B}_{xy}$ is odd. They are also symmetrical to the transformation $x \leftrightarrow y$ ($\mathcal{B}_{xx}(x,y,Vt)=\mathcal{B}_{yy}(y,x,Vt)$ and $\mathcal{B}_{xy}(x,y,Vt)=\mathcal{B}_{yx}(y,x,Vt)$). These symmetries are preserved by (4.44)–(4.46) which reduce to

$$L_{xx}(\mathbf{x},t;\tau)=T^{\mathrm{C}}_{xx}(\mathbf{x},t-\tau;\tau)+D_{xx}(\mathbf{x},t;\tau)\,T_2(\tau), \tag{4.47}$$

$$L_{xy}(\mathbf{x},t;\tau)=T^{\mathrm{C}}_{xy}(\mathbf{x},t-\tau;\tau)+D_{xy}(\mathbf{x},t;\tau)\,T_2(\tau), \tag{4.48}$$

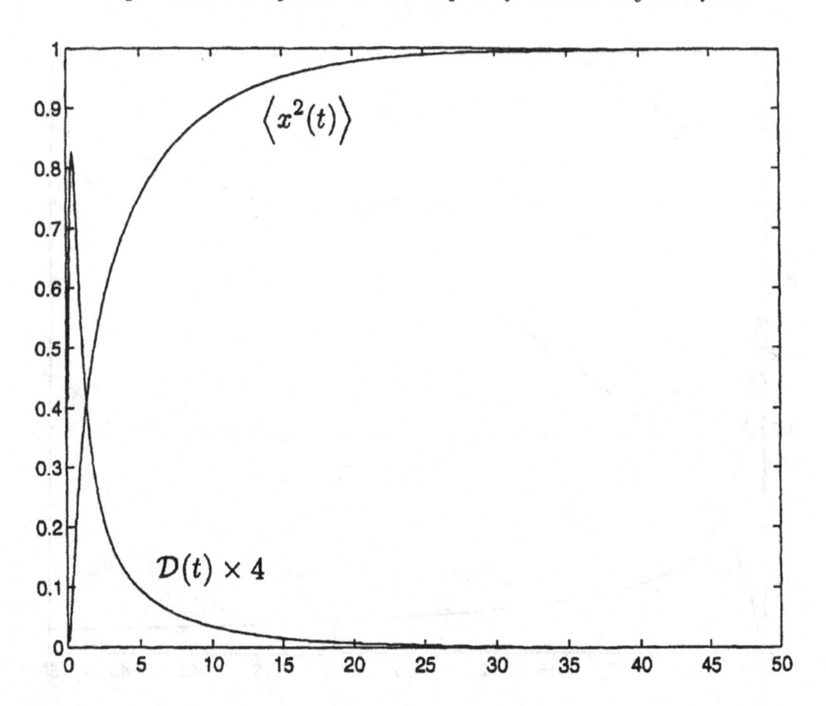

FIGURE 4. The same as in figure 3 but for the one-dimensional version of (4.39), with $b = b(x)$.

where $T_{ij}^{\mathrm{C}}(\mathbf{x}, t-\tau; \tau)$ is the Corrsin term (first term on the r.h.s. of (4.44)) and

$$T_2(\tau) = \iint_{-\infty}^{\infty} \mathrm{d}x'\mathrm{d}y' \left(D_{xx}(\mathbf{x}', \tau; \tau) \frac{\partial^2 \gamma(\mathbf{x}', \tau)}{\partial x'^2} + D_{xy}(\mathbf{x}', \tau; \tau) \frac{\partial^2 \gamma(\mathbf{x}', \tau)}{\partial x' \partial y'} \right). \quad (4.49)$$

The set of equations (4.47)–(4.49) is closed by giving the expression of the mean square displacement $\langle x^2(t)\rangle$ in terms of the Lagrangian correlations. We show that the usual expression $\langle x^2(t)\rangle = 2\int_0^t \mathrm{d}\tau\ D(0,0;\tau)$ is not valid in the nonlinear case and that $D(0,0;\tau)$ does not represent the running diffusion coefficient. The diffusion coefficient $\mathcal{D}(t)$ can be calculated as

$$\mathcal{D}(t) = \tfrac{1}{2}\frac{\mathrm{d}}{\mathrm{d}t}\langle x^2(t)\rangle = V\langle b_x(\mathbf{x}(t), Vt)\, x(t)\rangle. \quad (4.50)$$

The average in (4.50) can be determined with a method similar to that used in deriving (4.44). The following expression is finally obtained for the diffusion coefficient:

$$\mathcal{D}(t) = \langle x^2(t)\rangle\, T_2(t) + \iint_{-\infty}^{\infty} \mathrm{d}x'\mathrm{d}y' D(\mathbf{x}', \tau; \tau)\gamma(\mathbf{x}', \tau). \quad (4.51)$$

It can easily be shown that in the quasilinear limit $\alpha \to 0$ (4.51) reduces to the usual diffusion coefficient $D(0,0;\tau)$.

We have solved numerically (4.47)–(4.49) and (4.51) in the highly nonlinear limit where $\mathbf{b}$ is a function of $\mathbf{x}$ only. We have considered a Gaussian correlation function for the potential Ψ : $\langle \Psi(\mathbf{x})\Psi(\mathbf{x}')\rangle = \exp\left(-(\mathbf{x}-\mathbf{x}')^2/2\right)$ in these calculations. The results are presented in figure 3. One can see that the running diffusion coefficient $\mathcal{D}(t)$ has a maximum which is followed by a continuous decrease to zero. The mean square displacement $\langle x^2(t)\rangle$ grows slower than linearly. Thus, the nonlinear terms in (4.47), (4.48) have a strong influence on the solution, which is quite different from the result obtained with the Corrsin approximation. It appears that the new terms take into account particle trapping on the constant $\Psi(\mathbf{x})$ lines. This can be seen more clearly in the one-dimensional case, see figure 4. The running diffusion coefficient has a similar maximum but its decay is very fast in this case. Thus the mean square displacement saturates (at the square of the correlation length of the potential). These results are in agreement with the qualitative image of the problem.

This work was performed in collaboration with Madalina Vlad, partly at the Centre d'Etudes Cadarache, France and at the Université Libre de Bruxelles, Belgium. Important contributions of J. H. Misguich and of R. Balescu are gratefully acknowledged.

REFERENCES

BALESCU, R. 1988 *Transport Processes in Plasmas,* vol. 1 and 2. North Holland.

ISICHENKO, M. B. 1991 *Plasma Phys. Control. Fusion* **33**, 795.

KADOMTSEV, B. B. & POGUTSE, O. P. 1979 In *Proc. 7th Intl. Conf. on Plasma Phys. & Control. Nucl. Fusion Res.*, vol. 1, p. 649. International Atomic Energy Agency.

KROMMES, J. A., OBERMAN, C. & KLEVA, R. G. 1983 *J. Plasma Phys.* **30**, 11.

MARUYAMA, K. & SHIBATA, F. 1988 *Physica* A **149**, 447.

MCCOMB, W. D. 1990 *The Physics of Fluid Turbulence.* Clarendon Press.

MISGUICH, J. H., VLAD, M., SPINEANU, F. & BALESCU, R. 1995 *Comments on Plasma Phys. Control. Fusion* **17**, 45.

ROCHESTER, A. B. & ROSENBLUTH, M. N. 1978 *Phys. Rev. Lett.* **40**, 38.

RUBIN, R. J. & WEISS, G. H. 1982 *J. Math. Phys.* **23**, 250.

SPINEANU, F. & VLAD, M. 1994 *J. Plasma Phys.* **51**, 113.

SPINEANU, F. & VLAD, M. 1995 *J. Plasma Phys.* **54**, 333.

VLAD, M., SPINEANU, F., MISGUICH, J. H. & BALESCU, R. 1996 *Phys. Rev.* E **53**, 5302.

WANG, H.-D., VLAD, M., VANDEN EIJNDEN, E., SPINEANU, F., MISGUICH, J. H. & BALESCU, R. 1995 *Phys. Rev.* E **51**, 4844.

See also

BALESCU, R., WANG, H.-D. & MISGUICH, J. H. 1994 *Physics of Plasmas* **1**, 3826.

Non-equilibrium statistical mechanics and ergodic theory

By LEONID A. BUNIMOVICH

Center of Dynamical Systems and Nonlinear Studies, Georgia Institute of Technology, Atlanta, USA
and
Institute of Oceanology, Russian Academy of Sciences, Moscow, Russia

The goal of non-equilibrium statistical mechanics is to explain how the erratic (chaotic) motion of particles on the small (microscopic) scale appears to be regular and organized on a large (macroscopic) scale. The general belief is that macroscopic dynamics emerges in a rigorous mathematical way in the idealized limit of infinite scale separation. We give a brief review of rigorous mathematical results that justify this idea.

---

1. Introduction

Ergodic theory was founded by Boltzmann and Gibbs to elucidate how, from random motion of particles on the small scale, a deterministic motion of a fluid emerges on the large scale. The important issue is to understand what causes a randomness of the motion of particles. According to the ideas of the founders of statistical mechanics, a state of a system of particles is random because any system in nature contains an enormous number of particles. Therefore, it does not make sense to speak about an exact state of such systems, but rather one should consider a probabilistic ensemble of such states. Thus the randomness emerges because of a large number of degrees of freedom (particles) in a system.

The other point of view has not that long a history. It was created by the development of the modern theory of hyperbolic dynamical systems. It has been demonstrated that a system of only two elastically colliding particles demonstrates stochastic behavior, see Sinai (1970). The reason for that is an extremely strong internal instability of motion of this system that causes an exponential divergence in time of trajectories initially close to each other. Dynamical systems with such exponential internal local instability are called hyperbolic.

The remarkable fact is that the knowledge of local behavior in hyperbolic systems often allows us to give a global description of their dynamics. Such a global description gives the possibility in some cases of deriving the properties of macrodynamics from the properties of microdynamics. An extremely important global property is mixing (decay of time correlations) that was introduced by Gibbs. It is the strong dynamical mixing that provides the mechanism that pushes a system to equilibrium.

A system of erratically moving particles can afford to maintain a state close to a local equilibrium because of the local conservation laws. There are three fundamental conservation laws of mass, momentum and energy that constitute constrains not immediately visible on the microscopic scale. Thus, the fundamental problem is to prove for a deterministic mechanical system (system of particles) the existence of (nondegenerate) transport coefficients. These coefficients (diffusion, viscosity, thermal conductivity) appear in the macroscopic (hydrodynamical) equations that govern the evolution of the corresponding systems. Transport coefficients can be expressed through the characteristics of microdynamics via Green–Kubo formulas (see Balescu (1975), Spohn (1991)). (The corresponding relations are also called fluctuation–dissipation theorems.)

We discuss here rigorous results on the existence of diffusion and viscosity coefficients for the simplest models of interacting particles where these coefficients are nondegenerate.

2. Transport coefficients for periodic fluids

We consider an infinitely extended one component fluid in thermal equilibrium. The average with respect to the equilibrium distribution is denoted by $\langle\cdot\rangle$. In three dimensions, the fluid has five locally conserved fields: the particle density $n^{(0)}(x,t)$, the three components of the momentum density $n^{(\alpha)}(x,t)$, $\alpha = 1,2,3$, and the energy density $n^{(4)}(x,t)$, which depend on location $x \in \mathbb{R}^3$ and time $t \in \mathbb{R}$. They are distributions on phase space indexed by x,t.

By the local conservation law we have, in a distributional sense,

$$\frac{\partial}{\partial t} n^{(i)}(x,t) + \operatorname{div} j^{(i)}(x,t) = 0, \tag{2.1}$$

$i = 0,\ldots,4$, with the local currents $j^{(i)}$. (Since the interaction between particles has some range, the local currents are not uniquely defined. However, the space averaged currents are, see Bunimovich & Spohn (1996).)

The Green–Kubo formula for the transport coefficients reads as

$$\Gamma^{(ik)}_{\alpha\beta} = \frac{1}{2k_{\mathrm{B}}T}\int \mathrm{d}t \int \mathrm{d}^3x \left(\langle j^{(i)}_\alpha(x,t) j^{(k)}_\beta(0,0)\rangle - \langle j^{(i)}_\alpha(0,0)\rangle\langle j^{(k)}_\beta(0,0)\rangle\right), \tag{2.2}$$

$\alpha,\beta = 1,2,3$, $i,k = 0,\ldots,4$, with k_{B} the Boltzmann constant and T the temperature of the fluid, where we used the stationarity of the equilibrium measure in space-time. By rotation invariance and time reversal symmetry, in fact only three out of the 15×15 coefficients in (2.2) survive. They can be expressed by a linear combination of shear and bulk viscosity and thermal conductivity, see Spohn (1991).

It has been noted within molecular dynamics, see Erpenbeck & Wood (1977) and Hoover (1986), that, at least in principle, transport coefficients are also well defined for systems with a finite number of particles. The main observation is that, for a periodically repeated fluid the total (i.e. space averaged) currents are meaningful and transport coefficients can still be defined via the finite volume version of (2.2). Then (2.2) equals the covariance matrix of the time-integrated local currents, normalized by $\sqrt{t}$. Thus the mathematical issue is to prove a central limit theorem (CLT) for the time integrated currents. To derive the noninvertible diffusion (macro-) equation from the deterministic invertible microdynamics, one has to prove a statement stronger than CLT. It is called an invariance principle (see e.g. Billingsley (1969)).

It is natural to try first to prove the existence of nonvanishing transport coefficients for the simplest models where they are nontrivial.

The simplest (yet nontrivial) model to study a diffusion should contain at least one moving particle. It is the classical Lorentz gas, see e.g. Balescu (1975), Spohn (1991) and Erpenbeck & Wood (1977). The existence of diffusion for a periodic Lorentz gas was proved in Bunimovich & Sinai (1981) for two dimensions and in Chernov (1994) for any number of dimensions.

A model with nontrivial viscosity should have at least two particles, see Hoover (1986) and Bunimovich & Spohn (1996). The existence of nonvanishing shear and bulk viscosities was proved in Bunimovich & Spohn (1996) and Bunimovich (in preparation) for the periodic two-disk fluid. We will briefly discuss these results in subsequent sections.

The Lorentz gas and the periodic two-sphere fluid could be reduced to dispersing billiards, see Sinai (1970). The afore-mentioned results were obtained by making use of the theory of such billiards, see Bunimovich *et al.* (1990) and Bunimovich *et al.* (1991).

We divide the physical space $\mathbb{R}^d$ into cells (hypercubes) of linear dimension l and consider a fluid periodically repeated over all of $\mathbb{R}^d$ with N particles per unit cell. They interact via a central force, $-\nabla V$. Equivalently, we consider N particles, position q_i, momentum p_i, mass m, on the d-dimensional torus $\Lambda = [0, l]^d$. They interact via the pair force $F(q) = -\sum_{n \in \mathbb{Z}^d} \nabla V(q - nl)$, which is constructed as a sum over periodic images. In our case (a short-range potential) the sum consists of only a single non-zero term.

The total energy current for this system turns out to be equal to, see Bunimovich and Spohn (1996),

$$j_E(t) = \sum_{j=1}^{N} \frac{1}{2m} p_i(t)^2 \frac{1}{m} p_i(t) + \tfrac{1}{2} \sum_{i,k=1\,(i \neq k)}^{N} \big(q_i(t) - q_k(t)\big) \frac{1}{m} p_i(t) F\big(q_i(t) - q_k(t)\big).$$

If $N = 2$ then $p_1 + p_2 = 0$. Therefore

$$m j_E = \frac{1}{2m}(p_1^2 p_1 + p_2^2 p_2) + \tfrac{1}{2}(q_1 - q_2)(p_1 + p_2) F(q_1 - q_2) = 0. \tag{2.3}$$

Thus the thermal conductivity vanishes for this system, and to have a nontrivial thermal conductivity one would have to consider a system of $N \geq 3$ particles. Such systems can also be reduced to billiards. However, the corresponding billiards are not dispersing. They belong to the class of semi-dispersing billiards (see, e.g. Bunimovich *et al.* (1990) and Bunimovich *et al.* (1991)). Such billiards do not have the strong chaotic properties of dispersing billiards. Therefore, it is necessary to further develop the theory of nonuniformly hyperbolic billiards, see Bunimovich (1985), in order to prove the existence of the thermal conductivity for a periodic three disk fluid. It has already been explained (see also Hoover (1986) and Bunimovich & Spohn (1996)) that this is the simplest system that can have a nontrivial thermal conductivity.

3. Diffusion in a periodic Lorentz gas

Consider the dynamical system which corresponds to the motion of a single particle between fixed scatterers in $\mathbb{R}^d$, $d \geq 2$. Outside all scatterers, the particle moves with constant velocity and at reflections it changes its velocity according to the usual law of elastic reflections. This model was introduced by Lorentz in 1905 to describe the dynamics of electrons in metals.

For the sake of simplicity we will only discuss the case $d = 2$. We assume that scatterers are disks of an arbitrary diameter and the configuration of scatterers is periodic.

The phase space $\mathcal{M}$ of our dynamical system consists of points $x = (q, v)$ where $q = (q^1, q^2)$ are coordinates, $v = (v^1, v^2)$ is the velocity of the particle. Without any loss of generality we can consider $|v| = \sqrt{(v^1)^2 + (v^2)^2} = 1$. The dynamics will be denoted by $\{S^t\}$.

We shall denote by Π a fundamental domain of the periodic configuration of scatterers. (It can always be chosen as a semi-open set the closure of which is a rectangular.) There are two essentially different classes of periodic Lorentz gases. They differ by the possibility for a moving particle to have or not to have an arbitrarily long free path. A periodic Lorentz gas has a finite horizon if there exists a constant L such that the length of any straight segment which avoids all scatterers cannot exceed L. Otherwise a periodic Lorentz gas has an infinite horizon. The simplest configuration with finite (infinite) horizon corresponds to a triangular (square) lattice of scatterers (see figure 1).

Let μ be a probability measure concentrated on the set $\mathcal{M} \cap (\Pi \times S^1)$ which is absolutely continuous with respect to the Lebesgue measure on $\mathcal{M}$ and its density $p(x) \in C^1$. Thus x

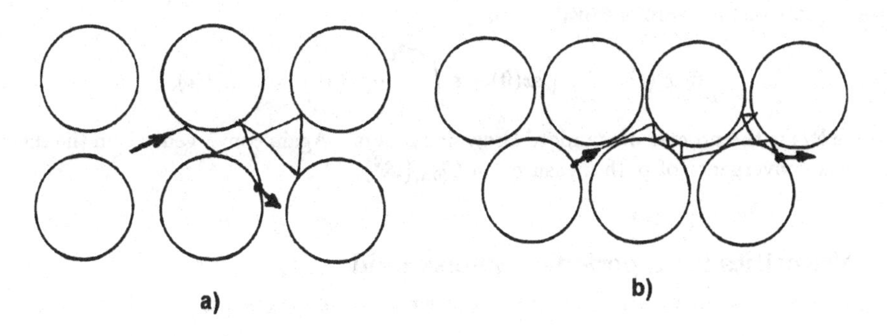

FIGURE 1. Periodic Lorentz gas with (a) an infinite and (b) a finite horizon.

is a random variable distributed according to the measure μ. If $S^t x = x(t) = \big(q(t), v(t)\big)$ then $q(t), v(t)$ are also random variables.

THEOREM 3.1 (BUNIMOVICH & SINAI (1981)). *There exists a non-degenerate two-dimensional Gaussian probability distribution with density g such that*

$$\lim_{t\to\infty} \mu\Big\{x : \frac{q(t)}{\sqrt{t}} \in B\Big\} = \int_B g(q^1, q^2)\, \mathrm{d}q^1 \mathrm{d}q^2. \tag{3.4}$$

Here B is a bounded open subset of the plane, the boundary of which has area equal to zero. Theorem 3.1 proves the existence of a nondegenerate diffusion for a periodic Lorentz gas with a bounded free path.

The next theorem is a stronger version of theorem 3.1. For every t we put $q_t(s) = (1/\sqrt{t})\ q(st)$, $0 \le s \le 1$. Thus we consider the diffusive scaling of orbits of the moving particle.

The measure μ induces the probability distribution on the set of all possible orbits $q_t(s)$, which are points of the space $C_{[0,1]}\big(\mathbb{R}^2\big)$ of continuous functions defined on the segment $[0,1]$ with values in $\mathbb{R}^2$. We shall denote this measure by μ_t.

THEOREM 3.2 (BUNIMOVICH & SINAI (1981)). *The measures μ_t converge weakly to a Wiener measure.*

The Wiener measure is the measure on the space of trajectories of the random process of Brownian motion. This process was axiomatically introduced by Wiener in 1921 to describe a process of a random walk with continuous time. Hence theorem 3.2 shows that the motion of a particle in the periodic Lorentz gas with a bounded free path is similar on large space and time scales to Brownian motion. It was the first mechanical model (and still is the only non one-dimensional, i.e. nondegenerate model) for which it has been proven.

Theorem 3.2 also provides the first rigorous derivation of the noninvertible diffusion (macro-) equation from invertible microdynamics. Indeed, it is well known that the transition probability function for a Wiener process coincides with the fundamental solution of the diffusion equation.

Theorem 3.2 also means that a noninvertible macroscopic dynamics can be rigorously derived from an invertible microscopic dynamics in a limit of infinite scale separation. Indeed theorems 3.1 and 3.2 ensure that the diffusion matrix $D_{ij} = \frac{1}{2}\int \mathrm{d}t\, \langle v_i(t) v_j(0)\rangle$

has at least one nonzero eigenvalue and

$$\lim_{\epsilon\to 0} x^\epsilon(t) = \lim_{\epsilon\to 0}\left(\epsilon x(0) + \epsilon\int_0^{\epsilon^{-2}t} \mathrm{d}s\, v(s)\right) = \sqrt{2D}\, b(t),$$

where $b(t)$ is the process of standard Brownian motion. Again convergence is in the sense of weak convergence of path measures on $C_{[0,t]}(\mathbb{R}^2)$.

4. Viscosities for a periodic two-disk fluid

Since the total momentum is conserved, we canonically transform as

$$q = q_1 - q_2, \quad p = p_1 - p_2, \quad q_c = \tfrac{1}{2}(q_1 + q_2), \quad p_c = \tfrac{1}{2}(p_1 + p_2). \tag{4.5}$$

Then the relative motion is governed by

$$m\frac{\mathrm{d}}{\mathrm{d}t} q(t) = p(t), \quad \frac{\mathrm{d}}{\mathrm{d}t} p(t) = 2F(q(t)), \tag{4.6}$$

where $q \in \left[-\frac{l}{2}, \frac{l}{2}\right]^d$ with periodic boundary conditions. Imposing the center of mass momentum $p_c = 0$, we obtain for the stress tensor (see Bunimovich & Spohn (1996) for details)

$$\tau_{\alpha\beta}(t) = \frac{1}{2m} p_\alpha(t) p_\beta(t) + q_\alpha(t) F_\beta(q(t)), \tag{4.7}$$

$\alpha, \beta = 1, \ldots, d$.

For the central limit theorem we need the dynamics (4.6) to have good mixing properties, which we ensure by taking V as a hard core potential. To simplify even further we choose $d = 2$. We then have a periodically repeated fluid in the plane consisting of two hard disks of diameter R per unit cell. They interact through perfectly elastic collisions. Certainly, a periodic two-disk fluid is the simplest model to study viscosity. Indeed, to have a sheared flow one needs at least two particles that move with different velocities.

Two disks on the two-torus are dynamically equivalent. We set $p = mv$ and the mass $m = 1$. To have relative speed $|v| = 1$, we set $E = \frac{1}{2m}(p_1^2 + p_2^2) = \frac{1}{4}$. The relative velocity is then specified by an angle θ, with $0 \le \theta < 2\pi$. We take a unit two-torus, i.e. $l = 1$. Then the only parameter left is the hard core radius R, which can be used to label the reduced density, i.e. density/(closed packing density), according to $\rho^* = \sqrt{3}R^2$.

We distinguish two dynamically very different cases (see figure 2). If $0 < R < 1/2$, then the fluid particles can enter the next cell and may even move an arbitrarily long distance without collision. It is easy to see from (4.5) that the corresponding one-particle billiard (4.6) has an infinite horizon. The central limit theorem fails for the periodic Lorentz gas with infinite horizon because of the 'strong' memory on the long free paths, see Bunimovich (1985) and Blecher (1992). However, for sufficiently latge diameter, the integrand $v_\alpha(t)v_\beta(t)$ suppresses phase volume (see details in Bunimovich & Spohn (1996)) and viscosities might still exist. This problem remains open.

If $1/2 < R < 1/\sqrt{2}$, the disks are confined and cannot pass each other. The one-particle billiard has a 'finite horizon' and its domain has a 'diamond' shape (see figure 3(b)). (Note that, because of the imposed quadratic symmetry, close packing cannot be reached.)

We have $\tau_{\alpha\beta} = \tau_{\beta\alpha}$. Denote the time-integrated stress tensor by $\tau_{\alpha\beta}([0,t])$. The central limit theorem has to be proved for

$$\frac{1}{\sqrt{t}}\tilde{\tau}_{\alpha\beta}([0,t]) = \frac{1}{\sqrt{t}}\left[\sum_{n=1}^{\infty} \chi(t_n \le t)\left(\tfrac{1}{2} - \theta\right) q_\alpha(t_n) \tfrac{1}{2}\left(v'(t_n) - v(t_n)\right)_\beta - \delta_{\alpha\beta} t p_D\right], \tag{4.8}$$

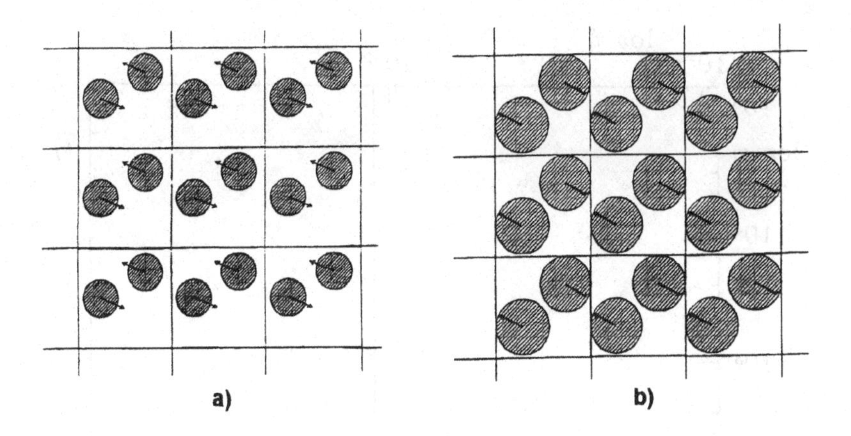

FIGURE 2. Two-disk fluid.

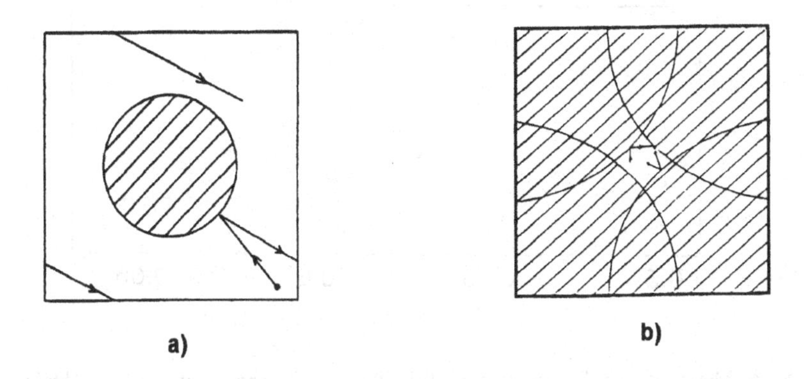

FIGURE 3. Corresponding billiard problems: (a) infinite horizon, (b) finite horizon.

where θ is the standard step function,

$$p_D = \left(1 - \sqrt{4R^2 - 1}\right)\left(4\left(1 - \sqrt{4R^2 - 1} - \pi R^2\left(1 - \frac{4}{\pi}\arccos\frac{1}{2R}\right)\right)\right)^{-1},$$

t_n is the time of nth collision, $v'(t_n)$ is post-collisional and $v(t_n) = v'(t_{n-1})$ is pre-collisional velocity, $\chi(\cdot)$ is a characteristic function (for details see Bunimovich & Spohn (1996)).

To simplify notation we set $\tau = (\tau_1, \tau_2, \tau_3)$ with

$$\tau_1([0,t]) = \tfrac{1}{2}(\tilde{\tau}_{11}([0,t]) + \tilde{\tau}_{22}([0,t])) - p_D t,$$

$$\tau_2([0,t]) = \tfrac{1}{2}(\tilde{\tau}_{11}([0,t]) + \tilde{\tau}_{22}([0,t])),$$

$$\tau_3([0,t]) = \tilde{\tau}_{12}([0,t])).$$

The covariance matrix for τ has the form

$$D = \begin{bmatrix} \sigma_1 & 0 & 0 \\ 0 & \sigma_2 & 0 \\ 0 & 0 & \sigma_3 \end{bmatrix}.$$

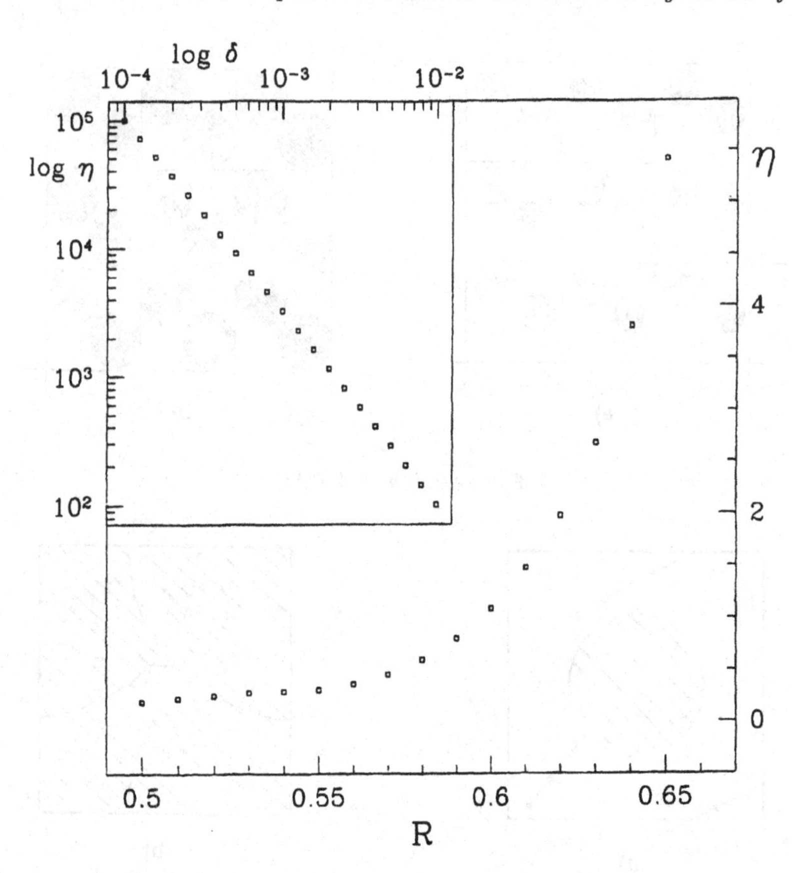

FIGURE 4. Shear viscosity η against hard disk diameter R. The insert is a logarithmic plot.

Then, up to dimensional factors, $\sigma_1 = \xi$, the bulk viscosity, $\sigma_3 = \eta$, the shear viscosity, and $\sigma_2 = \tilde{\eta}$ which becomes equal to η for a rotationally invariant fluid. To establish $D > 0$ as a matrix one only has to prove that for $i = 1, 2, 3$

$$\lim_{t\to\infty} \frac{1}{t} \langle \tau_i([0,t])^2 \rangle = \sigma_i \tag{4.9}$$

exists and is strictly positive, where the average is with respect to $\mu = C\,\mathrm{d}q_1\,\mathrm{d}q_2\,\mathrm{d}v$. Here C is a normalization factor.

THEOREM 4.1 (BUNIMOVICH & SPOHN (1996), BUNIMOVICH (IN PREPARATION)). *For $i = 1, 2, 3$ there exists a variance σ_i, $0 < \sigma_i < \infty$ such that for any real number z,*

$$\lim_{t\to\infty} \mu \left\{ \frac{1}{\sqrt{t}} \tau_i([0,t]) < z \right\} = \frac{1}{\sqrt{2\pi\sigma_i}} \int_{-\infty}^{z} \mathrm{d}u\, \mathrm{e}^{-u^2/(2\sigma_i)}.$$

In particular, $0 < \eta, \xi < \infty$. For every bounded open set $A \subset \mathbb{R}^3$ with $\mu(\partial A) = 0$ we have

$$\lim_{t\to\infty} \mu \left(\left\{ \frac{1}{\sqrt{t}} \tau([0,t]) \in A \right\} \right) = \int_A \prod_{i=1}^{3} \mathrm{d}u_i \frac{1}{\sqrt{2\pi\sigma_i}} \mathrm{e}^{-u_i^2/(2\sigma_i)}.$$

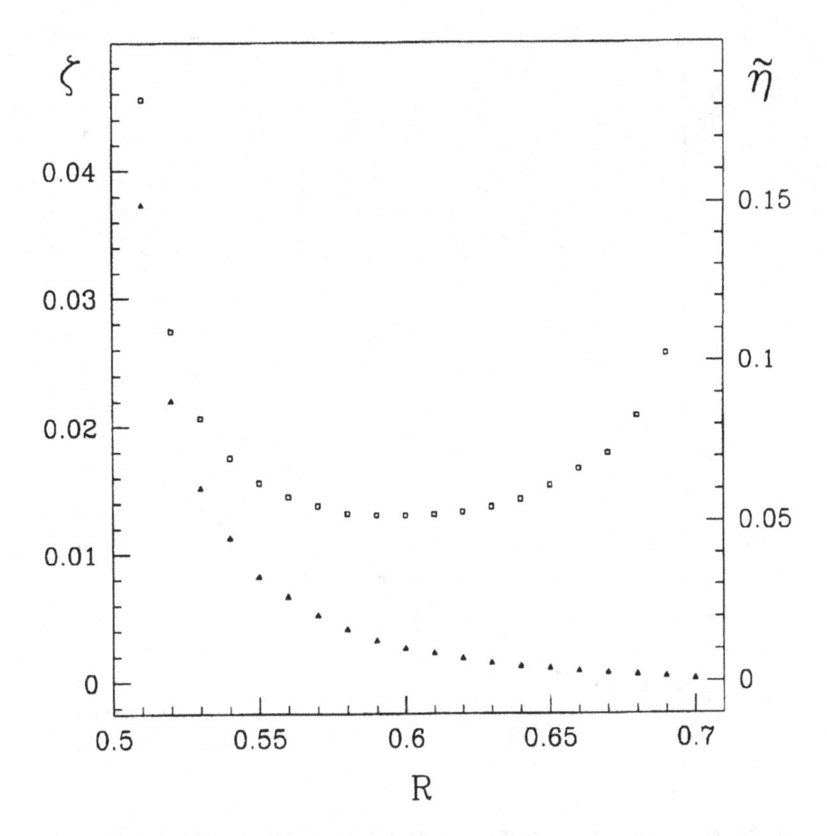

FIGURE 5. Bulk viscosity ξ (□) and viscosity $\tilde{\eta}$ (△) against hard disk radius R.

Theorem 4.1 establishes the existence of a strictly positive viscosity for a periodic two-disk fluid. This result was proven in Bunimovich & Spohn (1996) for a generic value of the parameter R (radius of disks). However, it certainly should hold for all R. This has been proven by Bunimovich (in preparation). The most delicate part of the proof is the analysis of singularities in the corresponding billiard problem.

Singularities work against hyperbolicity (local exponential instability). A trajectory 'remembers' its visits to a small neighborhood of singularities.

The important fact is that there are singularities of only one type in the periodic Lorentz gas, while there are two different types of singularities in the periodic two-disk fluid (in the case of a finite horizon). These singularities correspond to trajectories tangent to the boundary (the same occurs in the Lorentz gas) and to trajectories that hit the singularities of the boundary ∂Q (see figure 3(b)). It can be shown, see Bunimovich *et al.* (1991), that in the periodic Lorentz gas, the number of singular manifolds that intersect at a point cannot grow faster than linearly in time. On the other hand, the different types of singularities can in a sense 'help' each other. Therefore the estimate obtained in Bunimovich (in preparation) is much weaker, but is sufficient to prove the CLT.

The proof of theorem 4.1 gives very little information on the actual value of the viscosity, if any. For this, one has to determine numerically the averages $\langle \tau_j([0,t])^2 \rangle$, $j = 1, 2, 3$. The corresponding calculations were performed by Martin Fließer, see Bunimovich &

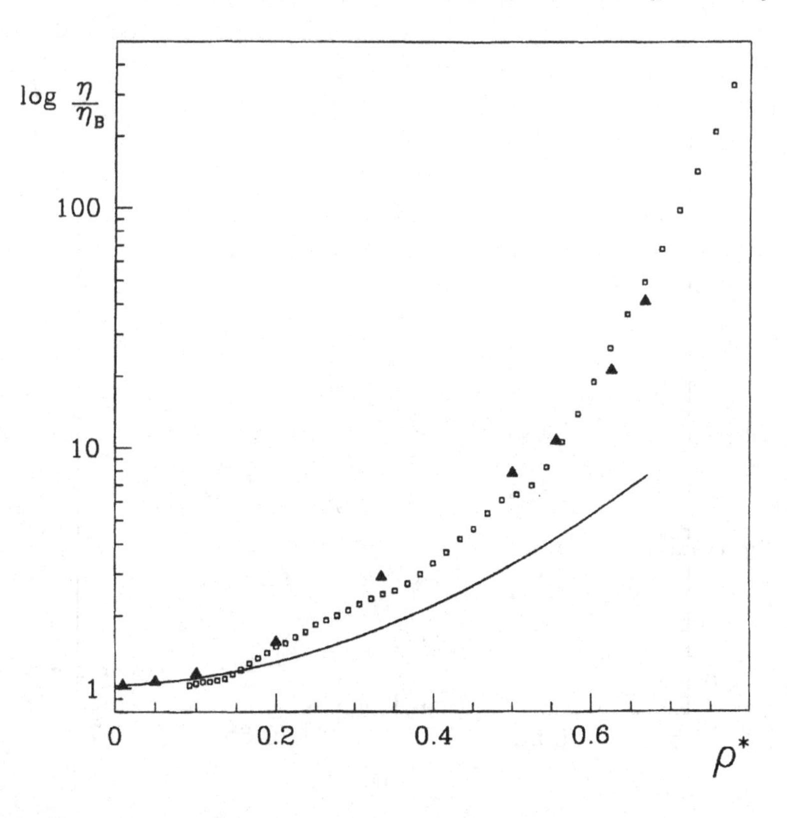

FIGURE 6. Shear viscosity η (▫) against density ρ^*; solid line represents the Enskog approximation and ▵ are the data from Alder *et al.* (1970). The vertical scale is in units of ($d = 2$) Boltzmann shear viscosity η_B.

Spohn (1996). By Monte Carlo 10^5 initial phase points uniformly distributed on the phase space have been generated. For given initial condition, we follow the dynamical trajectory. Since the Lyapunov exponent is positive, numerical accuracy is lost after 9–30 collisions depending on the value of R. The viscosity is determined from a linear least square fit over the last 5–10 collision times. Since the fit turns out to be essentially perfect, we expect to already be in an asymptotic regime (see figures 4 and 5).

The shear viscosity increases with R, the denser the fluid the stronger the internal friction, and diverges as $\left(1/\sqrt{2} - R\right)^{-3/2}$ for $R \to 1/\sqrt{2}$, as can be seen from the insert in figure 4.

It is interesting to compare the $N = 2$ results with those for larger systems, see Erpenbeck & Wood (1977) and Alder *et al.* (1970). To do so we have to write the viscosity in dimensionless form. Such numerical calculations were performed by M. Fließer (see Bunimovich & Spohn (1996)). In figures 7 and 8 we compare our data ($N = 2$, $d = 2$) with molecular dynamics for $d = 3$, $N = 108$, 500, 4000, given in Erpenbeck & Wood (1977) and Alder *et al.* (1970). (No numerical results for $d = 2$ seem to be available.)

As usual, we define the reduced density for $\rho^* = \rho/\rho_{cr}$ where ρ_{cr} is the density of close packing. The variables η/η_B and ξ/η_B are dimensionless and depend only on ρ^* and N.

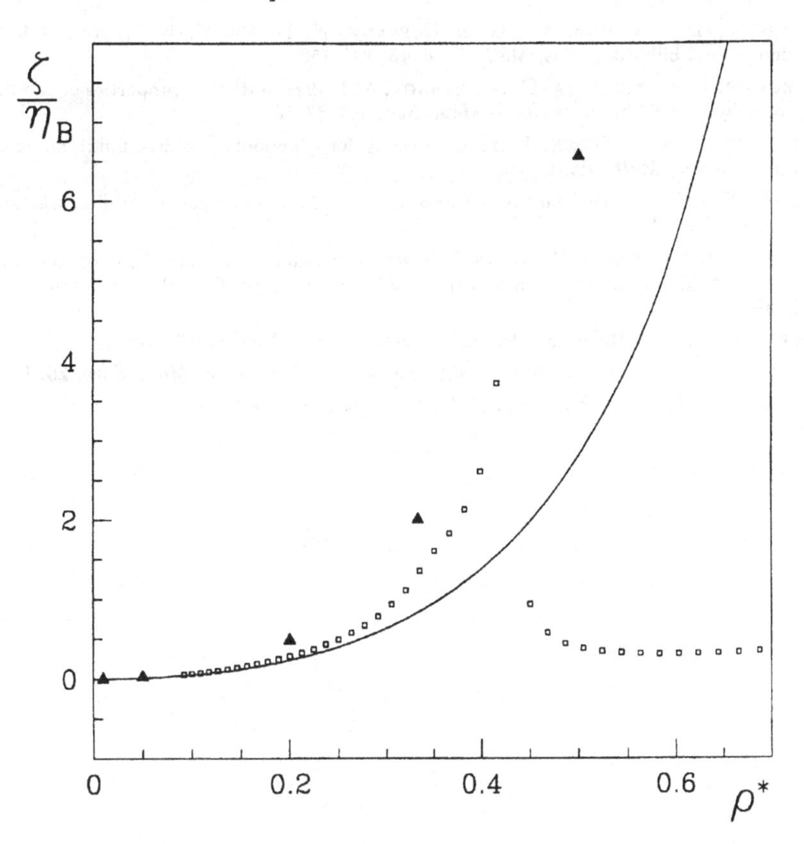

FIGURE 7. Bulk viscosity ξ (□) against density ρ^*; solid line represents the Enskog approximation and △ are the data from Alder *et al.* (1970). The vertical scale is in units of ($d = 2$) Boltzmann shear viscosity η_B.

We emphasize that figures 7 and 8 contain no fit parameters. Thus, it seems that low dimensional hyperbolic dynamics mimics the dynamics of multiparticle systems.

This work was partially supported by U.S. NSF grant DMS93-03769.

REFERENCES

ALDER, B. J., GASS, D. M. & WAINWRIGHT, T. E. 1970 Studies in molecular dynamics. VIII. The transport coefficients for a hard-sphere fluid. *J. Chem. Phys.* **53**, 3813–3826.

BALESCU, R. 1975 *Equilibrium and Nonequilibrium Statistical Mechanics.* Wiley.

BILLINGSLEY, P. 1969 *Convergence of Probability Measures.* Wiley.

BLECHER, P. M. 1992 Statistical properties of two-dimensional periodic Lorentz gas with infinite horizon. *J. Stat. Phys.* **66**, 315–373.

BUNIMOVICH, L. A. 1985 Decay of correlations in dynamical systems with chaotic behavior. *J. Exp. Theor. Phys.* **89**, 1452–1471.

BUNIMOVICH, L. A. Singularities in billiards and transport coefficients. In preparation.

BUNIMOVICH, L. A. & SINAI, YA. G. 1981 Statistical properties of the Lorentz gas with periodic configuration of scatterers. *Comm. Math. Phys.* **78**, 479–497.

BUNIMOVICH, L. A., SINAI, YA. G. & CHERNOV, N. I. 1990 Markov partitions for two-dimensional billiards. *Russ. Math. Surv.* **45**, 105–152.

BUNIMOVICH, L. A., SINAI, YA. G. & CHERNOV, N. I. 1991 Statistical properties of two-dimensional hyperbolic billiards. *Russ. Math. Surv.* **46**, 47–106.

BUNIMOVICH, L. A. & SPOHN, H. 1996 Viscosity for a periodic two-disk fluid: an existence proof. *Comm. Math. Phys.*

CHERNOV, N. I. 1994 Statistical properties of the periodic Lorentz gas. Multidimensional case. *J. Stat. Phys.* **74**, 11–53.

ERPENBECK, J. J. & WOOD, W. W. 1977 Molecular dynamics techniques for hard core systems. In *Statistical Mechanics. Modern Theoretical Chemistry* (ed. B. J. Berne), vol. 6, pp. 1–40. Plenum.

HOOVER, W. G. 1986 *Molecular Dynamics.* Lect. Notes in Physics, vol. 258. Springer.

SINAI, YA. G. 1970 Dynamical systems with elastic reflections. *Russ. Math. Surv.* **25**, 137–189.

SPOHN, H. 1991 *Large Scale Dynamics of Interacting Particles.* Springer.

Pseudochaos in statistical physics

By BORIS CHIRIKOV

Budker Institute of Nuclear Physics, Novosibirsk, Russia

A new generic dynamical phenomenon of *pseudochaos* and its relevance to statistical physics, both modern and traditional, are discussed in some detail. Pseudochaos is defined as a statistical behaviour of the dynamical system with *discrete* energy and/or frequency spectrum. The statistical behaviour, in turn, is understood as a time-reversible but nonrecurrent relaxation (at average) to some steady state, superimposed with irregular fluctuations. Our main attention is payed to the most important and universal example of pseudochaos, so-called *quantum chaos*, that is dynamical chaos in bounded mesoscopic quantum systems. Quantum chaos as a mechanism for implementation of the fundamental correspondence principle is also discussed.

The quantum relaxation localization, a peculiar characteristic implication of pseudochaos, is reviewed in both time-dependent and conservative systems, with special emphasis on the *dynamical decoherence* of quantum chaotic states. Recent results on the peculiar global structure of the energy shell, Green function spectra and eigenfunctions, both localized and ergodic, in a generic conservative quantum system are presented.

Examples of pseudochaos in classical systems are given, including linear oscillators and waves, digital computers and completely integrable systems. A far-reaching similarity between the dynamics of a quantum system with few degrees of freedom at high energy levels ($n \to \infty$), and that of many degrees of freedom ($N \to \infty$) is also discussed.

---

1. Introduction: the second birth of pseudochaos

The concept of *pseudochaos* was first introduced explicitly by Chirikov (1991b) in an attempt to interpret the very controversial phenomenon of *quantum chaos*, and to understand its mechanism and physical meaning. The term itself was borrowed from the theory of *pseudorandom number generators* in digital computers. Even though such imitations of 'true' random quantities are widely used in many *numerical experiments*, e.g. those employing Monte Carlo techniques, this pseudorandomness was always considered a very specific mathematical model, of no general interest for fundamental physics. However, in recent numerous attempts to understand quantum chaos, which is attracting the ever growing attention of researchers (see e.g. Casati (1985), Giannoni *et al.* (1991), Heiss (1992) and a collection of papers in Casati & Chirikov (1995a)), it is becoming more and more clear that this *specific mechanism* provides, in fact, a *typical* chaotic behaviour in physical systems.

Moreover, from the viewpoint of fundamental physics, pseudochaos is in principle the only kind of chaos possible in physical systems of finite dimensions. In infinite macroscopic systems of traditional statistical mechanics (TSM), both *classical and quantum*, particularly in the principal TSM concept of the thermodynamic limit, $N \to \infty$, where N is the number of degrees of freedom, this situation is not the case. Namely, it has been rigorously proved (see e.g. Kornfeld *et al.* (1982)) that, roughly speaking, 'true' chaos is a generic phenomenon in this limit even if for any finite N the system is completely integrable!

The discovery of dynamical chaos in finite (and even low-dimensional) *classical* systems, a fundamental breakthrough in recent decades, has crucially changed classical statistical mechanics. By now, this new mechanism for the statistical laws is well understood (but still not very well known), and has acquired firm mathematical foundations in modern ergodic theory, see Kornfeld *et al.* (1982).

In spite of the success of this new mechanism a 'minor' problem still remains: such a mechanism does not work in finite *quantum* systems, whose motion is bounded in *phase space* and, hence, whose energy and frequency spectra are discrete.

The simplest solution to this problem, which nowadays seems to be almost commonly accepted, is that dynamical chaos in such systems is simply impossible. However, this seemingly obvious 'solution' is in fact a trap, as it immediately leads to a sharp and very profound contradiction with the fundamental correspondence principle, see Casati & Chirikov (1995b). We need to choose what to sacrifice, this principle or else 'true' (classical) chaos. I prefer to drop the latter. If the phenomenon of quantum chaos really did violate the correspondence principle, as some physicists suspect, it would indeed be a great discovery, since it would mean that classical mechanics is not the limiting case of quantum mechanics, but a different (separate) theory. 'Unfortunately', there exists a less radical (but also interesting and important) resolution of this difficulty, pseudochaos, which is the main topic of my talk.

Within such a philosophical framework the central physical problem is to understand the nature and mechanism of dynamical chaos in quantum mechanics. In other words, we need the *quantum* theory of dynamical chaos, including the transition to the classical limit. Certainly, quantum chaos is a new dynamical phenomenon, see Casati & Chirikov (1995b), related but not identical to classical chaos. We call it *pseudochaos*, the term *pseudo* intending to emphasize the difference from the 'classical' chaos in the ergodic theory. From the physical point of view, which I accept here, the latter, strictly speaking, does not exist in Nature. So, within the common philosophy of universal quantum mechanics *pseudochaos is the only true dynamical chaos.* Classical chaos is but a limiting pattern which is, nevertheless, very important, both in theory, to compare with real (quantum) chaos and in applications, as a very good approximation in a macroscopic domain, as is the whole of classical mechanics. Ford (1995) calls it *mathematical chaos*, as contrasted to *real physical chaos* in quantum mechanics.

I emphasize again that classical chaos is impossible in *finite and closed* quantum systems to which my talk is restricted. Particularly, I am not going to discuss here quantum measurement in which, necesserily, macroscopic (infinite-dimensional) processes are involved, see e.g. Casati & Chirikov (1995b).

Thus, the physical meaning of the term *pseudochaos* is principally different from (and even opposite to) that of *pseudorandom numbers* in a computer. The reason for the original term *pseudo* was twofold. At the beginning, the first and only meaning was related to the common belief that (by definition) no dynamical, deterministic, system like a computer can produce anything random. This delusion has been overcome in the theory of dynamical chaos in the field of *real* numbers. However, the digital computer works on a finite lattice of *integers*. This is qualitatively similar to quantum behaviour, see Chirikov *et al.* (1981). Computer numbers, like quantum variables, can at most be *pseudorandom* only, in contrast to the 'true' random classical quantities represented by real numbers. But then, a very special notion of *pseudorandom* was scrambling up to the level of a new fundamental concept in physics.

Quantum chaos is a part of quantum dynamics which, in turn, is a particular class of dynamical systems. It became a real physical problem upon discovery and understanding of classical dynamical chaos. To explain the problem I need to briefly remind you of the main peculiarities of classical chaos, especially those that are crucial in quantum theory.

2. Asymptotic chaos in classical mechanics

There are two equivalent descriptions of classical mechanics or, more generally, of any finite-dimensional dynamical system: via individual trajectories, and via a distribution function, or phase-space density, for Hamiltonian systems (most fundamental).

The trajectory obeys the equations of motion, which in general are nonlinear. It desribes a particular realization of a system's dynamics depending on the initial conditions. The phase density satisfies the Liouville equation, which is always *linear*, whatever the equations of motion, and which usually represents the typical (generic) dynamical behaviour of a given system. In particular, all zero-measure sets of special trajectories are automatically excluded.

Notice, however, that in some special cases, the phase density may display properties absent for trajectories. An interesting example, see Courbage & Hamdan (1995), is the correlation decay (and, hence, continuous spectrum) for a special initial phase density in a completely integrable system. The point is that such decay is related to the correlation between *different* trajectories rather than the behaviour on a given trajectory. The trajectory spectrum remains descrete, and the corresponding correlation persists. In particular, this explains a surprising phenomenon known as an 'echo' which is the revival of such correlations upon velocity reversal. An interesting open question concerns the exact conditions for a phase density to represent the *trajectory* properties. This is the primary problem in dynamics.

The strongest statistical properties of a dynamical system are related to the *local* exponential instability of trajectories, as described by the *linearized* equations of motion, provided the motion is *bounded* in phase space. These two conditions are *sufficient* for a rapid mixing of trajectories by the mechanism of 'stretching and folding'. For the linear equations of motion the combination of both conditions is impossible unless the whole phase space of the system is finite. A well-known example of the latter situation is the model described by the *linear* 'Arnold cat map', see Arnold & Avez (1968):

$$\begin{aligned} \overline{p} &= (p + x) \pmod 1 \\ \overline{x} &= (x + \overline{p}) \pmod 1 \end{aligned} \tag{2.1}$$

on a unit torus. The motion is exponentially unstable with (positive) Lyapunov exponent $\Lambda = \ln[(3+\sqrt{5})/2] > 0$, and is bounded due to the operation (mod 1). Notice that the linearized motion is described by the same map but *without* (mod 1), that is in the *infinite* plane ($-\infty < \mathrm{d}p, \mathrm{d}x < \infty$). It is unbounded and *globally* unstable but perfectly regular. We have so-called hyperbolic motion:

$$\mathrm{d}p = a\exp(\Lambda t) + b\exp(-\Lambda t), \quad \mathrm{d}x = c\exp(\Lambda t) + d\exp(-\Lambda t), \tag{2.2}$$

where the constants a, b, c, d depend on initial conditions and on Λ, and the integer t is the discrete map time. Remarkably, the motion (2.2) is time-reversible but *unstable* in both senses ($t \to \pm\infty$). This implies time reversibility of all statistical properties for the main system (2.1). This is a surprising conclusion which is still confusing some researchers (see e.g. Misra & Prigogine (1983)).

A nontrivial part of the relation between instability and chaos is in that the instability must be *exponential*. A power law instability is insufficient for chaos. For example, if we replace the first of equations (2.1) by $\overline{p} = p$, the model becomes completely integrable with oscillation frequency depending on the integral of the motion p (nonlinear oscillation). This produces *linear* (in time) instability, but the motion remains regular (with discrete spectrum). This is a typical property of completely integrable nonlinear oscillations, see Casati *et al.* (1980), which leads to a confusing difference in dynamical behaviour between the trajectories and phase densities, as mentioned above. Another open question is how to choose the correct time variable for a particular dynamical problem, see Casati & Chirikov (1995b). A change of time may convert the exponential instability into a power one, and *vice versa* (see e.g. Blümel (1994) for discussion).

The two above conditions for dynamical chaos can be realized in very simple systems (e.g. low-dimensional) like the model of (2.1). Another simple example, to which I will refer below, is the so-called 'kicked rotator' described by the *standard map*, see Chirikov (1991b), Casati & Chirikov (1995b), Chirikov (1979) and Chirikov *et al.* (1981):

$$\overline{p} = p + k \sin x, \quad \overline{x} = x + T\,\overline{p}, \tag{2.3}$$

also on either a torus $((x,p) \pmod{2\pi})$ or a cylinder $(x \pmod{2\pi},\ -\infty < p < \infty)$. This model is also well studied, and has many physical applications. The motion on a cylinder is bounded in one variable only. This, however, is sufficient for chaos.

The exponential instability implies a continuous spectrum of the motion which is equivalent, roughly speaking, to the mixing, or temporal *correlation decay*. Apparently, this is the most important characteristic property in statistical mechanics, underlying the principal and universal statistical phenomenon of *relaxation* to some steady state, or statistical equilibrium.

Aperiodic relaxation is especially clear in the Liouville picture for phase density behaviour (see e.g. Arnold & Avez (1968)). Consider a basis for Liouville's equation, for example

$$\varphi_{mn} = \exp[2\pi\mathrm{i}(mx + np)], \tag{2.4}$$

where m, n are any integers, in a simple example of model (2.1). In other words, we represent the phase density as a Fourier series:

$$f(x,p,t) = \sum_{m,n} F_{mn}(t)\varphi_{mn}(x,p) = \sum_{m,n} F_{mn}(0)\exp[2\pi\mathrm{i}(m(t)x + n(t)p)]. \tag{2.5}$$

Each term in this series, except φ_{00}, has zero total probability, and characterizes the spatial correlation in the phase density. The map (2.1) induces a map for the Fourier amplitudes and for harmonic numbers:

$$\overline{F}_{mn} = F_{\overline{m}\overline{n}}, \quad \overline{n} = n + m, \quad \overline{m} = m + \overline{n}. \tag{2.6}$$

Remarkably, the variables $m(t)$ and $n(t)$ obey the same map as that for the *linearized* equations of motion in variables $\mathrm{d}x, \mathrm{d}p$, and with the same instability rate Λ on the *infinite* lattice (m,n). The dynamics of the phase density in the Fourier representation, described by the same equation (2.2) (upon substitution of m, n for $\mathrm{d}x, \mathrm{d}p$), is also unbounded, globally unstable, and regular. This is not surprising, as both representations describe the *local* structure of the motion. Dynamical chaos is a *global* phenomenon determined, nevertheless, by the microdetails of the initial conditions, due to the exponential instability of the motion, see Casati & Chirikov (1995b) and Chirikov (1994). Accordingly, in the original phase space the temporal density fluctuations are chaotic, as are almost all trajectories of the map (2.1).

The only *stationary* mode $m = n = 0$ represents, in this picture, the statistical steady state, while all the others describe *nonstationary* fluctuations. The latter are another characteristic property of statistical behaviour. These higher modes can be separated from the average statistical relaxation by so-called coarse-graining, or spatial averaging, which is a projection of the phase density on a finite (and arbitrarily fine) partition of the phase space. The kinetic (particularly, diffusive) description of the statistical relaxation is restricted to such a coarse-grained projection only, while the fluctuations work as a dynamical *generator of noise*.

Another elegant method of separating out the average relaxation is a suppression of the fluctuations using Prigogine's Λ operator, see Misra & Prigogine (1983), which provides an invertible smoothing of the exact phase density, see Kumicak (1996). True, the inverse operator is an improper one, yet this method could be efficiently used in some theoretical constructions. Contrary to common belief, it has nothing to do with time irreversibility, see Misra & Prigogine (1983) and Goodrich *et al.* (1980). Moreover, unlike the coarse-grained projection, the Λ-smoothed phase density is as reversible as the exact one (in principle but not in practice, of course). The origin of the misunderstanding concerning *irreversibility* is apparently related to the necessary restriction on the initial smoothed density, which was missed in the theory of Goodrich *et al.* (1980). Such a density is a technical rather than actual property of the system, and hence it does not need to be arbitrary. A similar operation is often used in quantum mechanics (for different purposes) to convert the Wigner function (the counterpart of exact classical phase density) into the so-called Husimi distribution, which is an expansion in the coherent states (see e.g. Casati & Chirikov (1995b)).

Nonstationary fluctuations/correlations of the phase density form a *stationary* flow into higher modes $|m|, |n| \to \infty$ (see Prigogine (1963)), and keep the memory of the exact initial conditions (see the first equation of (2.6)) providing time reversibility for the exact density. A stationary correlation flow is only possible for the *continuous* phase space, which is a characteristic feature of classical mechanics. This allows for an asymptotic formulation of the ergodic theory ($t \to \pm\infty$). Notice that both the trajectories and the full density are *time-reversible*. However, the latter, unlike the former, is *nonrecurrent*. Reversed relaxation, and in particular 'antidiffusion', describe the growth of a large fluctuation, which is eventually (as $t \to -\infty$) followed by standard relaxation in the opposite direction of time, see Chirikov (1994).

3. Quantum pseudochaos: a new dimension in ergodic theory

Dynamical chaos is one limiting case of the modern general theory of dynamical systems which describes statistical properties of the deterministic motion (see e.g. Kornfeld *et al.* (1982)). No doubt, this theory has been developed on the basis of classical mechanics. Yet, as a general mathematical theory, it does not need to be restricted to classical mechanics only. In particular, it can be applied to quantum dynamics, and indeed it was, with a surprising result. Namely, it had been found at the beginning, see Casati *et al.* (1979), and was subsequently well confirmed, see e.g. Chirikov (1991b), Casati & Chirikov (1995b), Chirikov *et al.* (1981), Cohen (1991) and Casati *et al.* (1986), that quantum mechanics does not typically permit 'true' (classical-like) chaos. This is because in quantum mechanics the energy (and frequency) spectrum of any system, whose motion is bounded in phase space, is discrete, and its motion is almost periodic. Hence, according to the existing ergodic theory, such a quantum dynamics belongs to the limiting case of regular motion which is the opposite of dynamical chaos. The ultimate origin of quantum almost-periodicity is the discreteness of the phase space itself (or, in a more formal language, the noncommutative geometry of this space). This property is the basis of quantum physics, directly related to the fundamental uncertainty principle. Nevertheless, another fundamental principle, the correspondence principle, requires the transition to classical mechanics in all cases, including dynamical chaos with all its peculiar properties.

Now, the principal question to be answered is: where is the expected quantum chaos in ergodic theory? The answer, see Casati & Chirikov (1995b), Chirikov *et al.* (1981) and Chirikov (1994), (not commonly accepted as yet) was concluded from a simple observation (in principle well-known but never fully understood) that the sharp border between the discrete and continuous spectrum is physically meaningful in the limit $|t| \to \infty$ only, the condition actually assumed in ergodic theory. Hence, to understand quantum chaos, the existing ergodic theory needs some modification by introducing a new 'dimension': time. The *finite-time statistical properties* of a dynamical system, both quantum and classical, become a new and central problem in ergodic theory.

Within a finite time, the discrete spectrum is dynamically equivalent to the continuous one, thus providing much stronger statistical properties of the motion than was (and still is) expected in the ergodic theory in the case of a discrete spectrum. It turns out that the motion with discrete spectrum may exhibit *all* statistical properties of classical chaos, but only on *finite* time scales.

The absence of classical-like chaos in quantum dynamics apparently contradicts not only the correspondence principle but also the fundamental statistical nature of quantum mechanics. However, even though the random element in quantum mechanics ('quantum jumps') is unavoidable, it can be singled out and separated from the proper quantum processes. Namely, the fundamental randomness in quantum mechanics is related only to a very specific event – the *quantum measurement* – which, in a sense, is foreign to the proper quantum system itself. This allows us to divide the whole problem of quantum dynamics into two qualitatively different parts:

- Proper quantum dynamics, as described by a very specific dynamical variable, the wavefunction $\psi(t)$, obeying some deterministic equation, for example the Schrödinger equation. The discussion in what follows will be limited to this part only.
- Quantum measurement, including the registration of the result, and hence, the collapse of the function ψ, which still remains a very vague issue in view of the fact that there is no common agreement even on whether this is a real physical problem or an ill-posed one, so that the Copenhagen interpretation of quantum mechanics answers all 'admissible' questions. In any case, up to now there is no dynamical description of the quantum measurement, including the collapse of ψ.

The recent breakthrough in the understanding of quantum chaos has been achieved particularly due to the above-mentioned philosophy of separating out the dynamical part of quantum mechanics. This philosophy is accepted, explicitly or more often implicitly, by most researchers in this field.

3.1. *Time scales of pseudochaos*

The existing ergodic theory is asymptotic in time, and thus involves no explicit time scales.† There are two reasons for it. One is technical: it is much easier to derive the asymptotic relations than to obtain rigorous finite-time estimates. The second reason is more profound. All statements in the ergodic theory hold up to measure zero, that is excluding some peculiar nongeneric sets of measure zero. Even this minor imperfection of the theory did not seem completely satisfactory but has been 'swallowed' eventually, and is now commonly tolerated by both physicists and mathematicians. In a finite-time theory all these exceptions acquire a *small but finite* measure, which would be apparently 'unbearable' (for mathematicians). Yet, there is a standard mathematical 'trick' for avoiding both these difficulties.

The most important time scale t_R in quantum chaos is given by the general estimate, see Casati & Chirikov (1995b) and Chirikov *et al.* (1981):

$$\ln(\omega t_R) \sim \ln Q, \quad t_R \sim \frac{Q^\alpha}{\omega} \sim \rho_0 \le \rho_H, \tag{3.7}$$

† Asymptotic statements in the ergodic theory should not always be understood literally, to avoid physical misconceptions (see e.g. addendum to Prigogine (1963)). Actually, classical chaos has also its time scales, for example a dynamical one ($\sim \Lambda^{-1}$), see Casati & Chirikov (1995b).

where ω and $\alpha \sim 1$ are system-dependent parameters, and $Q \gg 1$ stands for some large quantum parameter (in semiclassical region). It can be, e.g. a quantum number $Q = I/\hbar$, related to a characteristic action variable I, or the total number of states for the bounded quantum motion in the phase space domain of volume Γ: $Q \approx \Gamma/(2\pi)^N$.† The time scale t_R is called the *relaxation time scale*, referring to one of the principal properties of chaos, *statistical relaxation* to some steady state. The physical meaning of this scale is simple; it is directly related to the fundamental uncertainty principle ($\Delta t\, \Delta E \sim 1$) as implemented in the second equation (3.7), where ρ_H is the *full* average energy level density (also called the Heisenberg time). For $t \lesssim t_R$, the discrete spectrum is not resolved, and statistical relaxation follows the classical (limiting) behaviour. This indeed is the 'gap' in the ergodic theory (supplemented with time as the additional dimension) where pseudochaos, and in particular quantum chaos, dwells. A more accurate estimate relates t_R to a *part* ρ_0 of the level density. This is the density of the so-called *operative eigenstates* only (those which are present in a particular quantum state ψ, and which actually control its dynamics).

The formal trick mentioned above is not to consider finite-time relations, which we really need in physics, but rather a special *conditional limit*:

$$t, Q \to \infty, \quad \tau_R = \frac{t}{t_R(Q)} = \text{const}, \tag{3.8}$$

where τ_R is a new dimensionless time. The *double* limit (3.8) (unlike the single one $Q \to \infty$) is *not* the classical mechanics which holds, in this representation, for $\tau_R \lesssim 1$, and with respect to statistical relaxation only. For $\tau_R \gtrsim 1$, the behaviour becomes essentially quantum (even in the limit $Q \to \infty$!) and nowadays is called *mesoscopic*. In particular, the quantum steady state is in general quite different from classical statistical equilibrium, in that the former may be *localized* (under certain conditions), that is *nonergodic*, in spite of classical ergodicity.

Another important difference is in *fluctuations* which are also a characteristic property of chaotic behaviour. In comparison with classical mechanics, the quantum $\psi(t)$ plays, in this respect, an intermediate role between the classical trajectory with large relative fluctuations (~ 1), and the coarse-grained classical phase density with no fluctuations. Unlike both, the fluctuations of $\psi(t)$, or rather those of averages on a quantum state $\psi(t)$, are typically $\sim d_H^{-1/2}$, where d_H is the number of operative eigenstates associated with the quantum state ψ (d_H can be called the *Hilbert dimension* of the state ψ). In other words, the chaotic $\psi(t)$ represents statistically a *finite ensemble* of some number ($\sim d_H$) of independent systems, even though formally $\psi(t)$ describes a *single* system. The fluctuations clearly demonstrate the difference between physical time t and the auxiliary variable τ: in the double limit $(t, Q \to \infty)$ the fluctuations vanish, and one needs a new 'trick' to recover them for a finite Q.

The relaxation time scale should not be confused with the *Poincaré recurrence time* t_P ($\gg t_R$) which is typically much longer, and which sharply increases with decreasing recurrence domain. The time scale t_P characterizes large fluctuations (for both the classical trajectory and the quantum ψ, but not the classical phase density). On the contrary, t_R characterizes the average relaxation process. Rare recurrences (the more rare the larger quantum parameter Q) make quantum relaxation similar to classical (nonrecurrent) relaxation.

Stronger statistical properties (than relaxation and fluctuations) are related in the ergodic theory to the exponential instability of motion. The importance of those stronger properties for statistical mechanics is not completely clear, see Farquhar (1964). Never-

† Here and in what follows I put $\hbar = 1$.

theless, in accordance with the correspondence principle, these stronger properties are also present in quantum chaos, but on a *much shorter* time scale t_r:

$$\Lambda\, t_r \sim \ln Q. \tag{3.9}$$

Here Λ is the classical Lyapunov exponent. This time scale was discovered and partly explained by Berman & Zaslavsky (1978) (see also Casati & Chirikov (1995b) and Chirikov *et al.* (1981)). We call it a *random time scale.* Indeed, according to Ehrenfest's theorem, the motion of a narrow wave packet follows the beam of classical trajectories as long as the packet remains narrow, and hence it is as random as in the classical limit. Even though the random time scale is very short, it grows indefinitely as $Q \to \infty$. Thus, temporary, finite-time quantum pseudochaos turns into classical dynamical chaos in accordance with the correspondence principle.

Again, we may consider the *conditional limit:*

$$t, Q \to \infty, \quad \tau_r = \frac{t}{t_r(Q)} = \text{const.} \tag{3.10}$$

Notice that the new scaled time τ_r is different from that entering equation (3.8) (τ_R).

If we fix the time t, then in the limit $Q \to \infty$ we obtain the transition to the classical instability in accordance with the correspondence principle, while for Q fixed, and $t \to \infty$, we get the proper quantum evolution in time. For example, the quantum Lyapunov exponent satisfies

$$\Lambda_q(\tau_r) \to \begin{cases} \Lambda, & \tau_r \ll 1 \\ 0, & \tau_r \gg 1. \end{cases} \tag{3.11}$$

Quantum instability ($\Lambda_q > 0$) was observed in numerical experiments, see Casati & Chirikov (1995b) and Toda & Ikeda (1987). What terminates the instability for $t \gtrsim t_r$? A simple explanation is suggested by the classical picture of the phase density evolution on the integer Fourier lattice m, n discussed above for model (2.1). Classical Fourier harmonics m, n are of a kinematical nature, without any *a priori* dynamical restriction. In particular, they can be arbitrarily large (and actually are for a chaotic motion), which corresponds to a continuous classical phase space. On the contrary, the quantum phase space is discrete. At first glance, quantum wave packet stretching/squeezing, as for a classical system, does not seem to be principally restricted, since only a two-dimensional area (per degree of freedom) is bounded in quantum mechanics. However, Fourier harmonics of the quantum phase density (Wigner function) are directly related to quantum dynamical variables, in particular to the action variables, whose values are restricted by the quantum parameter Q, which leads to estimate (3.9). In the simple model (2.1) this is related to a finite size of the whole phase space. In general, in a conservative system, even with an infinite phase space, the restriction is imposed by the energy conservation. Numerical experiments reveal that the original wave packet, after a considerable stretching similar to the classical one, is rapidly destroyed. Namely, it splits into many new small packets, see Casati & Chirikov (1995b) and Toda & Ikeda (1987). The mechanism of this sharp 'disrupture' of the classical-like motion is not quite clear (for a possible explanation see Casati & Chirikov (1995b) and Chirikov (1993)). The resulting picture is qualitatively similar to that for the classical phase density, the main difference being in the spatial fluctuation scale, now bounded from below by $1/Q$. Nevertheless, the quantum phase density can also be decomposed into a coarse-grained average, and fluctuations. An important implication of this picture for the wave packet time evolution is the rapid and complete destruction of the so-called generalized coherent states in quantum chaos, see Perelomov (1986).

In the quasiclassical region ($Q \gg 1$) the scale is $t_r \ll t_R$. This leads to the surprising conclusion that quantum diffusion and relaxation are *dynamically stable*, contrary to the classical behaviour. This, in turn, suggests that, in general, the instability of the motion is not important *during* statistical relaxation. Nevertheless, the *foregoing* correlation decay on the short time scale t_r is crucial for the statistical properties of quantum dynamics.

Dynamical stability of quantum diffusion has been proved in striking numerical experiments with time reversal, see Shepelyansky (1983). In a classical chaotic system the diffusion is immediately recovered due to numerical 'errors' (not random!) amplified by the local instability. On the contrary, the quantum 'antidiffusion' proceeds until the system passes near the initial state to a very high accuracy. Only then is normal diffusion restored. The stability of quantun chaos on a relaxation time scale is comprehensible, as the random time scale is much shorter. Nevertheless, the accuracy of the reversal (up to $\sim 10^{-15}$ (!)) is surprising. Apparently this is explained by the relatively large size of the quantum wave packet as compared to the unavoidable rounding-off errors, unlike the classical computer trajectory whose size is comparable to these errors, see Chirikov (1992b). In the standard map (2.3) (upon quantization) the size of the optimal, least-spreading, wave packet is $\Delta x \sim \sqrt{T}$, see Chirikov *et al.* (1981). On the other hand, any quantity in the computer must well exceed the rounding-off error δ ($\ll 1$), e.g. $T \gg \delta$, and $(\Delta x)^2/\delta^2 \gtrsim (T/\delta)\delta^{-1} \gg 1$.

3.2. *Classical-like relaxation and residual fluctuations*

The relaxation time scale t_R is more important than the time scale t_r for two reasons. First, it is much longer than t_r, and second, it is related to the principal process of statistical relaxation which is the basis of statistical mechanics. The short scale t_r was interpreted by Berman & Zaslavsky (1978) (see also Zaslavsky (1981)) as a limit for the classical-like behaviour of chaotic quantum motion. Subsequently, it was found that the *method* of quasiclassical quantization can be extended to much longer times, see Chirikov *et al.* (1981) and Sokolov (1984). However, the *physics* on both time scales is qualitatively different: dynamical instability on the scale t_r, and statistical relaxation on t_R.

The discrete pseudochaos spectrum is not resolved on the whole scale t_R, and relaxation follows the classical law. Consider, for example, model (2.3), the standard map on a torus with total number of quantum states Q, and p, x being the action–angle variables.

If the perturbation parameter $k \gtrsim Q$, the relaxation to ergodic steady state in this model, as well as in model (2.1), is very quick, with characteristic relaxation time $t_e \sim 1$ (iterations). Such regimes often take place in physical systems. Here I consider another case, more interesting for the problem of pseudochaos, namely *diffusive relaxation*, which occurs for a sufficiently weak perturbation

$$k \ll Q. \tag{3.12}$$

In the classical limit this relaxation is descibed by the standard diffusion equation

$$\frac{\partial f(p,t)}{\partial t} = \frac{1}{2}\frac{\partial}{\partial p} D(p) \frac{\partial f(p,t)}{\partial p}, \tag{3.13}$$

where $f(p,t) = \langle f(p,x,t)\rangle_x$ is a coarse-grained phase density (averaged over x), and

$$D = \frac{\langle(\Delta p)^2\rangle}{t} \approx \frac{k^2}{2} \tag{3.14}$$

is the diffusion rate. Equation (3.14) holds for the standard map if $K \equiv kT \gg 1$, which is also the condition for global chaos in this model, see Chirikov (1979). The relaxation to the ergodic steady state $f_s = 1/Q$ is exponential with characteristic time

$$t_e = \frac{Q^2}{2\pi^2 D} \approx \frac{Q^2}{\pi^2 k^2}. \tag{3.15}$$

In the diffusive regime ($k \ll Q$) this time is $t_e \gg 1$. This average relaxation is stable and regular, in spite of the underlying chaotic dynamics.

The quantized standard map $\overline{\psi} = \hat{U}\psi$ is described by the unitary operator

$$\hat{U} = \exp\left(-i\frac{T\hat{p}^2}{2}\right)\exp(-ik\cos\hat{x}) \tag{3.16}$$

on a cylinder ($Q \to \infty$), see Casati *et al.* (1979), where $\hat{p} = -i\partial/\partial x$, and by a similar but somewhat more complicated expression on a torus, see Izrailev (1990).

There are three quantum parameters in this model: the perturbation k, the period T and the size Q, but only two classical combinations remain: the perturbation $K = kT$ and the classical size $M = TQ/(2\pi)$, which is the number of classical resonances over the torus. Notice that the quantum dynamics is in general richer than the classical, as the former depends on an extra parameter. It is, of course, another representation of Planck's constant which I have set equal to one. That is why, in the quantized standard map, we need both parameters k and T, and cannot combine them in a single classical parameter K.

The quasiclassical region, where we expect quantum chaos, corresponds to $T \to 0$, $k \to \infty$, and $Q \to \infty$, while the classical parameters $K = \text{const}$ and $M = \text{const}$ are fixed.

A technical difficulty in evaluating t_R for a particular dynamical problem is that the density ρ_0 depends on the dynamics. So, we have to solve a self-consistent problem. For the standard map the answer is known (see Casati & Chirikov (1995b)):

$$t_R = \rho_0 = 2D. \tag{3.17}$$

This is a remarkable relation, as it connects essentially *quantum* characteristics (t_R, ρ_0) with the *classical* diffusion rate D, see (3.14).

The quantum diffusion rate depends on the scaled (dimensionless) time τ_R (3.8), and is given by

$$D_q = \frac{D}{1+\tau_R} \to \begin{cases} D, & \tau_R = t/t_R \ll 1 \\ 0, & \tau_R \gg 1. \end{cases} \tag{3.18}$$

This is an example of scaling in a discrete spectrum which eventually stops quantum diffusion.

The character of the steady state crucially depends on the ratio t_R/t_e. Define the *ergodicity parameter* λ (see Casati & Chirikov (1995b)):

$$\lambda = \frac{D}{Q} \sim \left(\frac{t_R}{t_e}\right)^{\frac{1}{2}} \sim \frac{k^2}{Q} \sim \frac{K}{M}k. \tag{3.19}$$

Consider first the case $\lambda \gg 1$, when the time scale t_R is long enough to allow for the completion of the classical-like relaxation. In that case the final steady state, as well as all the eigenfunctions, are ergodic, i.e. the corresponding Wigner functions are close to the classical microcanonical distribution in the phase space. This region is inevitably reached if the classical parameter K/M is kept fixed, while the quantum parameter $k \to \infty$, in agreement with the Shnirelman theorem, or else with a physical generalization of this theorem, see Shnirelman (1974). This region is called *far quasiclassical asymptotics*.

A principal difference between the quantum ergodic state and the classical state is the existence of *residual fluctuations* in the former. In the quasiclassical region, the chaotic quantum steady state is a superposition of a large number of eigenfunctions. As a result, almost any physical quantity fluctuates in time. Even in the discrete spectrum we are considering here, these fluctuations are very irregular. In the case of a classical-like ergodic steady state all Q eigenfunctions essentially contribute to the fluctuations. Moreover, we would expect their contributions to be statistically almost independent. Hence, the fluctuations should scale $\sim Q^{-1/2} = d_{\rm H}^{-1/2}$, where $d_{\rm H}$ is the Hilbert dimension of the ergodic state. That, indeed, is the case, according to numerical experiments, see Casati *et al.* (in preparation). For example, the energy fluctuations were found to follow a simple relation

$$\frac{\Delta E_{\rm s}}{E_{\rm s}} \approx \frac{1}{\sqrt{Q}}, \tag{3.20}$$

where

$$(\Delta E_{\rm s})^2 = \overline{E^2(t)} - E_{\rm s}^2, \quad E_{\rm s} = \overline{E(t)}, \quad E(t) = \frac{\langle p^2 \rangle}{2}. \tag{3.21}$$

Here a bar indicates time averaging over a sufficiently long time interval ($\gg t_{\rm e}$), and brackets denote the usual average over the quantum state.

Equation (3.20) suggests the complete quantum decoherence in the final steady state for any initial state, even though the steady state is formally a pure quantum state. For $Q \gg 1$ the fluctuations are small, so that *statistically* the quantum relaxation is *nonrecurrent*. The decoherence of a chaotic quantum state is also confirmed by the independence (up to small fluctuations) of the final steady state energy $E_{\rm s}$ of the initial $E(0)$. Since any particular initial quantum state is strongly coherent, the decoherence is a result of the quantum chaos. It is called *dynamical decoherence.* This is one of the most important results in quantum chaos.

3.3. *Mesoscopics: quantum behaviour in the quasiclassical region*

If the ergodicity parameter $\lambda \ll 1$ is small, all the eigenstates and the steady state are non-ergodic, or localized. That is because the scale $t_{\rm R}$ is not long enough to support classical-like diffusion, which stops before classical relaxation is completed. For this reason this situation is also called *quantum diffusion localization.* As a result the structure of eigenfunctions and of the steady state remains essentially quantum, no matter how large the quantum parameter $k \to \infty$. This is called *intermediate quasiclassical asymptotics* or the *mesoscopic* domain. In particular, it corresponds to $K > 1$ fixed, $k \to \infty$ and $M \to \infty$, while $\lambda \ll 1$ remains small.

The popular term 'mesoscopic' means here some intermediate behaviour between classical and quantum. In other words, in mesoscopic phenomenona both classical and quantum features occur. Again, the correspondence principle requires the transition to completely classical behaviour. That, indeed, is the case, as mesoscopic phenomena occur in the region where the quantum parameter $k \gg 1$ is large, but still less than a certain critical value (corresponding to $\lambda \sim 1$) which determines the border of transition to fully classical behaviour (far quasiclassical asymptotics).

If $\lambda \ll 1$ is very small, the shape of localized eigenstates is asymptotically exponential, see Casati & Chirikov (1995b), and can be approximately described by a simple expression, see Casati *et al.* (to be published):

$$f_m(p) = \langle |\psi_m(p)|^2 \rangle \approx \frac{2}{\pi l \cosh[2(p - p_m)/l]}. \tag{3.22}$$

The localized steady state has a similar but somewhat more complicated shape, see Chirikov (1981) and Izrailev *et al.* (to be published). This is a simple approximation superimposed with large fluctuations. The parameter l is called the *localization length.* Interestingly, the two localization lengths ($l_{\rm s}$ for steady state and l for eigenfunctions) are different, see Chirikov *et al.* (1981):

$$l_s \approx D \quad \text{while} \quad l \approx \frac{D}{2}, \tag{3.23}$$

because of large fluctuations.

In terms of localization length, the region of mesoscopic phenomena is defined by

$$1 \ll l \ll Q. \tag{3.24}$$

The inequality on the left reflects classical features of the state, while the right hand one refers to quantum effects. The combination allows for a classical description, at least in the standard map, of statistical relaxation to a quantum steady state by a phenomenological diffusion equation for the Green function, see Casati & Chirikov (1995b) and Chirikov (1991a):

$$\frac{\partial g(\nu,\sigma)}{\partial\sigma} = \tfrac{1}{4}\frac{\partial^2 g}{\partial\nu^2} + B(\nu)\frac{\partial g}{\partial\nu}. \tag{3.25}$$

Here $g(\nu,0) = |\psi(\nu,0)|^2 = \delta(\nu-\nu_0)$ and

$$\nu = \frac{p}{2D}, \quad \sigma = \ln(1+\tau_R), \quad \tau_R = \frac{t}{2D}. \tag{3.26}$$

The additional drift term in the diffusion equation, with

$$B(\nu) \approx \mathrm{sgn}(\nu-\nu_0) = \pm 1, \tag{3.27}$$

describes the so-called quantum coherent back-scattering, which is the dynamical mechanism of localization.

The solution of (3.25) is, see Casati & Chirikov (1995b),

$$g(\nu,\sigma) = \frac{1}{\sqrt{\pi\sigma}}\exp\left[-\frac{(\Delta+\sigma)^2}{\sigma}\right] + \exp(-4\Delta)\,\mathrm{erfc}\left(\frac{\Delta-\sigma}{\sqrt{\sigma}}\right), \tag{3.28}$$

where $\Delta = |\nu-\nu_0|$.

Asymptotically, as $\sigma \to \infty$, the Green function $g(\nu,\sigma) \to 2\exp(-4\Delta) \equiv g_s$ approaches the localized steady state g_s, exponentially in σ but only as a power-law in physical time τ_R or t ($g - g_s \sim 1/\tau_R$). This is the effect of a discrete motion spectrum. Numerical experiments confirm prediction (3.28), at least up to the logarithmic accuracy $\sim \sigma \approx \ln\tau_R$, see Casati & Chirikov (1995b) and Cohen (1991).

A physical example of localization is the quantum suppression of the diffusive photoeffect in a hydrogen atom, see Casati *et al.* (1987). Depending on parameters, the suppression can occur, no matter how large the atomic quantum numbers. This is a typical mesoscopic phenomenon which had been predicted by the theory of quantum chaos, and was subsequently observed in laboratory experiments.

One might expect that, in the case of localization ($D \ll Q$), the fluctuations scale like $l^{-1/2} \sim k^{-1}$, as the number of coupled eigenfunctions in the localized steady state is $\sim l$. This, however, is *not* the case as was found in the first numerical experiments (Casati & Chirikov (1995b)). According to more accurate data, see Casati *et al.* (in preparation), the fluctuations are described by

$$\frac{\Delta E_s}{E_s} = \frac{A}{k^\gamma} = \frac{a}{d_H^{\gamma/2}}, \tag{3.29}$$

with fitting parameters $\gamma = 0.55$, $a = 0.65$. For a nonergodic state, the Hilbert dimension can be defined as (Casati & Chirikov (1995b))

$$d_H^{-1} = \tfrac{1}{3}\int |\psi(p)|^4\,dp = \int f^2(p)\,dp, \tag{3.30}$$

where $f(p)$ is a smoothed (coarse-grained) density, and the factor $\frac{1}{3}$ accounts for the ψ fluctuations, see Flambaum *et al.* (1994). In the case of exponential localization (3.22), $d_{\rm H} \approx \pi^2 l/4$. The most important parameter here, γ, is about half the expected value $\gamma = 1$. This result suggests some fractal properties of localized eigenfunctions and/or their spectra. To put it another way, a slow fluctuation decay (3.29) implies *incomplete* quantum decoherence, which can be characterized by the number d_s of statistically independent components in the steady state, see Casati & Chirikov (1995b). Then, from (3.20) and (3.29) we obtain in the two limits:

$$\frac{d_s}{d_{\rm H}} \approx \begin{cases} 1, & \lambda \gg 1 \\ d_{\rm H}^{\gamma-1}/a^2 = 2.4\, d_{\rm H}^{-0.45}, & \lambda \ll 1. \end{cases} \tag{3.31}$$

This result was confirmed by Izrailev *et al.* (to be published) for a band random matrix model.

The phenomenon of quantum diffusion localization also explains the limitation of quantum instability in systems with infinite phase space, like the standard map on a cylinder. Indeed, the maximum number of coupled states here is determined by the localization length, whatever the total number of states in the system. Hence, we should substitute the quantum parameter $Q \sim l \sim k^2$ in estimate (3.9). Even if there is no localization (e.g. for a standard map with the parameter $k(p)$ depending on p, see Casati & Chirikov (1995b) and Chirikov *et al.* (1981)), so that the quantum diffusion goes on and the quantum spectrum becomes continuous, the number of coupled states increases with time as a power law only ($\Delta p \sim \sqrt{t}$), and hence the quantum Lyapunov exponent vanishes on the relaxation time scale, $\Lambda_{\rm q} \to 0$. Only if the action variables grow exponentially, the instability rate $\Lambda_{\rm q}$ remains finite, and quantum chaos becomes asymptotic, as in the classical limit (see Chirikov *et al.* (1981) and Weigert (1990) for such 'exotic' models).

3.4. *Examples of pseudochaos in classical mechanics*

Pseudochaos is a new generic dynamic phenomenon missed in ergodic theory. No doubt, the most important particular case of pseudochaos is quantum chaos. Nevertheless, pseudochaos occurs in classical mechanics as well. A few examples of classical pseudochaos are given below, which may help in understanding the physical nature of quantum chaos. Furthermore, they unveil new features of classical dynamics.

- *Linear waves* are an example of pseudochaos close to quantum mechanics (see e.g. Chirikov (1992a)). Here, only a part of quantum dynamics is discussed, that described, e.g. by the Schrödinger equation, which is a linear wave equation. For this reason, sometimes quantum chaos is called wave chaos, see Šeba (1990). Classical electromagnetic waves are used in laboratory experiments as a physical model for quantum chaos, see Stöckmann & Stein (1990). The 'classical' limit corresponds here to geometrical optics, and the 'quantum' parameter $Q = L/\lambda$ is the ratio of a characteristic size L of the system to the wave length λ. As is well known from optics, no matter how large the ratio L/λ, the diffraction patterns prevail at sufficiently large distances $R \gtrsim L^2/\lambda$. This gives a sort of relaxation scale: $R/\lambda \sim Q^2$.
- *A linear oscillator* (multi-dimensional) is also a particular representation of waves. A broad class of quantum systems can be reduced to this model, see Eckhardt (1988). Statistical properties of linear oscillators, particularly in the thermodynamic limit ($N \to \infty$), were studied by Bogolyubov (1945) in the framework of TSM. On the other hand, the theory of quantum chaos suggests a richer behaviour for large but finite N, in particular, characteristic time scales for harmonic oscillations, see Chirikov (1986), the number of the degrees of freedoms N playing the role of 'quantum' parameter.

• ***Completely integrable nonlinear systems*** also reveal pseudochaotic behaviour. An example of statistical relaxation on the Toda lattice was given in Ford *et al.* (1973) much before the problem of quantum chaos arose. Moreover, the strongest statistical properties in the limit $N \to \infty$, including that equivalent to exponential instability (a so-called K-property), were rigorously proven just for completely integrable systems for any finite N, see Kornfeld *et al.* (1982).

• ***A digital computer*** is a very specific classical dynamical system, whose properties are extremely important in view of ever growing interest in numerical experiments, now covering all branches of science. The computer is an 'overquantized' system, in which *any* quantity is *discrete*, while in quantum mechanics only the product of two conjugated variables is. The 'quantum' parameter here is $Q = M$, which is the largest computer integer, and the short time scale (3.9), $t_r \sim \ln M$, represents the number of digits in the computer word, see Chirikov *et al.* (1981). Due to discreteness, any dynamical trajectory in a computer eventually becomes periodic, an effect well known in the theory and practice of pseudorandom number generators. One should take all necessary precautions to exclude such computer artifacts in numerical experiments (see e.g. Maddox (1994) and references therein). As for mathematics, the periodic approximations in dynamical systems are also considered in ergodic theory, apparently without any relation to pseudochaos in quantum mechanics or a computer, see Kornfeld *et al.* (1982).

Computer pseudochaos seems to be the most convincing argument for researchers who are still reluctant to accept quantum chaos as at least a kind of chaos. They insist that only classical-like (asymptotic) chaos deserves this name. But this is the same chaos which was (and still is) studied to a large extent on a computer and so is chaos inferred from pseudochaos!

4. Statistical theory of pseudochaos: random matrices

A complete solution of a dynamical quantum problem can be obtained via diagonalization of the Hamiltonian, to find the energy (or quasi-energy) eigenvalues and eigenfunctions. The evolution of any quantity is then expressed as a sum over the eigenfunctions. For example, the energy time dependence is

$$E(t) = \sum_{m,m'} c_m c^*_{m'} E_{mm'} \exp[\mathrm{i}(E_m - E_{m'})t], \tag{4.32}$$

where $E_{mm'}$ are matrix elements, and the initial state in momentum representation is $\psi(n,0) = \sum_m c_m \varphi_m(n)$. For chaotic motion this dependence is in general very complicated, but statistical properties of the motion can be inferred from the statistics of the eigenfunctions $\varphi_m(n)$ (i.e. of the matrix elements $E_{mm'}$) and of the eigenvalues E_m.

Nowadays there exists a well developed random matrix theory (RMT, see e.g. Brody *et al.* (1981)) which describes average properties of a typical quantum system with a given symmetry of the Hamiltonian. At the beginning, the object of this theory was assumed to be a very complicated (in particular multi-dimensional) quantum system as a representative of a certain statistical ensemble. In the course of understanding the phenomenon of dynamical chaos, it became clear that the number of degrees of freedom of the system is irrelevant. Instead, the number of quantum states (quantum parameter Q) is of importance, providing dynamical chaos in the classical limit.

This approach to the theory of complex quantum systems like atomic nuclei, was adopted by Wigner (1955) 40 years ago, much before the problem of quantum chaos was formulated. He introduced the so-called band random matrices (BRM), which were most suitable to account for the structure of conservative systems. However, due to severe mathematical difficulties, RMT immediately turned to a much simpler case of statistically homogeneous (full) matrices, for which impressive theoretical results have been achieved, see Brody *et al.* (1981). The price was that full matrices describe the local chaotic structure only, a limitation especially inacceptable for atoms, see Flambaum *et al.* (1994) and Chirikov (1985). Only recently, interest returned to the original Wigner BRM, see Casati *et al.* (1993) and Feingold *et al.* (1991).

One of the main results in studies of quantum chaos was the discovery of quantum diffusion localization. This mesoscopic, quasiclassical phenomenon discussed above, has been well studied and confirmed by many researchers for dynamical models described by maps. Contrary to common belief, the maps describe not only time-dependent systems, but also conservative ones in the form of Poincaré maps, see e.g. Bogomolny (1992). Nevertheless, to my knowledge, up to now no direct studies of quantum localization in conservative systems have been undertaken, either in the laboratory or in numerical experiments. Moreover, the very existence of quantum localization in conservative systems is challenged, see Bogomolny (1994). Here, I briefly descibe recent results concerning the structure of localized quantum chaos in the momentum space of a generic conservative system with few degrees of freedom, which is classically strongly chaotic, in particular ergodic on a compact energy surface, see Casati *et al.* (to be published).

In general, RMT is a statistical theory of systems with discrete energy spectra. This is the principal property of quantum dynamical chaos, see Casati & Chirikov (1995b). Thus, RMT turned out to become (accidentally!) a statistical theory for the incoming quantum chaos. Remarkably, this statistical theory does not include any time-dependent noise, that is any coupling to a thermal bath, the standard element of most statistical theories. Moreover, a *single matrix* from a given statistical ensemble represents a typical (generic) *dynamical* system of a given class characterized by the matrix parameters. This makes an important bridge between dynamical and statistical descriptions of quantum chaos. A similarity between the problem of quantum diffusion localization in momentum space, and the well known dual problem of Anderson localization in configurational space of disordered solids, see Anderson (1978) and Fishman *et al.* (1984), is especially clear and instructive in matrix representation.

Consider real Hamiltonian matrices of a general type

$$H_{mn} = H_{nn}\,\delta_{mn} + v_{mn}, \quad m,n = 1,\ldots,N, \tag{4.33}$$

where off-diagonal matrix elements $v_{mn} = v_{nm}$ are random and statistically independent, with $\langle v_{mn}\rangle = 0$ and $\langle v_{mn}^2\rangle = v^2$ for $|m-n| \le b$, and are zero otherwise. The most important characteristic of these Wigner band random matrices (WBRM) is the average energy level density ρ defined by

$$\frac{1}{\rho} = \langle H_{mm} - H_{m'm'}\rangle\,, \tag{4.34}$$

where $m' = m-1$. The averaging here and below is understood as performed either on a disorder, i.e. over many random matrices, or within a single sufficiently large matrix. These are equivalent, due to the assumed statistical independence of matrix elements. In other words, many matrices are statistically equivalent to a large one. In general, the quantum numbers m and n are arbitrary, but we will have in mind those related to the action variables, thus considering the quantum structure in momentum space. The basis in which the matrix elements are calculated is usually assumed to correspond to a completely integrable system with N quantum numbers, where N is the number of the degrees of freedom. By ordering the basis states in energy, we can represent N quantum numbers by a single one related to the energy, which is also an action variable.

In the classical limit the definition of WBRM (4.33) corresponds to the standard Hamiltonian $H = H_0 + V$, where the perturbation V is usually assumed to be sufficiently small, while the unperturbed Hamiltonian H_0 is completely integrable.

The quantum model (4.33) is defined by 3 independent physical parameters: ρ, v and b. The fourth parameter, the matrix size Q, is considered to be technical in this model provided $Q \gg d_e$ (see (4.38) below) is large enough to avoid boundary effects.

In terms of the unperturbed energy E_0, the classical chaotic trajectory of a given total energy E = const fills up the *energy shell* $\Delta E_0 = \Delta V$ with ergodic (microcanonical) measure w_e depending on a particular perturbation function V. In the quantum system this measure characterizes the shape (distribution) of the ergodic eigenfunction (EF) in the unperturbed basis. Conversly, if we keep the unperturbed energy fixed, E_0 = const, the measure w_e describes the band of energy surfaces E = const whose trajectories reach the unperturbed energy E_0. In a quantum system, the measure w_e in the latter case corresponds to the energy spectrum of a Green function (GS) with initial energy E_0. This characteristic was originally introduced also by Wigner (1955) as the 'strength function', the term still in use in nuclear physics. Nowadays, it is also called the 'local density of (eigen)states'.

For a typical perturbation represented by WBRM, w_e depends on the Wigner parameter $q = (\rho v)^2/b$, see Wigner (1955) and also Casati & Chirikov (1995b) and Fyodorov *et al.* (1996),

$$w_e(E) = \begin{cases} \dfrac{2}{\pi E_{SC}^2}\sqrt{E_{SC}^2 - E^2}, & |E| \le E_{SC}, \quad q \gg 1 \\[2ex] \dfrac{\Gamma/(2\pi)}{E^2 + \Gamma^2/4}\,\dfrac{\pi}{2\arctan[1/(\pi q)]}, & |E| \le E_{BW}, \quad q \ll 1, \end{cases} \tag{4.35}$$

provided $\eta = \rho v \gtrsim 1$, which is the condition for coupling neighbouring unperturbed states by the perturbation. In the opposite case, $\eta \ll 1$, the impact of the perturbation is negligible, and this fact is called ***perturbative localization.*** The latter is a well known *quantum* effect, but not one we are interested in (for chaotic phenomena it was first considered by Shuryak (1976)). What is less known is that for the coupling of *all* unperturbed states within the Hamiltonian band, a stronger condition is required, i.e.

$$\eta \gtrsim \sqrt{b} \quad \text{or} \quad q \gtrsim 1. \tag{4.36}$$

This is a simple estimate in first order perturbation theory. Indeed, the coupling is $\sim V/\delta E$. Within the band in question, energy detuning is $\delta E \sim b/\rho$, while the total random perturbation is $V \sim v\sqrt{b}$, leading to the estimate (4.36). In the opposite case, $q \lesssim 1$, a ***partial perturbative localization*** takes place. This is also a *quantum* phenomenon, but again not what we have in mind when speaking of quantum localization. The mechanism of perturbative localization is relatively simple and straightforward. This quantum effect is completely absent only in the semicircle (SC) limit of (4.35), where the width of the energy shell $\Delta E = 2E_{SC} = 2\sqrt{8bv^2} = 4\sqrt{2q}E_b \gg E_b$, and $E_b = b/\rho$ is the half-width of the Hamiltonian matrix band in energy. The last inequality allows for *diffusive* quantum motion within the energy shell, as a single random jump is $\sim b \ll \rho\Delta E$. The quantum localization under consideration is just related to the localization (suppression) of quantum diffusion by the interference effects in discrete spectrum, see e.g. Casati & Chirikov (1995b). Notice that the SC width immediately follows from the above estimate for perturbative localization: $\delta E \sim \Delta E \sim v\sqrt{b}$.

In the lower limit of (4.35), that of Breit–Wigner (BW), the full size of the energy shell $\Delta E = 2E_{BW} = 2E_b$ is equal to that of the Hamiltonian band. However, due to the partial perturbative localization explained above, the main peak of the quantum ergodic measure is considerably narrower, with width $\Gamma = 2\pi\rho v^2 = 2\pi q E_b \ll E_b$. This is again in accordance with the simple estimate: $\delta E \sim \Gamma \sim v\sqrt{\rho\Gamma}$.

To my knowledge, the quantum distributions (4.35) were theoretically derived and studied for GSa only. Classically, the measure w_e seems to be the same for both $E =$ const and $E_0 =$ const, as determined by the same perturbation V. One of the main recent results in studies of WBRM, see Casati *et al.* (to be published), is that the classical symmetry between EFs ($E =$ const) and GSa ($E_0 =$ const) is in general lost in quantum mechanics. Namely, in the ergodic case, such a statistical symmetry still persists. However, quantum localization drastically violates the symmetry, producing a very intricate and unusual global structure of quantum chaos.

In a sense, a conservative system is always localized (finite ΔE), even for ergodic motion. This sometimes is the origin of misunderstandings (see e.g. Feingold & Piro (1995)). In fact, such a *classical localization* is a trivial consequence of energy conservation, as was explained above. It persists, of course, in the classical limit as well. Here we are interested in *quantum localization* as explained above. In what follows it will simply be called localization.

As for maps, the localization in conservative systems also depends on the ergodicity parameter λ (see 3.19), where now

$$\lambda = a\frac{b^2}{d_e} = \frac{ab^{3/2}}{4\sqrt{2}c\eta}. \tag{4.37}$$

Here the ergodicity is not related to the total number of states Q, as in maps (3.19), but to that within the energy shell of width ΔE:

$$d_e = c\rho(\Delta E)_{SC} = 4c\eta\sqrt{2b}. \tag{4.38}$$

The Hilbert dimension d_e is also called the *ergodic localization length*, as it is a measure of the maximum number of basis states (BS) coupled by the perturbation in the case of ergodic motion. The numerical factor $c \approx 0.92$ is directly calculated from the limiting expression (4.35) for a particular definition of d in (3.30). Formally equation (4.38) is only valid in the SC region ($q \gg 1$), but was shown by computations to hold, within the accuracy of a few per cent, down to $q \approx 0.4$.

The parameter λ (4.37) was found in Feingold *et al.* (1991), and implicitely used there (without any reference to ergodicity). Its meaning was explained in detail in Casati *et al.* (1993), where the factor $a \approx 1.2$ was also calculated numerically.

Localization is characterized by the parameter

$$\beta_d = \frac{d}{d_e} \approx 1 - e^{-\lambda} < 1. \tag{4.39}$$

Here d stands for the actual average localization length of EFs measured according to the same definition (3.30). The empirical relation (4.39) was found in numerical experiments, see Casati *et al.* (1993), to hold in the whole interval $\lambda \leq 2.5$, and even up to $\lambda \approx 7$, see Casati *et al.* (to be published).

In the BW region $d_e = \pi\rho\Gamma = 2\pi^2 bq$ and $\lambda \approx ab/(2\pi^2 q) \gg 1$, as $q \ll 1$ in (4.35) and $b \gg 1$ in a quasiclassical region.

Hence, localization is only possible in the SC domain which was studied in Casati *et al.* (to be published).

The numerical results, see Casati *et al.* (to be published), were obtained from *two* individual matrices: the main one for the localized case, with parameters

$$\lambda = 0.23,\ q = 90,\ Q = 2400,\ v = 0.1,\ b = 10,\ \rho = 300,\ \eta = 30,\ d_e = 500,$$

and an additional one for the ergodic case, with parameters

$$\lambda = 3.6,\ q = 1,\ Q = 2560,\ v = 0.1,\ b = 16,\ \rho = 40,\ \eta = 4,\ d_e = 84.$$

All results are entirely contained in the EF matrix c_{mn}, which relates the eigenfunctions ψ_m to the unperturbed basis states φ_n:

$$\psi_m = \sum_n c_{mn}\varphi_n, \quad w_m(n) = |\psi(n)|^2 = c_{mn}^2 = w_{mn}, \tag{4.40}$$

in momentum representation, and with the eigenvalues $E_m \approx m/\rho$. From the matrix c_{mn} the statistics of EFs and GSa was evaluated. In order to suppress large fluctuations in individual distributions, an averaging over 300 in the central part of the matrix was performed in two different ways: with respect to the energy shell center ('global average', localization parameter (4.39) $\beta_d = \overline{\beta}_g$), and with respect to the centers of individual distributions ('local average', $\beta_d = \overline{\beta}_l$). Furthermore, the average $\langle\beta_d\rangle$ over β_d values from individual distributions was computed.

In the ergodic case, $\lambda = 3.6$, both average distributions for EFs are fairly close to the SC law. This is a remarkable result, as that law was theoretically predicted for *another* distribution, i.e. GSa. More precisely, the bulk ('cap') of the distributions is very close to the limiting SC (4.35), except for the vicinity of the SC singularities. Numerical values of the localization parameter ($\overline{\beta}_g = 1.08$, $\overline{\beta}_l = 0.94$, $\langle\beta_d\rangle = 0.99$) are in reasonable agreement with the scaled one $\beta_d = 0.97$ for $\lambda = 3.6$, see (4.39).

As expected, GS structure is similar: $\overline{\beta}_g = 1.07$, $\overline{\beta}_l = 1.06$, $\langle\beta_d\rangle = 0.98$.

For finite q, all distributions are bordered by two symmetric, steep tails which apparently fall off even faster than a simple exponential with characteristic width $\sim b$. A physical mechanism for tail formation is a specific quantum tunneling via intermediate BSs, see Flambaum *et al.* (1994). An asymptotic theory of the tails was developed by Flambaum *et al.* (1994), Wigner (1955) and Silvestrov (1995). Surprisingly, it works reasonably well even near the SC borders.

The structure of the matrix c_{mn} is completely different in the localized case, $\lambda = 0.23$. The EF *local* average is a clear evidence for exponential localization with $\overline{\beta}_l = 0.24$, which is again close to the scaled $\beta_d = 0.21$ for $\lambda = 0.23$. However, the *global* average reveals a nice SC (with tails) in spite of localization ($\overline{\beta}_g = 0.98$). It shows that, on average, the EFs homogeneously fill up the whole energy shell. In other words, their centers are randomly scattered over the shell.

Unlike the ergodic case, the localized GS structure is quite different from that of EFs. Both averages now well fit the SC distribution ($\overline{\beta}_g = 0.98$, $\overline{\beta}_l = 0.96$ as compared to $\overline{\beta}_g = 0.99$, $\overline{\beta}_l = 0.24$ for EFs). Thus, even though GSa look extended, they are localized! This is immediately clear from the third average $\langle\beta_d\rangle = 0.20$. The explanation of this apparent paradox is that even though the GSa are extended over the shell, they are *sparse*, i.e. contain many 'holes'.

In the analysis of the WBRM structure, the theoretical expression (4.38) for the ergodic localization length d_e (the energy shell width) was used. In more realistic and complicated physical models this might impede the analysis. In this respect the new method for direct empirical evaluation of d_e, and hence the important localization parameters β_d and λ in (4.39), from both the average distributions for GSa and the global average for EFs, looks very promising. It was elaborated by Casati *et al.* (to be published).

A physical interpretation of this structure, based upon the underlying chaotic dynamics, is the following. Spectral sparsity decreases the level density of the operative EFs, which is the main condition for quantum localization via decreasing the relaxation time scale, see e.g. Casati & Chirikov (1995b). However, the initial diffusion and relaxation are still classical, similar to the ergodic case, which requires extended GSa. On the other hand, EFs are directly related to the steady state density, see Chirikov *et al.* (1981), both being solid because of a homogeneous diffusion during the statistical relaxation.

This picture allows us to conjecture that for a classically *regular* motion the EFs also become sparse, so that EF/GS symmetry is apparently restored.

5. Conclusion: pseudochaos and traditional statistical mechanics

Quantum chaos is a particular but most important example of a new generic dynamical phenomenon, *pseudochaos*, in almost periodic motion. Statistical properties of motion with discrete spectra is not a completely new subject of research. It goes back to the time of intensive studies on the mathematical foundations of statistical mechanics *before* dynamical chaos was discovered or, say, understood (see e.g. Kac (1959)). This early stage of the theory, as well as the whole TSM, was equally applicable to both classical and quantum systems. As for the problem of pseudochaos, one of the most important rigorous results with far-reaching implications was the *statistical independence* of oscillations with incommensurate (linearly independent) frequencies ω_n, such that the only solution of the resonance equation

$$\sum_{n=1}^{N} m_n \omega_n = 0 \tag{5.41}$$

in integers is $m_n \equiv 0$ for all n. This is a generic property of real numbers. In other words, the resonant frequencies (5.41) form a set of zero Lebesgue measure. If we define $y_n = \cos(\omega_n t)$, the statistical independence of y_n means that the trajectory $y_n(t)$ is ergodic in the N-cube $|y_n| \leq 1$. This is a consequence of ergodicity of the phase trajectory $\phi_n(t) = \omega_n t \pmod{2\pi}$ on the torus $|\phi_n| \leq \pi$.

Statistical independence is the basic property of a set to which the probability theory is to be applied. In particular, the sum of statistically independent quantities,

$$x(t) = \sum_{n=1}^{N} A_n \cos(\omega_n t + \phi_n), \tag{5.42}$$

which is the motion with discrete spectrum, is a typical object of this theory.

However, familiar statistical properties, like Gaussian fluctuations, postulated (directly or indirectly) in TSM, are reached in the thermodynamic limit ($N \to \infty$) only, see Kac (1959). In TSM, this limit corresponds to infinite-dimensional models, see Kornfeld (1982), which provide a very good approximation for macroscopic systems, both classical and quantum.

What is really necessary for good statistical properties of the almost periodic motion (5.42) is a large number of frequencies N_ω, which makes the discrete spectrum continuous (as $N_\omega \to \infty$). In TSM this condition is satisfied by setting $N_\omega = N \to \infty$. The same holds for quantum fields which are infinite-dimensional. In the finite-dimensional quantum mechanics *another* mechanism, independent of N, works in the quasiclassical region $Q \gg 1$. Indeed, if the quantum motion (5.42) (with $x(t)$ replaced by $\psi(t)$) is determined by many ($\sim Q$) eigenstates, we can set $N_\omega = Q$, which is independent of N. The actual number of terms in expansion (5.42) depends on the particular state $\psi(t)$. For example, if it is just an eigenstate, the sum reduces to a single term. This is analogous to some special peculiar trajectories of classical chaotic motion, whose total measure is *zero*. Similarly, in quantum mechanics we have $N_\omega \sim Q$ for *most* states, if the system is *classically chaotic*.

If the motion is regular in the classical limit, the quantity $N_\omega(\ll Q)$ becomes considerably smaller. For example, in the standard map $N_\omega = Q$ in the ergodic case, $N_\omega \sim k^2$ in the case of localization (both cases being classically chaotic, $K > 1$) but only $N_\omega \sim k \ll k^2 \lesssim Q$ for classically regular motion ($K < 1$). The quantum chaos-to-order transition is not as sharp as the classical one, but the ratio $N_\omega(K > 1)/N_\omega(K < 1) \sim k \to \infty$ increases with the quantum parameter k.

Thus, as far as the mechanism of quantum chaos is concerned, we essentially *come back* from the ergodic theory to an old TSM, with the replacement of the number of the degrees of freedom N by the quantum parameter Q. However, in quantum mechanics, unlike TSM, we are not interested in the limit $Q \to \infty$, which is simply the *classical* mechanics. Here, the central problem is in the statistical properties for *large but finite* Q. This problem does not really exist in TSM, describing macroscopic systems. In *finite-Q* (or finite-N) pseudochaos we have to introduce the basic concept of a *time scale*, see Chirikov *et al.* (1981). This allows for interpretation of quantum chaos as a *new* dynamical phenomenon, related to but not identical with classical dynamical chaos. Hence, the term *pseudochaos*, emphasizing its difference from time asymptotic chaos in the ergodic theory.

In my opinion, the fundamental importance of quantum chaos is precisely in that it reconciles two apparently opposite regimes, regular and chaotic, in the general theory of dynamical systems. Study of quantum chaos helps us to better understand the old mechanism of chaos in multi-dimensional systems. In particular, the existence of characteristic time scales similar to those in quantum systems was conjectured in Casati & Chirikov (1995b).

Is pseudochaos really chaos?

Until recently, even the concept of classical dynamical chaos was rather incomprehensible, especially to physicists. I know that some researchers actually observed dynamical chaos in numerical or laboratory experiments. But... did their best to get rid of it as some artifact, noise or other interference! Now the situation in this field is upside down: many researchers insist that if an apparent chaos is not like that in classical mechanics (and in existing ergodic theory), then it is not chaos at all. Hence, sharp disputes over quantum chaos. The peculiarity of the current situation is that in most studies of 'true' (classical) chaos, a digital computer is used. But there only *pseudochaos* is possible, similar to that in *quantum* (not classical) mechanics!

Hopefully, this 'infant disease' of quantum chaos will be over before long.

The concept of quantum chaos presented above has been developed in long-term collaboration with G. Casati, J. Ford, I. Guarneri, F. M. Izrailev and D. L. Shepelyansky.

REFERENCES

ANDERSON, P. W. 1978 *Rev. Mod. Phys.* **50**, 191; see also LIFSHITS, I. M., GREDESKUL, S. A. & PASTUR, L. A. 1988 *Introduction to the Theory of Disordered Systems.* Wiley; FYODOROV, YA. V. & MIRLIN, A. D. 1994 *Intl. J. Mod. Phys.* **8**, 3795.

ARNOLD, V. I. & AVEZ, A. 1968 *Ergodic Problems of Classical Mechanics.* Benjamin.

BERMAN, G. P. & ZASLAVSKY, G. M. 1978 *Physica* A **91**, 450; see also BERRY, M., BALAZS, N., TABOR, M. & VOROS, A. 1979 *Ann. Phys.* **122**, 26.

BLÜMEL, R. 1994 *Phys. Rev. Lett.* **73**, 428; see also SCHACK, R. 1995 *Phys. Rev. Lett.* **75**, 581.

BOGOLYUBOV, N. N. 1945 *On Some Statistical Methods in Mathematical Physics*, p. 115; see also 1970 *Selected Papers*, vol. 2, p. 77. Naukova Dumka. In Russian.

BOGOMOLNY, E. B. 1992 *Nonlinearity* **5**, 805.

BOGOMOLNY, E. B. 1994 Private communication.

BRODY, T., FLORES, J., FRENCH, J., MELLO, P. PANDEY, A. & WONG, S. 1981 *Rev. Mod. Phys.* **53**, 385; see also MEHTA, M. 1991 *Random Matrices.* Academic Press.

CASATI, G., ED. 1985 *Chaotic Behaviour in Quantum Systems.* Plenum; see also SELIGMAN, T. & NISHIOKA, H., EDS. 1986 *Quantum Chaos and Statistical Nuclear Physics.* Springer; PIKE, E. & SARKAR, S., EDS. 1987 *Quantum Measurement and Chaos.* Plenum; CERDEIRA, H. *et al.*, EDS. 1991 *Quantum Chaos.* World Scientific; CVITANOVIĆ, P., PERCIVAL, I. & WIRZBA, A., EDS. 1992 *Quantum Chaos - Quantum Measurement.* Kluwer; CASATI, G., GUARNERY, I. & SMILANSKY, U., EDS. 1993 *Quantum Chaos.* North-Holland.

CASATI, G. & CHIRIKOV, B. V., EDS. 1995a *Quantum Chaos: Between Order and Disorder.* Cambridge Univ. Press.

CASATI, G. & CHIRIKOV, B. V. 1995b The legacy of chaos in quantum mechanics. In Casati & Chirikov (1995a), p. 3; see also *Physica* D **86**, 220.

CASATI, G., CHIRIKOV, B. V., FORD, J. & IZRAILEV, F. M. 1979 Lecture Notes in Physics, vol. 93, p. 334. Springer.

CASATI, G., CHIRIKOV, B. V. & FORD, J. 1980 *Phys. Lett.* A **77**, 91.

CASATI, G., FORD, J., GUARNERI, I. & VIVALDI, F. 1986 *Phys. Rev.* A **34**, 1413.

CASATI, G. *et al.* 1987 *Phys. Reports* **154**, 77; see also CASATI, G., GUARNERI, I. & SHEPELYANSKY, D. L. 1988 *IEEE J. of Quantum Electr.* **24**, 1420.

CASATI, G., CHIRIKOV, B. V., GUARNERI, I. & IZRAILEV, F. M. 1993 *Phys. Rev.* E **48**, R1613.

CASATI, G., CHIRIKOV, B. V., FUSINA, G. & IZRAILEV, F. M. Quantum steady state: relaxation and fluctuations. In preparation.

CASATI, G., CHIRIKOV, B. V., GUARNERI, I. & IZRAILEV, F. M. Quantum ergodicity and localization in generic conservative systems: the Wigner band random matrix model. To be published.

CHIRIKOV, B. V. 1979 *Phys. Reports* **52**, 263.

CHIRIKOV, B. V. 1985 *Phys. Lett.* A **108**, 68.

CHIRIKOV, B. V. 1986 *Foundations of Physics* **16**, 39.

CHIRIKOV, B. V. 1991a *Chaos* **1**, 95.

CHIRIKOV, B. V. 1991b Time-dependent quantum systems. In Giannoni *et al.* (1991), see below, p. 443.

CHIRIKOV, B. V. 1992a Linear chaos. In *Nonlinearity with Disorder.* Springer Proc. in Physics, vol. 67. Springer.

CHIRIKOV, B. V. 1992b The problem of quantum chaos. In Heiss (1992), see below, p. 1.

CHIRIKOV, B. V. 1993 The uncertainty principle and quantum chaos. In *Proc. 2nd Intl. Workshop Squeezed States and Uncertainty Relations*, p. 317. NASA.

CHIRIKOV, B. V. 1994 Natural laws and human prediction. In *Proc. Intl. Symposium Law and Prediction in (Natural) Science in the Light of our New Knowledge from Chaos Research*, Salzburg; see also Linear and nonlinear dynamical chaos. In *Proc. Intl. School Let's Face Chaos through Nonlinear Dynamics*, Ljubljana.

CHIRIKOV, B. V., IZRAILEV, F. M. & SHEPELYANSKY, D. L. 1981 *Sov. Sci. Rev.* C **2**, 209; see also *Physica* D **33**, 77 (1988).

COHEN, D. 1991 *Phys. Rev.* A **44**, 2292.

COURBAGE, M. & HAMDAN, D. 1995 *Phys. Rev. Lett.* **74**, 5166; see also COURBAGE, M. 1996 Unpredictability in some non-strongly chaotic dynamical systems. In *Proc. Intl. Conf. on Nonlinear Dynamics, Chaotic and Complex Systems* (ed. E. Infeld *et al.*). Zakopane, Poland, November 1995. Contributed papers. *J. Technical Phys.* **37**.

ECKHARDT, B. 1988 *Phys. Reports* **163**, 205.

FARQUHAR, I. 1964 *Ergodic Theory in Statistical Mechanics.* Wiley; see also PENROSE, O. 1979 *Rep. Prog. Phys.* **42**, 1937.

FEINGOLD, M., LEITNER, D. & WILKINSON, M. 1991 *Phys. Rev. Lett.* **66**, 986; see also *J. Phys.* A **24**, 1751.

FEINGOLD, M. & PIRO, O. 1995 *Phys. Rev.* A **51**, 4279; see also FEINGOLD, M. & LEITNER, D. 1993 Semiclassical localization in time-independent K-systems. In *Proc. Intl. Conference Mesoscopic Systems and Chaos*, Trieste.

FISHMAN, S., GREMPEL, D. & PRANGE, R. 1984 *Phys. Rev. Lett.* **29**, 1639; see also SHEPELYANSKY, D. L. 1987 *Physica* D **28**, 103.

FLAMBAUM, V. V, GRIBAKINA, A. A., GRIBAKIN, G. F. & KOZLOV, M. G. 1994 *Phys. Rev.* A **50**, 267.

FORD, J. 1995 Private communication.

FORD, J. *et al.* 1973 *Prog. Theor. Phys.* **50**, 1547.

FYODOROV, YA. *et al.* 1996 *Phys. Rev. Lett.* **76**, 1603.

GIANNONI, M., VOROS, A. & ZINN-JUSTIN, J., EDS. 1991 *Chaos and Quantum Physics.* North-Holland.

GOODRICH, K., GUSTAFSON, K. & MISRA, B. 1980 *Physica* A **102**, 379.

HEISS, D., ED. 1992 *Chaos and Quantum Chaos.* Springer.

IZRAILEV, F. M. 1990 *Phys. Reports* **196**, 299.

IZRAILEV, F. M. *et al.* Quantum diffusion and localization of wave packets in disordered media. To be published.

KAC, M. 1959 *Statistical Independence in Probability, Analysis and Number Theory.* Math. Ass. of America.

KORNFELD, I., FOMIN, S. & SINAI, YA. 1982 *Ergodic Theory.* Springer; see also KATOK, A. & HASSELBLATT, B. 1994 *Introduction to the Modern Theory of Dynamical Systems.* Cambridge Univ. Press.

KUMICAK, J. 1996 Physical meaning of the Λ operator of the Brussels school theory – explicit construction of Markov processes corresponding to the baker's transformation. In *Proc. Intl. Conf. on Nonlinear Dynamics, Chaotic and Complex Systems* (ed. E. Infeld *et al.*). Zakopane, Poland, November 1995. Contributed papers. *J. Technical Phys.* **37**.

MADDOX, J. 1994 *Nature* **372**, 403.

MISRA, B. & PRIGOGINE, I. 1983 Time, probability and dynamics. In *Long-Time Prediction in Dynamics* (ed. C. Horton *et al.*), p. 3. Wiley; see also PRIGOGINE, I. & DRIEBE, D. J. Time, chaos and the laws of nature. These proceedings.

PERELOMOV, A. M. 1986 *Generalized Coherent States and Their Applications.* Springer.

PRIGOGINE, I. 1963 *Non-equilibrium Statistical Mechanics.* Wiley; see also BALESCU, R. 1975 *Equilibrium and Non-equilibrium Statistical Mechanics.* Wiley.

ŠEBA, P. 1990 *Phys. Rev. Lett.* **64**, 1855.

SHEPELYANSKY, D. L. 1983 *Physica* D **8**, 208; see also CASATI, G. *et al.* 1986 *Phys. Rev. Lett.* **56**, 2437; DITTRICH, T. & GRAHAM, R. 1990 *Ann. Phys.* **200**, 363.

SHNIRELMAN, A. I. 1974 *Usp. Mat. Nauk* 6, **29**, 181; see also BERRY, M. 1977 *J. Phys.* A **10**, 2083; VOROS, A. 1979 Lecture Notes in Physics, vol. 93, p. 326. Springer; SHNIRELMAN, A. I. 1993 On the asymptotic properties of eigenfunctions in the regions of chaotic motion. Addendum in LAZUTKIN, V. F. *KAM Theory and Semiclassical Approximations to Eigenfunctions.* Springer.

SHURYAK, E. V. 1976 *ZhETF* **71**, 2039.

SILVESTROV, P. G. 1995 *Phys. Lett.* A **209**, 173.

SOKOLOV, V. V. 1984 *Teor. Mat. Fiz.* **61**, 128; see also HELLER, E. *et al.* 1991, 1992 *J. Chem. Phys.* **94**, 2723 (1991); *Phys. Rev. Lett.* **67**, 664 (1991); **69**, 402 (1992); *Physica* D **55**, 340 (1992).

STÖCKMANN, H. & STEIN, J. 1990 *Phys. Rev. Lett.* **64**, 2215; see also WEIDENMÜLLER, H. *et al.* 1992 *Phys. Rev. Lett.* **69**, 1296.

TODA, M. & IKEDA, K. 1987 *Phys. Lett.* A **124**, 165; see also BISHOP, A. *et al.* 1989 *Phys. Rev.* B **39**, 12423.

WEIGERT, S. 1990 *Z. Phys.* B **80**, 3; see also WEIGERT, S. 1993 *Phys. Rev.* A **48**, 1780; BERRY, M. 1990 True quantum chaos? An instructive example. In *Proc. Yukawa Symposium*;

Benatti, F. *et al.* 1991 *Lett. Math. Phys.* **21**, 157; Narnhofer, H. 1992 *J. Math. Phys.* **33**, 1502.

Wigner, E. 1955 *Ann. Math.* **62**, 548; see also **65**, 203 (1957).

Zaslavsky, G. M. 1981 *Phys. Reports* **80**, 157.

Foundations of non-equilibrium statistical mechanics

By J. P. DOUGHERTY

Department of Applied Mathematics and Theoretical Physics, University of Cambridge, U.K.

In this paper, some basic ideas in the foundations of non-equilibrium statistical mechanics (NESM) will be reviewed. The work will be presented in the context of classical mechanics, as attempts to combine classical and quantum formulations, by alternating between them, tend to be confusing. I shall consider only a closed system. Some writers work with open systems, but it seems to me that the additonal terms inserted to represent the interactions between an open system and its environment are merely empirical, so the theory loses all claim to fundamental status as it no longer starts from first principles.

The eventual aim will be to derive, from those first principles, macroscopic equations such as the heat conduction equation, etc. The connection with experiment can only be made when the derived equations are, in turn, solved. The derived equations are therefore only an intermediate stage between the microsopic laws and macroscopic observation.

---

1. Unstable dynamics

We note the following the features of classical microscopic physics. The equations are of Hamiltonian (symplectic) form. They are deterministic, meaning that solutions for given initial values exist and are unique. They are measure-preserving in phase space. Also, importantly, they are reversible. Closely allied to this is the point that there is no concept of equilibrium. They describe correctly, to within the limitations of classical mechanics, what would actually be seen microscopically if the means were available.

Hamilton's equations are generally non-integrable. To see what this means, consider first the exception, the 2-body problem. Phase space is 12-dimensional, (6 for each particle). There are 6 invariants that express the simple motion of the mass-centre, 3 expressing the constancy of the vector angular momentum, and 1 for the energy. These 10 first integrals reduce the problem to one in the 2 remaining dimensions, which is easily dealt with. For the 3-body problem, phase space is 18-dimensional. There are the same 10 invariants of the motion, with the obvious modification of their definitions to include all 3 particles. The initial data thus selects a particular 8-dimensional manifold. It also selects an initial point within that manifold. But the dimensionality cannot generally be further reduced. This is often reflected in the fact that the orbit can reach, in due time, any open set within the manifold. More seriously, the orbit is commonly unstable with respect to the initial values within the manifold, so small errors are magnified. Although the system is still deterministic, microscopic predictability has been lost, in the sense that no further progress is possible, at the foundational level, either with analytic or computational work.

Such dynamical systems, considered within the manifolds defined by all the available invariants, can be classified by properties listed here, in rising order of the severity of the description, in the following list: quasi-ergodic; ergodic; weak mixing; mixing; Lebesgue; Kolmogorov; Bernoulli. Many of these properties correspond to features of the spectrum of the Liouvillian, but the details are omitted here.

For a macroscopic system whose molecules have a given form and a given law of interaction, it is found to be extremely difficult to establish such properties rigorously.

For the present one has to assume, guided by circumstantial observational evidence, that a property at about the level of 'mixing', holds. This gap in the argument means that NESM does not yet form a completely logical branch of theoretical physics.

2. Observational considerations

Experiments deal with macroscopic variables, such as pressure, magnetisation, fluxes, etc. These may be simple discrete quantites, like total energy, or may involve new independent variables, such as position. They are not such as to identify individual particles.

An important concept, discussed by Jaynes (1983), p.297, is that of *macroscopic reproducibility*. We conceive an experiment as the preparation of some specimen of matter and the observation of macroscopic variables, perhaps evolving in time. The preparation does not control all the microsopic dynamics. If the experiment can be repeated with consistent macroscopic results it is macroscopically reproducible. This stands in contrast to the loss of microscopic predictability for large systems, but it undoubtedly occurs. Such experiments are the ones selected for publication and theoretical interpretation, and they indicate what will be a good choice of macroscopic variables to use for a given situation. If molecular physics had been such that macroscopic reproducibility never occurred, then the science of matter would take a very different form, statistical mechanics would not exist, and we would not be holding this meeting!

Macroscopically reproducible experiments are characterised by irreversibility, and the thermodynamic arrow of time. This means that prediction is possible, while retrodiction would be difficult, at any rate over a long period. This is borne out by the form of the macroscopic equations. They are well posed, in the sense of Poincaré, for future time but not for past time. The thermodynamic arrow coincides with the physiological arrow (which we experience directly) but that is really tautological.

States of macroscopic equilibrium enter naturally as the final outcome of irreversible evolution of isolated systems. Their existence, and irreversibility itself, are in apparent contradiction to the microscopic reversibility of the underlying dynamics. That paradox has been under discussion for more than a century!

3. Probability

It is widely agreed that probability theory must provide the solution to the problems just raised. So we write

$$\rho(X,t)\,\mathrm{d}\Omega = \text{probability that } X \text{ is in } \mathrm{d}\Omega, \tag{3.1}$$

where $X \in \Gamma$ is the representative point for the whole system, Γ is phase space. We also write $\rho \in E$ where E is an appropriate function space. Liouville's equation is

$$\dot{\rho} = L\rho, \tag{3.2}$$

where L is the Liouvillian operator.

One possible motivation for the introduction of probability in this context is to say that owing to the instability with respect to initial data, and the imperfection of observations, error bars have to be attached to the initial position in phase space. So one considers the evolution of a small set; the indicator function for that set evolves according to Liouville's equation, and more generally, since the equation is linear and homogeneous, any other solutions may be reached just by superposition. Of course, the equation also preserves positivity and the normalisation, as required for probability distributions.

Here we shall simply accept that one then embarks on the mathematical manipulation of probability, and the outcome is sufficient to explain NESM. If required to reveal what underlying philosophy of probability is being adopted, we would reply that it is the frequency interpretation, in accord with the idea of repetition of experiments already used in the concept of macroscopic reproducibility. Others (for example Jaynes) have argued for the Bayesian approach, about which we comment later. (A third possibility, based on the ergodic theorem relation between time- and phase-averages is completely out of place in NESM, and is in our opinion irrelevant to equilibrium statistical mechanics also.)

4. Macroscopic observables

Given the microscopic state, X, any quantity of the form $a(y) = M(y, X)$ can be regarded as an observable, where y is some new free variable. The possible values of $a(y)$ will generally lie in some new function space, A. A macroscopic observable is of this form, with the understanding that all that can be obtained is the expectation value of a, which by abuse of notation we also write as $a(y,t)$. (By allowing ρ to be a δ-function the microscopic version of a may be recovered.)

$$a(y,t) = \int_\Gamma M(y,X)\rho(X,t)\,\mathrm{d}\Omega. \tag{4.3}$$

This creates a map $M : E \to A$ which we shall also write $a = M\rho$. As is common in the theory of integral equations, the same symbol, M denotes both the map, and the kernel which expresses it in detail.

The problem is how to combine this with Liouville's equation $\dot{\rho} = L\rho$. A first step to this is to define a split of E in the form $E = E_1 \oplus E_2$, so $\rho_1 + \rho_2$. Here E_2 is the null space of M and E_1,

$$E_2 = \{\rho : M\rho = 0\}. \tag{4.4}$$

The choice of E_1 is not unique but the eventual theory should be independent of the choice made.

The map $P : E \to E$ defined by setting $\rho_2 = 0$ is a projection operator. The fundamental problem of NESM is that P does not in general commute with L. Again, as P is not invertible, a choice of $a(y)$ implies a unique value of ρ_1 but not of ρ_2.

5. Brussels calculus

The split of E implies a split of Liouville's equation in the form

$$\begin{bmatrix} \dot{\rho}_1 \\ \dot{\rho}_2 \end{bmatrix} = \begin{bmatrix} L_{11} & L_{12} \\ L_{21} & L_{22} \end{bmatrix} \begin{bmatrix} \rho_1 \\ \rho_2 \end{bmatrix}. \tag{5.5}$$

Supposing initial data $a(y, t_0)$ given, we wish to solve (5.5) in the interval (t_0, t) and attempt at the same time to eliminate ρ_2. Formal manipulation leads to

$$\left\{ \frac{\partial}{\partial t} - L_{11}(t) \right\} \rho_1(t) = D(t - t_0)\rho_2(t_0) + \int_{t_0}^{t} \psi(t - t')\rho_1(t')\,\mathrm{d}t'. \tag{5.6}$$

Here, $D(\cdot)$ and $\psi(\cdot)$ are new kernels with complicated expressions in terms of L_{ij} and the Greens function of L_{22}. This equation could, in principle, be used to compute the evolution of ρ_1 in a way that involves its values at intermediate times t', reminiscent of hysteresis, but the initial value of ρ_2 is still involved.

To make progress, we introduce a *fading hypothesis*, according to which 'macroscopic reproducibility' $\Leftrightarrow$ 'D and ψ are sufficiently rapidly fading'.

If (and only if) these properties hold, it is possible to eliminate ρ_2 by allowing $t_0 \to -\infty$, leading to results summarised in the next section.

6. Subdynamics

The equation obtained as just described is

$$\left\{\frac{\partial}{\partial t} - L_{11}(t)\right\}\rho_1(t) = \int_0^\infty \psi(\tau)\rho_1(t-\tau)\,\mathrm{d}\tau. \tag{6.7}$$

Note that although ρ_2 has been eliminated, there is no assumption that it is negligible, or zero.

Equation (6.7) is a linear, homogeneous Volterra integro-differential equation with infinite delay. It is time-transitive but not time-reversible. Its solutions (if any) form a vector space, associated with the particular choice of observable M, of the space of all solutions of Liouville's equation. The irreversibility has been introduced by the 'user' in the attempt to solve for future time. There is a large literature of mathematical investigations of equations of this type carried out of the last 30 years, see Gripenberg *et al.*(1990).

It may turn out that (6.7) does not possess any solutions, owing to the divergence of the integral. If that happens, it signifies that an impracticable choice of observation M has been selected for the particular system.

The investigation of (6.7) constitutes the subject of *subdynamics*, created by Prigogine and co-authors. The method used is the natural one of taking the Laplace Transform, the equation being of the convolution type. Actually they derive the Laplace transform of (6.7) by conducting the eliminations starting from the Laplace transform of (5.5).

7. More about subdynamics

If asked to nominate a result to be the analogue in NESM of the canonical distribution, we would suggest equation (6.7).

In practice, to use (6.7) for a particular system involves immense difficulties owing to the very complicated structure of ψ. Progress can only be made by embarking on perturbation expansions and other approximations. The book of Balescu (1975) offers a large amount of work following those lines.

For the system to approach equilibrium, as observed at the level M, some analytic conditions are required; for example, the two-body system would not approach equilibrium. It is believed that L needs to possess a continuous spectrum. Clearly, if L has a discrete eigenvalue, then its eigenfunction, if other than $\rho =$ constant, will prevent such approach to equilibrium, unless it lies entirely in E_2.

In practice, however, if L has a very closely packed but discrete spectrum, such that $\Delta\omega$ is a typical separation between the frequencies, then the distinction between that and a continuous spectrum cannot be detected if $\Delta\omega \ll$ (the age of the universe)$^{-1}$.

For a system where ψ is such as to imply an extremely short memory, equation (6.7) becomes effectively markovian, but still irreversible. For example, for a dilute gas with short-range interactions, it becomes the Boltzmann equation. But if this is not so, the system explores a bigger space of solutions, and exhibits hysteresis.

Some progress in exploring the mathematics of (6.7) has been made by replacing the real dynamical system with models like the Baker transformation, or elaborations of it, see Penrose & Coveney (1994).

8. Entropy

Two aspects of the preceding line of thought and development deserve close attention.

(i) It seems not to correspond with that of most physicists.

(ii) It makes no use, or even mention, of entropy. See Lebowitz (1993) for an exposition of the more familiar approach.

These two points are closely related, and will now be discussed. The (Gibbs) entropy is of course

$$S_{\mathrm{G}}[\rho] = -\int_{\Gamma} \rho \ln \rho \, \mathrm{d}\Omega \tag{8.8}$$

and, as is well known, S_{G} is invariant if ρ satisfies Liouville's equation.

Entropy is thus an intrinsic property of ρ and can be introduced by a consideration of the number of ways in which the distribution can be constructed, beginning with a division of phase space into cells (S is actually the logarithm of the number). If S is low, the distribution is in some sense an unlikely one. In a rather similar argument, the same concept occurs in information theory, where $-S$ appears as the information content of a distribution. These arguments appear in the first instance in the context of discrete distributions, and their use in the continuous case raises other questions. A further property of S is that it is additive for disjoint systems. Also, it can be identifed with the thermodynamic entropy, which is the quantity that increases according to the second law. Finally, the canonical distribution of equilibrium statistical mechanics is such as to maximise S subject to the macroscopic data.

This collection of ideas has created a widespread expectation that NESM can be founded on them. Irreversibility occurs because a system will progress from a less likely to a more likely state, and equilibrium represents the most likely state of all, with maximum entropy. Hence the law of increasing entropy, but this must be reconciled with the theorem that S is invariant. To do this one introduces the idea of 'coarse-graining', explained by the discarding of information.

Although many physicists think of statistical mechanics in those terms, it is difficult to see how, then, to formulate NESM from first principles and develop methodology for the construction of macroscopic equations. One author (Jaynes (1983), p. 287 ff) has had the courage to attempt it in detail. His methods, with some applications, have appeared in a book by Grandy (1988). We shall refer to this as the 'Maxent' approach.

9. Maxent development of NESM

The following is my own presentation of this material. We may take it in two stages. The notation is the same as that used earlier.

• *First stage.* We suppose $a(0)$ given and wish to predict $a(t)$. As in our previous discussion, there is no unique $\rho(0)$, but we settle the choice by requiring $\rho(0)$ to maximise $S[\rho]$ subject to $M(\rho) = a(0)$. We shall call this the 'quasi-equilibrium' distribution, ρ_q associated with a. These distributions take the form of the exponential of a linear expression that contains Lagrange undetermined multipliers, similar to the canonical distribution. Having done this, compute $\rho(t)$ at a later time from Liouville's equation, from which $a(t)$ follows.

This cannot be correct. It is not time-transitive: if used for the time intervals (t_1, t_2) then (t_2, t_3), the results disagree with those obtained by going directly from t_1 to t_3. Also, the quasi-equilibrium distribution does not exhibit dissipative processes.

• *Second stage.* Modify the above rule as follows. Although the procedure is to start at time $t = 0$, accept that the system has already been evolving for $t < 0$, and that data $a(t)$ from an interval $t_1 < t < 0$ is to be included. Add the additional constraint that ρ should maximise $S[\rho]$ with respect to that earlier data also. As S is actually constant, we can set up the maximisation procedure at a single time, $t = 0$; the influence of the earlier times is imposed by (formally) translating the data from the time it is recorded to the time $t = 0$ using Heisenberg-type operators (in a non-quantum context). Lagrange multipliers are then required for all that earler data, and the expression involving the multipliers becomes an integral over past time. The resulting formulation is difficult and nonlinear. Some progress can be made in the near-equilibrium case exploiting the convenient form of the canonical distribution in calculations. We note that no mention is made of unstable dynamics, and it is unclear how t_1 is to be assigned.

10. Equivalence with subdynamics

To ensure transitivity in the macroscopic equations, we need $t_1 \to -\infty$. If the resulting integrals diverge, the failure should be identified with a lack of macroscopic reproducibility. This is beginning to look reminiscent of subdynamics, though in its integral equation (rather than Laplace transform) version. On closer examination, it emerges that the imposition of the constraints is actually carrying out the same work as the subdynamics elimination of ρ_2 using different variables. But in that case what is the rôle of the maximisation of S, which is absent in subdynamics?

An easy analogy for the above occurs in the elementary case of discrete probabilities, such as the dice problem. Here, the set of probabilities $\{p_i\}, 1 \leq i \leq 6$, are to be assigned on the basis of some rather limited data, to be handled as constraints under which S is to be maximised. Normally the number of constraints is less than the number of probabilities to be determined, and one can solve for the probabilities and the undetermined multipliers. The entropy is then indeed made stationary. If the number of constraints equals the number of probabilities, one can solve the constraints alone to determine the probabilities, and S. Nearby values are inaccessible, so S is not stationary (though the solution might not be unique). But, not noticing the situation, one might embark on some alternative route through the algebra of solving for the probabilities and the undetermined multipliers. Although no error would result, the outcome would no longer represent a variational calculation.

The methodology associated with the two approaches, in the sense of perturbations, etc., looks very different, and it would seem that both make useful contributions.

11. Law of increasing entropy

As mentioned before, entropy did not even appear in the description of subdynamics, and $S[\rho]$ is invariant if ρ satisfies Liouville's equation. S is the idea behind the Maxent approach, but would also be invariant. Can the familiar law of increasing entropy be incorporated into NESM, in either or both of these approaches?

A possible answer is the following. Let $a(t)$ be the macroscopic evolution obtained from these methods, and let $\rho_q(t)$ be $\rho_q[a(t)]$ defined as previously. Although we discarded this as a basis for evolution equations, here it is merely defined for the present purpose, and does not satisfy Liouville's equation. Now define an *observational entropy* $S_{\text{obs}}(t) = S[\rho_q(t)]$. Then it is plain that $S_{\text{obs}}(t)$ generally increases, assuming macroscopic reproducibility, and eventually arrives at its maximum value, subject to the physical invariants, when thermal equilibrium is reached. This does not prove that the increase would be *monotonic*, though in practice we know that it almost always is so. Exceptions are known: spin echoes and plasma wave echoes. It remains unclear why the exceptions are so rare. However, on this explanation, the second law would appear not to have the fundamental status that is usually ascribed to it; in fact its association with the concept of macroscopic reproducibility may actually lead one to regard it as tautological!

More details of some of the material reviewed here may be found in my recent articles, see Dougherty (1993) and Dougherty (1994).

REFERENCES

BALESCU, R. 1975 *Equilibrium and Non-equilibrium Statistical Mechanics.* Wiley.

DOUGHERTY, J. P. 1993 Explaining statistical mechanics. *Studies in History and Philosophy of Science* **24**, 843–866.

DOUGHERTY, J. P. 1994 Foundations of non-equilibrium statistical mechanics. *Phil. Trans. Royal Soc.* A **346**, 259–305.

GRANDY, W. T. 1988 *Statistical Mechanics*, vol. 2.

GRIPENBERG, G., LONDEN, S.-O. & STAFFANS, O. 1990 *Volterra Integral and Functional Equations.* Cambridge.

JAYNES, E. T. 1983 *Collected Papers* (ed. R. D. Rosenkrantz). Reidel.

LEBOWITZ, J. L. 1993 Macroscopic laws, microscopic dynamics, time's arrow and Boltzmann's entropy. *Physica* A **194**, 1–27.

PENROSE, O. & COVENEY, P. V. 1994 Is there a 'canonical' non-equilibrium ensemble? *Proc. Royal Soc.* A **447**, 631–646.

Thermomechanical particle simulations

By W. G. HOOVER[1–3], H. A. POSCH[1], CH. DELLAGO[1], O. KUM[2,3], C. G. HOOVER[2], A. J. DE GROOT[2,3] AND B. L. HOLIAN[4]

[1]Institut für Experimentalphysik, Universität Wien, Wien, Austria

[2]Lawrence Livermore National Laboratory, Livermore, USA

[3]Department of Applied Science, University of California at Davis/Livermore, Livermore, USA

[4]Los Alamos National Laboratory, Los Alamos, USA

Thermomechanics is a generalization of classical mechanics in which heat transfer and temperature play essential roles. It is specially useful away from equilibrium. Here we review thermomechanical particle simulation techniques and their applications, both small-scale and large. We illustrate the impact of thermomechanical methods on irreversible statistical mechanics and on the interpretation of macroscopic irreversible phenomena. Pressing problems for the future are also outlined and discussed.

---

1. Introduction

To simulate open-system flow processes involving heat transfer and dissipation it is necessary to augment conventional Newtonian mechanics with explicit treatments of temperature and heat reservoirs. In § 2 we review the definition of the ideal-gas temperature, and show that it is appropriate for thermomechanical simulations. In § 3 we use Hamilton's Principle of Least Action to incorporate this definition of temperature, with corresponding temperature controls, into particle-based simulations.

Thermomechanical simulations carried out in this way have already led to a variety of results. On the most fundamental microscopic level, a description of the flow within the many-body phase space, the simulations establish interconnections linking the spectrum of Lyapunov instability exponents to the information dimension of time-reversible multifractal repellors and attractors, and to the rate of external entropy production, all in accord with the Second Law of Thermodynamics. These relationships are discussed generally, and illustrated for many-body field-driven flows, in § 4 and § 5. We then consider, in § 6, these fundamental microscopic concepts for the simplest possible problem, the two-dimensional isokinetic Galton Board. Here, the phase space is relatively simple, so that the multifractal nature of the nonequilibrium phase-space distributions can be visualized completely.

Thermomechanical particle simulations have so far provided not only all the relatively simple properties, such as the equilibrium mechanical and thermal equations of state and the nonequilibrium transport coefficients, but also a quantitative treatment of some of the more complicated macroscopic flow processes, such as plasticity and fracture, for which no satisfactory macroscopic explanation exists. Some recent results are described in § 7.

On the macroscopic level, matter is described in terms of its constitutive properties, rather than in terms of interatomic forces. Because particle methods tend to be unusually stable, they are desirable tools for solving macroscopic problems too. But the microscale time and spatial limitations of atomistic particle methods mean that macroscopic problems require macroscopic particles. For such macroscopic problems, Smooth Particle Applied Mechanics is a useful method with some interesting connections to molecular dynamics. In § 8 we discuss the application of Smooth Particle Applied Mechanics to the simulation of hydrodynamic instabilities. Finally, we comment on some of the puzzling loose ends awaiting a better understanding in the near future.

2. Temperature far from equilibrium

In equilibrium thermodynamics there are many alternative definitions of temperature. The most familiar are:

(i) the ideal-gas temperature scale defined by the pressure of an equilibrium ideal-gas thermometer,

$$T_{\text{ideal gas}} \equiv \frac{PV}{Nk},$$

where P is the isotropic pressure exerted by an N-particle ideal gas in a volume V,

(ii) the entropic temperature defined by the isochoric entropy derivative of the internal energy,

$$T_{\text{entropic}} \equiv \left(\frac{\partial E}{\partial S}\right)_V.$$

The lack of a satisfactory definition of entropy far from equilibrium, as well as the relative simplicity of the ideal-gas temperature scale, both suggest that the ideal-gas thermometer be selected to define temperature away from equilibrium. A convincing case for this choice can be based on a detailed calculation, using kinetic theory.

If we consider a heavy particle, with mass M and velocity U, able to collide with representative light ideal-gas particles, of mass m and with velocities $\{u\}$ chosen from the Maxwell–Boltzmann velocity distribution characterizing an equilibrium ideal gas at a fixed temperature T, the average momentum transfer $\langle\Delta(MU)\rangle$ and energy transfer $\langle\Delta(MU^2/2)\rangle$ per collision are simple averages, weighted with the mean speed, $|u-U|$, over the Maxwell–Boltzmann distribution of gas-particle velocities. The results are (i) that the heavy-particle velocity decays exponentially to zero while (ii) the average energy, $\langle MU^2/2\rangle$, approaches the equilibrium value, $DkT/2$, in D dimensions. The number of collisions required for the velocity and energy changes is of order M/m.

Our theoretical knowledge of dilute-gas behavior is summarized by the Boltzmann Equation, which leads again to these same results for the effects of collisions on the heavy-particle velocity and energy. A one-line derivation of them, given by Salmon (1980), is particularly instructive. He takes the point of view that gas-phase momenta change in such a way as to maximize the Boltzmann entropy. He represents this tendency toward equilibration by introducing a generalized entropic force, defined as a gradient in momentum space and taken in the direction of increasing Boltzmann entropy. The magnitude of the characteristic force is determined by the kinetic temperature T. The time required for the force to become effective defines a collision time, τ. Salmon's simple picture leads to the following *ad hoc* recipe for the entropic force:

$$\left(\frac{\mathrm{d}p}{\mathrm{d}t}\right)_S \equiv -\frac{mkT}{\tau}\nabla_p \ln\left(\frac{f}{f_{\text{eq}}}\right), \quad f_{\text{eq}} \propto \exp\left(-\frac{p^2}{2mkT}\right),$$

which then gives directly the Fokker–Planck equation for the time evolution the one-particle distribution function:

$$\frac{\mathrm{d}f}{\mathrm{d}t} = -\nabla_p\cdot\left[f\left(\frac{\mathrm{d}p}{\mathrm{d}t}\right)_S\right] \equiv \frac{1}{\tau}\nabla_p\cdot(pf) + \frac{mkT}{\tau}\nabla_p^2 f.$$

The Fokker–Planck equation is of course nothing other than a weak-collision limit of the Boltzmann equation. But its interpretation as the result of an entropic force is interesting.

This point of view associates a common origin for (i) the decay of the velocity toward zero together with (ii) the diffusive randomizing force of collisions, leading to the equilibrium value of the kinetic temperature.

3. Motion algorithms from Least Action

Gillilan & Wilson (1992) showed that the simplest equilibrium application of the Principle of Least Action to conservative particle dynamics, minimizing the action integral for a trajectory from $-\mathrm{d}t$ to $+\mathrm{d}t$, gives rise to the familiar Størmer integration algorithm, which we write here explicitly as a set of particle motion equations, one for each degree of freedom:

$$\{x_- - 2x_0 + x_+ = (\mathrm{d}t)^2 F_0/m\}.$$

The subscripts indicate the times $\{-\mathrm{d}t, 0, +\mathrm{d}t\}$. In the variational calculation the end-point coordinates $\{x_-, x_+\}$ are fixed. The set of intermediate coordinates $\{x_0\}$ is then varied to minimize the trapezoidal-rule integration of the Lagrangian. The equilibrium motion algorithm which results, the Størmer algorithm, has been used by generations of scientists for solving Newton's conservative equations of motion. The only technical details required to ensure an efficient simulation are (i) short-ranged forces vanishing linearly at a cutoff, so as not to degrade the second-order integrator accuracy, and (ii) a linked-list catalog of near neighbors, so as to make the simulation time proportional to N rather than to N^2.

The equilibrium Størmer algorithm is patently time-reversible. Levesque & Verlet (1993) pointed out that this reversibility is exact, to the very last bit, provided that the algorithm is applied in an *integer* state space, with all coordinates $\{x(n\mathrm{d}t)\}$ integers and with the coordinate changes due to the forces rounded off to integers also. There is very likely no way to apply this idea of exact 'bit reversibility' away from equilibrium.

The least action principle itself *can* be applied away from equilibrium, where it is necessary to control the temperature, or the internal energy, or other dynamical variables. Consider the simplest case, temperature control. To keep the kinetic temperature fixed during the time interval from $-\mathrm{d}t$ to $+\mathrm{d}t$ requires a Lagrange multiplier to impose the constraint $\mathrm{d}K/\mathrm{d}t = 0$. The resulting equations of motion:

$$\left\{\frac{\mathrm{d}p}{\mathrm{d}t} = F - \zeta_K p\right\}, \quad \zeta_K = -\frac{1}{2K}\frac{\mathrm{d}E}{\mathrm{d}t},$$

can be solved with a slight modification of the Størmer algorithm or, if better accuracy is desired, by using a conventional fourth-order Runge–Kutta integrator.

It is noteworthy that these isokinetic equations of motion are still time-reversible. In the reversed motion both the momenta $\{p\}$ and any friction coefficients $\{\zeta_K\}$ change sign. It is straightforward to generalize the least-action derivation of the motion algorithm to treat time-dependent constraints or to constrain the internal energy in the presence of external fields. We explore both the constrained and unconstrained algorithms in the two following sections.

4. Conservative nonequilibrium simulations

The difficulties involved in simulating nonequilibrium flows with Newtonian mechanics can best be appreciated by considering a simple example, a fluid in which half the particles are accelerated to the right, and half to the left, by an external field of fixed strength F. We choose to study a two-dimensional square 36-particle system of particles interacting

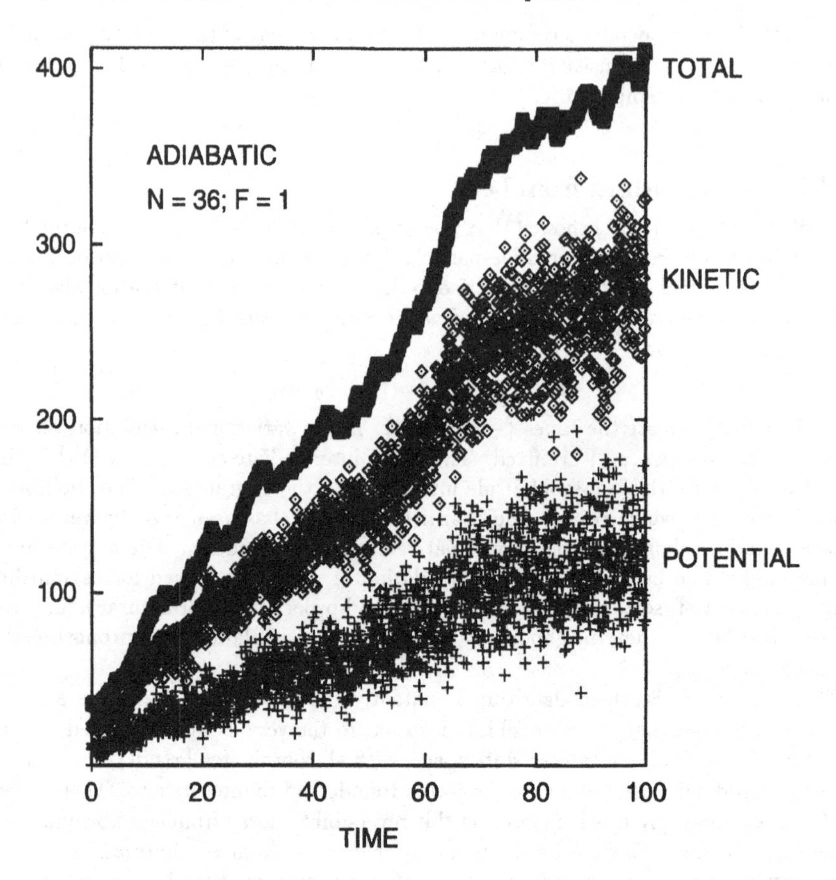

FIGURE 1. Time variation of the total, kinetic, and potential energies for a system of 36 particles driven by a field of unit strength. The mechanics is purely Hamiltonian, so that the system heats up with no increase in phase volume.

with a short-ranged repulsive potential,

$$\phi = 100\,\epsilon\left[1-(r/\sigma)^2\right]^4.$$

The overall reduced density is unity, so that $V = N\sigma^2$. For convenience we choose reduced units with the energy ϵ, the collision diameter σ, and the particle mass m all set equal to unity. We further choose the forces due to the external field F equal to $\{(\pm 1, 0)\}$.

With the periodic boundaries usual in simulations of bulk properties, intuition, bolstered by the numerical results shown in figure 1, suggests that such a field-driven system has no nonequilibrium steady state. If the field is reasonably strong, currents develop and the system heats up. The rate at which work is done by the field, $F \cdot JV$, where J is the current density, then causes a proportional heating of the fluid. As the fluid heats up, the heating effect of the field becomes relatively less important, until, at sufficiently high temperature, the motion more and more closely approximates that of an equilibrium system. Evidently such an adiabatically heated system never achieves a steady state and instead only proceeds, with gradually increasing fluctuations, in the general direction of equilibrium.

It is interesting to see that, even with a very strong external field, the resulting nonequi-

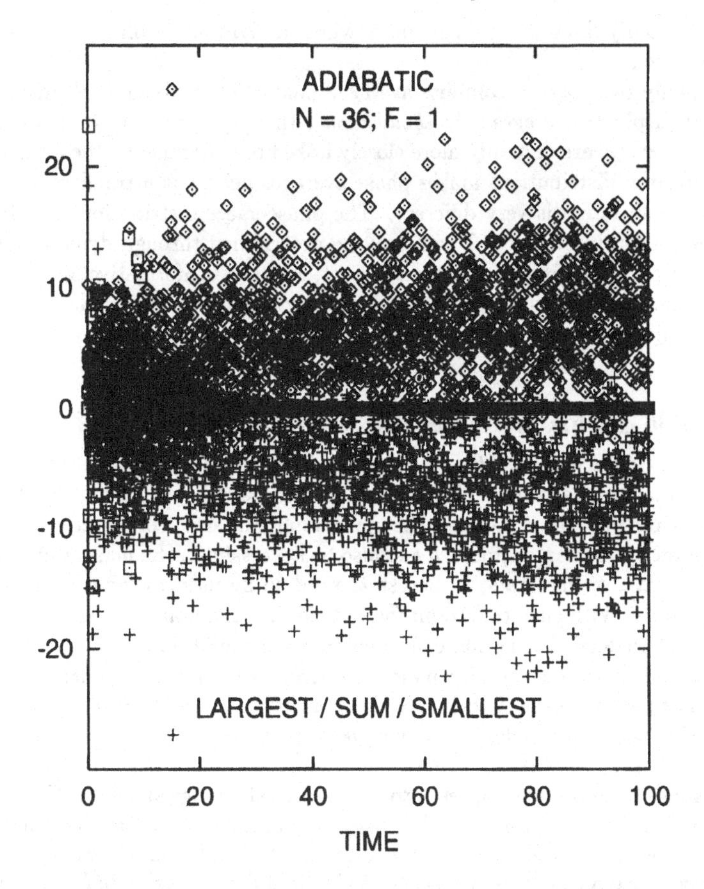

FIGURE 2. Time variation of the largest and smallest local Lyapunov exponents in the 36-particle adiabatic simulation of figure 1. The sum of the local exponents is also plotted, and shows that the instantaneous values become paired, following the decay of transients, at a reduced time of approximately 40.

librium motion remains adiabatic and its phase-space description is straightforward. Such a phase-space flow obeys Liouville's Theorem, taking place at constant phase volume. Thus, if we imagine the motion of a small phase-space hypervolume surrounding our trajectory's initial state, the density associated with that volume is carried along forever, without change. Figure 1 shows the increase of energy and temperature for a reduced time of 100. Figure 2 shows the pairing of the local (instantaneous) Lyapunov exponents that occurs, as it must for any solution of Hamilton's equations of motion. The vanishing of the summed exponents shows that there is no change in the comoving phase-space volume.

This example, like the more familiar free-expansion one, shows the inadequacy of Gibbs' statistical entropy away from equilibrium. Entropy, unlike pressure and temperature, has no mechanical analog, and so is not a useful concept far from equilibrium. There is no analog of Zermélo's Recurrence Paradox here because the periodic boundaries make energy a multiple-valued function of the particle coordinates. The Lyapunov spectrum for such a time-reversible flow, two exponents for each degree of freedom, is composed of rapidly fluctuating Smale pairs which individually sum to zero, as shown, for one pair, in figure 2.

There are only two ways to compute averaged quantities, as dynamical time averages or as statistical phase averages. *At equilibrium* either approach can be followed. Although time averages are certainly more closely linked to experiments, the availability of Gibbs' equilibrium distributions makes phase averages nearly as attractive. *Away from equilibrium* the situation is very different. The phase-space distribution is multifractal and is *never* known in advance. It can only be determined through dynamical simulation. Thus, away from equilibrium, the simplest, and most informative, approach is to analyze a single long trajectory generated by special steady-state boundary conditions. We illustrate this approach in the following section.

5. Reversible dissipative nonequilibrium simulations

In his stimulating lectures *On the Character of Physical Law*, Feynman repeatedly emphasized the fundamental nature of Hamilton's Principle of Least Action. We have seen that this principle provides a basis for the most commonly used conservative integration algorithm. Exactly the same approach can also be successful in the nonequilibrium case, where thermostats, or ergostats, or barostats need to be used to induce a nonequilibrium steady state. With such constraint forces field-driven heating no longer leads to an unsteady state. Instead the motion converges to a nonequilibrium steady state. Under such constrained conditions, the Lyapunov spectrum is no longer symmetric, though a modified Smale pairing of the time-averaged exponents can persist. Sarman *et al.* (1992) established the conditions under which 'conjugate pairs' of exponents undergo identical shifts.

In phase space the motion collapses onto a multifractal strange attractor. The information dimension of this attractor is the dimensionality of the largest comoving phase-space object which does not get smaller as time goes on. As Kaplan and Yorke suggested, it can be estimated from the Lyapunov spectrum by finding the maximum number of exponents which can be summed, with the sum remaining positive. In the case of field-driven currents the numerical results indicate an information dimension which decreases below the equilibrium value in proportion to the square of the field strength.

To achieve an isoenergetic steady state, the equations of motion derived from the Least Action Principle are the following:

$$\left\{\frac{\mathrm{d}p}{\mathrm{d}t} = -\nabla\Phi \pm F - \zeta_E p\right\}, \quad \zeta_E = \sum \frac{\pm F\cdot p}{2mK}.$$

Figure 3 shows results for the same 36-particle field-driven system discussed earlier, but with an isoenergetic constraint imposed for reduced times greater than 50. The kinetic and potential energies continue to fluctuate normally. The instantaneous values of the largest and smallest Lyapunov exponents, and their sum, is shown in figure 4. Though Sarman *et al.* (1992) showed that the pairing rule must be satisfied for the global averaged exponent pairs, a close examination of the data here show that there is no exact pairing for the local exponents. Considerably larger deviations from instantaneous pairing occur if an isokinetic, rather than isoenergetic, constraint is imposed. In either constrained nonequilibrium case the overall sum of the global exponents is negative, reflecting the collapse of the comoving phase volume onto a strange attractor. This collapse, which is qualitatively unlike the adiabatic constant-volume flow, corresponds to a steady divergence of the Gibbs entropy, which approaches $-\infty$ at long times.

The special time-reversible equations of motion used here have interesting consequences

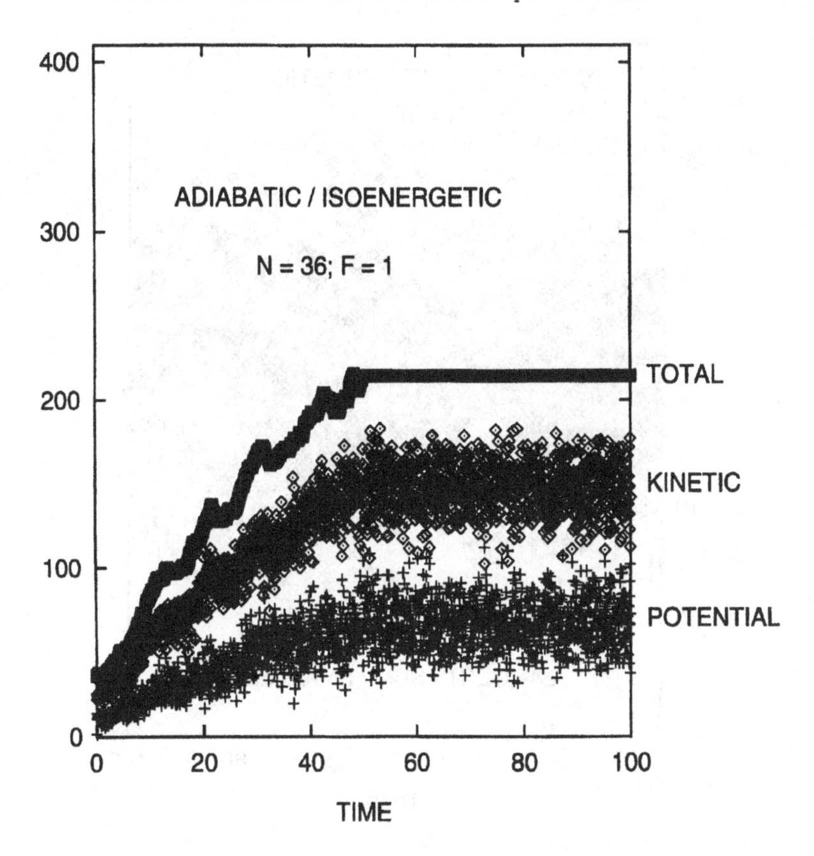

FIGURE 3. Time variation of the total, kinetic, and potential energies for a system of 36 particles driven by a field of unit strength. The total energy is constrained, and the comoving phase volume begins to collapse to a strange attractor, at a reduced time of 50.

for the phase-space flow. Liouville's Theorem for the flow establishes that the constrained comoving phase volume changes with time according to the instantaneous values(s) of the friction coefficient(s):

$$\frac{\mathrm{d}\ln f}{\mathrm{d}t} \equiv \sum \zeta_E,$$

where the sum includes all ergostatted degrees of freedom. Conservation of the comoving phase-space probability, $f\otimes$, where $\otimes$ is the small comoving element of phase volume, occupied by f, relates the sum of Lyapunov exponents to the sum of the friction coefficients:

$$\sum \lambda \equiv \left\langle \frac{\mathrm{d}\ln\otimes}{\mathrm{d}t} \right\rangle \equiv \left\langle -\frac{\mathrm{d}\ln f}{\mathrm{d}t} \right\rangle = \left\langle -\sum \zeta_E \right\rangle,$$

where the angular brackets $\langle\cdot\rangle$ indicate a time average. Finally, the external heat exchanged through the friction coefficients, divided by the temperature of the exchange, gives the rate of external entropy production, $\mathrm{d}S/\mathrm{d}t$. As a direct consequence of the least-action approach to deterministic temperature and heat transfer there results the exact chain of relations linking the *instantaneous* friction coefficients, the *instantaneous*

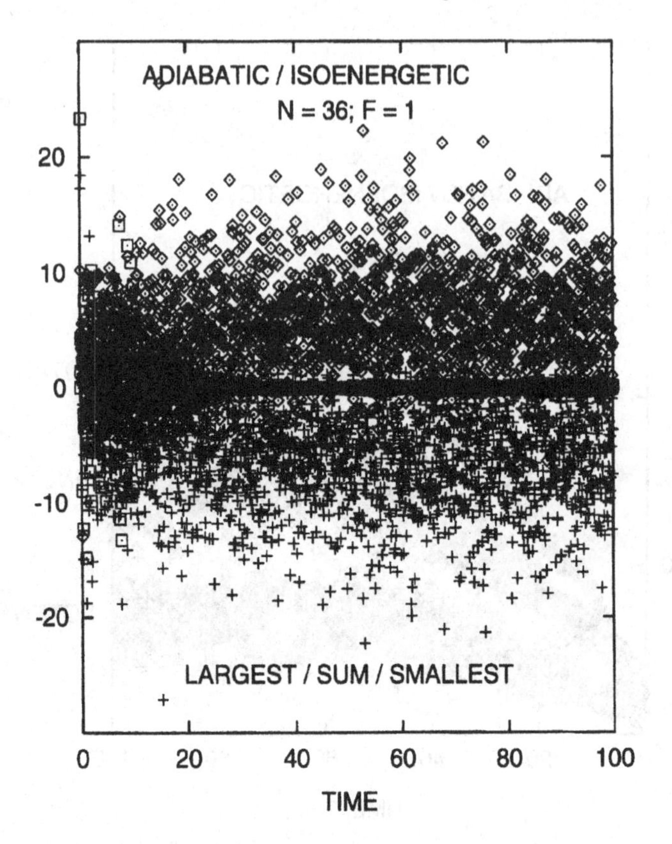

FIGURE 4. Time variation of the largest and smallest local Lyapunov exponents in the 36-particle constrained isoenergetic simulation of figure 2. In this case the exact instantaneous pairing of pairs of exponents is destroyed by the isoenergetic constraint.

Lyapunov exponents, and the *instantaneous* entropy production:

$$\sum \zeta_E \equiv -\sum \lambda \equiv \frac{\mathrm{d}(S/k)}{\mathrm{d}t}.$$

Of course, this chain of instantaneous equalities applies just as well to the time-averaged quantities.

The qualitative phenomenon illustrated here, the breaking of the symmetry of the Lyapunov spectrum, with the summed spectrum reproducing the rate of external entropy production, is a perfectly general consequence of the time-reversible equations of motion based on the Principle of Least Action. Because the geometric features associated with irreversibility are much more easily demonstrated in small systems, we consider the two-body version of this nonequilibrium problem in the next section.

6. Quantitative results from the Galton board

Historically, the Galton Board was the first system for which the generic multifractal time-reversible phase-space structure just discussed was established, see Machta & Zwanzig (1983), Ladd & Hoover (1985), Hoover *et al.* (1985), Moran *et al.* (1987), Hoover *et al.* (1988), Hoover & Moran (1989), Hoover & Moran (1992), Vance (1992), Hoo-

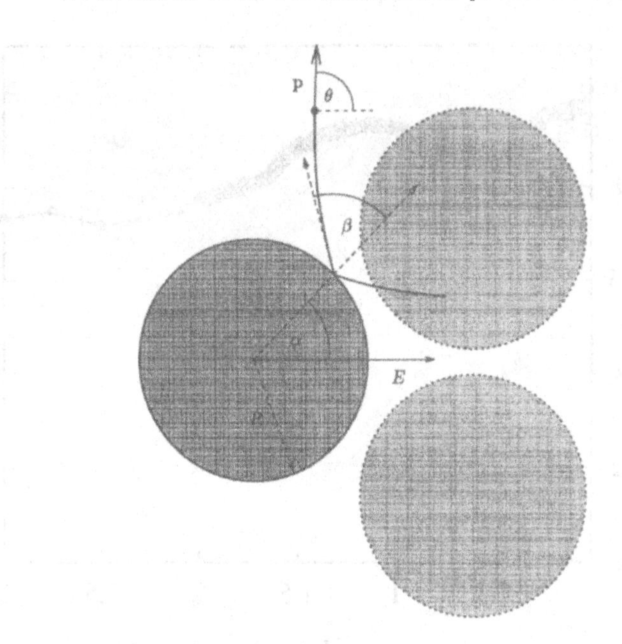

FIGURE 5. Definition of the angles α and β defining a hard-disk collision in the Galton Board. θ gives the direction of the particle velocity relative to the field direction.

ver *et al.* (1992), Chernov *et al.* (1993), Petravic *et al.* (1994), Lloyd *et al.* (1995) and Dellago *et al.*. The first nonequilibrium simulations, see Machta & Zwanzig (1983), incorporated shear, and were a caricature of viscous flow. By 1985, studies of field-driven diffusion became more common than studies of shear, mainly because diffusion is a slightly simpler problem, with fewer phase-space dimensions. These two viscous and diffusive Galton Board models are representative of the computer revolution in physics, in which models are chosen for study because they are both instructive and computationally feasible, though not necessarily analytically tractable. Thus the Galton Board is more informative than is the Baker Map though the Baker Map is more easily analyzed. The Galton Board occupies an interesting middle ground in which many of the results established numerically could also be deduced from a purely theoretical basis. Such a model is very useful for the progress of mathematical physics. Simulations stimulate theoretical advances which then feed back new ideas for simulation. That the Galton Board has been specially fruitful for nonequilibrium statistical mechanics is clear from the references cited by Dellago *et al.* (1995).

The Galton Board problem is a special two-body version of the simple field-driven constrained problem of the last section. The periodic boundaries are chosen to be consistent with a particular lattice structure, here the triangular lattice, and the driving field is chosen in a particular direction, here perpendicular to one of the three directions of closest packing. To simplify the dynamics, the interparticle interaction is further simplified to an impulsive hard-disk interaction. The field-dependent conductivity is defined in the usual way, as the ratio of the current to the field strength. Though the problem is at first glance complicated, being described by eight phase-space variables, $\{x, y, p_x, p_y\}$, fixing the center of mass and the kinetic energy gives five constraints, so that the motion occurs in a three-dimensional phase space. The description can be further simplified, reducing

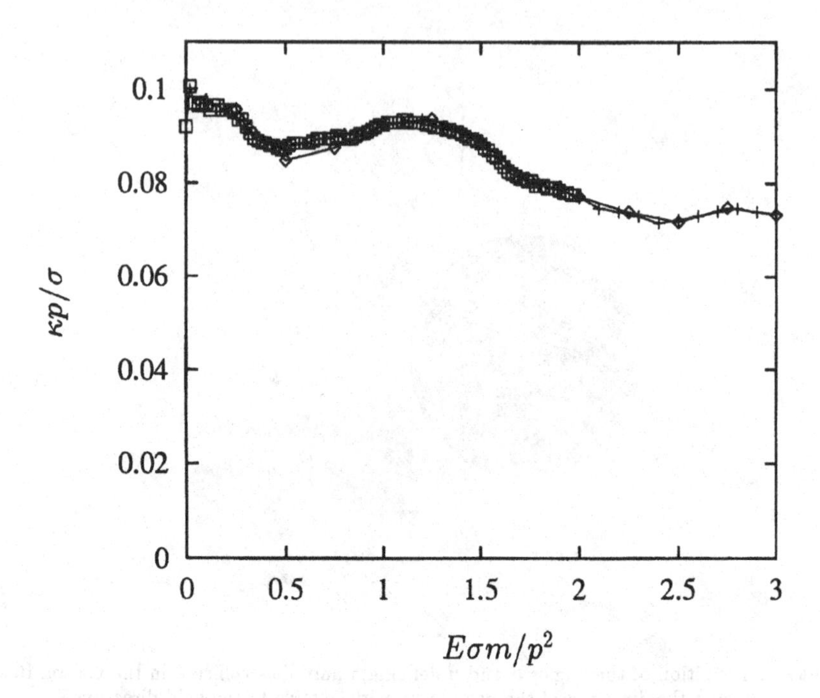

FIGURE 6. Variation of Galton Board conductivity as a function of field strength at 4/5 the close-packed density. The points shown are taken from Moran *et al.* (1987), Dellago *et al.* (1995) and Lloyd *et al.* (1995). The agreement among these three data sets is excellent.

to a list of successive collisions, each of which can be described by two angles, α and β, as defined in figure 5.

The conductivity and the phase-space densities for this model were known in 1986. The Lyapunov exponents have just very recently been calculated. There are only two, with the sum negative, equal to the rate at which field energy is dissipated by the ergostat. The variation of the conductivity, Lyapunov exponents, and a Poincaré section of the phase-space density with field strength are shown in the figures 6–8. The multifractal phase-space structure, with an information dimension strictly less than the equilibrium value, and the positive transport coefficient, despite time-reversible equations of motion, are both characteristic of all nonequilibrium simulations employing time-reversible thermostats or ergostats. The Galton Board is the first model for which these results were established, both numerically and theoretically.

It is noteworthy that the time-reversed attractor is the unstable phase-space repellor, on which the Second Law of Thermodynamics is violated. Because this repellor has a space-filling ergodic character, with Hausdorff dimension equal to the embedding dimension, there is a repellor point arbitrarily close to any point in space. Thus this chaos could easily be 'controlled' to get a current violating the Second Law by converting heat to work.

7. Microscopic plasticity and fracture simulations

Atomistic simulations of the progress of dislocations and cracks through crystals have been carried out for 20 years. These simulations are motivated by the failure of continuum

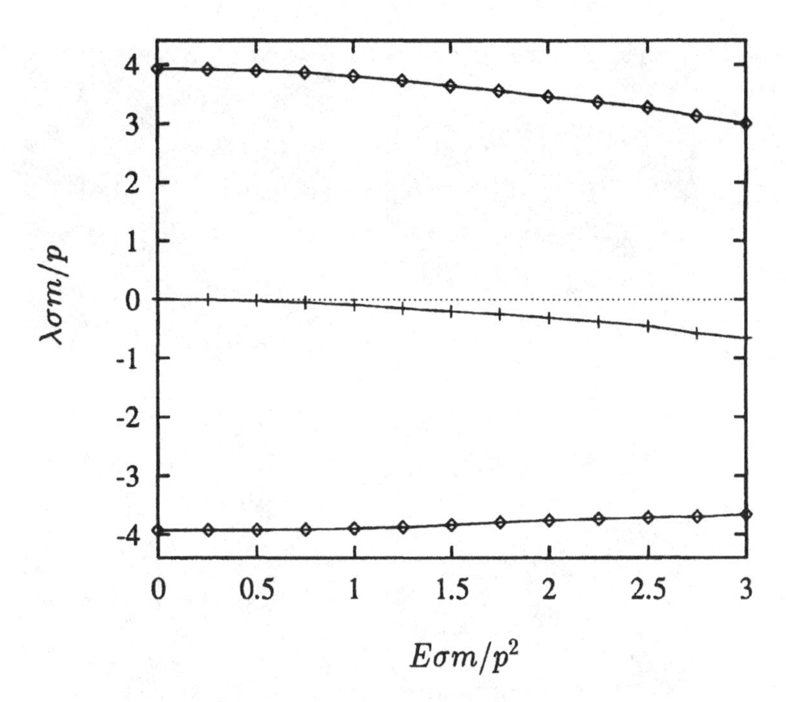

FIGURE 7. Variation of Galton Board Lyapunov exponents as a function of field strength at 4/5 the close-packed density. The sum of the exponents is also plotted and is equal to $-\mathrm{d}(S_{\mathrm{external}}/k)/\mathrm{d}t$.

mechanics to deal satisfactorily with atomic-scale failure processes. Of course, the forces used in the atomistic simulations are oversimplified and speculative, so that neither a microscopic nor a macroscopic picture can provide a quantitative understanding of 'real' plasticity and fracture. The embedded-atom potential provides the necessary flexibility for dealing with metals, which are only poorly described by pair potentials.

Some of the largest-scale molecular dynamics simulations to date, with more than ten million particles, have been carried out at Los Alamos. Goals of this work are, first, to characterize both brittle and ductile crack growth and bifurcation, and then, to understand the influences of temperature and interparticle forces on these features. Figure 9 shows a bifurcated crack, propagating through a cold two-dimensional crystal of a quarter million atoms. To prevent sound waves from reflecting at the system boundaries, special viscoelastic boundaries were developed and implemented in regions with a width of twenty atomic diameters.

The first large-scale nonequilibrium simulations of plastic flow in three dimensions were devoted to the deformation of models of crystalline and amorphous silicon. Apart from the complexity of dealing with three-body forces, the main problem involved developing boundary conditions describing the interaction of an indentor with the silicon workpiece. Eventually these plasticity simulations were carried out with over a million atoms, and at sufficiently low deformation rates to provide a size-independent yield strength in accord with experiment. A specimen simulation is shown in figure 10. Such simulations revealed both a shear-induced phase transformation and the formation mechanism for new surfaces adjacent to the indentor. Both the fracture and plasticity simulations would have been

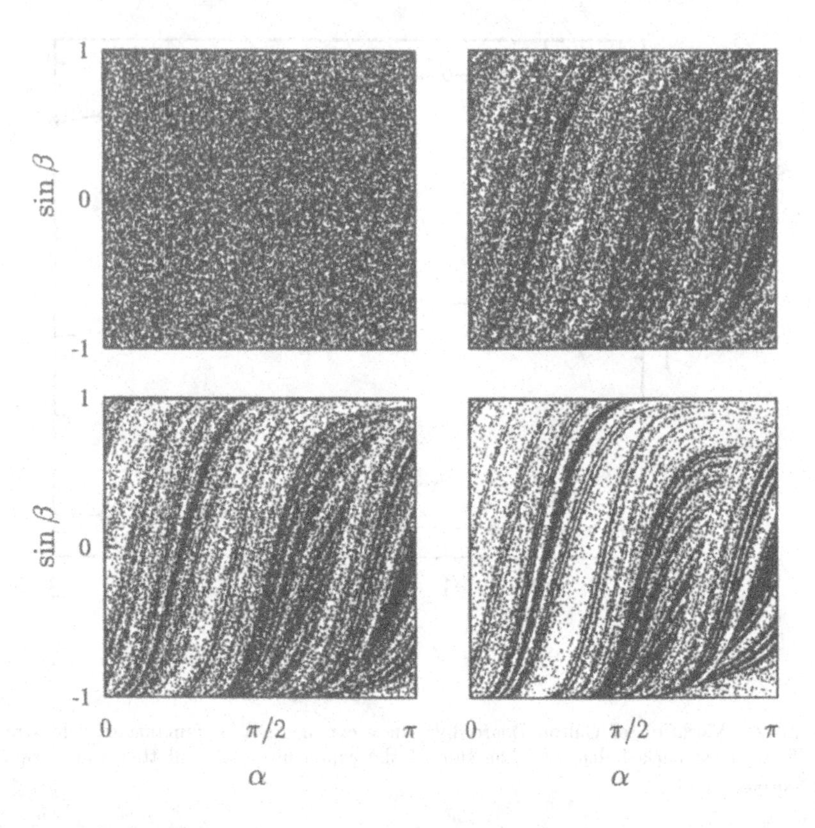

FIGURE 8. Typical phase-space sections showing the distribution of collisions as a function of field strength, all at 4/5 the close-packed density. The field strengths shown are 0, 1, 2, and 3, in units of $p^2/(m\sigma)$, where a point mass m, with momentum p, is scattered by disks of diameter σ.

impossibly time-consuming ten years ago. Both required parallel processors, linked lists of neighbors, and special boundary conditions for extracting energy and momentum from the system.

8. Smooth Particle Applied Mechanics

Particle methods can be applied to macroscopic problems too, using Smooth Particle Applied Mechanics, a technique invented by Lucy and Monaghan in 1977, see Lucy (1977) and Monaghan (1992). The individual smooth particles represent smoothed distributions of material, with their own masses, energies, velocities, stresses, heat fluxes, etc. The mass distribution of each particle in space is described by a weighting function $w(r)$ which is usually, but not invariably, fixed in form and range. Typically, the first and second derivatives of $w(r < h)$ vanish at the cutoff radius, h, which is chosen sufficiently large that each particle interacts with thirty or forty others. Values of the various continuum flow variables are obtained as sums of particle contributions. Density is an important example. The density at a location r is calculated as a sum of contributions from the

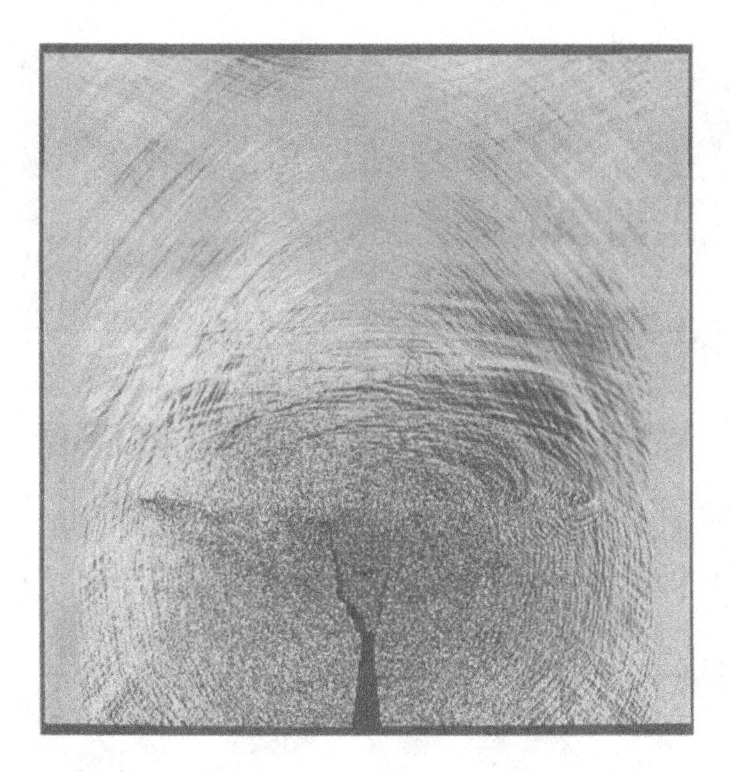

FIGURE 9. Propagation of a crack through a cold two-dimensional crystal of Lennard–Jones spline particles with quiet boundaries. Note the bifurcation, or branching of the crack. The vertical boundaries contain particles interacting with dashpot forces which increase linearly in strength over a region twenty atomic diameters in width.

weights of all nearby smooth particles:

$$\rho(r) \equiv \sum m\, w(r - r_i),$$

This simple formulation makes it unnecessary to solve the continuity equation.

The motion equations for the smooth particles involve the individual stresses as well as the gradients of the weighting, or smoothing, functions which describe the spread of the particles in space:

$$\left\{ \frac{\mathrm{d}v_i}{\mathrm{d}t} = \sum m \left[(\sigma/\rho^2)_i + (\sigma/\rho^2)_j \right] \cdot \nabla_i w_{ij} \right\}.$$

To the extent that the stress and density are slowly varying in space, the smooth particle trajectories look just like particle trajectories governed by a pair potential proportional to $w(r)$.

Smooth particles have mostly been used to solve difficult problems in astrophysics, but there are beginning to be more applications to familiar problems in fluid mechanics. Figure 11 shows a single frame from a two-dimensional smooth-particle simulation of the Rayleigh–Bénard instability, the formation of convective rolls when a thermally-expanding fluid is exposed, simultaneously, to a temperature gradient and a parallel gravitational field. A detailed investigation shows that systems of a few thousand smooth particles are sufficient to reproduce the kinetic energy and the time development of

FIGURE 10. Indentation pit in a model of silicon. The tetrahedral indentor which created the pit moved at 1/5 the speed of sound. The simulation shown here used 373,248 silicon atoms.

the exact continuum flow field within a few percent, see Rapaport (1988), Puhl *et al.* (1989), Posch *et al.* (1995) and Kum & Hoover (1995). Figure 12 compares density and temperature contours calculated with 5000 smooth particles with exact contours obtained from a grid-based solution of the Navier–Stokes equations.

The similarity of the smooth particle motion equations to the microscopic ones makes smooth particles a promising component of hybrid models spanning a range of space and time scales. In solutions of the continuum equations, boundary conditions often present a major difficulty. In the smooth particle case, two different types of boundaries have proved very useful. Mirror boundaries, shown in figure 11, associate image particles with those bulk particles within range of the boundary. An explicit treatment of surfaces, based on a smooth particle analog of surface tension, has proved useful for simulating both static and rotating liquid drops and for studying the instability of the interface between materials of differing densities.

The accurate treatment of material surfaces is a research frontier in numerical continuum mechanics. In simulations of automobile collisions or the unstable buckling of structures composed of struts, it is essential to find all material contacts in a complex mesh, quickly and reliably. This can be done by using a generalization of the linked-list approach, with an additional space-fixed grid devoted to the detection and prevention

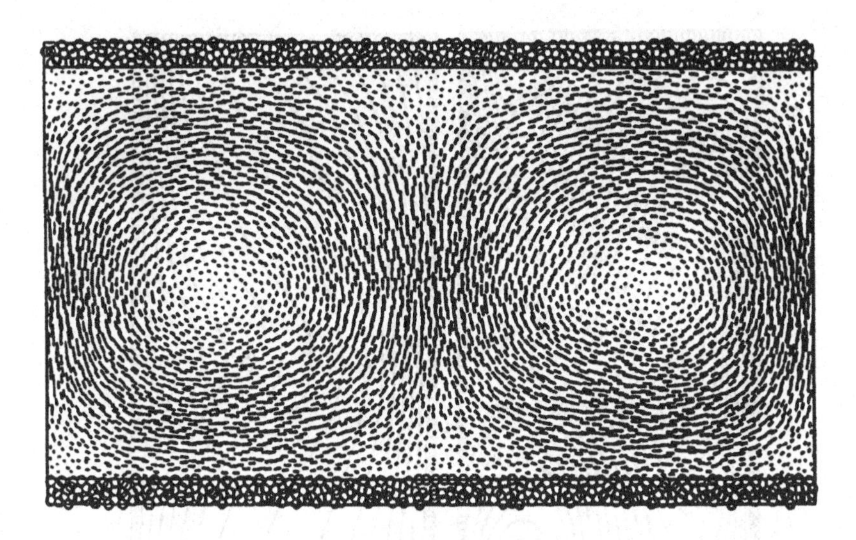

FIGURE 11. Typical smooth-particle velocity distribution for Rayleigh–Bénard flow; 5000 smooth particles were used. Note the reflected image particles at the top and bottom boundaries.

of intersurface contacts. Figure 13 shows the deformation of a sheet made up of elastic-plastic shell elements. The sheet was crumpled, to form a wad, by simultaneously squeezing the original structure between three pairs of mutually-orthogonal moving plane walls. The resulting structure has several interesting multifractal aspects.

9. Puzzles for the future

One of the stated goals of the conference organizers is to identify current problems important for the future growth of understanding and applying Chaotic Complex Systems. From the perspective of particle simulations there are several areas which need more attention. In outline form these are

(i) characterizing multifractal objects,
(ii) understanding ergodic strange attractors,
(iii) developing smooth boundaries,
(iv) excluding unwanted fluctuations,
(v) improving our description of turbulence,
(vi) relating Nosé's mechanics to mechanical variational principles.

Let us consider each of these problem areas in more detail.

The most basic need is for better characterization of the fractal objects themselves. For these objects, the fractal dimension, or dimensions, or its spectrum, are not sufficiently detailed to distinguish a crumpled sheet of paper from a puffy cumulus cloud. More quantifiable descriptors of fractal geometry are sorely needed to help us find our way around in the zoo of these objects.

The fractal objects generated by nonequilibrium particle simulations are mathematically subtle. It seems evident that the twin notions of (a) ubiquitous shrinking of phase volume onto a strange attractor which is simultaneously (b) ergodic are dangerously close to nonsense. For an integer (or rational) state space these two things cannot both be true.

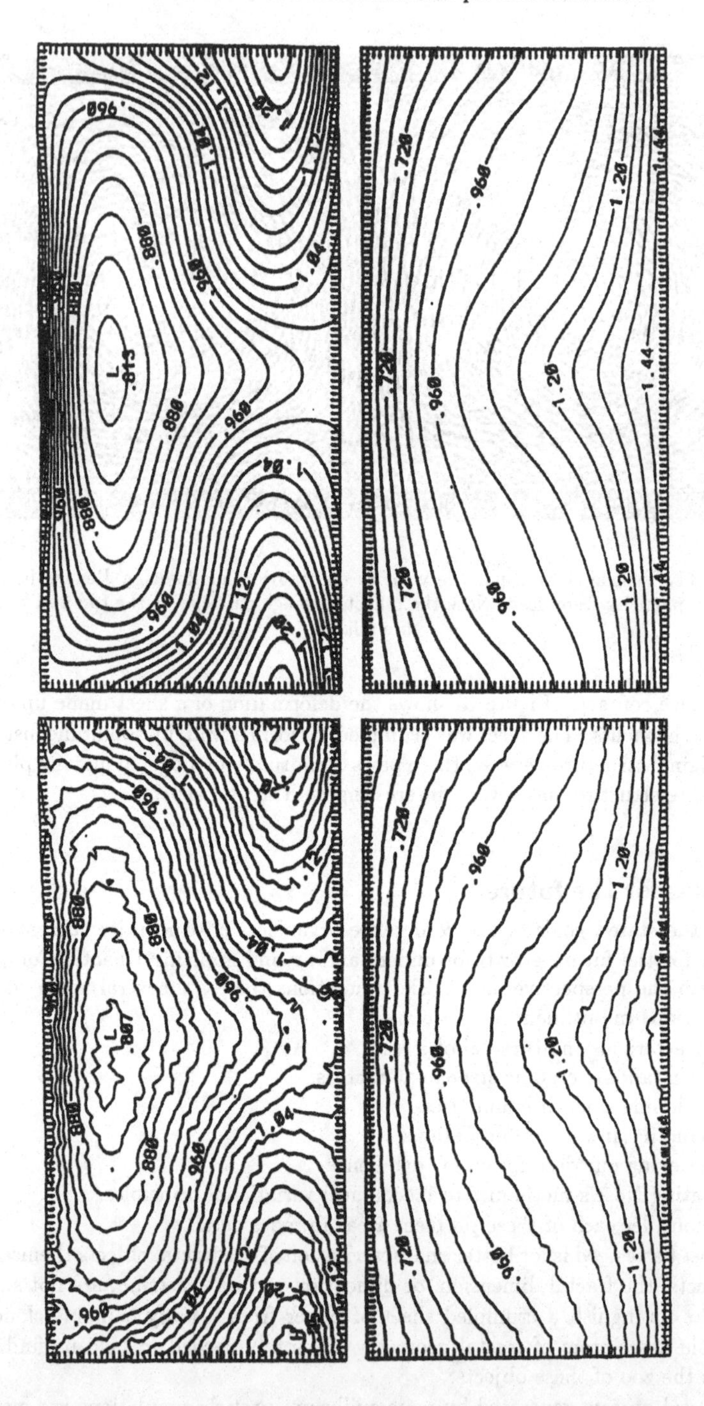

FIGURE 12. Density and temperature profiles for the smooth-particle simulation of figure 11 are compared to exact profiles from a highly-accurate solution of the continuum equations.

FIGURE 13. Simulation of the crumpling of a structure composed of elastic-plastic shell elements using the computer program DYNA3D.

Nevertheless, the shrinkage in phase space and the ergodicity are both characteristic of the fractal objects of most interest for physicists.

The simultaneous shrinking and recurrence bring to mind some paradoxical mathematics from two generations ago. The Banach–Tarski paradox refers to that disease of set theory, when augmented by the Axiom of Choice, which makes it possible to assemble two identical objects from the cut-up pieces of another. This paradox has been specially nicely illustrated, using oranges, by French (1988). This failure of augmented set theory to conserve mass is likely related to our present conceptual difficulties in understanding multifractal objects.

Other immediate needs of particle simulators are more special and have to do with smooth and quiet boundaries, the number of degrees of freedom, and fluctuations. Simulation boundaries need to be made as smooth as is possible. An atomistic adaptation of the smooth-particle image approach might be useful. It is also desirable to reduce the number of extraneous degrees of freedom. All particle models are saddled with fluctuations (called 'heat', in the case of molecular dynamics) which can be both realistic and interesting, but with a cost which greatly exceeds their benefit. A means of discarding irrelevant degrees of freedom would be welcome. Finding an efficient route to the Kolmogorov spectrum of turbulence using smooth particles is a related interesting open problem.

For simplicity we have discussed here only isoenergetic and isokinetic forms of equilibrium and nonequilibrium mechanics. Nosé (1991) discovered a generalization of mechanics, designed to replicate Gibbs' canonical distribution in the equilibrium case, now called Nosé–Hoover mechanics. So far the link of that mechanics to the variational principles is missing.

WGH thanks the Universities of California and Vienna, as well as the organizers of this meeting, for making it possible to present this work at Zakopane. Work at the Lawrence Livermore National Laboratory and at the Los Alamos National Laboratory was performed under the auspices of the United States Department of Energy, according to Contracts W-7405-Eng-48 and W-7405-Eng-36. Work at the University of Vienna was supported by the Fonds zur Förderung der Wissenschaftlichen Forschung, Grant P9677. Though this paper primarily reflects the views of WGH and HAP, it was made possible by the dedicated work of all the authors.

REFERENCES

CHERNOV, N. I., EYINK, G. L., LEBOWITZ, J. L. & SINAI, Y. G. 1993 Derivation of Ohm's law and a deterministic mechanical model. *Phys. Rev. Letts.* **70**, 2209. The reader is cautioned that the multifractal dimension referred to in this reference as a 'Hausdorff dimension' is more often referred to as the 'information dimension'.

DELLAGO, C., GLATZ, L. & POSCH, H. A. 1995 *Phys. Rev.* E **52**, 4817.

FRENCH, R. M. 1988 *Mathematical Intelligencer* **10(4)**, 21.

GILLILAN, R. E. & WILSON, K. R. 1992 *J. Chem. Phys.* **97**, 1757.

HOOVER, W. G. 1992 In *Microscopic Simulations of Complex Hydrodynamic Phenomena* (ed. M. Mareschal & B. L. Holian). Plenum. See particularly figure 4.

HOOVER, W. G. & MORAN, B. 1989 *Phys. Rev.* A **40**, 5319.

HOOVER, W. G. & MORAN, B. 1992 *Chaos* **2**, 599.

HOOVER, W. G., MORAN, B., HOOVER, C. G. & EVANS, W. J. 1988 *Phys. Lett.* A **133**, 114.

HOOVER, W. G, WINER, K. A. & LADD, A. J. C. 1985 *Intl. J. Eng. Sci.* **23**, 483.

KUM, O., HOOVER, W. G., & POSCH, H. A. 1995 *Phys. Rev.* E **52**, 4899.

LADD, A. J. C. & HOOVER, W. G. 1985 *J. Stat. Phys.* **38**, 973.

LEVESQUE, D. & VERLET, L. 1993 *J. Stat. Phys.* **72**, 519.

LLOYD, J., NIEMEYER, M., RONDONI, L. & MORRISS, G. P. 1995 *Chaos* **5**, 536.

LUCY, L. B. 1977 *Astron. J.* **82**, 1013.

MACHTA, J. & ZWANZIG, R. W. 1983 *Phys. Rev. Letts.* **50**, 1959.

MONAGHAN, J. J. 1992 *Ann. Rev. Astron. & Astrophys.* **30**, 543.

MORAN, B., HOOVER, W. G. & BESTIALE, S. 1987 *J. Stat. Phys.* **48**, 709.

NOSÉ, S. 1991 *Prog. Theor. Phys. Suppl.* **103**, 1.

PETRAVIC, J., ISBISTER, D. J. & MORRISS, G. P. 1994 *J. Stat. Phys.* **76**, 1045.

POSCH, H. A., HOOVER, W. G., & KUM, O. 1995 *Phys. Rev.* E **52**, 1711.

PUHL, A., MANSOUR, M. M. & MARESCHAL, M. 1989 *Phys. Rev.* A **40**, 1999.

RAPAPORT, D. C 1988 *Phys. Rev. Lett.* **60**, 2480.

SALMON, J. 1980 *Hadronic Journal* **3**, 1080.

SARMAN, S., EVANS, D. J. & MORRISS, G. D. 1992 *Phys. Rev.* A **45**, 2233.

VANCE, W. N. 1992 *Phys. Rev. Lett.* **69**, 1356.

See also (numerals in parantheses following references indicate related sections):

ABRAHAM, F. F., BRODBECK, D., RAFEY, R. A. & RUDGE, W. E. 1994 *Phys. Rev. Lett.* **73**, 272. (7).

ASHURST, W. T. & HOOVER, W. G. 1976 *Phys. Rev.* B **14**, 1465. (7).

BERENDSEN, H. J. C. & VAN GUNSTEREN, W. F. 1986 In *Molecular-Dynamics Simulation of Statistical-Mechanical Systems* (ed. G. Ciccotti & W.G. Hoover). North-Holland. (2,3).

FOILES, S. M., BASKES, M. I. & DAW, M. S. 1986 *Phys. Rev.* B **33**, 7982. (2,3).

GRAY, S. K., NOID, D. W. & SUMPTER, B. G. 1994 *J. Chem. Phys.* **101**, 4062. (2,3).

HOLIAN, B. L., DE GROOT, A. J., HOOVER, W. G. & HOOVER, C. G. 1990 *Phys. Rev.* A **41**, 4552. (2,3).

HOLIAN, B. L., HOOVER, W. G., & POSCH, H. A. 1987 *Phys. Rev. Letts.* **59**, 10. (6).

HOLIAN, B. L. & RAVELO, R. 1995 *Phys. Rev.* B **51**, 11275. (7).

HOLIAN, K. S. & HOLIAN, B. L. 1989 *Intl. J. Impact Eng.* **8**, 115. (7).

HOOVER, W. G. 1988 Reversible mechanics and time's arrow. *Phys. Rev.* A **37**, 252. This paper was written for a special issue of the Journal of Statistical Physics, honoring Professor Prigogine (September, 1987), but was rejected. (5,6).

HOOVER, W. G. 1993 *Physica* A **194**, 450. (6).

HOOVER, W. G. 1995 *Phys. Letts.* A **204**, 133. (2,3).

HOOVER, W. G., HOLIAN, B. L. & POSCH, H. A. 1993 *Phys. Rev.* E **48**, 3196. (2,3).

HOOVER, W. G. & POSCH, H. A. 1994 *Phys. Rev.* E **49**, 1913. (6).

HOOVER, W. G., POSCH, H. A., HOLIAN, B. L., GILLAN, M. J., MARESCHAL, M. & MASSOBRIO, C. 1987 *Mol. Sim.* **1**, 79. (6).

KUSNEZOV, D., BULGAC, A. & BAUER, W. 1990 *Annals of Physics* **204**, 155. (2,3).

KUSNEZOV, D., BULGAC, A. & BAUER, W. 1992 *Annals of Physics* **214**, 180. (2,3).

POSCH, H. A. & HOOVER, W. G. 1992 In *Molecular Liquids: New Perspectives in Physics & Chemistry* (ed. Teixeira-Dias). NATO Advanced Science Institute, vol. 20. Kluwer Academic Publishers.

Quantum dynamics on a Markov background and irreversibility

By BORIS PAVLOV

St Petersburg University, St Petersburg, Russia
and
University of Auckland, Auckland, New Zealand

1. Introduction

Using operator theory, we construct a class of models of quantum systems on a stochastic background, which exhibit irreversibility in a straightforward way. Assuming that the Hamiltonian of the quantum system depends on time via some Markov process, we examine the evolution operators $U[x(t)]$ which correspond to the trajectories $x(t)$ of the process. Averaging the evolution operators over all trajectories with fixed ends $x(0) = x_0$, $x(t) = x_t$, we get a family of *averaged evolution operators* $\mathcal{Z}(t)$ acting in the corresponding *quantum-stochastic space* which are represented by generalized kernels

$$\{\mathcal{Z}(t)\}_{x,y} \equiv \langle U\,[x(t)]\rangle_{x_0=x,\,x_t=y}\,.$$

This family turns out to be a contracting semigroup, generated by some dissipative operator, which in general is not a weak perturbation of a self-adjoint operator. In many interesting cases it has a complex branch of an absolutely continuous spectrum, or there may even exist a domain in the complex plane filled with the spectrum. The spectral approach to the description of quantum dynamics on a Markov background was suggested for some problems in Mössbauer spectroscopy in Belozerskij & Pavljuhin (1976) and Belozerskij & Pavljuhin (1977), and developed for mathematical model Hamiltonians in Pavlov & Ryshkov (1987), Pavlov & Ryshkov (1989) and Cheremshantsev (1990).

In this paper we develop the described approach for Liouville's operator and investigate the properties of the corresponding irreversible dynamics. We show that for the generator of this dynamics a *positivity-preserving self-adjoint dilation* can be constructed in some extended Hilbert space. Moreover, the corresponding *scattering matrix* can be defined; this connects the asymptotic behaviour of the dilated unitary dynamics and the spectral properties of the original contracting dynamics, similarly to the classical characteristic function of weakly-perturbed dissipative operators. In the last section the case of the classical dynamics on a stochastic background is discussed.

2. Averaged evolution, irreversibility and quantum chaos

Let us consider a Hamiltonian H which depends on time via a Markov process $x(t)$ taking values in the space X of stochastic states. A natural description of the correpond-ing evolution is given by Liouville's equation for the density matrix $\rho(t)$, which is an element of the space Q of operators of the Hilbert–Schmidt type, with the inner product $\langle\rho_1, \rho_2\rangle = \mathrm{Tr}\,(\rho_2^+\rho_1)$:

$$\frac{1}{\mathrm{i}}\frac{\mathrm{d}\rho}{\mathrm{d}t} = H\rho - \rho H \equiv H^\times\rho.$$

The time evolution of the initial data for the last equation is given by the unitary T-product $U(t) : \rho(0) \to \rho(t)$, which is generated by the the superoperator $H^\times$ satisfying

$$\frac{1}{\mathrm{i}}\frac{\mathrm{d}U}{\mathrm{d}t} = H^\times U,$$

$$U(0) = I.$$

The evolution operators $U(t)$ along a fixed trajectory of the Markov process obviously do not form a group or even a semigroup. Nevertheless, averaging $U(t)$ over all trajectories with fixed ends

$$x(0) = x_0, \quad x(t) = x_t$$

yields a *contracting* semigroup generated by a *dissipative operator*. In the case of a discrete stochastic space the spectrum of the corresponding generator B may have several one-dimensional branches in the upper half-plane of the spectral parameter, which are isomorphic to the absolutely continuous spectrum of the time independent Liouvillian with a 'frozen' stochastic process.

THEOREM 2.1. *Assume that the stochastic space X is finite-dimensional and that the evolution of the transition probabilities of $x(t)$ is governed by the Fokker–Planck differential equation for transition probabilities in X with a positive matrix M:*

$$\frac{\mathrm{d}P}{\mathrm{d}t} + MP = 0.$$

Then the generator of the quantum evolution averaged over all trajetories with fixed ends is a dissipative operator in quantum stochastic space $K \times X$ given by the construction:

$$L_M = \left\{\mathrm{diag}_X H^{\times}(x)\right\} + \mathrm{i}M \times I_Q.$$

Here I_Q is the unit operator in the quantum space Q of density matrices.

This theorem can be used for the description of irreversible quantum dynamics as a limiting case of the dynamics averaged over all trajectories with fixed ends, when the dependence of the Hamiltonian on the Markov process is suppressed. The limiting Liouvillian is in general irreducible, so in general it cannot be represented as a superoperator $H^{\times}$ generated by any self-adjoint Hamiltonian H.

A comprehensive analysis of irreversible quantum dynamics on a stochastic background is presented in Davies (1976). In our case many characteristic features of this dynamics can be deduced directly from the spectral analysis of the averaged evolution. Moreover, we can indicate some cases where the averaged evolution exhibits a chaotic behaviour in the sense of I. Prigogine (see Prigogine & Stengers (1988)).

It is known that the spectral subspace of the Liouvillian corresponding to the zero eigenvalue consists of linear combinations of the orthogonal projections $\phi_k\rangle\langle\phi_k$ onto pure quantum eigenstates ϕ_k of the corresponding Hamiltonian. These projections remain constant in the course of evolution. On the other hand, the off-diagonal elements of the density matrix (correlations) $\mathrm{e}^{\mathrm{i}(\lambda_k-\lambda_l)}\phi_k\rangle\rho_{kl}\langle\phi_l$ are changing in time. For quantum systems with absolutely continuous spectra, some correlations can go to zero because of the Riemann–Lebesgue Lemma. If *all* correlations of the given quantum system go to zero in the course of evolution, the limiting density matrix $\rho_\infty = \lim_{t\to\infty}\rho(t)$ is diagonal, and thus *should be constant in the course of evolution!* I. Prigogine suggested to call the quantum systems which exhibit this paradoxical behaviour, *chaotic quantum systems.* The dynamics of such quantum systems is obviously irreversible.

We present here a simple example of a quantum system on the Markov background. It exhibits the described behaviour. Actually even the exponential decay of all correlations is admissible, if the dynamics of the system fulfil the conditions of the Lax–Phillips scattering scheme.

We assume that the discrete spectrum of the Liouvillian in each stochastic state consists of the zero eigenvalue, which corresponds to the subspace of pure quantum states in the space of all density matrices, and two real eigenvalues $\pm\gamma$ corresponding to stochastic states $\pm$, *associated with the same eigenfunction.* The absolutely continuous spectrum of constant multiplicity μ and the corresponding eigenfunctions are supposed to be unaffected by the stochastic process. In this case, the generators $L_\pm$ of the quantum evolution in every stochastic state are represented as the orthogonal sum of the discrete component

$$L_{\mathrm{d}}^{\pm} = \pm\gamma \begin{bmatrix} 1 & 0 \\ 0 & -1 \end{bmatrix},$$

the zero-component, corresponding to the zero eigenspace N_0, and the matrix multiplication operator on the real axis of the spectral variable k in the space $L_2(Q)$ (counting multiplicity $\mu = \dim Q$):

$$L_\pm = L_{\mathrm{d}}^{\pm} \oplus \{0 \times P_0\} \oplus k.$$

Averaging the corresponding quantum evolution over the trajectories of the Markov process with two-point stochastic space and the stochastic matrix M:

$$M = \frac{\kappa^2}{2}\begin{bmatrix} 1 & -1 \\ -1 & 1 \end{bmatrix}, \quad \kappa > 0,$$

we get the following generator of the averaged evolution:

$$\mathbf{L} \equiv \begin{bmatrix} L_+ & 0 \\ 0 & L_- \end{bmatrix} + \mathrm{i}\frac{\kappa^2}{2}\begin{bmatrix} I_{\mathrm{q}} & -I_{\mathrm{q}} \\ -I_{\mathrm{q}} & I_{\mathrm{q}} \end{bmatrix},$$

where I_{q} is the unit operator in the quantum space of the Hilbert-Schmidt operators. Actually, the spectrum of $\mathbf{L}$ has two absolutely continuous branches $\{\mathbf{R}\}$ and $\{\mathrm{i}\kappa + \mathbf{R}\}$. The corresponding eigenfunctions are combinations of eigenfunctions δ_{k-s} of the absolutely continuous spectrum of $L_\pm$ in the form $\Psi_{\mathrm{sym}}(s) \equiv (\delta_{k-s}, \delta_{k-s})$ for the real branch $\{\mathbf{R}\}$, and in the form $\Psi_{\mathrm{asym}}(s) \equiv (\delta_{k-s}, -\delta_{k-s})$ for the complex branch $\{\mathrm{i}\kappa + \mathbf{R}\}$. The discrete spectrum of $\mathbf{L}$ consists of two eigenvalues, 0 and $\mathrm{i}\kappa$, of infinite multiplicity, which correspond to the eigenspaces

$$N_{\mathrm{sym}} = \{(n, n),\ n \in N_0\} \quad \text{and} \quad N_{\mathrm{asym}} = \{(n, -n),\ n \in N_0\},$$

and two multiple eigenvalues $\lambda_\pm$ (of multiplicity 2).

Let us consider the Cauchy problem for the nonstationary Liouville equation with the operator $\mathbf{L}$:

$$\frac{1}{\mathrm{i}}\frac{\mathrm{d}\wp}{\mathrm{d}t} = \mathbf{L}\wp.$$

Fixing any initial condition $\wp_0$, we can represent the solution $\wp(t)$ in the form of the decomposition into eigenfunctions of $\mathbf{L}$:

$$\begin{aligned} \wp(t) &= \mathrm{e}^{\mathrm{i}t\mathbf{L}}\wp_0 \\ &= P_{N_{\mathrm{sym}}}\wp_0 + \int_{-\infty}^{\infty} \mathrm{e}^{\mathrm{i}kt}\psi_{\mathrm{sym}}\langle\wp(0), \psi_{\mathrm{sym}}\rangle\,\mathrm{d}k \\ &\quad + \int_{-\infty+\mathrm{i}\kappa}^{\infty+\mathrm{i}\kappa} \mathrm{e}^{\mathrm{i}kt}\psi_{\mathrm{asym}}\langle\wp(0), \psi_{\mathrm{asym}}\rangle\,\mathrm{d}k \\ &\quad + \sum_\pm \mathrm{e}^{\mathrm{i}\lambda t}\langle\wp_0, \psi\rangle\psi + \mathrm{e}^{-\kappa t}P_{N_{\mathrm{asym}}}\wp_0. \end{aligned}$$

It is obvious that the asymptotic behaviour of the component of the density matrix $\wp(t)$ in the subspace of the discrete spectrum corresponds exactly to the case of quantum chaos, described in Prigogine & Stengers (1988): the correlation component $\sum_\pm \mathrm{e}^{\mathrm{i}\lambda t}\langle\wp_0, \psi\rangle\psi$, and the component corresponding to N_{asym} decrease exponentially as $t \to \infty$, and the component $P_0\wp_0$, corresponding to pure states, remains constant. The behaviour of the continuous spectrum is more interesting. The antisymmetric branch just gives a standard exponentially decreasing term, but the symmetric branch $\{\mathbf{R}\}$ of the spectrum seems to produce a nontrivial contribution, which tends to zero with polynomial speed as $t \to \infty$, due to standard oscillation arguments (Riemann–Lebesgue lemma). Nevertheless, some special situations can be found where this term decreases exponentially. These situations correspond to the case when Lax–Phillips theory can be applied to $\mathbf{L}$, see Lax & Phillips (1967).

3. Self-adjoint dilation and scattering matrix

The eigenvalues of the generator of the averaged evolution can be interpreted as resonances of some dilated unitary dynamics. Actually, using the construction of the self-adjoint dilation of the dissipative operator $\mathbf{L}$, we can construct a self-adjoint operator $\mathcal{L}$ in a larger Hilbert space $\mathcal{H}$, such that its compression onto the quantum-stochastic space $H \times X \subset \mathcal{H}$ coincides with the averaged evolution:

$$\mathrm{e}^{\mathrm{i}t\mathbf{L}} = P_{\mathbf{H}} \mathrm{e}^{\mathrm{i}t\mathcal{L}}|_{\mathbf{H}}.$$

It is easy to see that the constructed dilation is *positivity preserving* in the following sense: for every density matrix $\rho \in \mathbf{H}$, the result of the corresponding dilated evolution is a completely positive matrix with respect to the vectors $\{\varphi\}$ of the initial space $\mathbf{H}$. Indeed, for every finite family of vectors φ_l, $0 < l < \infty$, from the Hilbert space of wave-functions, and for any set ξ_l of complex numbers, the quadratic form of the density matrix $\sum_{l,n} \langle \rho\, \varphi_l, \varphi_n \rangle \xi_l \bar{\xi}_n \equiv \mathrm{Tr}\,(\wp_\varphi^+ \rho)$, $\rho_\varphi^+ \in \mathbf{H}$, is positive. Hence it is also positive when replacing the test-matrix ρ_φ by the result of the corresponding quantum evolution; it automatically remains positive after averaging the evolution; and it is invariant under the replacement of the contraction semigroup $\mathrm{e}^{\mathrm{i}t\mathbf{L}}$ by the corresponding unitary dilation $\mathrm{e}^{\mathrm{i}t\mathcal{L}}$:

$$\mathrm{Tr}\,(\rho_\varphi^+ \mathrm{e}^{\mathrm{i}t\mathbf{L}} \wp) = \mathrm{Tr}\,(\rho_\varphi^+ \mathrm{e}^{\mathrm{i}t\mathcal{L}} \wp),$$

since $\wp, \rho_\varphi^+ \in \mathbf{H}$.

The standard procedure for construction of the self-adjoint dilation (see Nagy & Foias (1967)) is applicable to weak dissipative perturbations of self-adjoint operators. Here we use another approach (see Pavlov (1983) and Pavlov (1995)) which proves to be valid for strong perturbations when the dissipative operator has its essential spectrum in the complex plane.

We consider the case of the two-point stochastic space $(\pm)$, when the Markov process is governed by the matrix

$$\kappa^2 P_a \equiv \frac{\kappa^2}{2} \begin{bmatrix} 1 & -1 \\ -1 & 1 \end{bmatrix}.$$

The dissipative generator of the averaged dynamics, represented in matrix form, is

$$\begin{bmatrix} L_+ & 0 \\ 0 & L_- \end{bmatrix} + \mathrm{i}\frac{\kappa^2}{2} \begin{bmatrix} I_\mathrm{q} & -I_\mathrm{q} \\ -I_\mathrm{q} & I_\mathrm{q} \end{bmatrix},$$

where $L_\pm$ are quantum Liouvillians corresponding to the stochastic states $\pm$, and I_q is the unit operator in the quantum space of density matrices. Actually, the second term coincides, up to a factor κ^2, with the orthogonal projection P_a in the quantum-stochastic space onto the subspace E_a of elements which are antisymmetric with respect to the stochastic variables. Due to the idempotency of the projections $\left((P_\mathrm{a})^2 = P_\mathrm{a}\right)$ this generator has the form

$$A + \mathrm{i}G^2,$$

where $G = \kappa P_a$.

The dilation of this operator is constructed by attaching to the quantum-stochastic space two orthogonal 'incoming' and 'outgoing' channels $D_\pm$

$$D_+ = E \times L_2(0,\infty), \quad D_- = E \times L_2(-\infty,0),$$

with momentum operators

$$2\mathrm{i}\frac{\mathrm{d}u_\pm}{\mathrm{d}\xi}, \quad \xi \in (0,\pm\infty),$$

acting on $D_\pm$. In our case the perturbation $\mathrm{i}G^2$ is not a relatively weak operator with respect to the real part A. The construction of the minimum self-adjoint dilation is described by the following general statement:

THEOREM 3.1. *Consider a dissipative operator L, represented in the form of a sum of a self-adjoint operator A acting in a Hilbert space K, and an imaginary part $\mathrm{i}G^2$, constructed as the square of a bounded nonnegative operator G (we assume that* $\operatorname{Range} G = E$*):*

$$L = A + \mathrm{i}G^2.$$

Then the operator, which is defined in the orthogonal sum of the space K and the incoming and outgoing channels $D_\pm = E \times L_2(0, \pm\infty)$ by the formula

$$\mathcal{L}\begin{bmatrix} u_- \\ u \\ u_+ \end{bmatrix} = \begin{bmatrix} 2\mathrm{i}\frac{\mathrm{d}u_-}{\mathrm{d}\xi} \\ Au + G[u_-(0) + u_+(0)] \\ 2\mathrm{i}\frac{\mathrm{d}u_+}{\mathrm{d}\xi} \end{bmatrix}$$

with boundary condition

$$u_+(0) - u_-(0) = \mathrm{i}Gu,$$

is the minimum self-adjoint dilation of L. It has an absolutely continuous spectrum filling the real axis (generally with varying multiplicity). If the generalized limits of the resolvent of the real part A in some dense linear subspace E', $E' \subset E$, exist, the eigenfunctions of the dilation can be represented through these limits.

In particular, this operator has two orthogonal systems $\psi_\pm$ of eigenfunctions of the scattered-wave type, which form a basis of the double-invariant subspaces $H_\pm$ generated by $D_\pm$, respectively, and two orthogonal systems of eigenfunctions of absolutely continuous spectra $\psi^\rangle$, $\psi^\langle$ in complementary subspaces $H^\rangle, H^\langle$.

For instance

$$\psi_- = \begin{bmatrix} \mathrm{e}^{-\mathrm{i}\frac{1}{2}\lambda\xi}\nu \\ u \\ \mathrm{e}^{-\mathrm{i}\frac{1}{2}\lambda\xi}S^+(\lambda)\nu \end{bmatrix}, \quad \nu \in E'.$$

The transmission coefficient S^+ is nontrivial only in the subspace E, and is given there, in terms of the generalized limit $R^A(\lambda - \mathrm{i}0)$ of the resolvent of the real part A, by the formula

$$S^+(\lambda) = \frac{I_E - \mathrm{i}GR^A(\lambda - \mathrm{i}0)G}{I_E + \mathrm{i}GR^A(\lambda - \mathrm{i}0)G}.$$

The central component u of ψ_- is given by the formula

$$u = -R^A(\lambda - \mathrm{i}0)G\left[S^+\nu + \nu\right].$$

The eigenfunctions $\psi^\langle$ of the complementary component of $\mathcal{L}$ in $H \ominus H_-$ vanish on D_- and are represented through the eigenfunctions of the operator L (which play the role of *the central component) and exponentials in the outgoing space:*

$$\psi^{(} = \begin{bmatrix} 0 \\ u \\ \mathrm{i}Gue^{-\frac{\mathrm{i}}{2}\lambda\xi} \end{bmatrix}.$$

Here u is the properly normalized eigenfunction of the operator L:

$$Au + \mathrm{i}G^2 u = \lambda u.$$

In the special case when $G = \kappa^2 P_{\mathrm{a}}$, an elegant formula can be derived for the zeroes and root vectors of the transmission coefficient. Denoting by $R_\pm$ the resolvents of the Liouvillians corresponding to both stochastic states, we can write down the restriction of the transmission coefficient to the antisymmetric subspace $E = P_{\mathrm{a}}\{K \times X\}$ in the form

$$\frac{I_q - \mathrm{i}\kappa^2(R_+ + R_-)/2}{I_q + \mathrm{i}\kappa^2(R_+ + R_-)/2},$$

where $R_\pm = (L_\pm - \lambda I)^{-1} \equiv (L \pm \Gamma - \lambda I)^{-1}$. Hence the roots of the transmission coefficient are defined by the nontrivial solutions of the homogeneous equation

$$\left[L_+ - (\lambda + \mathrm{i}\kappa^2)I\right]\nu - \mathrm{i}\kappa^2\Gamma R_-\nu \equiv 0.$$

In our case, the difference $2\Gamma = L_+ - L_-$ is a two-dimensional operator, hence it is relatively weak with respect to the Liouvillian L_-. Then the root vectors ν are just the result of a perturbation of the eigenfunctions of A, corresponding to the real spectrum $\sigma_{\mathrm{a}} = \{\lambda_{\mathrm{a}}\}$ of A,

$$\lambda = \lambda_{\mathrm{a}} - \mathrm{i}\kappa^2,$$

and can be found as solutions of the corresponding Lippmann–Schwinger equation, assuming that this spectrum is absolutely continuous, and the products $\Gamma[L_\pm - (\lambda_{\mathrm{a}} - \mathrm{i}0)I]^{-1}$ have generalized limits on it. If $\psi_{\mathrm{a}}(\lambda_{\mathrm{a}})$ are the eigenfunctions of the absolutely continuous spectrum of A, the root-vectors satisfy the equation

$$\nu(\lambda) = \psi_{\mathrm{a}}(\lambda_{\mathrm{a}}) + [L_+ - (\lambda_{\mathrm{a}} - \mathrm{i}0)I]^{-1}\kappa^2\Gamma R_-\nu,$$

which can be solved explicitely in our case. Thus the transmission coefficient in this case has a *continuous* branch of complex zeroes which is complex conjugate to the branch of the absolutely continuous spectrum of the dissipative generator of the averaged dynamics, similar to the usual situation for weak dissipative perturbations of self-adjoint operators, where only the discrete spectrum of resonances is present in the complex domain $\operatorname{Im}\lambda > 0$. Similarly to the case of weak perturbations, the adjoint transmission coefficient S has a branch of roots in the upper half-plane, which coincide with the complex, absolutely continuous spectrum of the dissipative generator. The adjoint transmission coefficient plays the role of the corresponding *characteristic function.* The functional model of the contraction semigroup $e^{\mathrm{i}t\mathbf{L}}$ arises when we write it down in the incoming spectral representation of the self-adjoint dilation $e^{\mathrm{i}t\mathcal{L}}$.

4. Irreversiblility of classical dynamics on the Markov background

The approach used for quantum systems can be applied to classical systems as well. Let us consider a classical dynamical system on a smooth n-dimensional manifold, governed by the Hamilton function depending on time via a Markov process $\alpha(t)$ with discrete stochastic space $X = \{\alpha_s\}$

$$H = H\big(\alpha(t), q, p\big).$$

The piecewise smooth solution of the system of Hamilton equations along the given trajecory of the process is composed of pieces of the solutions of the corresponding equations with the 'frozen' Hamilton function $H(\alpha, q, p)$, $\alpha(t) = \alpha_s$, $t \in (t_s, t_{s+1})$:

$$\frac{\mathrm{d}q}{\mathrm{d}t} = \frac{\partial H}{\partial p}(\alpha_s),$$
$$\frac{\mathrm{d}p}{\mathrm{d}t} = -\frac{\partial H}{\partial q}(\alpha_s),$$

$$q(t_s + 0) = q(t_s - 0); \quad p(t_s + 0) = p(t_s - 0).$$

To apply the approach described above we should reformulate the classical dynamics first in dual terms of the densities $\rho(t, q, p)$ carried by the flow $q(t)$, $p(t)$ defined above:

$$\frac{\partial \rho}{\partial t} = \{H_{\alpha(t)}, \rho\} = H^{\times}_{\alpha(t)}\rho.$$

Here $H^{\times}_{\alpha_s}$ is a skew-symmetric operator in the Hilbert space $L_2(\mathrm{d}p\,\mathrm{d}q)$ of square-integrable densities, defined by the bracket-operation (see Guillemin & Sternberg (1984))

$$H^{\times}\rho = \frac{\partial \rho}{\partial q}\frac{\partial H}{\partial p} - \frac{\partial \rho}{\partial p}\frac{\partial H}{\partial q}.$$

The use of the aproach described above for the stochastic background defined by the Markov process $\alpha(t)$ with the Fokker–Planck equation for the transition probabilities P,

$$\frac{\mathrm{d}P}{\mathrm{d}t} + MP = 0,$$

now gives the following differential equation for the averaged evolution in the product space $C = X \times L_2(\mathrm{d}q\,\mathrm{d}p)$:

$$\frac{\mathrm{d}\wp}{\mathrm{d}t} = \left\{\mathrm{diag}_X H^{\times}_{\alpha}\right\}\wp + M \times I_C\,\wp,$$

with proper initial conditions. Multiplication by i gives the dissipative operator on the right hand side, which can be analysed by the methods described above. In particular, the corresponding scattering matrix can be defined. It can serve as a tool for investigation of the asymptotic behaviour of the averaged dynamics.

The author is grateful to Professor I. R. Prigogine for directing his attention to mathematical problems concerning irreversibility of quantum dynamics and quantum chaos. The author is also grateful to Professor I. Antoniou for stimulating discussions on the problem of existence of positivity-preserving dilations, and to Professor N. Levenberg, who read and commented on the text.

This work was partly supported by the grant 'ESPRIT' of the EC commission for Education and Research and a Staff Research Grant of the University of Auckland, New Zealand.

REFERENCES

BELOZERSKIJ, G. & PAVLJUHIN, Y. 1976 *ZhETF* 2, **70**, 717.

BELOZERSKIJ, G. & PAVLJUHIN, Y. 1977 *Fizika Tverdogo Tela* 5, **19**, 1279.

CHEREMSHANTSEV, S. 1990 Proceedings of Steklov Institute of Acad. Sci. USSR, vol. 184.

DAVIES, E. B. 1976 *Quantum Theory of Open Systems.* Academic Press.

GUILLEMIN, V. & STERNBERG, S. 1984 *Symplectic Techniques in Physics,* p. 468. Cambridge University Press.

LAX, P. & PHILLIPS, R. 1967 *Scattering Theory.* Academic Press.

NAGY, B. SZ. & FOIAS, C. 1967 *Analyse Harmonique des Operators de l'Espace de Hilbert.* Masson, Acad Kiado.

PAVLOV, B. 1983 *Proceedings of the International Congress of Mathematicians, Warsaw,* vol. 2, pp. 1011–1025.

PAVLOV, B. 1995 Spectral analysis of a dissipative singular Schrödinger operator in terms of a functional model. *Encyclopedia of Mathematical Sciences,* vol. 65, p. 65. Springer.

PAVLOV, B. & RYSHKOV, A. 1987 *Problemy Matematicheskoj Fiziki* **12**, 54.

PAVLOV, B. & RYSHKOV, A. 1989 Scattering on a random point potential. In *Applications of Self-Adjoint Extensions in Quantum Physics* (ed. P. Exner & P. Šeba). Lecture Notes in Physics, vol. 324. Springer.

PRIGOGINE, I. & STENGERS, I. 1988 *Entre le Temp et l'Eternite.* Fayard.

See also

CHUESHOV, I. 1978 In *Differential Equations and Methods of Functional Analysis.* pp. 133–141. English translation in *Selecta Mathematica Sovietica* **4**.

Time, chaos and the laws of nature

By ILYA PRIGOGINE AND DEAN J. DRIEBE

Center for Studies in Statistical Mechanics and Complex Systems, The University of Texas at Austin, Austin, USA
and
International Solvay Institutes for Physics and Chemistry, Free University of Brussels, Brussels, Belgium

The constructive role of irreversible processes has motivated an effort to incorporate them in the basic dynamical laws. Dynamics may be described on either the individual level, corresponding to trajectories or wavefunctions, or on the statistical level as probability distributions. For unstable dynamical systems the statistical description is irreducible to the individual description when probability distributions are considered in extended functional spaces. In this setting, irreversibility is made apparent and the measurement problem of quantum mechanics is avoided. The laws of nature then express possibilities instead of certainties, as is appropriate for the evolving world we live in.

In recent years, a radical change of perspective has been witnessed in science following the realization that large classes of systems may exhibit abrupt transitions, a multiplicity of states, coherent structures or a seemingly erratic motion characterized by unpredictability often referred to as deterministic chaos. Classical science emphasized stability and equilibrium; now we see instabilities, fluctuations and evolutionary trends in a variety of areas ranging from atomic and molecular physics through fluid mechanics, chemistry and biology to large scale systems of relevance in environmental and economic sciences, see Prigogine (1980), Nicolis & Priogine (1989). Concepts such as 'dissipative structures' and 'self-organization' have become quite popular. The distance from equilibrium, and therefore the arrow of time play an essential role in these processes, somewhat like temperature in equilibrium physics. When we lower the temperature we have various states of matter in succession . In nonequilibrium physics and chemistry, when we change the distance from equilibrium the observed behavior is even more varied.

How can these findings be interpreted from the point of view of the basic laws of physics? Newtonian dynamics, as well as relativity and quantum physics, do not include any distinction between past and future. There is no flow of time. This puzzle has led to an unending series of controversies. This may be called the time paradox. It is interesting that the time paradox was only identified in the second half of the 19th century. It was then that the Viennese physicist Ludwig Boltzmann tried to emulate what Charles Darwin had done in biology and to formulate an evolutionary approach to physics. But at that time, the laws of Newtonian physics had been accepted as expressing the ideal of objective knowledge. As they imply the equivalence between past and future, any attempt to confer to the arrow of time a fundamental meaning was resisted as a threat to the ideal of objective knowledge. Newton's laws were considered final in their domain of application, somewhat as quantum mechanics is today considered to be final by many physicists. How then can we introduce unidirectional time without destroying these amazing achievements of the human mind?

In October 1994, there appeared a special issue of *Scientific American* devoted to 'Life in the Universe'. There is an article, written by an eminent physicist, Steven Weinberg, which is relevant to our discussion, see Weinberg (1994). He writes: 'But as much as we would like to take a unified view of nature, we keep encountering a stubborn duality in the role of intelligent life in the universe, as both subject and student. ... On one hand, there is the Schrödinger equation, which describes in a perfectly deterministic way how the wavefunction of any system changes with time. Then, quite separate, there is a set of principles that tells how to use the wavefunction to calculate the probabilities of various possible outcomes when someone makes the measurement'.

The measurement? Is it suggested that *we*, through our measurements, would be at the origin of the cosmic evolution? Weinberg speaks of a stubborn duality. This is a point of view which we find in many recent publications, for example, in the famous book by Steven Hawking: *A Brief History of Time*, see Hawking (1988). There Hawking advocates a geometrical interpretation of cosmology. In short, time would be an "accident" of space. But Hawking understands that it is not enough. We need an arrow of time to deal with 'intelligent life'. Therefore, as do many other cosmologists, Hawking introduces, in addition, the so-called 'anthropic principle'. But no indication is given as to how the anthropic principle could ever emerge from a static geometrical universe. With the anthropic principle we are back to Cartesian dualism: on one side matter 'res extensa' described by geometry, on the other, the 'mind' associated with 'res cogitans'. Descartes described in this way the striking difference which exists between the behavior of simple physical systems such as a frictionless pendulum and the functioning of the human brain.

This dualism is difficult to accept for the modern mind. But the 19th century has left us a double heritage. We have 'the laws of nature', which involve time symmetry; but we also have an evolutionary description associated with entropy. Entropy introduces the distinction between time reversible and time irreversible processes. Irreversible processes produce entropy. But again, how to relate the existence of irreversible processes to the basic laws? A popular interpretation – similar in its spirit to Weinberg's statement – is that it would be us, through our approximations, who would be responsible for the 'apparent' observation of irreversible processes. To make such an argument plausible the first step is to present the consequences of the second law as trivial, as self-evident. For a 'well-informed' observer, such as the demon imagined by Maxwell, the world would appear as perfectly time reversible. We would be the fathers of evolution and not the children. But recent developments in nonequilibrium physics and chemistry point in the opposite direction.

1. Irreversible processes in nature

Let us briefly summarize the present situation. At equilibrium one of the thermodynamic potentials (i.e. the free energy) is minimum. As a result, fluctuations of external or internal origin are damped as they are followed by processes which bring the system back to the minimum of the potential. Near equilibrium it is the entropy production per unit time which is minimum. This again implies stability, but there is a new factor: irreversibility may become a source of order. This is already clear in classical experiments such as thermal diffusion. We heat one wall of a box containing two components and cool the other. The system evolves to a steady state in which one component is enriched in the hot part and the other in the cold part. We have an ordering process that would be impossible in equilibrium. Far from equilibrium there is no longer in general any extremum of a potential and stability is not assured. In steady states, fluctuations may then appear and lead to new spatio-temporal structures which one of us (I.P.) named dissipative structures as they are conditioned by a critical value of the distance from equilibrium, and therefore also by a minimum amount of entropy production associated with dissipation. Dissipative structures are characterized by a new coherence associated with long-range interactions and symmetry breaking (well-known examples are chemical clocks and so-called Turing structures). The appearance of dissipative structures occurs at 'bifurcation points', where new solutions of the nonlinear equations of evolution become stable. We have in general a succession of bifurcations which leads to an historical dimension. At bifurcations there are generally many possibilities open to the system out of which one is randomly realized. As a result, determinism breaks down even on the macroscopic scale. The conclusions are clear: irreversible processes (associated with the arrow of time) are as real as the reversible processes described by mechanics. Irreversible processes therefore cannot correspond to approximations we would introduce in the basic laws.

Therefore we come back to our problem. How to extend the basic laws of physics to include both reversible and irreversible processes? How to overcome the time paradox? Here interesting progress has been realized over the last years. The main aim of this paper is to give a simple introduction to these new ideas. A first remark is that we need an extension of dynamics only for classes of systems where we expect irreversible processes to arise. A well-documented example is 'deterministic chaos'. These are unstable systems. Trajectories corresponding to different initial conditions diverge exponentially in time. We shall in the next section present a new formulation of the laws of dynamics valid for deterministic chaos. Then we shall consider the more general case of unstable Hamiltonian classical or quantum systems.

The basic idea is common to the two situations. It is well known since the pioneering work of Gibbs and Einstein that we can describe dynamics from two points of view. On the one hand, we have the individual description in terms of trajectories in classical dynamics, or of wavefunctions in quantum theory. On the other hand, we have the description in terms of ensembles described by a probability distribution, ρ (called the density matrix in quantum theory). For Gibbs and Einstein, the ensemble point of view was merely a convenient computational tool when exact initial conditions were not available. In their view, probabilities express ignorance, lack of information. Moreover, it has always been admitted that, from the dynamical point of view, the consideration of individual trajectories and of probability distributions were equivalent problems. We can start with individual trajectories and then derive the evolution of probability functions or *vice versa.* The probability distribution ρ indeed corresponds to a superposition of trajectories. It was therefore natural to assume that the two levels of description: the 'individual' level (corresponding to single trajectories) and the 'statistical' level (corresponding to ensembles) would be equivalent.

Is this always so? For stable systems where we do not expect any irreversibility, this is indeed true. Gibbs and Einstein were right. The individual point of view (in terms of trajectories) and the statistical point of view (in terms of probabilities) are then indeed equivalent. But for unstable dynamical systems, such as associated with deterministic chaos, this is no longer so. At the level of distribution functions we obtain a new dynamical description which permits us to predict the future evolution of the ensemble including characteristic time scales. This is impossible at the level of individual trajectories or wavefunctions. The equivalence between the individual level and the statistical level is then broken. We obtain new solutions for the probability distribution which are 'irreducible' as they do not apply to single trajectories. In this new formulation the symmetry between past and future is broken.

2. The laws of chaos

We start with chaotic maps because they are the simplest systems to illustrate how irreversibilty emerges from unstable dynamics. A map, which is a discrete-time dynamical process, may arise from a continuous-time system or may describe a process that acts at certain time intervals with free motion in between. One may also consider a map simply as a model which can be used to illustrate essential features of the dynamics.

The simplest chaotic system is known as the Bernoulli map, see Schuster (1988). We have a variable x defined on the interval from 0 to 1. This interval is the 'phase space' of the system. The map is given by the rule that the value of x at some given time step is twice the value at the previous time step. In order to stay in the interval from 0 to 1 though, if the new value exceeds 1, only the fractional part is kept. The rule for the map is thus concisely written as $x_{n+1} = 2x_n \pmod 1$, where n represents time, which takes integer values. Here the map may only be run forward in time since the rule for the map cannot be inverted. This is because two distinct points at time n may yield the same point at time $n+1$, e.g. 1/6 and 2/3 both yield 1/3. Typical records of trajectories for this system are shown in figure 1.

This very simple system has the remarkable property that even though successive values of x are completely determined, they also have quite random properties. If x is written in binary notation, successive values are obtained simply by removing the first digit in the expansion and shifting the remaining digits. This means that after m time steps information about the initial value to an accuracy of 2^{-m} is now amplified to give whether the value of x is between 0 and 1/2 or 1/2 and 1. This amplification in any initial uncertainty of the value of x makes following trajectories for more than a few time steps a practical impossibility.

A generic initial value of x would be an irrational number with an infinite non-repeating expansion. This would lead to a trajectory that forever wanders throughout the phase space. But rational numbers, with repeating or terminating binary expansions, thus leading to periodic or fixed-point trajectories, are densely distributed among irrational numbers. This means that qualitatively different behavior, in the sense of trajectory dynamics, arises from initial conditions that are infinitesimally close. This kind of complicated microstructure of phase space for chaotic systems is in contrast to systems with regular dynamics where initial conditions throughout large regions of phase space lead to similar behavior, see Prigogine (1980).

These facts suggest that a much more natural way to consider the time evolution in chaotic systems is in terms of ensembles of trajectories defined by probability distributions. The evolution of an ensemble, determined by a probability distribution, is given by superposing trajectories. Then the probability distribution evolves through the application of an operator, usually denoted by U, known as the Frobenius–Perron operator, see Lasota & Mackey (1994). To obtain the distribution $\rho(x, n)$ at some time n, we apply the operator successively n times to the initial distribution $\rho(x, 0)$. Thus, $\rho(x, n) = U^n\rho(x, 0)$. In contrast to the unpredictable trajectory behavior, that of the probability distribution is completely predictable and furthermore, for all 'smooth' initial distributions approaches an equilibrium state as is shown in figure 2. By a smooth distribution, we mean one that does not just represent a trajectory, which would be a distribution localized at a single point. Then we would just get back to problems with trajectories. A point distribution localized at $x = \tilde{x}$ is written in terms of a Dirac delta function as $\delta(x - \tilde{x})$.

In order to understand how an operator acts on a class of functions one calculates the spectral decomposition of the operator. The spectral decomposition of an operator depends not just upon how the operator acts on a function but also on the type of function the operator is considered to act on. In quantum mechanical problems, the operators are considered to act on 'nice' normalizable functions that are members of a Hilbert space. Time evolution operators, even in classical mechanics, have traditionally

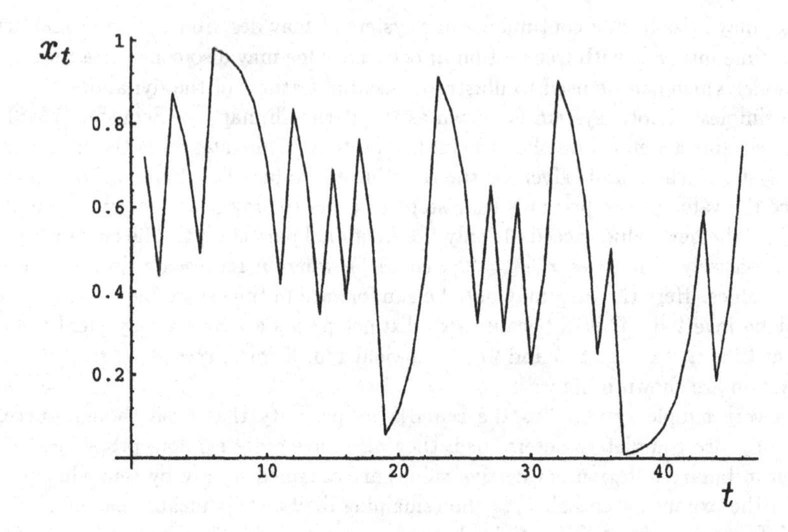

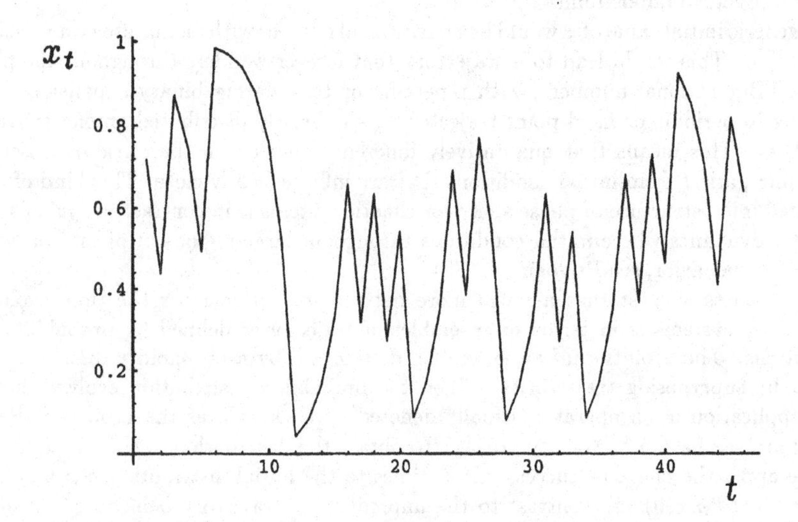

FIGURE 1. Records of two trajectories with nearby initial conditions under evolution of the Bernoulli map are shown. For the first few time steps the records are indistinguishable but after about ten time steps they follow completely different paths.

been analyzed in Hilbert space. A class of operators known as Hermitian operators plays a special role. These operators have only real eigenvalues in Hilbert space. The time evolution is then expressed as $e^{i\omega t}$, which is a purely oscillating function because ω is a real number. In order to have an explicit approach to equilibrium expressed by decay modes as $e^{-\gamma t}$ it is necessary to go outside the Hilbert space where Hermitian operators may have complex eigenvalues.

Physicists are used to working in Hilbert space since John Von Neumann employed it

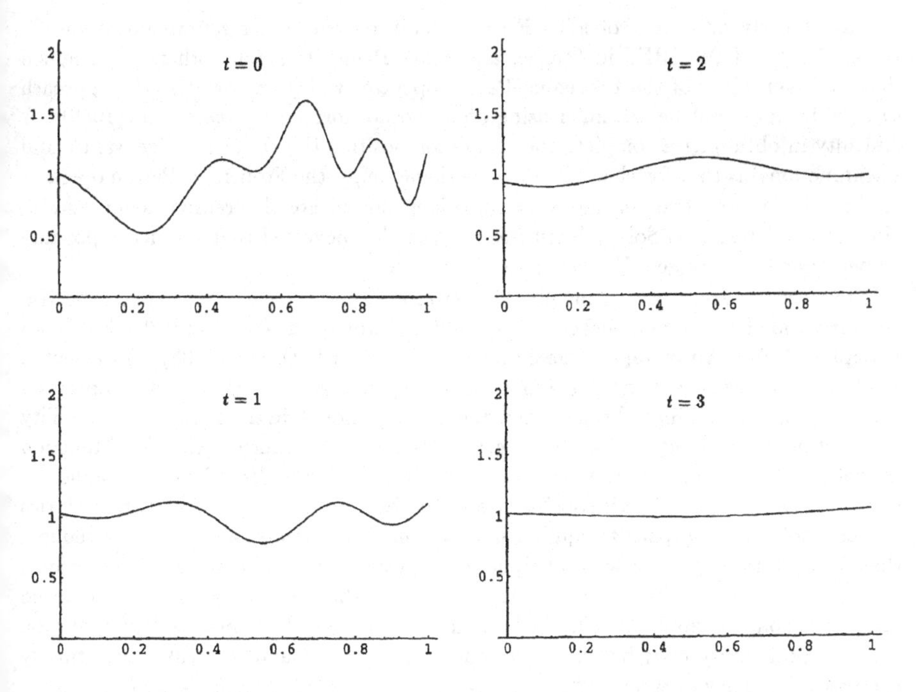

FIGURE 2. The evolution of a nonequilibrium probability distribution in the Bernoulli map is shown. At the initial time, $t = 0$, the distribution is quite inhomogeneous. At successive time steps the distribution becomes more smooth and by the third time step is very close to the equilibrium uniform distribution.

in the early 1930s as the mathematical framework for the then new quantum mechanics. Around the same time, Koopman (1931) proposed that classical mechanics could also be formulated in Hilbert space so that it could have a structure parallel to quantum mechanics.

One of the fundamental characteristics of Hilbert space is that it is 'self-dual', i.e. the dual space is the same Hilbert space. In Hilbert space, Hermitian operators act in the same way to the left and the right. Thus they only have real eigenvalues. This seemed to be the correct framework for quantum mechanics, because in this theory physical observables are represented as operators and the values the observables take (real numbers) are the eigenvalues of the operator.

For stable or periodic systems such as a stable atom or an harmonic oscillator, the Hilbert space setting works fine. For unstable atoms, and general time-dependent processes, Hilbert space is insufficient. This is because the decay rates and other quantities characterizing the approach to equilibrium correspond to *complex* eigenvalues. Complex energies have been used in an ad hoc way almost from the beginning of quantum mechanics to account for decaying states. But complex eigenvalues may be obtained rigorously if the operators act in a generalized functional space, often called a rigged Hilbert space, see Böhm & Gadella (1989). A rigged Hilbert space is obtained when the Hilbert space is restricted, which simultaneously causes the dual space to be enlarged. The dual space will usually become a generalized functional space. To include irreversible processes in the fundamental laws of physics, a rigged Hilbert space is the appropriate mathematical setting.

Only recently has the Frobenius–Perron operator come under serious investigation. David Ruelle, of the IHES in France, and Mark Pollicott, among others, determined that the eigenvalues of the Frobenius–Perron operator which characterize the approach to equilibrium could be obtained using an elaborate formalism, see Ruelle (1986). A difficulty in obtaining a complete spectral decomposition (i.e. both the eigenvalues and eigenfunctions) is that for chaotic one-dimensional maps, the Frobenius–Perron operator does not have a spectral decomposition in Hilbert space. For this reason, Ioannis Antoniou, of the Internatial Solvay Institutes in Brussels, suggested that the decomposition be performed in a 'rigged' Hilbert space.

In the last few years, several members of our group have calculated the explicit spectral decomposition for several model chaotic maps, see Antoniou & Tasaki (1992b), Hasegawa & Saphir (1992), Antoniou & Tasaki (1992a), Hasegawa & Driebe (1993). An essential result of this work is that the eigenfunctions corresponding to the spectral decomposition with decay modes are indeed generalized functions. As noted, in the sense of a probability distribution, a trajectory is written as a generalized function. Since a generalized function cannot be multiplied by another, the representation with generalized functions cannot be applied to trajectories. This is consistent with the fact, previously noted, that trajectories do not display an approach to equilibrium, but smooth distributions do. Recall though, that the evolution operator for probability distributions, the Frobenius–Perron operator, is defined by superposing trajectories. It is remarkable that the solution in the generalized functional space cannot be reduced back to trajectories. It is in this representation that the probability distribution is the fundamental object and its dynamics naturally expresses the arrow of time.

The Frobenius–Perron operator for the Bernoulli map, U_{B}, acts on a probability distribution as

$$U_{\mathrm{B}}\rho(x) = \tfrac{1}{2}\left[\rho\left(\frac{x}{2}\right) + \rho\left(\frac{1+x}{2}\right)\right].$$

The operator U_{B} has polynomial eigenstates, which coincidentally are called Bernoulli polynomials. Corresponding to the Bernoulli polynomial of degree m is the eigenvalue $1/2^m$.

In order to construct a spectral decomposition of U_{B}, we need to consider how it acts on functions to its left. This operation is the adjoint operator of U_{B} denoted as $U_{\mathrm{B}}^{\dagger}$. It acts on a function of x simply by replacing x by the rule for the Bernoulli map: $2x$ (mod 1). The operator $U_{\mathrm{B}}^{\dagger}$ in Hilbert space only has eigenvalues of absolute value 1, since it is an isometric operator. Thus, in Hilbert space, there are no eigenstates of $U_{\mathrm{B}}^{\dagger}$ corresponding to the eigenvalues $1/2^m$ (except for $m = 0$) and so no spectral decomposition of U_{B} is possible there. If we allow $U_{\mathrm{B}}^{\dagger}$ to act on generalized functions, we can obtain the eigenstates corresponding to the eigenvalues $1/2^m$. They are given by derivatives of the Dirac delta function as $(-1)^m[\delta^{(m-1)}(x-1) - \delta^{(m-1)}(x)]$.

Thus, the spectral decomposition of U_{B} is in a rigged Hilbert space: the right eigenstates are polynomials which span a subspace of the Hilbert space of square integrable functions; the left eigenstates are generalized functions which span a 'larger' functional space than Hilbert space, see Antoniou & Tasaki (1992b), Hasegawa & Saphir (1992), Antoniou & Tasaki (1992a), Hasegawa & Driebe (1993).

As noted, the trajectory dynamics of the Bernoulli map is irreversible from the beginning. By adding an extra dimension to the phase space, we may extend the Bernoulli map to obtain an invertible system known as the baker map, see Prigogine (1980). The dynamics associated with the new dimension is contracting (by a factor of 1/2) so the map preserves the area of an initial element of phase space. The two features of invertibility and preservation of phase space area are the essential features of Hamiltonian dynamics governing mechanical systems in nature.

Contrary to the Bernoulli map, there is a representation of the Frobenius–Perron operator of the baker map in Hilbert space, but there no irreversibility is apparent. All eigenvalues are of absolute value 1 so that no decay modes are determined. The generalized spectral representation of the baker map has also been recently constructed, see Hasegawa & Saphir (1992), Antoniou & Tasaki (1992a), Hasegawa & Driebe (1993). This representation contains eigenvalues of absolute value less than 1 (again $1/2^m$) so that the corresponding eigenfunctions decay. Both the left and right eigenfunctions are generalized functions and this representation again can only be applied to smooth probability distributions.

The original invertibility of the trajectory dynamics in the baker map leads to two distinct representations of the dynamics on the level of smooth probability distributions. The original group evolution in the Hilbert space representation splits into two semi-groups in the generalized representation. This means that there is a class of distributions which will always yield a natural approach to equilibrium in our future. This approach to equilibrium here is not due to any coarse-graining or projection of the dynamics but is an exact result which allows us to see that irreversibility already appears at the most basic level in this sytem.

Maps with even more realistic features, such as diffusion, have been considered by our group, see Hasegawa & Driebe (1993). Diffusion is usually modelled as due to underlying random forces. This is the theory of Brownian motion developed by Einstein, Langevin and others early in this century. The physical picture was motivated by the original experiments of the British botanist Robert Brown, who observed the erratic motion of pollen particles suspended in water. The explanation for the erratic motion is that, since the fluid is made up of so many particles with thermal motion, the pollen particle is in effect kicked randomly. The characteristic feature of diffusive motion is that the average of the square of the distance travelled by a diffusing particle is proportional to time. The proportionality factor defines the diffusion coefficient. This is in contrast to regular motion at a fixed velocity where the distance itself is proportional to time, with the proportionality factor being the velocity.

Traditional Brownian motion theory breaks direct contact with the underlying deterministic equations governing motion on microscopic scales by assuming a random force. However, we now understand that many particles, or many degrees of freedom, are not necessary to have complex dynamics that appears as random. Simple chaotic systems like the one-variable Bernoulli map generate such a seemingly random output because the dynamics is chaotic, see Lasota & Mackey (1994).

We may then construct a map by extending the Bernoulli map over many intervals, instead of just the one interval it is usually defined on. The extension is made in such a way that, in each interval, the dynamics is just like the Bernoulli map, but the map injects the iterates to the next interval to the right or left depending on where it lands in the original interval. This is a model of deterministic diffusion which is known as the multi-Bernoulli map. The multi-Bernoulli map is a simpler version of a similar map constructed from extending the baker transformation, known as the multi-baker map and introduced by Pierre Gaspard of the Free University of Brussels, see Gaspard (1992).

Generalized spectral representations of these models of deterministic diffusion have recently been constructed. The solutions have a surprisingly rich structure. First, the eigenvalues tell us that the diffusive mode is an exact mode in the system and is not obtained from a limiting procedure as is generally done to obtain diffusive modes, see

Hasegawa & Driebe (1993) and Gaspard (1992). The eigenvectors have interesting fractal properties, see Hasegawa & Driebe (1993).

These recent results on chaotic maps demonstrate that having many degrees of freedom is not responsible for irreversibility. It is the unstable microscopic dynamics that is responsible for the emergence of irreversibilty, as the results on the one- and two-degree of freedom systems we have studied indicate. When the trajectory dynamics is time reversible, there appears a time symmetry breaking associated with the choice of a dynamical semigroup on the level of the probability distribution. Since in nature we all share the same arrow of time, this choice expresses the unity found in nature.

3. Extension of classical and quantum mechanics

We come to classical and quantum mechanics. Here the central quantity is the Hamiltonian, which is the energy of the system expressed in terms of coordinates and momenta (in simple cases the momentum is given by the velocity multiplied by the mass). In these systems, time is acting continuously. We first consider classical mechanics. Again there are two descriptions: the 'individual' description in terms of trajectories in classical mechanics or wavefunctions in quantum mechanics; and the 'statistical' description in terms of probability distributions ρ. The probability distribution obeys the Liouville equation $\mathrm{i}\partial\rho/\partial t = L\rho$. This equation has the formal solution $\rho(t) = \mathrm{e}^{-\mathrm{i}Lt}\rho(0)$ so that the Frobenius–Perron operator is replaced by $\mathrm{e}^{-\mathrm{i}Lt}$. The operator L is a derivative operator in phase space and is called the Liouville operator, see Prigogine (1962).

Here we come to the problem of integrability at the center of the basic work of Henri Poincaré at the end of the 19th century. The Hamiltonian generally contains two terms corresponding respectively to the kinetic and the potential energy. Poincaré asked the question (we simplify somewhat): is it possible to eliminate the potential energy by an appropriate choice of variables? Then the system would become isomorphic to independent particles and integration of the equations of motion would be immediate. Poincaré has shown that this is in general impossible, and fortunately so. If the answer would have been affirmative there would be no possibility of coherence, no organization, and no life. The importance of Poincaré resonances is well-recognized today. It led to KAM theory, so called in honour of its founders Kolmogorov, Arnold and Moser.

Poincaré moreover identified the reason for nonintegrability: the existence of resonances between the various degrees of freedom. For each degree of freedom there is an associated frequency ω. Consider then a system characterized by two degrees of freedom. The corresponding frequencies are ω_1 and ω_2. Whenever $n_1\omega_1 + n_2\omega_2 = 0$, with n_1 and n_2 non-vanishing integers, we have resonance. These resonances lead to the problem of small denominators by showing up in perturbation calculations as $1/(n_1\omega_1 + n_2\omega_2)$. The resonances give rise to random trajectories. In this sense Poincaré resonances and non-integrability are also associated with chaos.

We shall be mainly interested in so-called 'Large Poincaré Systems' (LPS) in which the frequency ω_k depends continuously on the wave vector k. A simple example is provided by the interaction between an oscillator ω_1 and a field. Resonances appear when the field frequencies ω_k are equal to the oscillator frequency ω_1. Resonances in LPS are responsible for fundamental phenomena such as emission or absorption of light, decay of unstable particles and scattering of particles, to name a few. These resonances play a fundamental role in both classical and quantum physics.

We now come back to the problem of the equivalence between the 'individual' and 'statistical' descriptions already mentioned in the section on irreversible processes. Since the Liouville operator is an Hermitian operator, we cannot expect that the statistical description would lead to irreversible processes as long as we remain in Hilbert space.

Here we come to an essential point. In typical macroscopic situations, molecules collide continuously with each other. We have 'persistent' interactions. This is in contrast to 'transitory' interactions as considered e.g. in ordinary scattering experiments (described by so-called 'S-matrix' theory) in which we have free asymptotic 'in' and 'out' states. To describe persistent interactions we have to introduce 'delocalized' distribution functions spread out in space. For example, consider a distribution function $\rho(x,p)$, where x is the coordinate and p the momentum, and impose boundary conditions such as ρ going to a function independent of x when $x \to \pm\infty$. The classical theory of Fourier transforms tells us that a function independent of x requires singular Fourier transforms with a delta function singularity in the wave vector k. The distribution function in Fourier space is then the sum of a singular part and a regular part. Note that even in equilibrium statistical mechanics, where the probability distribution is a function of the Hamiltonian, it is a singular function in Fourier space, as the Hamiltonian contains the kinetic energy, which is independent of x and thus delocalized.

Our basic mathematical problem is the derivation of the spectral representation of the Liouville operator for LPS and delocalized distribution functions. This has been recently accomplished by Tomio Petrosky, of The University of Texas, and one of us, see Petrosky & Prigogine (1994). The eigenvalue problem has some very interesting features. The singular part satisfies a closed diffusion-type equation and the regular part is a well-defined functional of the singular part. But a diffusion equation leads to complex eigenvalues and therefore time symmetry is broken. Moreover, this solution cannot be expressed in terms of trajectories. As a result we again obtain spectral representations irreducible to trajectories. The trajectory now corresponds to stochastic realizations familiar from Brownian motion theory.

It is important to stress that irreversibility appears as an emergent property somewhat analogous from this point of view to phase transitions which can also only be defined with respect to ensembles of particles. Here also we need many molecules producing persistent interactions. The 'whole' is more than the sum of the 'parts'. Our theory recovers all known results in statistical mechanics (see for example Balescu (1975)) and goes beyond them as well. The usual methods only allow for the description of slow processes, of the order of the relaxation times. Our theory allows for the calculation of processes on much shorter time scales and enables the calculation of the appearance of spatial correlations, Driebe & Petrosky (1992). Moreover, delocalized distribution functions can be prepared in computer simulations and our theoretical predictions have been tested successfully. This is a first indication that, for the situations we have considered, classical physics has to be extended and irreversibility incorporated in its formulation.

Scattering of particles in both classical and quantum systems is one of the fundamental processes in nature. For free motion, the momentum p is an invariant of motion, i.e. it is independent of time. In the presence of scattering, classical theory leads to the definition of a new invariant of motion $I(x,p)$. As a result, its average value $\langle I\rangle_\rho$ taken with a *localized* distribution function (or density matrix) ρ is also invariant. Our theory predicts that this is no longer so for $\langle I\rangle_\rho$ taken with a delocalized distribution function, which corresponds to persistent interactions. Then the time derivative $\mathrm{d}\langle I\rangle_\rho/\mathrm{d}t$ satisfies a diffusion equation and varies strictly linearly with time. This prediction has been quantitatively verified by numerical simulations, see Petrosky & Prigogine (1993a), as shown in figure 3.

These results have been extended to quantum mechanics by Petrosky & Prigogine (1993b), see figure 4. Again, for LPS and delocalized distribution functions we have to

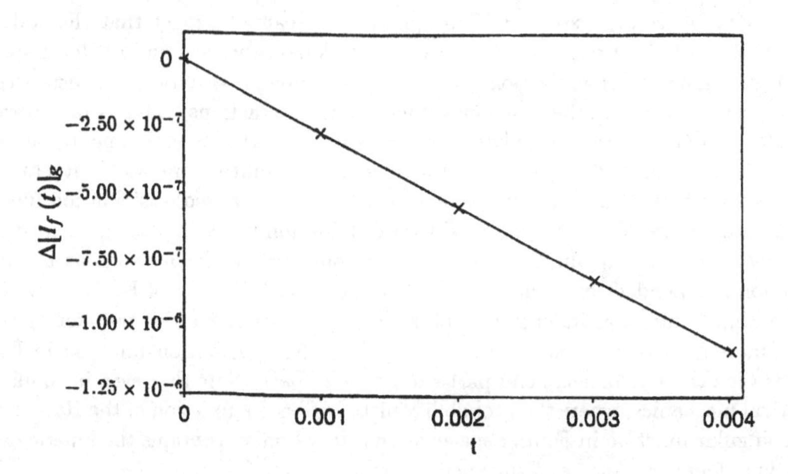

FIGURE 3. The result of a numerical simulation showing the linear change of the 'invariant' in classical two-body scaterring.

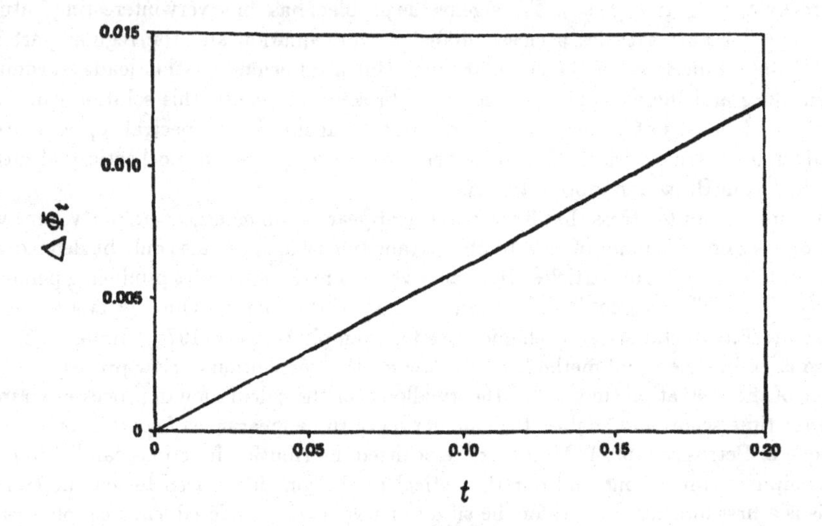

FIGURE 4. The result of a numerical simulation showing the linear change of the 'invariant' in quantum two-body scaterring.

extend the Liouville operator beyond Hilbert space. We then obtain new spectral decompositions which lead to complex eigenvalues and which are irreducible to wavefunctions. We may associate the new solutions with 'quantum chaos'.

In spite of the immense success of quantum mechanics, discussions about its foundations continue. The role of Newton's equation is now played by the Schrödinger equation which transforms the wavefunction $\psi(t_0)$, as given at initial time t_0, to the wavefunction $\psi(t)$ at time t. Schrödinger's equation is again time reversible and deterministic.

Generalizations of the standard formulation of quantum mechanics have been proposed. In the 60s and 70s Arno Böhm, Jean-Pierre Antoine and Manuel Gadella gave a

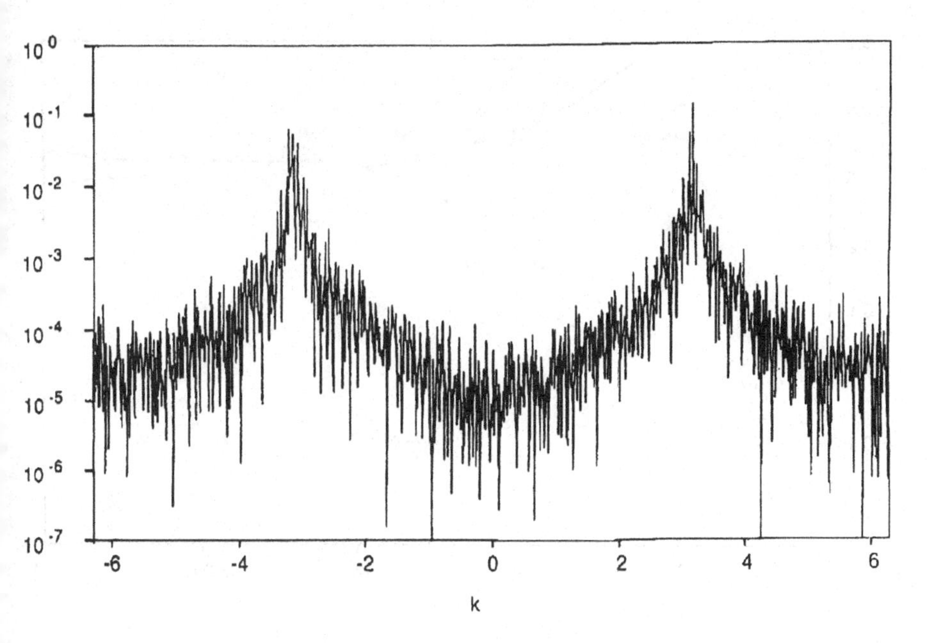

FIGURE 5. The probability $\rho_{k,k}(t)$ for finding the momentum at $t = 6t_r$.

formulation of quantum mechanics in rigged Hilbert spaces to deal with decay processes and radioactivity, see Böhm & Gadella (1989). An alternative method based on analytic continuation was proposed by Sudarshan *et al.* (1978). Both of these approaches can be seen as attempts to complete the early work of Nakanishi (1958) who constructed eigenvectors with complex eigenvalues for a model decay system. But these methods do not solve the problem of irreversibility associated with the approach to equilibrium; neither do they solve the 'quantum paradox'.

The wavefunction determines 'potentialities'. We therefore need an additional mechanism to go from potentialities to the actualities we measure. This introduces the 'collapse' of the wavefunction and leads to irreversibility. But we then obtain a dual structure at the basis of quantum theory. When to use the Schrödinger equation; when to introduce collapse? This leads to the quantum paradox. Many proposals to elucidate the conceptual foundations of quantum theory can be found in the literature. But as for the time paradox, they are mostly based on approximations we would introduce in the basic quantum laws. Also, none of these proposals leads to new predictions which could be tested. In contrast, our approach leads to a unified and testable formulation. The basic description is now on the statistical level. The collapse corresponds to situations where the initial state is outside Hilbert space (as for plane waves) and where Poincaré resonances lead to diffusive behaviour. Our theory permits us to describe the approach of quantum systems to equilibrium and eliminates the quantum paradox.

The quantum Lorentz gas is a model of a light particle, such as an electron, scattered by randomly distributed heavy particles, such as ions. A numerical simulation for a one-dimensional quantum Lorentz gas has been performed to verify the approach to equilibrium predicted by our complex spectral representation. In this model, N scatterers are distributed with density $c = N/L$ in a one-dimensional 'box' with size L. The box

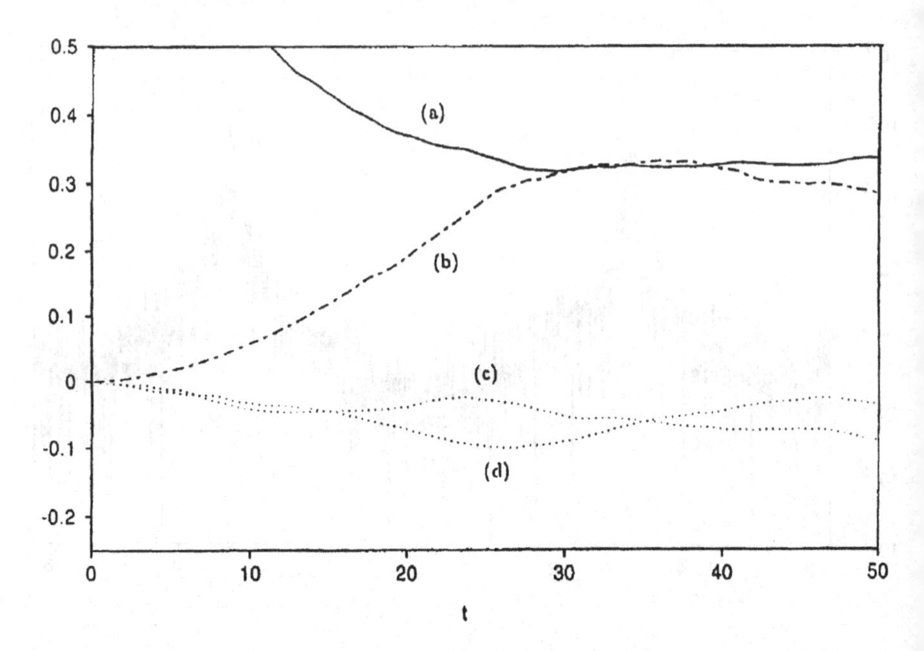

FIGURE 6. Time evolution of elements of the density matrix around $k = k_0$: (a) $\rho_{k,k}$, (b) $\rho_{-k,-k}$, (c) $\mathrm{Re}\,[\rho_{k,-k}]$, (d) $\mathrm{Im}\,[\rho_{k,-k}]$.

size has been chosen large enough to correspond to the thermodynamic limit (for the duration of the simulations).

The scatterers are distributed randomly in space. A light particle is scattered by these scatterers. The wave function for the light particle is represented by a plane wave, $\exp[ik_0x]$, at the initial time. The probability to find the particle at point x is then given according to the rules of quantum mechanics by: $\exp[ik_0x]\exp[-ik_0x]$; it is therefore independent of x. This corresponds to a delocalized situation outside Hilbert space. A repulsive delta function interaction between the light particle and the heavy scatterers is chosen. Our complex spectral representation predicts an approach to equilibrium with relaxation time $t_r = |k_0|/(8\pi^2\lambda^2c)$ for $\lambda \ll 1$. Here λ is the coupling constant measuring the strength of the interaction between the light particle and the scatterers.

In figure 5, the probability $\rho_{k,k}(t)$ for finding the momentum k on a logarithmic scale at $t = 6t_r$ is shown. The distribution has two sharp peaks around $k = \pm k_0$ which correspond to a 'microcanonical distribution' $\rho^{\mathrm{eq}}_{l,l} \sim \delta(\omega_l - \omega_{k_0})$ with $w_l = l^2/2$. The plane wave has therefore evolved to an equilibrium distribution (we have a 'collapse' of the wavefunction).

In figure 6, the time evolution of the diagonal elements, $\rho_{k,k}$ and $\rho_{-k,-k}$, and of the off-diagonal element $\rho_{k,-k}$, around $k = k_0$ is shown. The results clearly show the transformation of the pure state into a mixture (for a pure state the order of magnitude of the diagonal element and the off-diagonal element would be the same).

Quantum theory started with the observation that spectroscopic frequencies are the differences of two energy levels. But this is not true for the imaginary part of the eigenvalues. Quantum relaxation times, as observed or calculated, are not the differences between two levels. This already shows that irreversible processes cannot be described

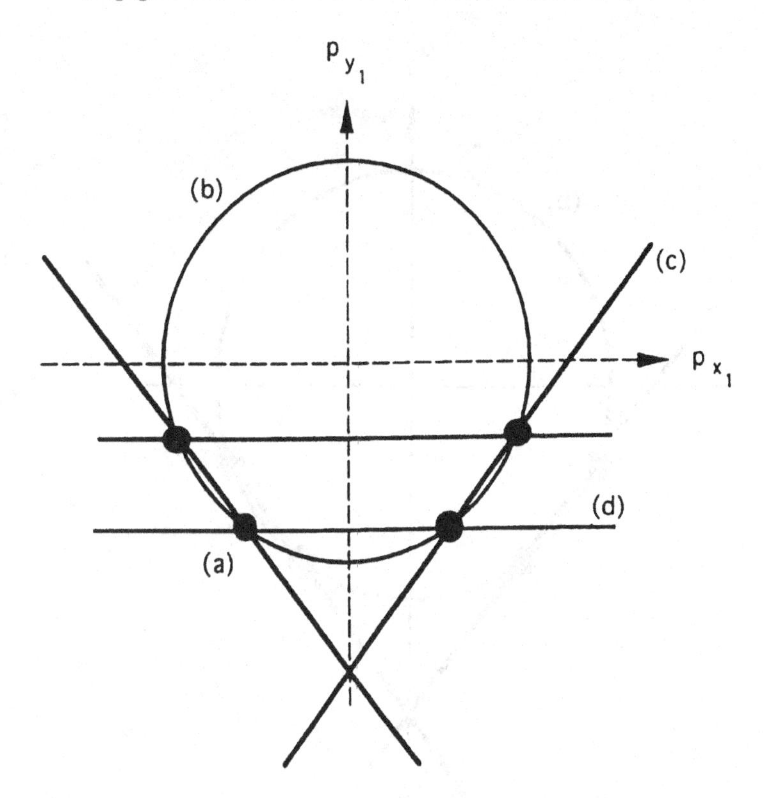

FIGURE 7. Theoretical prediction of the distribution of the momentum during the collision by our complex spectral representation.

in terms of wavefunctions in Hilbert space, but only in the Liouville space extended to generalized functions.

Usual scattering theory, i.e. the S-matrix approach, is an asymptotic theory. One considers a wave packet which was free for $t \to -\infty$, then interacts with a scatterer during some finite time and again becomes free for $t \to +\infty$. The time of observation is always considered much larger than any other time scale such as the duration of the collision. In contrast, our approach deals with persistent scattering associated with delocalized wave packets. Our approach permits an extension of scattering theory to *finite* time. Indeed, for a large but finite wave packet there exists a time scale where scattering can be described by delocalized wave packets (the system still does not 'know' that the wave packet is finite). This permits us to extend scattering theory to observation times much less than the collision time. This is the situation of interest in radioactive decay or in chemical reactions where observations are made *during* the collisions.

We shall consider here a simple three-body problem and show that our theory leads to a new effect not accessible by traditional scattering theory, see Petrosky *et al.* (1996). We assume quantum mechanics, and delta function repulsive interactions between each pair of particles. In the original coordinate system, there are three variables R_i which locate the particles with momentum P_i for $i = 1, 2, 3$. But if we go to the center-of-mass system we can eliminate one of these variables. This is achieved by so-called Jacobi coordinates defined by $r_{xi} = (R_j - R_k)/\sqrt{2}$ and $r_{yi} = \sqrt{3/2}\, R_i$, and similarly

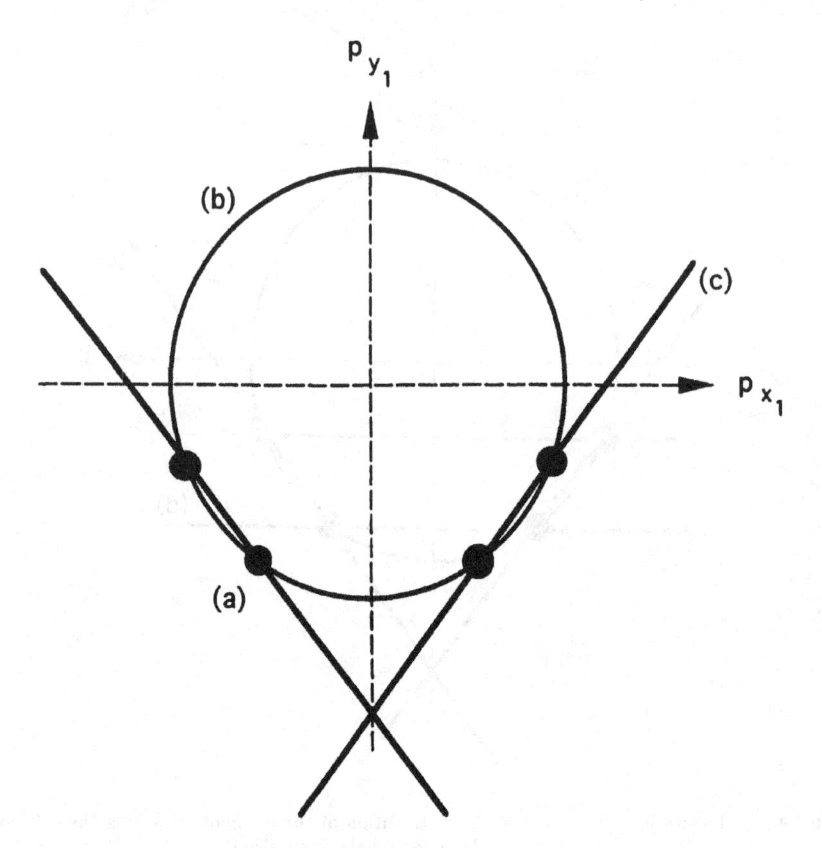

FIGURE 8. Theoretical prediction of the distribution of the momentum after the collision by the traditional S-matrix theory.

$p_{xi} = (P_j - P_k)/\sqrt{2}$ and $p_{yi} = \sqrt{3/2}\, P_i$, where (i, j, k) are chosen cyclically from $(1, 2, 3)$. In this coordinate system, the collision between, for example, particles 2 and 3 occurs at $r_{x1} = 0$ independent of the value of r_{y1}. As particle 1 does not participate in this collision, the momentum p_{y1} does not change. This means that one can interpret the process of three-body scattering as the evolution of a single 'particle' in two-dimensional space, interacting with three frictionless walls at $r_{xc} = 0$ (i.e. the r_{yc} axis), for $c = 1, 2, 3$, respectively.

Consider a specific scattering, where a collision between particles 1 and 2 is followed by a collision between particles 2 and 3. Our complex spectral representation is applicable for finite time scales, and predicts several secular effects during the collision. The strongest secular effect is proportional to t^2 and (a) the unperturbed energy is preserved between each two-body collision. This corresponds to the rescattering process. For a given initial momentum, the final momentum is distributed among the four points indicated in figure 7. There are three other secular effects which are proportional to t. They are the processes in which the energy is preserved (b) between the initial and final state, but not in the intermediate states; (c) between the final and intermediate states, but not between the intermediate and initial states, and (d) between the intermediate and initial states,

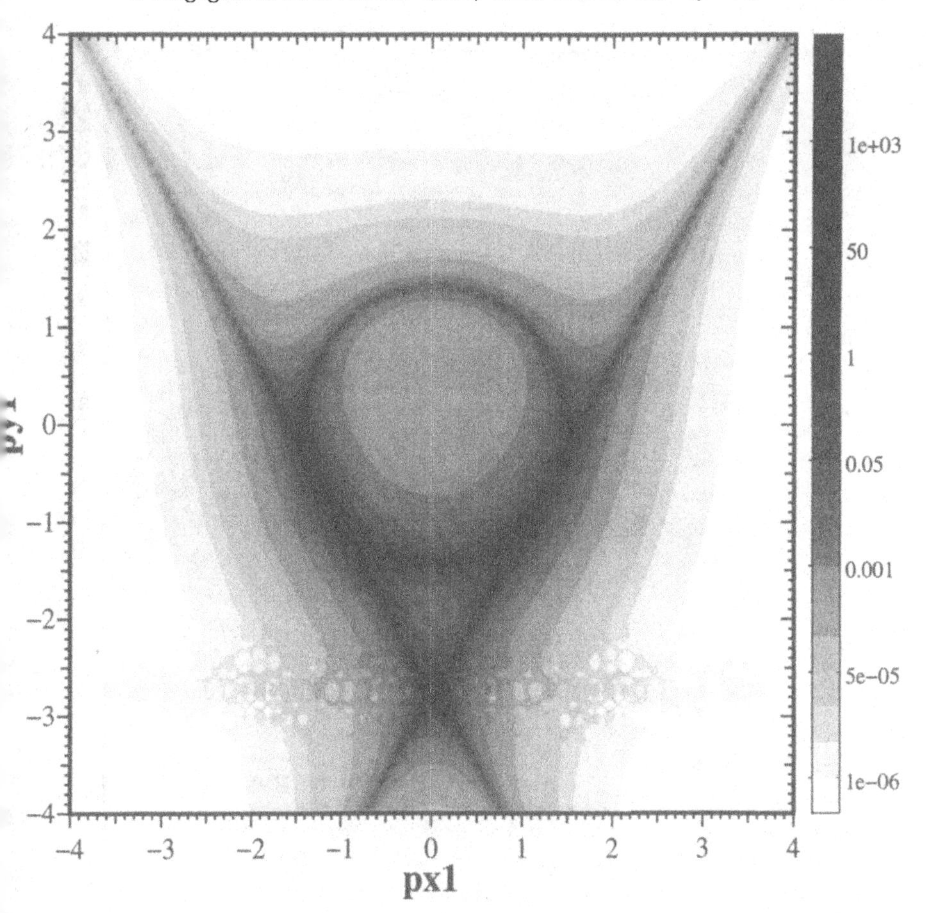

FIGURE 9. Numerical results of the distribution of the momentum during the collision.

but not between the final and intermediate states. Processes (c)–(d) are also shown in figure 7.

In contrast, the usual S-matrix theory predicts the distribution of the final momentum after the collision as shown in figure 8. The difference comes from process (d) which is transient and disappears after the collision. We display the results of a numerical simulation performed for an initial condition with a large wave packet, see Petrosky *et al.* (1996). Figure 9 shows the process during the collision, while figure 10 shows it after the collision. These results verify the prediction.

In our theory, the observer no longer plays a special role. We obtain a dynamical description of the measurement process. The measurement device has to be described by a semigroup presenting a broken time symmetry. It is the common arrow of time which is the necessary condition of our communication with the physical world, just as it is the condition of our communication with our fellow humans.

It is historically interesting that each basic physical phenomenon required an adequate mathematical formulation for its incorporation in a consistent theoretical scheme. Particle-wave duality required the operator calculus of quantum mechanics; gravitation required Riemannian geometry. Now we see that irreversibility requires generalized func-

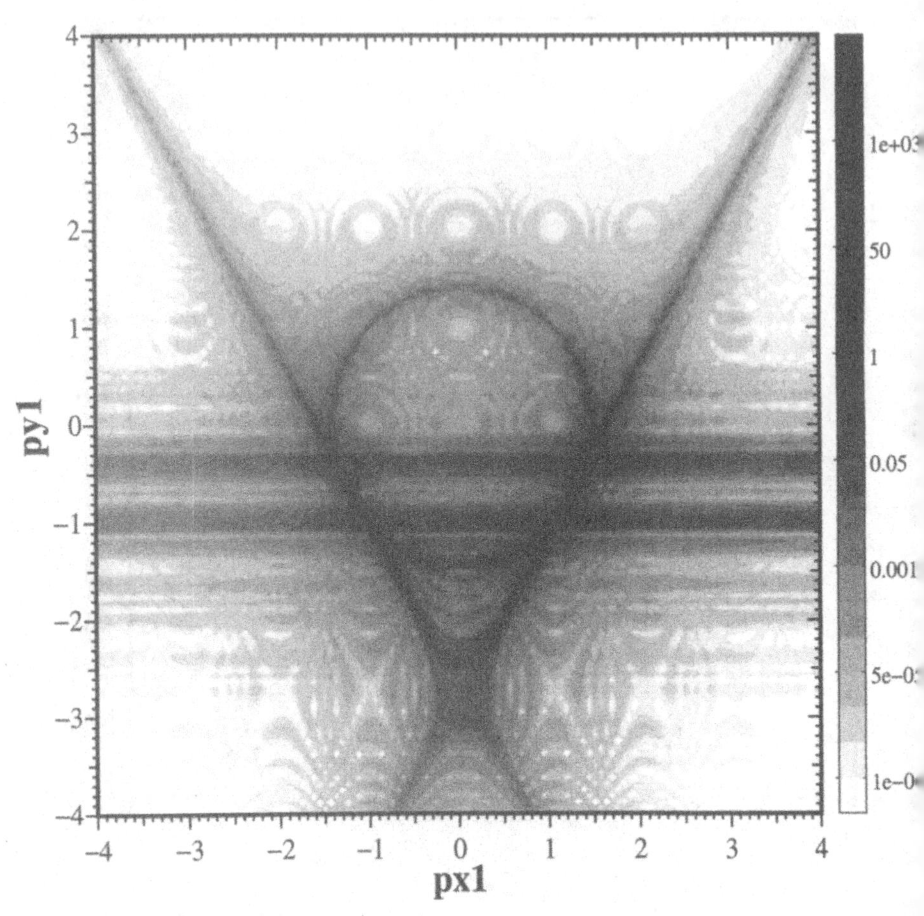

FIGURE 10. Numerical results of the distribution of the momentum after the collision.

tional analysis. This is because we need to go beyond Hilbert space to include irreversibility in the spectrum of the time evolution operators.

Statistical physics and thermodynamics have played an immense historical role in the evolution of physics in the 20th century. The divergences in the classical description of the specific heat of black-body radiation led to quantum theory. We are today in a somewhat similar situation. Instability and chaos force us to adopt a different point of view concerning the formulation of the laws of nature. They now express possibilities instead of certainties. There is no longer any contradiction between dynamical and thermodynamic descriptions of nature. Far from being a measure of our ignorance, entropy expresses a fundamental property of the physical world, the existence of a broken time symmetry leading to a distinction between past and future which is a universal property of both the nature we observe as well as a prerequisite for the existence of life and consciousness.

The work reported here is the outcome of a many year long effort by the Brussels–Austin group. We wish to especially acknowledge those who have made important contributions to the last phase: Tomio Petrosky, Suichi Tasaki, Ioannis Antoniou and Hiroshi

Hasegawa. We also wish to acknowledge support of the Welch Foundation, the United States Department of Energy, the European Commision and the Communauté Française de Belgique.

REFERENCES

ANTONIOU, I. & TASAKI, S. 1992a *Physica* A **190** 303.

ANTONIOU, I. & TASAKI, S. 1992b *J. Phys.* A**26** 73; see also GASPARD, P. 1992 *J. Phys.* A**25**, L483; ANTONIOU, I. & TASAKI, S. 1993 *Intl. Journal of Quantum Chemistry* **46**, 425.

BÖHM, A. & GADELLA, M. 1989 *Dirac Kets, Gamow Vectors and Gelfand Triplets.* Springer.

BALESCU, R. 1975 *Equilibrium and Non-equilibrium Statistical Mechanics.* Wiley.

DRIEBE, D. J. & PETROSKY, T. Y. 1992 *J. Stat. Phys.* **67**, 369.

GASPARD, P. 1992 *J. Stat. Phys.* **68**, 673.

HASEGAWA, H. H. & DRIEBE, D. J. 1993 *Phys. Lett.* A **176**, 193; see also TASAKI, S., ANTONIOU, I. & SUCHANECKI, Z. 1993 *Phys. Lett.* A **179**, 97; HASEGAWA, H. H. & DRIEBE, D. J. 1994 *Phys. Rev.* E **50**, 1781; GASPARD, P. 1993 *Chaos* **3**, 427.

HASEGAWA, H. H. & SAPHIR, W. C. 1992 *Phys. Rev.* A **46**, 7401.

HAWKING, S. 1988 *A Brief History of Time.* Bantam Books.

KOOPMAN, B. 1931 *Proc. Natl. Acad. Sci. USA* **17**, 315.

LASOTA, A. & MACKEY, M. 1994 *Chaos, Fractals, and Noise.* Springer.

NAKANISHI, N. 1958 *Prog. Theor. Phys.* **19**, 607.

NICOLIS, G. & PRIGOGINE, I. 1989 *Exploring Complexity.* Freeman.

PETROSKY, T. & PRIGOGINE, I. 1993a *Phys. Lett.* A **182**, 5.

PETROSKY, T. & PRIGOGINE, I. 1993b *Proc. Natl. Acad. Sci. USA* **90**, 9393.

PETROSKY, T. & PRIGOGINE, I. 1994 *Chaos, Solitons & Fractals* **4**, 311.

PETROSKY, T., ORDONEZ, G. & MIYASAKA, T. 1996 Extension of scattering theory for finite time: three-body scattering. *Phys. Rev.* A **53**, 4075.

PRIGOGINE, I. 1962 *Non-equilibrium Statistical Mechanics.* Wiley.

PRIGOGINE, I. 1980 *From Being to Becoming.* Freeman.

RUELLE, D. 1986 *Phys. Rev. Lett.* **56**, 405; see also POLLICOTT, M. 1990 *Ann. Math.* **131**, 331.

SCHUSTER, H. G. 1988 *Deterministic Chaos.* VCH Publishers.

SUDARSHAN, E. C. G., CHIU, C. B. & GORINI, V. 1978 *Phys. Rev.* D **18**, 2914.

WEINBERG, S. 1994 Oct. *Scientific American.*

Dynamic entropies and predictability of evolutionary processes

By WERNER EBELING

Institute of Physics, Humboldt-University, Berlin, Germany

We consider models of evolutionary processes and investigate symbolic sequences generated by evolution. The basic idea is that evolution 'likes' a criticality connected with the existence of long-range correlations, long time memory and historical behaviour (self-tuned criticality). In order to check this assumption, we investigate simple models of evolutionary growth as, e.g. the Fibonacci model of replication processes and the logistic map model for self-reproduction. We prove the existence of long-range correlations under certain (critical) conditions. We investigate the structure of sequences generated by evolutionary processes and, in particular, information carriers such as books and DNA strings. The higher order Shannon entropies and conditional entropies (dynamical entropies and their limits) are calculated, and characteristic scaling laws are found. Finally, other measures of long-range correlations, as e.g. algorithmic entropy, correlation function for distant letters, low frequency Fourier spectrum and the characteristic scaling of Hoelder exponents are discussed.

---

1. Introduction

All our experience with biological and social evolution processes is showing us that evolution is an historical process connected with long memory effects, see Ebeling *et al.* (1990), Feistel & Ebeling (1989), Kauffman (1993) and Ebeling & Nicolis (1991). One of the main reasons to expect long-range correlations is that evolution probably operates in regions of criticality. As is known from many examples in physics and in other natural sciences, critical conditions (i.e. operating near to transitions) imply the existence of long range correlations, see Bak *et al.* (1987) and Bak *et al.* (1989). Most systems show, for the critical value of control parameters, $a = a_{\rm cr}$, and in the neighbourhood of this value,

$$\frac{|a - a_{\rm cr}|}{a_{\rm cr}} \ll 1, \tag{1.1}$$

several peculiarities, as e.g. long relaxation times, long-range correlations, anomalous scaling laws, scale invariance and $1/f$ noise. In the following, structures occurring on the border between regular and chaotic regimes, or between two different regular or chaotic regimes, will be called critical structures. There exist far-reaching analogies between critical structures in non-equilibrium and critical phenomena known from equilibrium phase transitions. In order to get critical phenomena one has, in general, to fix the external parameters, as e.g. temperature and pressure in equilibrium or the pumping rate in non-equilibrium, at the critical values. However, there also exists an interesting class of systems showing 'self-organized criticality'. The notion of self-organized criticality (SOC) was introduced by Bak *et al.* (1987). The idea of SOC is that systems are kept in a critical state without any control parameter being set explicitly. Several authors expressed the idea that SOC may play a role in information processing systems, and especially in life phenomena, see Kauffman (1993), Bak *et al.* (1987), Bak *et al.* (1989), Langton (1990). Here we will use here the term 'self-tuned criticality' (STC) instead of SOC, defining it in a rather loose sense. In our notation, STC means that the system is in some sense critical by tuning its parameters by an adaptive process, and that it therefore shows long-range correlations. However, we will not include specific scaling laws in the definition of STC.

(This is often done in work on SOC.) In our view, too narrow a definition of SOC has led in the past to rather controversial discussions about the relevance of SOC in nature.

For reasons of simplicity, we consider here only discrete processes in time and state space which show rather long memory tails. We mention that, by using the methods of symbolic dynamics, any trajectory of a dynamic system may be mapped to a string of letters in an alphabet, see Ebeling & Nicolis (1992). Besides sequences generated by dynamical models of evolutionary processes such as the Fibonacci model and the logistic map, we shall also study long correlations in strings carrying information. Such sequences can be considered as frozen symbolic sequences. In particular, we study information carriers. The first examples are polynucleotids (DNA), the main information carriers in living systems which are sequences of nucleotides, see Gatlin (1972), Ebeling *et al.* (1990) and Feistel & Ebeling (1989). Further on we shall study examples of human messages such as books. For several reasons we expect the existence of long range structures in these sequences. Especially we are expecting correlations which are ranging from the beginning of a string up to the end of the string. Let us discuss some of those reasons:

(*a*) *Predictability.* We know from our every-day experience and from scientific research that the identification of the first hundred or thousand letters of the string already tells us a lot about the continuation. We often are making a decision in a book shop after reading just one page. Such expectations are justified only if there indeed exist long range correlations. This is the scientific expression of our intuitive expectations which are based on the experience that such sequences have a certain inherent predictability. As the basic quantity for estimating predictability we consider here the Shannon entropies for certain subtrajectories. Assuming that an observation has provided us with a certain trajectory of length n (an n-word), we may ask for the uncertainty of predicting the next state (letter). This is nothing else than the difference between the Shannon entropies for trajectories (words) of length $n+1$ and trajectories of length n:

$$h_n = H_{n+1} - H_n. \tag{1.2}$$

This conditional entropy measures the uncertainty of predicting a state one step in the future, provided a history consisting of n states, i.e. the present state and the previous $n-1$ states, is known, see Ebeling & Nicolis (1992). In this way the estimation of Shannons n-gram entropies, which are often called block entropies for a series of word length n, is our basic problem. Predictability is measured in this work by differences of Shannon entropies, in other words by conditional entropies. The existence of long correlations is expressed by long decreasing tails of the conditional entropies h_n. In general, our expectation is that any long-range memory decreases the conditional entropies and improves the chances for predictions.

(*b*) *Syntactical limitations.* Another heuristic reason to expect long-range correlations is the exponential explosion of the variety of possible subwords with increasing length for uncorrelated strings. Uncorrelated sequences generated on an alphabet of λ letters have a manifold of

$$N(n) = \lambda^n = \exp\big(n \ln(\lambda)\big) \tag{1.3}$$

different subwords of length n. A subword (word) here is any combination of letters, including the space, punctuation marks, etc. For $n > 100, \ldots, 1000$, the number $N(n)$ is extremely large. In other words, we need very sharp restrictions to select a meaningful subset from it. Long-range correlations provide such a strong selection criterion. Denoting the selected subset by $N^*(n)$ we must expect

$$\frac{N^*(n)}{N(n)} \to 0, \tag{1.4}$$

if $n \to \infty$. Bounds of this kind are given by the internal structure of texts and pieces of music especially by the rules of syntax. The syntactical rules do not allow for an arbitrary concatenation of words to sentences, most of them are forbidden. Furthermore we know that texts (and pieces of music) are formed by keywords (motifs) which are the raw material for the generation of a text (a piece of music). In fact all these rules lead to limitations of the growth with n.

In our earlier work we had made the conjecture that the number of allowed subwords scales like, see Feistel & Ebeling (1989),

$$N^*(n) \simeq n^{\alpha(\lambda)}. \tag{1.5}$$

This kind of scaling has recently been detected in musical structures, see Ebeling *et al.* (1995b). Furthermore it could be shown that strings generated by the Fibonacci construction and by the logistic map at the Feigenbaum point have this scaling property, see Ebeling & Nicolis (1992), Ebeling *et al.* (1995b) and Gramss (1994). For natural information carrying strings we have found in several cases the stressed exponential scaling, see Ebeling & Pöschel (1994) and Ebeling *et al.* (1995a),

$$N^*(n) \sim \exp(Cn^{\alpha}), \tag{1.6}$$

with $\alpha = \frac{1}{2}$. The relation between the effective numbers $N^*(n)$ and the block entropies which are studied below is an exponential one,

$$N^*(n) \sim \exp(H_n). \tag{1.7}$$

In this way, the scaling of the $N^*(n)$ is determined by the block entropies, which is another reason to study these quantities.

Our investigation is mainly based on the concepts of block entropies and conditional entropies, as well as on certain generalizations. Further on we will discuss a comparison to other measures of long-range relations based on methods of statistical physics. We mention, for example, transinformations, see Li & Kaneko (1992), Herzel *et al.* (1994b), Herzel *et al.* (1994a) and Herzel & Grosse (1995), algorithmic entropy, correlations functions and mean square deviations, see Ebeling *et al.* (1995a), $1/f^{\delta}$ noise, see Voss (1992), Anishchenko *et al.* (1994) and Ebeling & Neiman (1995), scaling exponents, see Peng *et al.* (1994), Stanley *et al.* (1994) and Ebeling *et al.* (1995a), and higher order cumulants, see Ebeling & Neiman (1995). Our working hypothesis which we formulated in earlier papers, see Ebeling & Nicolis (1992) and Anishchenko *et al.* (1994), is that texts and DNA show some structural analogies to strings generated by nonlinear processes at bifurcation points. This is demonstrated here by the analysis of the behaviour of dynamic entropies.

2. Conditional entropies

This section is devoted to the introduction of several basic terms stemming from information theory which were mostly already used by Shannon. Let us assume that the processes or structures to be studied are modelled by trajectories on discrete state spaces having total length L. Let λ be the length of the alphabet. Furthermore, let $A_1 A_2 \dots A_n$ be the letters of a given subtrajectory of length $n \leq L$. Let $p^{(n)}(A_1 \dots A_n)$ be the probability to find in the total trajectory a block (subtrajectory) with the letters $A_1 \dots A_n$. Then we may introduce the entropy per block of length n (the n-gram entropy):

$$H_n = -\sum p^{(n)}(A_1 \dots A_n) \log p^{(n)}(A_1 \dots A_n). \tag{2.8}$$

From the block entropies we derive conditional entropies (n-gram dynamic entropies) as the differences $h_n = H_{n+1} - H_n$. Furthermore we define

$$r_n = \log(\lambda) - h_n, \tag{2.9}$$

as the average predictability of the state following after a measured n-trajectory. We remember that $\log(\lambda)$ is the maximum of the uncertainty, so the predictability is defined as the difference between the maximal and the actual uncertainty. In other words, predictabilty is the information we get by exploration of the next state in the future in comparision to the available knowledge. In the following in most cases we shall use λ as the unit of the logarithms.

The limit of the dynamic n-gram entropies for large n is the entropy of the source (called dynamic entropy or Kolmogorov–Sinai entropy). The predictability of processes is closely connected with these dynamic entropies. Let us consider a certain section of length n of the trajectory, a time series, or another sequence of symbols $A_1 \dots A_n$, which is often denoted as a subcylinder. We are interested in the uncertainty of the predictions of the state following after this particular subtrajectory of length n. Again following the concepts of Shannon we now define the expression

$$h_n^{(1)}(A_1 \dots A_n) = -\sum p(A_{n+1}|A_1 \dots A_n) \log p(A_{n+1}|A_1 \dots A_n), \tag{2.10}$$

as the conditional uncertainty of the next state (1 step into the future) following behind the measured trajectory $A_1 \dots A_n$ ($A_i \in$ alphabet). Here and in the following all logarithms are measured in λ units. We note that in these units the following inequality holds:

$$0 \leq h_n^{(1)}(A_1 \dots A_n) \leq 1. \tag{2.11}$$

We define

$$r_n^{(1)}(A_1 \dots A_n) = 1 - h_n^{(1)}(A_1 \dots A_n), \tag{2.12}$$

as the predictability of the next state following after a measured subtrajectory, which is a quantity between zero and one. We note that the average of the local uncertainty

$$\begin{aligned} h_n = h_n^{(1)} &= \left\langle h_n^{(1)}(A_1 \dots A_n) \right\rangle \\ &= \sum p(A_1 \dots A_n) h_n^{(1)}(A_1 \dots A_n) \end{aligned}$$

leads us back to Shannon's uncertainty (n-gram dynamic entropy). A possible generalization concerns the case that we want to predict the state which deos not immediately follow the observed n-string, but only after k steps into the future, see Baraldi *et al.* (1995). We define

$$h_n^{(k)}(A_1 \dots A_n) = -\sum p(A_{n+k}|A_1 \dots A_n) \log p(A_{n+k}|A_1 \dots A_n),$$

as the uncertainty of the state which occurs k steps into the future after the observation of an n-block, or symbollically

$$[A_1 \dots A_n](k-1 \text{ states})[?].$$

Then we define accordingly the predictabilities

$$r_n^{(k)}(A_1 \dots A_n) = 1 - h_n^{(k)}(A_1 \dots A_n). \tag{2.13}$$

For $n = 1$ the predictability is closely related to the transinformation (mutual information) which may be expressed as, see Herzel *et al.* (1994b), Herzel *et al.* (1994a) and Herzel *et al.* (1995),

$$I(k) = r_1^{(k)} + (h_0 - 1).$$

For systems with long memory, it makes sense to study a whole series of predictabilities with increasing n-values (where n denotes the lower index)

$$r_1^{(k)}, r_2^{(k)}, r_3^{(k)}, \ldots, r_m^{(k)}.$$

Here m is an estimate for the length of the memory. Due to the inequality

$$r_{n+1}^{(k)} \geq r_n^{(k)},$$

the average predictability may be improved by taking longer blocks into account. In other words, one can gain an advantage for predictions by basing the predictions not only on actual states but on whole trajectory blocks which represent the actual state and its history.

3. The Fibonacci model and nonlinear map models for evolutionary processes

One of the first mathematical models of an evolutionary process was formulated in 1202 by the Italian mathematician Leonardo da Pisa, better known as Fibonacci, in his book *Liber Abaci.* Fibonacci considered the problem of how many rabbit pairs are generated after n breeding sessions, assuming the following simple rules, see Schroeder (1993):

- the game starts with an immature pair,
- rabbits mature in one season after birth,
- mature rabbit pairs produce one new pair every breeding session,
- rabbits never die.

This game produces the famous sequence of Fibonacci numbers 1, 1, 2, 3, 5, 8, 13, 21, 34, 55, ..., which appear in innumerable situations. The Fibonacci game may be encoded as a sequence of zeros and ones by using the rules $0 \to 1$ denoting 'young rabbits grow old', and $1 \to 10$ standing for 'old rabbits stay old and beget young ones'. Beginning with a single 0, continued iteration gives 1, 10, 101, 10110, ..., resulting finally in the infinite selfsimilar so-called Schroeder's rabbit sequence 1011010110110.... Alternatively we may formulate the rules by a grammar:

$$S_0 = 0, \tag{3.14}$$

$$S_1 = 1, \tag{3.15}$$

$$S_{n+1} = S_n S_{n-1}, \quad n = 1, 2, 3, \ldots. \tag{3.16}$$

In this way one gets Schroeder's rabbit sequence. We recall that the symbol 0 represents a young pair and the symbol 1 an old rabbit pair. In this language the generation rule $1 \to 10$ means that an old pair survives to the next 'year' $(n+1)$ and generates a young pair. Furthermore, the rule $0 \to 1$ means that the young pair is getting old. We note that the lengths of the generated sequences S_n are given by the Fibonacci numbers F_{n+1}. We can also mention a close connection to the so-called *golden mean* $\gamma = (\sqrt{5} - 1)/2$. Gramss (1994) calculated the entropy of Schroeder's sequence, and was able to show that dynamic entropies behave in the limit of large n as

$$h_n \propto n^{-1}. \tag{3.17}$$

This is based on the observation that the Fibonacci numbers show an exponential growth,

$$F_{n+1} = \frac{1}{\sqrt{5}}\left(\gamma^{-n} - \gamma^n\right) \propto \gamma^n \quad \text{as } n \to \infty. \tag{3.18}$$

Schroeder's sequence, which models the evolutionary process of 'self-reproduction of rabbits' is the first example of an evolutionary structure showing correlations of infinite length. Another well-studied simple model of evolution is a logistic map. This model of growth processes was studied in the last century by Verhulst and Pearl, and in a discrete form, recently by May, Feigenbaum and many others. We write this map in the form

$$x(n+1) = rx(n)\big(1 - x(n)\big), \tag{3.19}$$

$$0 \le r \le 4. \tag{3.20}$$

In order to generate a discrete string from this map we use the bipartition ($\lambda = 2$)

$$C_1 := \left[0, \tfrac{1}{2}\right) \to 0, \tag{3.21}$$

$$C_2 := \left[\tfrac{1}{2}, 1\right] \to 1. \tag{3.22}$$

The two states will be denoted by the symbols 0 and 1. In this way the evolution is mapped on binary strings. We denote these sequences as Feigenbaum strings. The rank-ordered word distributions for Feigenbaum strings were discussed by several authors, see e.g. Ebeling & Nicolis (1992) and Rateitschak (1995). For $r = 4$ all words of fixed length are equally distributed. For the Feigenbaum accumulation point $r = 3.5699456\ldots$, we also get rather simple word distributions consisting only of one or two steps in the dependence on the word length, see Ebeling & Nicolis (1992) and Rateitschak (1995). Furthermore we may give simple construction rules for the generation of these sequences. The grammatical rules are, see Rateitschak (1995),

$$S_0 = 1, \tag{3.23}$$

$$S_1 = 10, \tag{3.24}$$

$$S_{n+1} = S_n S_{n-1} S_{n-1}, \quad n = 1, 2, 3, \ldots . \tag{3.25}$$

This leads to

$$S_2 = 1011, \tag{3.26}$$

$$S_3 = 10111010, \tag{3.27}$$

$$S_4 = 1011101010111011, \quad \text{etc.} \tag{3.28}$$

Here the symbols S_n stand for sequences of length 2^n. This is the reason for the generation of infinite length correlations in the sequences. The rules described above generate selfsimilar structures. Accordingly, the n-gram block entropies satisfy

$$H_{n_{k+1}} = H_{n_k} + 1 = H_{n_1} + k. \tag{3.29}$$

By specialization of (3.29) we get for $n = 2^k$ a result first obtained by Grassberger (1986):

$$H_n = \log_2(3n/2). \tag{3.30}$$

In a similar way we obtain the H_n for all the other sequences. The block entropies show a logarithmic law of growth with n. For the dynamic entropies we get for the Grassberger numbers $n = 2, 4, 8, 16, \ldots$, the conditional entropies

$$h_n = \frac{4}{3n}. \tag{3.31}$$

In between two Grassberger numbers namely at $n = 3, 6, 12, 24, \ldots$, the dynamic entropies jump to a value according to the next Grassberger number. In this way a simple step function is obtained. We see that the dynamic entropy itself (which is the limit of infinite n) is zero. This correponds to a zero Lyapunov exponent and a zero algorithmic entropy, see Freund (1992)).

Long range correlations are also generated by intermittent processes, see Ebeling & Nicolis (1991). We now consider a simple model of intermittency, see Freund (1992) and

Rateitschak (1995):

$$x_{n+1} = f(x_n, \epsilon) = \begin{cases} x_n + x_n^2 + \epsilon, & x \in \left[-\frac{1}{2}, \frac{1}{2}\right) \\ -\frac{1}{2} + \xi, & x \in \left[\frac{1}{2}, \frac{3}{4} + \epsilon\right). \end{cases} \quad (3.32)$$

By again using a binary coding and denoting the laminar phase with a 0 and the turbulent phase with a 1 we get sequences of the following type

$$\ldots 000000001001000000000000100000001010000000000000010000001 \ldots \quad (3.33)$$

The characteristic quantity for intermittent processes (here type (i)) is the distribution of the length of the laminar phases $p_\epsilon(l)$, which depends of course on the parameter ϵ, the width of the channel. For $\epsilon \ll 1$ one can give an analytical estimate of the distribution and of the entropies, see Freund (1992) and Rateitschak (1995). In order to get a memory of infinite length laminar phases of infinite length have to be admitted (intermittency of type (ii) or type (iii)). A group of discrete maps of such type was investigated by Szepfaluzy & Gyorgyi (1986). The following scaling for the approach to the limit was found

$$h_n - h = n^{-\alpha}. \quad (3.34)$$

4. Average entropies and other correlation measures for evolutionary strings

As evolutionary strings we denote here the strings which were generated by evolutionary processes. In particular, we have in mind information carriers, as e.g. DNA and human writings. For biosequences, several authors have pointed out the existence of long-range correlations, see Li & Kaneko (1992), Herzel *et al.* (1994b) Herzel *et al.* (1994a), Voss (1992) and Peng *et al.* (1994). However, an analysis of the average uncertainties (the dynamic entropies) yields rather high values. Measured in bits, the limit uncertainty in most cases is larger than the 1.8 bit, i.e. larger than 0.9 in λ units, see Gatlin (1992) and Herzel *et al.* (1994b). For this reason, the average dynamic entropies do not seem to be an appropriate instrument to analyze DNA strings. However, as we show in the next section, local investigations of the entropy and the transinformation might be very powerful for the analysis of long correlations. Let us now consider several results available for texts. Originally, texts were generated by the writer as a dynamical process in real time. Nowadays, we find in books the frozen in results of this process in the form of a symbolic sequence. We have studied for example Melville's *Moby Dick* ($L \approx 1,170,200$) and Grimm's *Tales* ($L \approx 1,435,800$). Our methods for the analysis of the entropy of sequences were explained in detail elsewhere, see Pöschel *et al.* (1995). We have shown that, at least in a reasonable approximation, the scaling of the entropy against the word length is given by a square root law. Our best fit of the data obtained for texts on the 32–alphabet (measured in log(32) units) reads

$$H_n \approx 0.5\sqrt{n} + 0.05\,n + 1.7, \quad (4.35)$$

$$h_n \approx 0.25/\sqrt{n} + 0.05. \quad (4.36)$$

The dominating term is given by a square root law corresponding to a rather long memory tail. We mention that a scaling law of the square root type was first found by Hilberg who made a new fit for Shannon's original data, Hilberg (1990). We used our own data for $n = 1, \ldots, 26$, but also included Shannon's result for $n = 100$.

Let us now briefly summarize the results obtained from using other measures of correlations, see Ebeling & Neiman (1995) and Ebeling *et al.* (1995a). First we calculated the algorithmic entropy according to Lempel and Ziv which is introduced as the relation of the length of the compressed sequence (with respect to a Lempel–Ziv compression algorithm) to the original length. The results obtained for the Lempel–Ziv complexities (entropies) of several DNA sequences and for texts were compared with diffusion exponents. These scaling quantities were obtained by using the method proposed by Peng *et al.* (1994) and Stanley *et al.* (1994), and the invariant representation proposed by Voss (1992). Instead of the original string consisting of λ different symbols we generate λ strings on the binary alphabet (0,1) ($\lambda = 32$ for texts). In the first string we place a 1 on all positions where there is an 'a' in the original string and a 0 on all other positions. The same procedure is also carried out for the remaining symbols. Then we generate random processes corresponding to these strings moving one step upwards for any 1 and remaining on the same level for any 0. The resulting move over a distance l is called $y(k,l)$ where k denotes the symbol. Then by defining a λ-dimensional vector space considering $y(k,l)$ as the component k of the state vector at the (discrete) 'time' l we can map the text to a trajectory. The corresponding procedure is carried out for the DNA sequences which are mapped to a random walk on a ($\lambda = 4$)-dimensional discrete space. The power spectrum is defined as the Fourier transform of the correlation function $C(k,n)$ which measures the correlation of the letters of type k in a distance n, see Anishchenko *et al.* (1994). The results of spectral calculations for the original file of the *Bible,* for *Moby Dick* and for the same files shuffled on the word level or on the letter level correspondingly are presented in Ebeling & Neiman (1995). Furthermore, the power spectrum of *Moby Dick* shuffled on the chapter and on the page level was calculated, see Ebeling *et al.* (1995a). We have shown that the spectra of the original texts have a characteristic shape with a well expressed low frequency part. This shows the existence of long-range correlations in *Moby Dick.* An estimate for the exponent is 0.7. However, it is difficult to extract a precise value for the exponent corresponding to a power law. One should rather look at the shape of the whole curve which reminds us of a piecewise linear behaviour which is characteristic for multifractal structures. This point certainly needs further investigation but, as we believe, there is no reason to expect that a long text such as *Moby Dick* is a multifractal in a quantitative sense. On the other hand, the hierarchical character of the structure of texts is beyond a doubt, and we think that spectral curves may be characterized as a quantitative measure of this hierarchy. Shuffling on the page level destroys the low frequency branch of the spectrum. This clearly proves that the origin of the $\frac{1}{f}$ fluctuations is on a scale which exceeds the page level. The contribution of high frequencies corresponds to the structure on the word and sentence level. We also studied the anomalous diffusion coefficients which allow for a higher accuracy of the analysis, see Stanley *et al.* (1994). The mean-square displacement for the symbol k is determined as

$$F^2(k,l) = \langle y^2(k,l) \rangle - \left(\langle y(k,l) \rangle \right)^2, \tag{4.37}$$

where the brackets $\langle \cdot \rangle$ mean averaging over all initial positions. The behaviour of $F_2(k,l)$ for $l \gg 1$ is the focus of our interest. It is expected that $F_2(k,l)$ follows a power law, see Stanley *et al.* (1994),

$$F_1(k,l) \propto l^{\alpha(k)}, \tag{4.38}$$

where $\alpha(k)$ is the diffusion exponent for the symbol k. We note that the diffusion exponent is related to the exponent of the power spectrum, see Stanley *et al.* (1994). The case $\alpha(k) = 0.5$ corresponds to the normal diffusion or to the absence of long-range correlations. If $\alpha(k) > 0.5$, we have an anomalous diffusion which reflects the existence of long-range correlations. Besides the individual diffusion exponents for the letters we also get an averaged diffusion exponent α for the state space. The data are summarized in Ebeling & Neiman (1995) and Ebeling *et al.* (1995a). In the same way, we obtained other important statistical quantities: higher order moments and cumulants of $y(k,l)$, see Ebeling & Neiman (1995). By calculations of the Hölder exponents D_q, up to $q = 6$, we have shown that the higher order moments exhibit (in the limits of accuracy) the same scaling behaviour as the second moment. We repeated the procedure described above for the shuffled files. The results of the calculations in dependence on the shuffling level are shown in Ebeling *et al.* (1995a). We see from the analysis that the original DNA sequences and the texts show strong long-range correlations, i.e. the coefficients of anomalous diffusion are clearly different from $\frac{1}{2}$. After the shuffling below the page level, the sequences become practically Bernoullian in comparison with the original ones since the diffusion coefficients decrease to a value of about $\frac{1}{2}$. This decrease occurs in the shuffling regime between the page level and the chapter level. For DNA sequences the characteristic level of shuffling where the diffusion coefficient goes to $\frac{1}{2}$ is about 500–1000. Our result demonstrates that shuffling on the level of symbols, words, sentences or pages, or segments of length 500–1000 in the DNA case destroys the long range correlations which are felt by the mean square deviations.

5. Predictions based on a local analysis

As shown in the previous section, sometimes the analysis of average dynamic entropies fails to detect long-range correlations. An example is a DNA sequence where we know from the analysis of Fourier spectra, algorithmic entropies and diffusion exponents, that long correlations exist. On the other hand, the average uncertainty of predictions (the limit of the dynamic entropies) is in most cases higher than 0.9 (i.e. higher than 1.8 bits). Therefore, the average predictability of a state k steps into the future following after a trajectory (history) of length n,

$$r_n^{(k)} = \left\langle r_n^{(k)}(A_1 \dots A_n) \right\rangle = 1 - \left\langle h_n^{(k)}(A_1 \dots A_n) \right\rangle,$$

is always rather low.

As shown above, the average uncertainty decreases and the average predictability improves with increasing lenght of the 'history' $A_1 \dots A_n$ taken into account. This is true for DNA and for texts but, for DNA the saturation of the predictability for $n \ll 10$ reaches a plateau at about 0.9 and for texts the plateau is rather low (about .05–.10) and is reached for $n > 100$ only. Sometimes there seem to be several plateaus. For example in the case of meteorological data, after a plateau of about 7 days, a further increase of the predictabilities is observed, see Baraldi *et al.* (1995). This can be interpreted as a hint for higher order memory effects in the climate. Clearly, all the estimated quantities and all conclusions will depend on the length of the datastring L. In particular, the observed predictabilities

$$\hat{r}_n^{(k,L)} = \left\langle r_n^{(k,L)}(A_1 \dots A_n) \right\rangle, \quad k < L,$$

depend on the length of the datastring L and will increase monotonically with L. In many cases for $n > 10$–12 the relatively short length $L = 1000$–100000 of the datastring does not allow a reliable calculation of the average uncertainties. Therefore the question o how far the memory really goes is to be left open. For an algorithm to carry out length corrections we refer to earlier work, see Ebeling & Nicolis (1992), Ebeling *et al.* (1995b), Herzel *et al.* (1994a) and Pöschel *et al.* (1995). For practical applications, one is not so much interested in an average value but rather in a concrete prediction based on the

ℓ	string	h	r	ℓ	string	h	r
6305	agtga	0.74	0.26	10011	tgggg	0.75	0.25
6445	aaccc	0.72	0.28	10113	ccagc	0.79	0.21
6529	cccag	0.75	0.25	10154	ctcac	0.78	0.22
6618	gggct	0.68	0.32	10197	gcttg	0.77	0.23
7064	aatac	0.70	0.30	10223	gctgc	0.77	0.23
8393	gaaat	0.73	0.27	10246	gtgtg	0.74	0.26
9238	aatac	0.70	0.30	10287	gtgtt	0.77	0.23

TABLE 1. Examples of low uncertainties (h) – high predictabilities (r) of a letter at position $\ell+1$ following a given substring of length 5, ending at position ℓ. The data refer to the HIV-DNA (HIV2BEN-string).

observation of a concrete string of finite length n. In other words, one is more interested in concrete predictions than in 'average predictabilities'.

Therefore, we have also studied predictabilities of the states following the particular strings $A_1 \ldots A_n$,

$$r_n^{(k)}(A_1 \ldots A_n) = 1 - h_n^{(k)}(A_1 \ldots A_n). \tag{5.39}$$

This is a quantity which is local and fluctuates while going along the string, since the local 'history' $A_1 \ldots A_n$ changes from position to position. Another closely related fluctuating quantity is the transinformation, which is connected with the local predictability for $n = 1$ by the relation, see Herzel *et al.* (1994b), Herzel *et al.* (1994a) and Herzel & Grosse (1995),

$$I(k) = r_1^{(k)} + (h_0 - 1).$$

For DNA strings, extensive calculations of the transinformation are available, see Li & Kaneko (1992), Herzel *et al.* (1994b), Herzel *et al.* (1994a) and Herzel & Grosse (1995). Similar investigations are known for texts and for pieces of music, see Ebeling & Pöschel (1994) and Ebeling *et al.* (1995b). The transinformation is a strongly fluctuating quantity, its local peaks having the meaning of strong correlations at certain distances.

Let us now consider the local predictability based on longer histories $n \gg 1$. The local predictability of the symbol following a group of 8 symbols was studied by Herzel *et al.* (1994b). It was shown that at places where repetitions are found, the predictability is much higher than at normal places. In tables 1 and 2 several examples for the local uncertainty after certain substrings of various lengths are given.

We see that there is a noticeable dispersion of the uncertainty. Moving along the string, the local history $A_1 \ldots A_n$ changes from position to position. This makes $h_n(A_1 \ldots A_n)$ a local fluctuating quantity. Uncertainty and predictability fluctuate along the string; at some places the predictability is rather high, and the uncertainty rather low. The average is, in our case of the HIV-DNA, about 0.92 for the uncertainty, and 0.08 for the predictability.

6. Conclusions

Our results show that dynamic entropies are appropriate measures for studying the predictability of evolutionary processes. Of particular interest are local studies of the predictabilities of certain local histories. Long correlations are of special interest, since they improve the predictability. This means that one can in principle improve the predictions by basing them on longer observations. Furthermore, we can conclude that there

string	h	r	string	h	r
cta	0.82	0.18	atgct	0.66	0.34
cct	0.82	0.18	tgtcc	0.78	0.22
ata	0.83	0.17	ctgct	0.67	0.33
act	0.83	0.17	ttgcag	0.85	0.15
ggct	0.76	0.24	gagctg	0.77	0.23
cagct	0.64	0.37	ccagga	0.76	0.24
gagct	0.66	0.34	acgctg	0.83	0.17
ctgca	0.67	0.33	gatgatg	0.83	0.17
cagca	0.83	0.27	ggcgctg	0.80	0.21

TABLE 2. Several examples of uncertainties (h) and predictabilities (r) of a letter following a given substring, for the DNA of the λ virus.

are specific substrings, which are relatively seldom, where the uncertainty is less than 0.7, i.e. the predictability is better than 0.3. In other words, there are specific situations where the predictability is much better than average. It may be of practical importance to find all substrings which belong to this particular class. Due to the primary importance of long correlations for the improvement of predictions, we also considered here several other measures of long correlations. Our results show that, in particular the low frequency spectra and the scaling of the mean square deviations, are appropriate measures for the long-range correlations in symbolic sequences. However, as demonstrated by shuffling experiments, different measures operate on different length scales. The longest correlations found in our analysis comprise a few hundred or thousand letters, and may be understood as long-wave fluctuations of the composition. These correlations (fluctuations) give rise to the anomalous diffusion and to long-range $1/f$ fluctuations. These fluctuations may comprise several hundred or several thousand letters. There is some evidence that these correlations are based on the hierarchical organization of the sequences and on the structural relations between the levels. In other words, these correlations are connected with the grouping of the sentences into hierarchical structures as paragraphs, pages, chapters, etc. Usually inside a certain substructure the sequences show a greater uniformity on the letter level. In order to demonstrate this we have ivestigated, see Ebeling & Neiman (1995), the local frequency of the blanks and other letters (averaged over windows of length 4000) in the text *Moby Dick* depending on the position along the text. The original text shows a large-scale structure extending over many windows. This reflects the fact that, in some part of the texts we have many short words, e.g. in conversations (yielding the peaks of the space frequency), and in others we have more numerous long words, e.g. in descriptions and in philosophical considerations (yielding the minima of the space frequency). The shuffled text shows a much weaker non-uniformity of the text, the lower the shuffling level, the larger the uniformity. More uniformity means less fluctuations and more similarity to a Bernoulli sequence. For the case of DNA sequences, no analogies of pages, chapters, etc., are known. Neverthless, the reaction on shuffling is similar to those of texts. Our results clearly demonstrate that the longest-range correlations in information carriers are of structural origin. The entropy-like measures studied in § 2 operate on the sentence and on the word level. In some sense, entropies are the most complete quantitative measures of correlation relations. This is due to the fact that the entropies also include many-point correlations. On the other hand, the calculation of the higher order entropies is extremely difficult and at the present moment there is no hope for extending the entropy analysis to the level of thousands of letters.

Similar structures were recently obtained in sequences from time series, but this question is still under investigation. Possibly a more careful study of the long correlations in time series sequences may contribute to better predictions of evolutionary processes. In conclusion, we would like to express the hope that the analysis of entropies, power spectra and scaling exponents could be developed to be useful instruments for studies of the large scale structure of a rather broad class of time series and information carrying sequences.

The author thanks C. Baraldi, R. Feistel, J. Freund, I. Grosse, H. Herzel, L. Molgedey, A. Neiman, C. Nicolis, G. Nicolis, T. Pöschel, K. Rateitschak, H. Rose and A. Schmitt for many fruitful discussions and a collaboration on special aspects of the problems discussed here.

REFERENCES

ANISHCHENKO, V. S., EBELING, W. & NEIMAN, A. B. 1994 *Chaos, Solitons & Fractals* **4**, 69.

BAK, P., TANG, C., WIESENFELD, K., 1987 *Phys. Rev. Lett.* **59**, 381.

BAK, P., CHEN, L. & CREUTZ, M. 1989 *Nature* **342**, 780.

BARALDI, C., EBELING, W., NICOLIS, C. & NICOLIS, G. 1995 Preprint, Humboldt-University, Berlin.

EBELING, W., ENGEL, A. & FEISTEL, R. 1990 *Physik der Evolutionsprozesse.* Akademie-Verlag.

EBELING, W., FREUND, J. & RATEITSCHAK, K. 1996 *Intl. J. Bifurcation & Chaos.* In press.

EBELING, W. & NEIMAN, A. 1995 *Physica* A **215**, 233.

EBELING, W., NEIMAN, A. & POESCHEL, T. 1995a Dynamic entropies, long-range correlations and fluctuations in complex linear structures. In *Coherent Approach to Fluctuations* (Proc. Hayashibara Forum 1995). World Scientific.

EBELING, W., & NICOLIS, G. 1991 *Europhys. Lett.* **14**, 191.

EBELING, W. & NICOLIS, G. 1992 *Chaos, Solitons & Fractals* **2**, 635.

EBELING, W. & PÖSCHEL, T. 1994 *Europhys. Lett.* **26**, 241.

EBELING, W., PÖSCHEL, T. & ALBRECHT, K. F. 1995b *Intl. J. Bifurcation & Chaos* **5**, 51.

FEISTEL, R. & EBELING, W. 1989 *Evolution of Complex Systems.* Verlag der Wissenschaften; also Kluwer Academic Publ.

FREUND, J. 1992 Dissertation, Humboldt University, Institute of Physics, Berlin; see also FREUND, J. & HERZEL, H. 1995 *Chaos, Solitons & Fractals.* In press.

GATLIN, L., 1972 *Information Theory and the Living System.* Columbia University Press.

GRAMSS, T. 1994 *Phys. Rev.* E **50**, 2616.

GRASSBERGER, P. 1986 *Intl. J. Theor. Phys.* **25**, 907.

HERZEL, H. *et al.* 1994a *Chaos, Solitons & Fractals* **4**, 97.

HERZEL, H. & GROSSE, I. 1995 *Physica* A **216**, 518; see also HERZEL, H., EBELING, W. & GROSSE, I. In *Proc. Conf. Bioinformatics.* GBF Monographs, vol. 18.

HERZEL, H., SCHMITT, A. O. & EBELING, W. 1994b *Phys. Rev.* E **50**, 5061.

HILBERG, W. 1990 *Frequenz* **44**, 243.

LI, W. & KANEKO, K. 1992 *Europhys. Lett.* **17**, 655.

KAUFFMAN, S. A. 1993 *The Origins of Order.* Oxford University Press.

LANGTON, C. G. 1990 *Physica* D **42**, 12.

PENG, C.-K. *et al.* 1994 *Phys. Rev.* E **49**, 1685.

PÖSCHEL, T., EBELING, W. & ROSÉ, H. 1995 *J. Stat. Phys.* **80**, 1443.

RATEITSCHAK, K. 1995 Diploma Thesis, Humboldt-University, Institute of Physics, Berlin; see also RATEITSCHAK, K., EBELING, W. & FREUND, J. 1995 Nonlinear dynamical model for texts; RATEITSCHAK, K., FREUND, J. & EBELING, W. 1995 Dynamic entropy and long range correlations in nonlinear processes. In *Entropy and Entropy Generation* (ed. J. Shiner). Kluwer.

SCHROEDER, M. 1993 *Fractals, Chaos, Power Laws.* Freeman.

STANLEY, H. E., *et al.* 1994 *Physica* **A205**, 214.

SZEPFALUZY, P. & GYORGYI, G. 1986 *Phys. Rev.* A **33**, 2852.

VOSS, R. F. 1992 *Phys. Rev. Lett.* **68**, 3805; see also VOSS, R. F. 1994 *Fractals* **2**, 1.

Spatiotemporal chaos information processing in neural networks

By HAROLD SZU AND CHARLES HSU

Center for Advanced Computer Studies, University of Southwestern Louisiana, Lafayette, Louisiana, USA

We present a neuron model with a simple processor element. It is capable of generating chaos and can realize synaptic Hebbian learning in the framework of artificial neural networks (ANN). The model is a variant of classical McCullouch–Pitts neurons. Since it is analytically associated with a cubic polynomial, a mapping function with a single hump, it may be referred to as a sigmoidal N-shaped function. Thus, this model has been demonstrated with the bifurcation cascade toward chaos in the Feigenbaum sense. Furthermore, we have numerically investigated a large set of such neuron models. Snapshots of several hundred thousand neuronic outputs of a single layer, called neural images, illustrate the iterative neurodynamics. The fixed-point attractor dynamics, based on the Hebbian learning rule of the synaptic weight matrix among all chaotic neurons, has generated a mean field of the iteration feedback baseline from other neurons. It reveals a spatially coherent neural image as the information content. In the case of sigmoidal N-shape neurons, results of the neural images show psychologically the possibilities of fuzzy logic reasoning, misconception, perceptual habituation or adaptation, novelty detection and noise-generated hallucination. To achieve an exponentially fast pattern recognition inherited from the iterative mapping chaos, designs of electronic implementation of chaotic ANN chips are needed. A modified N-shaped function, named a piece-wise linear (PWL) N-shaped function, is designed and implemented by a voltage CMOS circuit. Also a VLSI chip with two chaotic neurons was fabricated through the MOSIS program. The chaotic behavior is analyzed and verified by Lyapunov exponents. The measurement diagnosis of the chip is demonstrated.

---

1. Introduction

This paper summarizes the biological and physical perspectives of neural network chaos for the purpose of efficient spatiotemporal information processing. Two major novel results are:

(i) a time-discrete class of a one-dimensional analog neuron mapping model is recursively demonstrated in an iterative fashion, without assuming replenishment delay, in order to exhibit Feigenbaum-like bifurcation cascades to chaos;

(ii) the equilibrium ensemble of such a large set of chaotic neural networks (CNN) is insured, because each neuron's input–output mapping baseline is changed according to a fixed-point Hebbian learning dynamics, under the surrounding influence averaged over all other neurons.

In § 1, the biological motivation of current trends of CNNs is itemized succinctly. In § 2, the potential applications and the reasons for investigating artificial neural network (ANN) chaos are given. In § 3, three mathematical models of chaotic ANN are presented. In § 4, the simulation results of neural images are given and discussed. In § 5, chaos chip implementation issues are addressed.

2. Biological background of chaos

Sufficient evidence exists, see Chialvo *et al.* (1990), Glass & Mackey (1988), Garfinkel *et al.* (1992), Hayashi *et al.* (1983) and Freeman (1986), to strongly support the hypothesis that brain systems operate in chaotic regions of state space. However, it is unclear how global chaos arises in the cortex, and whether it plays any essential role in perception and cognition. If it does, how might we discover and simulate this role?

How does globally coherent chaos arise? Single neurons are known to have chaotic domains of function. The characteristic Poisson interval histograms, flat autocovariances, and small co-variances between pairs of pulse trains may reflect that mode of operation. Alternatively, each single neuron may operate near a fixed point attractor and be driven by 1,000–10,000 others by re-entrant synaptic pathways with distributed delays. The model proposed here can provide a test bed for this question, because the elements simulating neurons can be easily moved parametrically through their point, limit cycle and chaotic basins of attraction. The spectra, dimensions, and phase portraits of the local and global chaotic outputs of the model can be compared to those with the known properties of units and EEGs from cortexes.

What is the optimal form of the sigmoidal nonlinearity? The most commonly used form is a monotonic, symmetric curve that contributes to the great stability. An asymmetric sigmoid, as revealed in cortical studies, provides a desired degree of instability for rapid state transitions. The N-shaped nonlinearity used in the proposed system is derived directly from studies on neurons, and its properties are consistent with Pavlovian paradoxical inhibition, cathodal block, and the N-shaped performance curves of several types of sensory receptors and central neurons. The present system will enable a systematic evaluation of this question.

Is chaos essential for the rapid and complete state transitions that are observed in the sensory cortex during serial perception? The present model can be trained on two or more classes of inputs to be discriminated, and can be parametrically shifted quickly and cleanly from a point or limit cycle attractor to one of several chaotic domains to test this hypothesis. Is chaos essential for learning novel stimuli? One hypothesis is that chaos provides unstructured and unpatterned neural activity needed to drive Hebbian synapses during the formation of nerve cell assemblies. These enable the cortex to generate novel patterns. The present system can test this by conducting learning in limit cycle versus chaotic states. Likewise, an essential part of learning is to habituate to irrelevant and ambiguous input (the context) while associative memories are laid down. The system can be used to explore the formation and operation of 'negative' images, which are already apparent in preliminary results using chaotic dynamics. Software simulation now in use can only give tentative answers to these questions, because the solution of the large number of equations involved by numerical integration is very slow, and the risk of being misled by numerical instabilities that might supplant or contaminate the desired chaotic solutions is high. Hardware embodiment is obviously needed, and it must be designed and constructed so that the parameters relevant to biological hypothesis testing are built in and easy to use. Furthermore, it is only by use of well designed chips that the serial architectures of corticocortical and corticothalamic operations can be simulated. Digital embodiments are relatively slow for *in-situ-in-vitro* investigation of more than one layer at a time.

In the near future, a class of 'cortical chips' is foreseen, in which the adjustable parameters are designed and built in, so as to simulate the synaptic and threshold effects of pharmacological agents known to affect the cortical function in specific ways, and to test specific hypotheses on how biochemical and genetic diseases of the brain can lead to neurological, perceptual, and cognitive dysfunctioning. That kind of simulation might assist in the development of palliative, remedial, and preventive therapeutic regimens by bridging the gulf between molecular and behavioral abnormalities of brain functions.

Why ANN modeling of chaos?

Based on the (saddle point) control of chaotic systems (e.g. in a magnetic ribbon) scientists, Garfinkel *et al.* (1992), Chialvo *et al.* (1990), Glass & Mackey (1988), Hayashi *et al.* (1983) and Freeman (1986), have recently demonstrated that the wide frequency band chaotic state of a dying rabbit's heartbeat can be restored to a healthy narrow band oscillation measured by the interbeat interval of the monophasic action potential of the heart. However, there remains an intriguing question – given biological chaos, whether in hearts or olfactory bulbs, see Freeman (1987), Freeman (1991), Yao & Freeman (1990) and Baird (1986). This is: *how is chaos useful for information processing in general?* Chaos, in the Poincaré sense, is usually exceedingly sensitive to the initial condition and intractable by rigorous mathematical analysis. This fact becomes even more difficult for a large interacting neural network. Mathematical solutions are based on two approaches:

(i) the Complex Processor Model, see e.g. Blum & Wang (1992), can be mathematically treated in the case of a few oscillator neurons with inhibition–excitation links,

(ii) the Complicated Communication Model, see e.g. Jackson (1989), Wang *et al.* (1990), Derrida & Meir (1988), Sompolinsky & Kanter (1986) and Hansel & Sompolinsky (1992), incorporates a bipolar version of a Hopfield network with second order Hebbian interconnects T_{ijk} having a stochastic switch-off probability.

While these mathematical analyses are interesting, the relevance to cortical information processing remains to be investigated. Our rationale, similar to the phase transition phenomenon, is that a coherent spatiotemporal structure might occur only in the limit of very large numbers of chaotic neurons, and that the present day computer is powerful enough to investigate chaotic ANNs based on any biologically meaningful neuron models. Thus, we wish to design chaotic neuron models in the framework of the McCullouch–Pitts (MP) neuron model and demonstrate in principle the spatiotemporal coherence due to the interaction of several hundred thousand chaotic neurons. However, the spatiotemporal evolutions reported here are computed for 4096 neurons iteratively for 15 time steps. This study of information processing capability is being pursued because:

- ANN Chaos may be an efficient knowledge representation of an external chaotic world.
- ANN Chaos may be a dynamic pattern recognition mechanism, since the chaos could represent a 'don't know' state while the other attractors, such as periodic or quasi-periodic attractors, could represent known or familiar states.
- Chaos is known to be exponentially fast to switch from chaos to one attractor basin and back to another.
- Does the individual neuron chaos become collectively more chaotic or more ordered, e.g. quasiperiodic orbits?
- Can we observe such a phase transition and in what sense can we quantify the degree of order or chaos?
- Does the nonlinear neurodynamics harvest both the advantage of being chaotic and being massively parallel and distributed?

Based on the same dynamics, we can take (2.1) as the fixed point. Then, we have derived the output dynamics of Grossberg and Pineda (note that we have taken the alphabetical order to denote **u**-input and **v**-output). We assume discrete neuronic mapping and always use alphabetical order to denote the input and output of a single ith neuron, e.g. u_i for the net input and v_i for the total output of the ith neuron. The sigmoidal outputs v_i are generally defined with an arbitrary slope $1/\tau_i$ as follows:

$$v_i = \sigma(u_i) = \frac{1}{1 + e^{-u_i/\tau_i}}, \tag{2.1a}$$

$$u_i = \sum_{j=1\,(j\neq i)}^{N} W_{ij}v_j - \Theta_i, \tag{2.1b}$$

where u_i is the net neuron input and Θ_i is the threshold value; equation (2.1*a*) is a sigmoid function; W_{ij} is the weight bewteen neuron i and neuron j, and v_i is the output of the ith neuron;

$$\frac{dv_i}{dt} = -\delta\left[v_i - \sigma\left(\sum_{j=1\,(j\neq i)}^{N} W_{ij}v_j - \Theta_i\right)\right]. \tag{2.2}$$

Alternatively, we can take the fixed point solution of the net input u_i, see (2.1):

$$\frac{du_i}{dt} = -\alpha\left[u_i - \sigma\left(\sum_{j=1\,(j\neq i)}^{N} W_{ij}v_j - \Theta_i\right)\right], \tag{2.3}$$

which has become known as the Hopfield net in terms of the Lyaponov energy gradient, the r.h.s. The two approaches are equivalent if W_{ij} is independent of time according to the Hopfield model. However, rapid updates of memory by chaotic dynamics imply, by setting both $\alpha = 0$ and $\delta = 0$, using directly a fixed point: the MP model (2.1), and N-shaped sigmoid as follows.

3. Neuron models

In the interest of further reducing the complexity of the implementation, we have considered three variations of neuron models: (i) an internal threshold dynamic model of three degrees of freedom, (ii) an analytical N-shaped input–output mapping based on a cubic polynomial, and (iii) a piecewise negative sigmoidal mapping model.

3.1. *Differential flow threshold dynamics model*

Recently, Szu & Rogers (see Szu & Rogers (1992a), Szu & Rogers (1992b), Aihara & Matsumoto (1986), Szu (1989) and Szu (1990)) have generalized the threshold value of the McCullouch–Pitts neuron model to be a vector function of *three* degrees of freedom, consistent with the differential flow requirement of the Kolmogorov–Arnold–Moses Theorem. Szu & Rogers have generalized the threshold to a vector $\Theta_i(t)$ of several components. They showed that when it has only two components, the MP model can produce pulses. This fact is consistent with the Hodgkin–Huxley model where two ionic channels exist for two degrees of freedom. Each degree of freedom may be simplified by employing first-order fixed-point dynamics, then together two first order equations can account for one second order equation that is mathematically necessary for any oscillation phenomenon involving the replenishment in firing pulses, see Szu & Rogers (1992a), Szu & Rogers (1992b), Aihara & Matsumoto (1986), Szu (1989) and Szu (1990). However, for a possible differential flow chaos, a third degree of freedom, a neural housekeeping activity, must be introduced. Its magnitude at the axon hillock Θ_i is assumed to be proportional to the value of the sigmoidal slope:

$$\Theta_i(v_i) = \gamma_i \frac{dv_i}{du_i} = 4\lambda_i v_i(1 - v_i), \tag{3.4}$$

where use is made of the slope of the sigmoid: $(dv_i/du_i) = v_i(1 - v_i)/\tau_i \leq 4$. This is a source of the initial-sensitive and unpredictable chaos in a single neuron model (see figure 1). Thus, Feigenbaum's λ-knob for the model is $4\lambda_i = \gamma_i/\tau_i$, which varies according to the slope τ_i of output firing rate v_i, equation (2.1*a*). The higher the slope value near the threshold, the more the change of the output v_i value. Thus, more is demanded of the

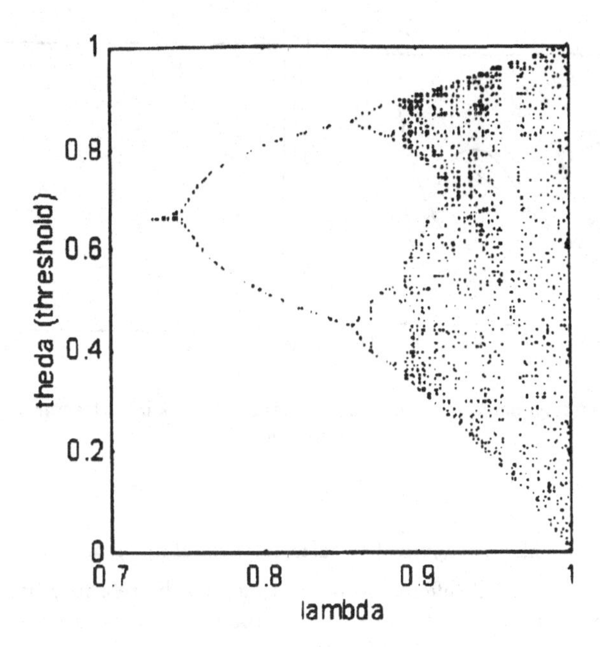

FIGURE 1. Threshold Dynamics Neuron Model assumes the threshold to be proportional to sigmoidal slope (3.4) giving Feigenbaum's map.

housekeeping logistics to maintain a regulated threshold function $\Theta_i(v_i)$, which can give a higher λ-knob value. When the λ-knob value exceeds $\lambda_c = 0.8924$, the output firing leads through the route of bifurcations to chaos. Similarly to the population growth of blowflies under a limited food supply, such a neuron model seems to be biologically meaningful, because a larger slope implies a larger change of firing rates and therefore larger demand on a housekeeping function.

Unfortunately, the three internal threshold dynamics consume extensive computing time, an hour per iteration on a Silicon Graphics workstation (see § 6 for details). This has motivated us to design a chaos chip for massively parallel ANN computing which also embodies a simplified internal dynamics with three adjustable parameters in the following N-shaped sigmoidal function. One is less certain about whether this mapping alone can produce chaos. However, chaos seems to follow when a time-delay is introduced to the discrete mapping, because the delay generates an iterative series solution which is composed of discrete mapping of all the previous time steps, including the sensitive initial condition. The phase plot of the mapping is quite different. The delay gives rise to a two-dimensional Hénon map, a sign of chaos, while without the delay it simply gives a one-dimensional Feigenbuam map, a strange attractor (see Hofstadter (1985), figure 17(*b*)). This is similar to the Nagumo–Fitzhugh binary model with delay that produces a measure of the Cantor triad, i.e. the devil's-staircase plotting the unpredictable frequency ratio jumps against the input strength. Following Harmon's tunneling diode implementation of a neuron, Harmon (1961), Caianiello & DeLuca (1966) and Nagumo & Sato (1972) (see also Yoshizawa *et al.* (1982)) studied a binary neuron model with delay for a *threshold-refractory period* that has also produced the measure of a Cantor triad.

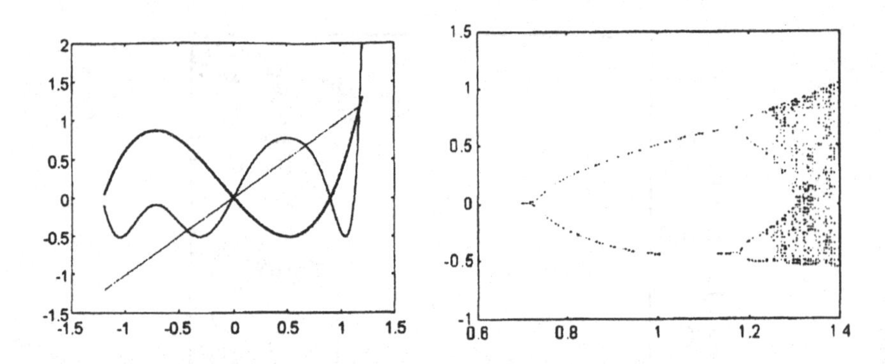

FIGURE 2. Cubic polynomial mapping: (*a*) four stable fixed point attractor, (*b*) bifurcation route to chaos.

3.2. *Antisymmetric cubic polynomial mapping*

We wish to simplify the 3D differential flow into an 1D discrete mapping, such as the Feigenbaum logistic map, as well as to preserve the piecewise negative sigmoidal logic. Thus, the following cubic polynomial function is proposed:

$$v_i = \frac{1}{c^2} u_i(u_i - b)(u_i + a) + v_{i0} = f(u_i), \tag{3.5}$$

where a, b, c are arbitrary real positive constants. The offset constant v_{i0} is introduced to keep the output positive.

The proposed cubic polynomial mapping is thoroughly explored and many of its interesting and useful properties are discovered. Simulation results for a particular case with $c^2 = \frac{2}{3}, b = 0.9$, and parameter a changing from 0.7 to 1.4 are shown in figure 2.

For example, an attractor of period four is found at $a = 1.215$ (figure 2(*a*)). Note that the fixed points are obtained at the intersections of function $g_4(u) = g_2(g_2(u))$, where $g_2 = f(f(u))$, and $v = u$. Figure 2(*b*) portrays the period doubling route to chaos and the onset of chaos at some critical value a_c. The more beautiful features are observed for the special case of the antisymmetric cubic polynomial function when $a = b$. A particular case for $a = b = 1.22$ ($c^2 = \frac{2}{3}$) is shown in figure 3(*a*). According to the bifurcation theory, see Hofstadter (1985), the attractor so obtained should have four stable fixed points. However, a cycle of two stable fixed points occurs. This is due to the different mapping function which is not fixed at the two ends as is the Feigenbaum mapping, and also has a third part which makes the curve antisymmetric. There are two cycles of frequency two, which never occur at the same time. Which cycle is chosen depends on the initial condition (i.e. on a seed of chaos) shown in figure 3. This feature may add more uncertainty in the bifurcation route to chaos as shown in figure 4. An example of a chaotic response is shown in figure 2(*b*). We have derived an asymptotic bifurcation route to chaos where Feigenbaum metric universality is generalized by the antisymmetric cubic polynomial function.

The use of the so explored cubic polynomial mapping model in the collective neural network is not adequate because of the unboundedness of the function, while the output of the unipolar neuron takes values from 0 to 1. Instead, a new model with an N-shaped sigmoidal function is proposed in the next section.

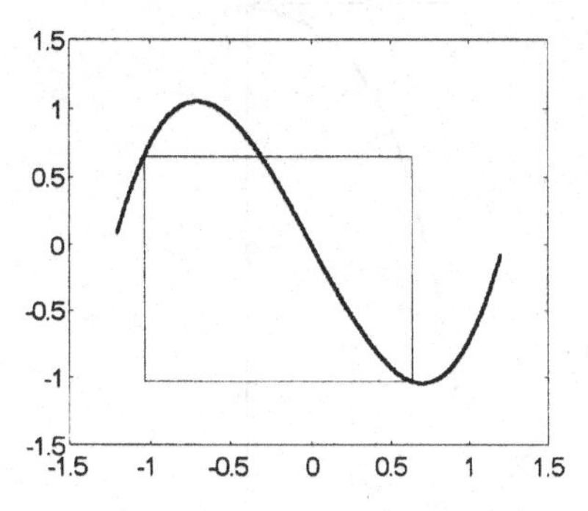

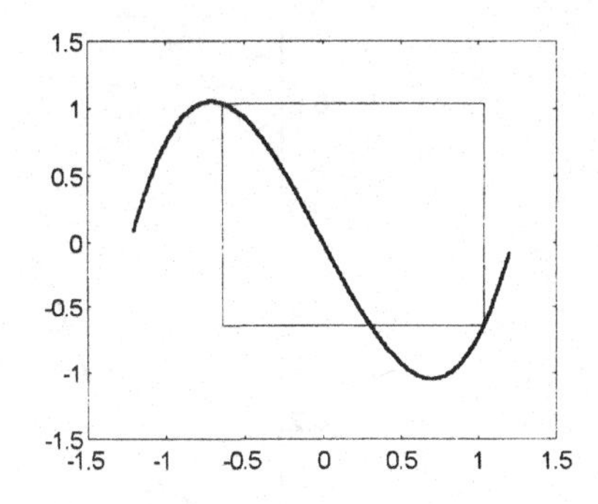

FIGURE 3. Symmetric Cubic Mapping: two different cycles with two fixed points. Only last 100 iterations are shown from 300 iterations. (a) $u = -0.1,\ a = b = 1.22$; (b) $u = 0.1,\ a = b = 1.22$.

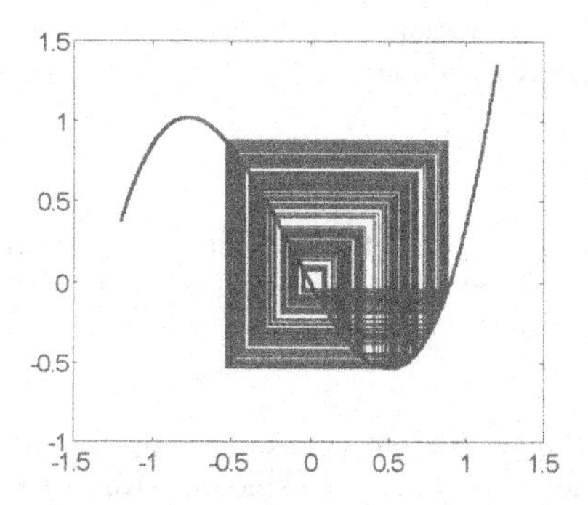

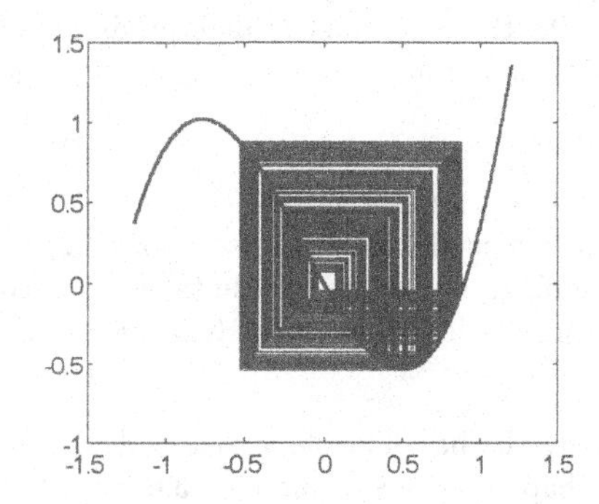

FIGURE 4. The antisymmetric cubic mapping bifurcation route to chaos; (a) seeds of chaos: $u = 0.1,\ a = 1.3,\ b = 0.9$; ($b$) seeds of chaos: $u = 0.5,\ a = 1.3,\ b = 0.9$.

3.3. *Piecewise negative logic model (N-shaped sigmoidal function)*

The N-shaped sigmoidal function that allows unbounded input, and yet has piecewise negative logic in the sense that more input gives less output. The bounded output from 0 to 1 is described by:

$$\sigma_{\mathrm{N}}(u) = \begin{cases} \dfrac{1}{1+\mathrm{e}^{-(u+0.5)/0.3}}, & u \leq -0.5 \\ -0.5u + 0.25, & -0.5 < u \leq 0.5 \\ \dfrac{2}{1+\mathrm{e}^{-(u-0.5)/0.3}} - 1, & 0.5 < u. \end{cases}$$

In terms of neurocomputing, this neuron model has a piecewise-negative logic of which the location is determined by three parameters similar to the threshold dynamics model. The N-shape activation transfer has three branches with a central piece of negative logic near the zero origin, where more input u implies less output v (see figure 5). The sick

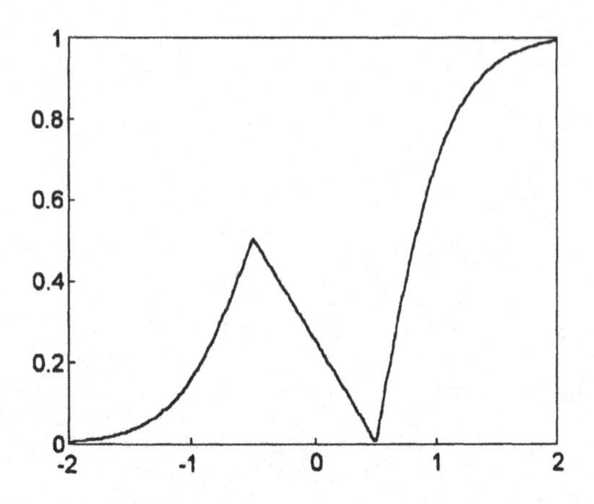

FIGURE 5. Analog N-shaped sigmoidal Neuron Model, $v = \sigma_N(u)$.

region of input is set within ± 0.5, and the recovery begins at $u = 0.5$ and $v = 0$. These are the three parameters determining the smooth N-shaped input–output function.

The set of iterative mapping for a single neuron is described by:

$$v_n = \sigma_N(u_n), \qquad (3.6)$$
$$u_{n+1} = v_n.$$

where n is the index of iterations. According to (3.6), the base-line of the iterative mapping, $u_{n+1} = v_n$, should take into account the weight changes following the Hebbian learning rule. Thus, a new base-line is introduced

$$v = \omega u,$$

where the base-line slope is $\omega = 1/W$ for a single neuron. The new mapping with the N-shape sigmoidal function (3.6) and the new base-line ω is explored with ω being a changing parameter. The results of such a mapping are shown in figure 6. The region of chaos is found for certain values of parameter ω. It is interesting that the mapping enters the chaotic region almost without going through the bifurcation process.

The iterative mapping with the N-shaped sigmoidal function in terms of the neuro-computing is considered. A collective neural network is described by a set of equations:

$$v_i = \sigma_N(u_i), \qquad (3.7)$$
$$u_i = \sum_{j=1\,(j\neq i)}^{N} W_{ij} v_j.$$

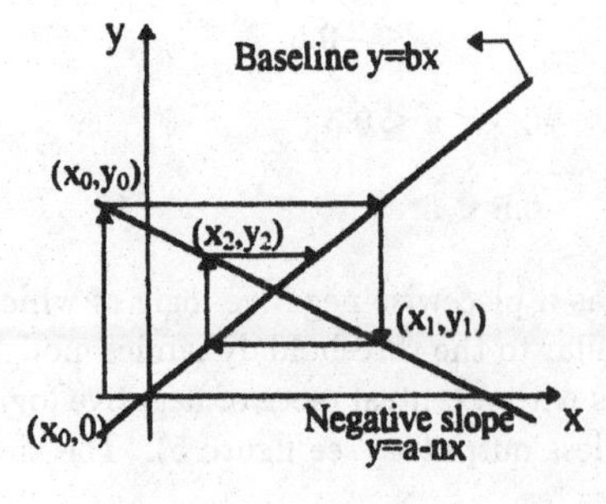

From the left graphics analysis, we have:
Iteration 0: $y_0 = a - nx_0$;
Iteration 1: $y_1 = a - na/b + n^2 x_0/b$;
Iteration 2: $y_2 = a - na/b + n^2 a/b^2 - n^3 x_0/b^2$;
Iteration 3: $y_3 = a - na/b + n^2 a/b^2 - n^3 a/b^3 + n^4 x_0/b^3$.

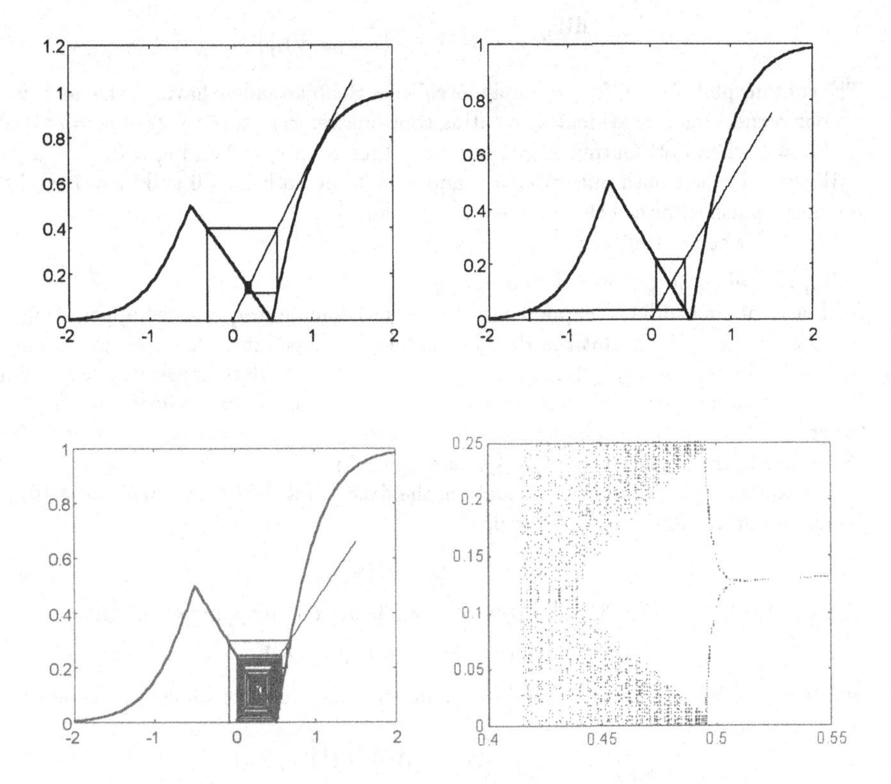

FIGURE 6. N-shape sigmoidal mapping with the changing slope of the base-line: (*a*) a fixed point response, (*b*) bifurcation is found only at one value of $\omega = 0.5$ (equal to the negative slope of the N-shape), (*c*) example of a chaos behavior, (*d*) chaos region.

Chaos behavior of two linear functions can be described as follows. The convergence of the intersection of the baseline function and the sigmoid N-shaped function σ_N is illustrated. The three results can be summarized:

(i) if $|n| = |b|$, the system is oscillating;
(ii) if $|n| < |b|$, the system is convergent;
(iii) if $|n| > |b|$, the system is divergent which could produce chaos. (n and b are the slopes of the negative segment of σ_N function and the baseline function, respectively.)

The characteristics of the three attractors can help to extract the images.

The effect and characterization of many-neuron iterative mapping hinges upon the iteration base-line whose slope changes due to the following Hebbian learning. Memory is given at equilibrium by the Hebbian product rule $v_i v_j$. Although the output v_i, due to the (3.4) mapping, is unipolar ($0 \le v_i \le 1$), the memory must be converted into bipolar analog values normalized arbitrarily between ± 1 ($-1 \le W_{ij} \le 1$) in order to provide excitation and inhibition. The diagonal term is set to zero for no self terms. Therefore, the Hebbian product $v_i v_j$ is changed to (no summation convention)

$$[x(\mathbf{v})_i x(\mathbf{v})_j] \equiv x(\mathbf{v})_i x(\mathbf{v})_j - \delta_{ij} x(\mathbf{v})_i x(\mathbf{v})_i, \quad x(\mathbf{v})_i \equiv 2v_i - 1. \tag{3.8}$$

Then, we postulate the simplest first order attractor dynamics as follows :

$$\frac{dW_{ij}}{dt} = -\beta\Big\{W_{ij} - [x(\mathbf{v})_i x(\mathbf{v})_j]\Big\}. \tag{3.9}$$

The contour plot of (3.9) for $i \neq j$ may be called a Hebbian saddle having extrema located at four corners (that is evident by rotating coordinates: $x_i x_j \equiv (\xi+\psi)(\xi-\psi) = \xi^2 - \psi^2$):

$W_{ij} = 1$ when both output $v_i = 1$ and $v_j = 1$, or both $v_i = 0$ and $v_j = 0$;

$W_{ij} = -1$ when both output $v_i = 1$ and $v_j = 0$, or both $v_i = 0$ and $v_j = 1$; and the memory vanishes toward the middle output value, $\frac{1}{2}$;

$W_{ij} = \frac{1}{4}$ when output $v_i = \frac{1}{4}$ and $v_j = \frac{1}{4}$;

$W_{ij} = 0$ when output $v_i = \frac{1}{2}$ and $v_j = \frac{1}{2}$;

Such a Hebbian saddle is mapped one to one onto domain $u_i u_j$, via the upper branch of σ_N, the equation (3.6). But the Hebbian saddle is mapped into three possible branches in the third quadrant ($v_i \leq 1/2, v_j \leq 1/2$) corresponding to dark image gray scale values. The exact saddle point has a zero memory at two possible input values: one from the lower branch at the logical breakdown point where $u_i = -1, u_j = -1$, and the other from the upper branch where $u_i \leq 1, u_j \leq 1$ (see figure 5).

To investigate the equilibrium situation, the fixed point Hebbian memory in (3.10) can be derived from (3.9) if $dW_{ij}/dt = 0$.

$$W_{ij}^{(0)} = [x(\mathbf{v})_i x(\mathbf{v})_j], \tag{3.10}$$

and substituting it into (3.5) to arrive at a single discrete mapping equilibrium of CNN:

$$k_i(n) u_i(n+1) = 2\sigma_N(u_i(n)) - 1, \tag{3.11}$$

where $n = 1, 2, 3, \ldots$ discrete time points, and the mean-field baseline slope is derived

$$\frac{1}{k_i(n)} \equiv \sum_{j=1\,(j\neq i)}^{N} \Big[2\sigma_N(u_j(n)) - 1\Big]\Big[2\sigma_N(u_j(n))\Big]. \tag{3.12}$$

Given a slope $k_i(n)$, the plot of (3.11) shows four intersection repulsive points and one potential bifurcation attractor point at the piece of negative logic.

4. Results of simulations

The behavior of a neural network composed of 28×35 fully connected N-shaped sigmoidal neurons governed by Hebbian learning is illustrated by image processing. We study the perception habituation. Simulation steps are given as follows:

1. *Initiation of Bipolar Memory:* Substitution of image **I** (row 1, column 1 in figure 7) into (3.12):

$$W_{ij} = \big(x(\mathbf{I})_i x(\mathbf{I})_j - \delta_{ij} x(\mathbf{I})_i x(\mathbf{I})_j\big) \equiv [x(\mathbf{I})_i x(\mathbf{I})_j] = W_{ij}^{t=1}. \tag{4.13a}$$

2. *N-shaped sigmoidal output* of unipolar image (row 1, column 2 in figure 7):

$$V_i^{t=1} = \sigma_N\big(U_i^{t=1}\big). \tag{4.13b}$$

3. *Updated* U_i^{t+1} by Hebbian learning algorithm in (3.12):

$$U_i^{t+1} = X_i^t \sum_{j=1\,(j\neq i)}^{N} (2X_j^t - 1)\, X_j^t. \tag{4.13c}$$

4. *N-shaped sigmoidal output* of image (top left 3rd in figure 7):

$$V_i^t = \sigma_N\big(U_i^t\big). \tag{4.13d}$$

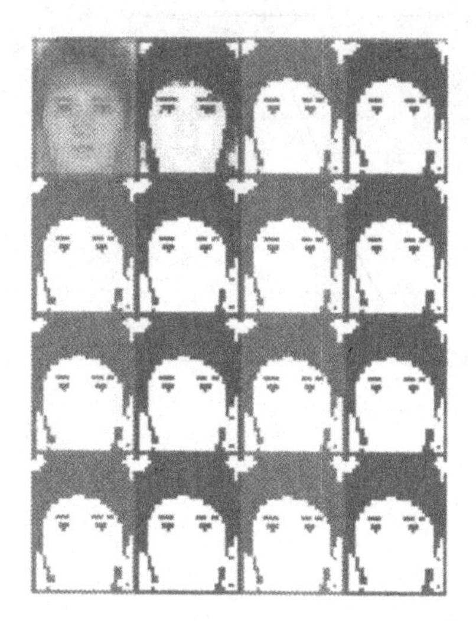

FIGURE 7. (a) 15 iterations plus the original image, (b) an original image.

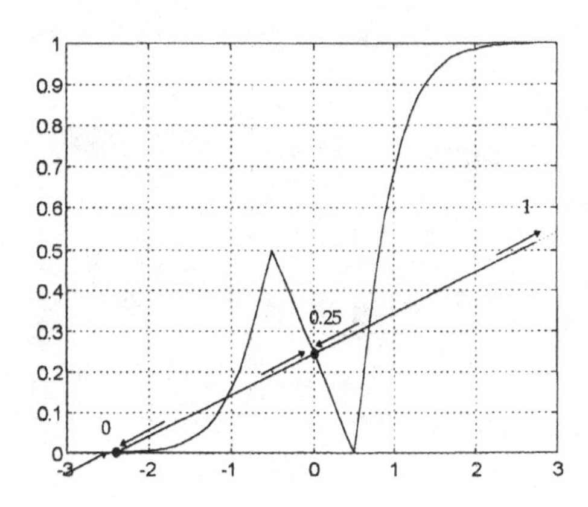

FIGURE 8. The characteristic of the three attractors in σ_N.

Repeat the above processes 3 and 4. Fifteen iterations plus the original image are displayed in figure 4. From the simulation it can be seen that (i) the image is bifurcated and oscillates and (ii) the output values of neurons are roughly among 0, 0.25 and 1. When we take a close look at the sigmoid N-shaped function, two attractors (0 and 1) of the iteration feedback baseline function in (4.13c) can be observed and are shown in figure 5. In addition, we discover that if $U = 0, V = \sigma_N(U) = \sigma_N(0) = 0.25$ from figure 8. If a bipolar sigmoid function shown in figure 9 is used instead of σ_N, the weights

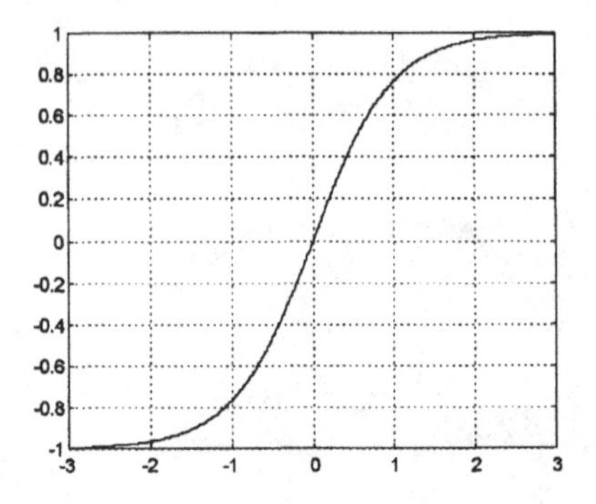

FIGURE 9. A bipolar sigmoid function.

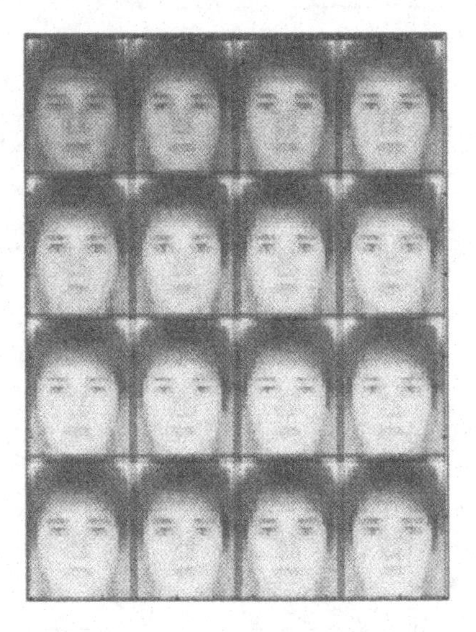

FIGURE 10. Demonstration of associative memory using a sigmoid function in figure 9.

in (4.13a) become self-organized associative memory. The image can be recalled from the associative memory. Figure 10 shows the 15 iterations of self organized associative memory. If part of an image is missing, the associative memory can still recall the original image. Figure 11 shows the result that althogh the partial image in iteration 9 is missing, the memory can recall the image.

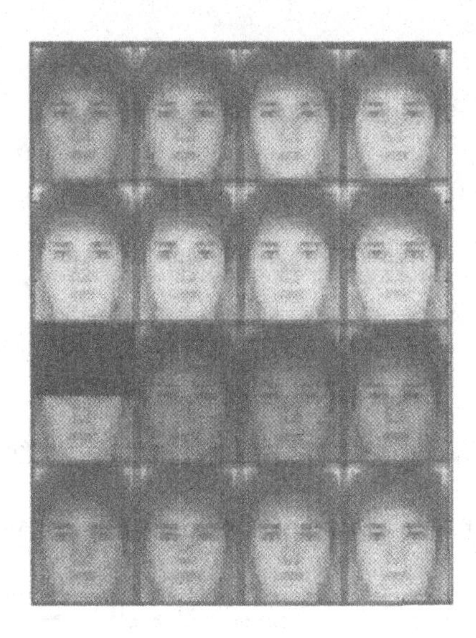

FIGURE 11. Recollection from a partial memory in the iteration 9.

5. Implementation issues of chaos chips

Hardware implementation is needed for several reasons. An obvious one is the slowness in software simulation, because each iteration step requires at least of the order of N^2 for N neuron global interactions. A SPICE simulation demonstrates that using a chip in studying such a chaotic neural network is feasible. Hardware can help designs of various chaos chips for fast processing applications. Details will be fully reported in IEEE Transaction on Neural Networks by November 1996.

We have shown through SPICE simulations that it is possible to implement chaotic neurons using a CMOS inverter and a linear resistor. A basic idea of the single neuron model implementation was presented in Szu *et al.* (1993). An ideal OP amplifier was used to implement the transfer function of the chaotic neuron model in Yamakawa *et al.* (1992). In this section, we give a complete hardware design and implementation of the chaotic neuron model. There are three components in a chaotic neuron: PWL N-shaped function, Sample-and-Hold, and the feedback *baseline* function. The piece-wise linear map $N(x)$ defined as in equation (5.14) can be realized by the composition of the transfer characteristics of a CMOS inverter circuitry and a linear resistor, see Szu *et al.* (1993). The graphical composition of those characteristics is shown in figure 12, see Szu *et al.* (1993). The SPICE simulation of this transfer characteristic is shown in figure 13. From figures 12 and 13 we note that the slopes and constants of the three piece-wise linear map $N(x)$ are functions of the physical parameter α:

$$N(x) = \begin{cases} \alpha x + 5, & x \leq 1.33 \\ (\alpha - 3.65)x + 9.85, & 1.33 < x \leq 2.7 \\ \alpha x, & 2.7 < x. \end{cases} \tag{5.14}$$

The neural model of the PWL N-shaped equation (5.15) is represented by the block

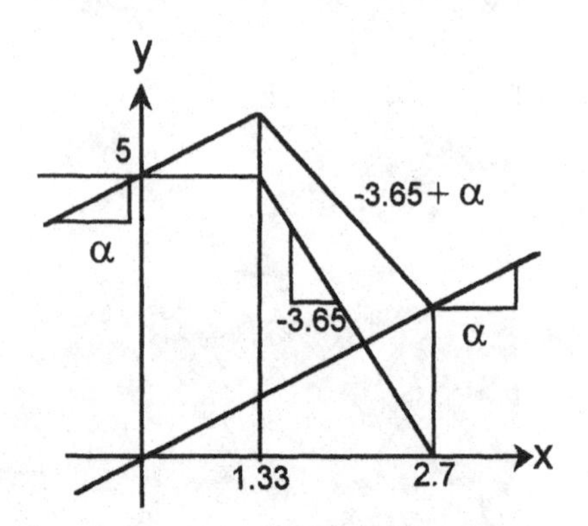

FIGURE 12. Graphical composition of PWL N-shaped function.

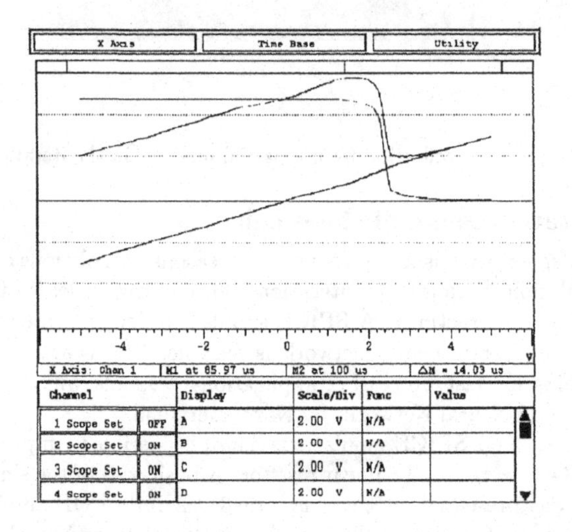

FIGURE 13. SPICE simulation of composition of PWL N-shaped function.

diagram of figure 14. Its possible realization using a CMOS inverter and several operational amplifiers and sample & hold circuitry is shown in figure 15, where V_n is the control voltage for changing α of the linear function; V_a and V_c are the control voltages to change the parameters a and c of the *baseline* function. The circuit is easily implemented in CMOS technology. A possible integration is shown in figure 16.

$$y_1(n+1) = N(x_1(n)), \tag{5.15a}$$

$$x_1(n+1) = w_{11}y_1(n+1) + A_1(n). \tag{5.15b}$$

In our circuit realization, the parameter a of the *baseline* function is fixed at one. The parameter c, which represents external inputs to the neuron from the other neurons, can be easily varied by the voltage V_c; V_n is the control voltage for parameter α, and α is realized as a ratio of resistances of a variable resistor and a passive resistor connected

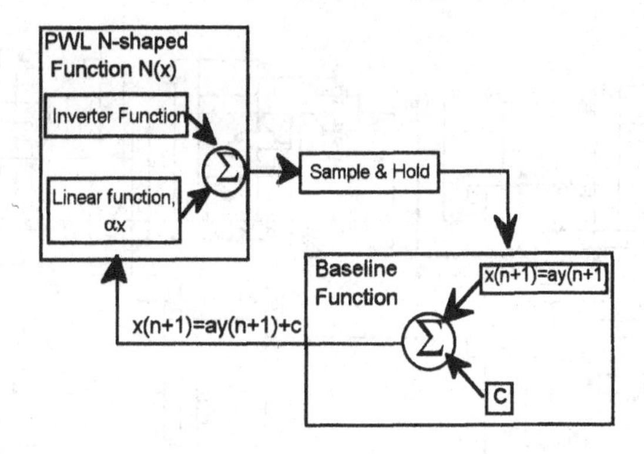

FIGURE 14. Conceptual block diagram of one chaotic neuron.

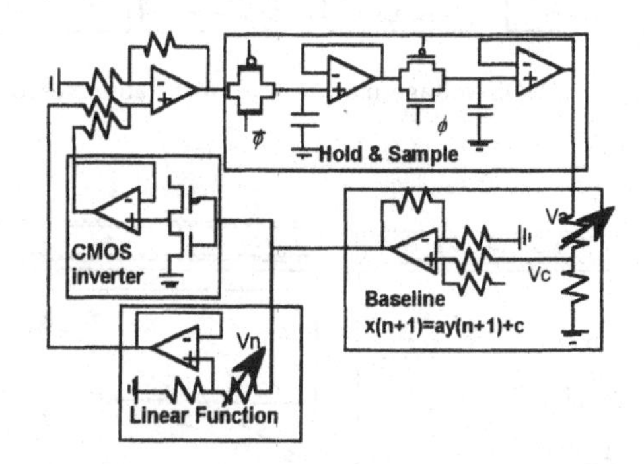

FIGURE 15. Circuit realization of one chaotic neuron, where V_a, V_n, V_c are control voltages.

in series. The variable resistor is controlled by V_n and its SPICE simulations are shown in figure 17. SPICE simulations of phase diagrams for some chosen cases are shown in figures 18 and 19. Figure 18 shows the SPICE simulation of a phase diagram of a chaotic neuron with $\alpha = 0.65$, $a = 1$, and $c = 0$, when $V_n = 6.2\,\text{V}$, and $V_c = 0.0\,\text{V}$. Figure 19 shows the phase diagram of a chaotic neuron with $\alpha = 0.5$, $a = 1$, and $c = 0$, when $V_n = 5.5\,\text{V}$, and $V_c = 0.0\,\text{V}$. The above circuit is designed using a voltage mode circuit for implementation of the chaotic neuron. A current mode circuit is also possible by CMOS technology.

The voltage-mode CMOS chaotic neuron was fabricated in a VLSI chip through the MOSIS program. The layout of the chip is shown in figure 20. The SPICE simulation was demonstrated in the previous section. The design of an active resistor is a dominant part of the circuit which is used for changing the PWL N-shaped function in this chip. Here α is the ratio of resistances of the active resistor and a passive resistor connected in series. The value of the active resistance, and therefore of the parameter α, is determined by the controlled voltage V_n.

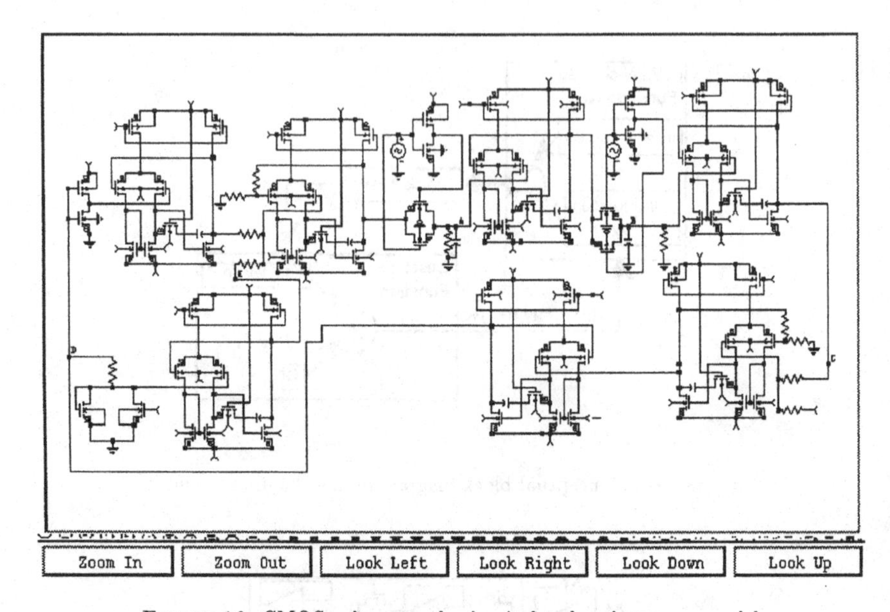

FIGURE 16. CMOS voltage mode circuit for chaotic neuron model.

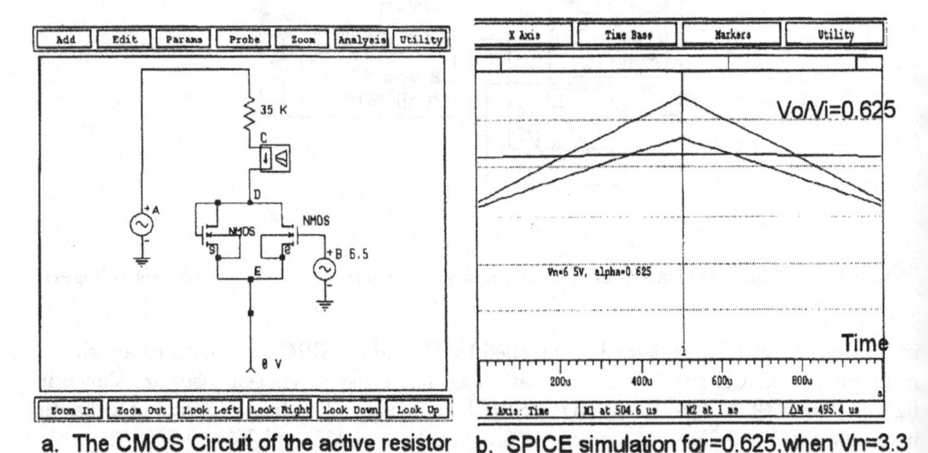

a. The CMOS Circuit of the active resistor b. SPICE simulation for=0.625,when Vn=3.3

FIGURE 17. The CMOS Circuit of the active resistor and SPICE simulations.

Three measurements for the chaotic chip are included in this section: (i) $V_n = 3.3\,\mathrm{V}$, $\alpha = 0.386$; the measurement of a square dance of the chaotic chip is shown in figure 21(a). (ii) $V_n = 6.2\,\mathrm{V}$, $\alpha = 0.645$; the measurement of chaotic behavior of a neuron is shown in figure 21(b). (iii) $V_n = 5.6\,\mathrm{V}$, $a = 0.51$; the measurement of three fixed points of the chaotic chip is shown in figure 21(c). The above measurements show that the chaotic neural network behaves as expected from theory.

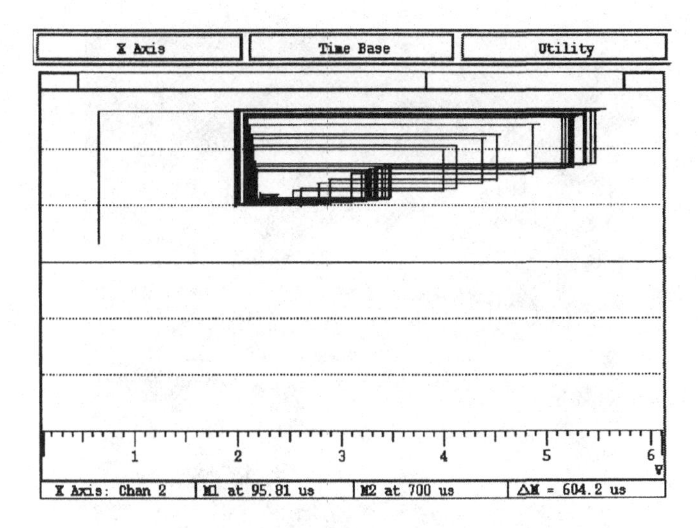

FIGURE 18. Phase diagram of a chaotic neuron in chaotic region with $\alpha = 0.65$, $a = 1$, and $c = 0$, where $V_n = 6.2\,\mathrm{V}$, and $V_c = 0\,\mathrm{V}$.

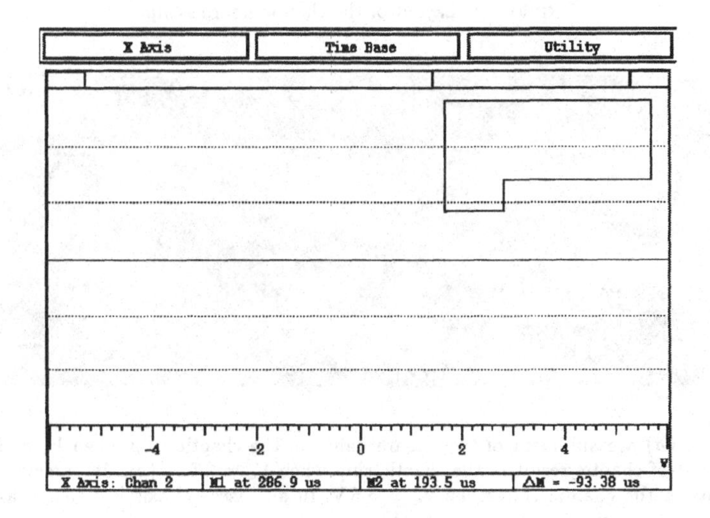

FIGURE 19. Phase diagram of a chaotic neuron in oscillation region with $\alpha = 0.5$, $a = 1$, and $c = 0$, where $V_n = 8.5\,\mathrm{V}$, and $V_c = 0\,\mathrm{V}$.

6. Comparison and conclusion

The Nagumo–Fitzhugh model, implemented by Yamakawa's chip and Aihara's circuit, is equivalent to a 2D map of the Hénon attractor:

$$x_{n+1} = f(x_n) - \alpha y_n, \tag{6.16a}$$

$$y_{n+1} = x_n, \tag{6.16b}$$

where the piecewise-linear discrete map $f(x_n)$ extended Hénon's quadratic function, a $x_n^2 - 1$, see Yamakawa *et al.* (1992). The sensitivity to the initial condition in the sense

FIGURE 20. Layout of the chaotic neuron chip.

FIGURE 21. (a) Measurement of two square dance in the chaotic chip when $V_n = 3.3\,\mathrm{V}$. (b) Measurement of chaotic result in the chaotic chip when $V_n = 6.5\,\mathrm{V}$. (c) Measurement of three fixed points in the chaotic chip when $V_n = 5.6\,\mathrm{V}$. In all cases, X-axis is $V(n)$, Y-axis shows $V(n+1)$ and one square is $500\,\mathrm{mV}\times 500\,\mathrm{mV}$.

of Poincaré chaos might be due to the fact that in this 2D case all the history is required to determine the present value:

$$x_{n+1} = \sum_{k=0}^{N} (-\alpha)^k f(x_{n-2k}). \tag{6.17}$$

On the contrary, the 1D map is due to the instability of the recursive solution, i.e. the geometric intersection of two curves,

$$y_{n+1} = f(x_n), \tag{6.18a}$$

$$y_n = \omega x_n, \tag{6.18b}$$

as the slope of the second curve, the iteration baseline, changes passing beyond the absolute value of the piecewise negative logic slope.

In ANN, the 1D mapping of ith neuron output is given by

$$v_i(t_{n+1}) = \sigma_N(u_i(t_n)), \tag{6.19a}$$

$$v_i(t_n) = [W_{ij} - \Theta_i \delta_{ij}]^{-1} u_j(t_n), \tag{6.19b}$$

where the recursive time index t_n is usually surpressed but understood in ANN. By comparison, the change of baseline slope ω is obviously caused by the inverse of the synaptic weight matrix, $[W_{ij} - \Theta_i \delta_{ij}]^{-1}$. The adjustment is done according to the first order fixed-point attractor equation:

$$\frac{dW_{ij}}{dt} = -\beta\Big\{W_{ij} - \big[x(\mathbf{v})_i x(\mathbf{v})_j\big]\Big\}, \tag{6.20}$$

where the Hebbian matrix learning rule is the analog bipolar traceless outer product (no summation convention):

$$W_{ij} = [x(\mathbf{v})_i x(\mathbf{v})_j] \equiv x(\mathbf{v})_i x(\mathbf{v})_j - \delta_{ij} x(\mathbf{v})_j x(\mathbf{v})_j.$$

The analog firing rate v_j is mapped to bipolar analog x_j: $x(\mathbf{v})_j \equiv 2v_j - 1$ for reasons of biological excitation and inhibition, and the diagonal self-talk $W_{ii} = O(\epsilon)$ is deleted to avoid the recursive fixed point along the unit diagonal term.

With this simplified neuron model, we are potentially capable of providing insight into drug effects upon the brain neural networks, see Malsburg & Bienenstock (1986), Principe & Lo (1991), Choi (1988), Amari (1983), Amari (1971), Amari & Maginu (1988), Kowalski *et al.* (1992), Bulsara *et al.* (1991) and Longtin *et al.* (1991), e.g. the intermittent symptoms of misconception, see Fitzhugh (1961), day dreaming and hallucination, see Ermentrout & Cowan (1979) and Szu *et al.* (1992), or other dynamics effects in brain, see Pribram (1971), and in neuroscience, see Science (1989) and Tsuda (1992). We expect that the spatiotemporal collective chaos is informative because of (i) an emergent property, as evident by the collective oscillations of neural images in a much lower dimensionality than the degrees of freedom of individual neurons, and (ii) random noise interference leading to jamming as evident by the collective blackening and whitening effect of neural images, as shown in IJCNN-92 Beijing, see Szu *et al.* (1992).

Our conclusions are:

- collective interactions of nonlinear dynamics can yield an emergent reduction in the dimensionality or degrees of freedom among individual chaos – the locking or quenching in a collective chaos,
- noise phenomena can produce a delocking of the 'coherence' effect of collective chaos – the switching off of the collective chaos.

The latter will be reported elsewhere. We have introduced (artificial) neural images as the graphical output of collective behaviors. Indeed, these figures have supported that goal of reduced dimensionality in collective chaos. Furthermore, work is planned to demonstrate technology applications based on the emergent collective property of neural chaos, such as real time chaos information processing, and the control of chaotic dynamics using chaos neurochips.

REFERENCES

AIHARA, K. & MATSUMOTO, G. 1986 Chaotic oscillations and bifurcations in squid giant axons. In *Chaos* (ed. A. V. Holden), pp. 257–269. Princeton Univ. Press.

AIHARA, K., TAKABE, T. & TOYODA, M. 1990 Chaotic neural network. *Phys. Lett.* A **144**, 333–340.

AMARI, S.-I. 1971 *Proc. IEEE* **59**, 35–46.

AMARI, S.-I. 1983 Field theory of self-organizing neural nets. *IEEE Trans. Sys. Man & Cyb.* SMC-**13**, 741–748.

AMARI S.-I. & MAGINU, K. 1988 Statistical neurodynamics of associative memory. *Neural networks* **1**, 63–73.

BAIRD, B. 1986 Nonlinear dynamics of pattern formation and pattern recognition in the rabbit olfactory bulb. *Physica* D **22**, 150–175.

BLUM, E. K. & WANG, X. 1992 Stability of fixed points and periodic orbits and bifurcations in analog neural networks. *Neural Networks* **5**, 577–587.

BULSARA, A., JACOBS, E. W., ZHOU, T., MOSS, F. & KISS, L. Stochastic resonance in a single neuron model theory and analog simulation. *J. Theor. Biol.* **152**, 531–555.

CAIANIELLO, E. R. & DELUCA, A. 1966 Decision equation for binary systems. Application to neuronal behavior. *Kybernetik* **3**, 33–40.

CHIALVO, D., GILMOUR, JR., D. & JALIFE, J. 1990 Low dimensional chaos in cardiac tissue. *Nature* **343**, 653–657.

CHOI, M. Y. 1988 Dynamic model of neural networks. *Phys. Rev. Lett.* **61**, 2809–2812.

DERRIDA, B. & MEIR, R. 1988 Chaotic behavior of a layered neural network. *Phys. Rev.* A **38**, 3116–3119.

ERMENTROUT, G. B. & COWAN, J. D. 1979 A mathematical theory for visual hallucination paterns. *Bio. Cyber.* **34**, 137.

FITZHUGH, R. 1961 Impulses and physiological states in theoretical models of nerve membrane. *Biophys. J.* **1**, 445–466.

FREEMAN, W. J. 1986 Petit mal seizure spikes in olfactory bulb and cortex caused by runaway inhibition after exhaustion of excitation. *Brain Res. Rev.* **11**, 259–284.

FREEMAN, W. J. 1987 Simulation of chaotic EEG patterns with a dynamic model of the olfactory system. *Biol. Cybern.* **56**, 139–1150.

FREEMAN, W. J. 1991 The physiology of perception. *Sci. Am.* **2**, 78–85.

GARFINKEL, A., SPANO, M., DITTO, W. & WEISS, J. 1992 Controlling cardiac chaos. *Science* **257**, 1230–1235.

GLASS, L. & MACKEY, M. 1988 From clocks to chaos. *The Rhythms of Life*, pp. 1–248. Princeton Univ. Press.

HANSEL, D. & SOMPOLINSKY, H. 1992 Synchronization and computation in a chaotic neural network. *Phys. Rev. Lett.* **68**, 718–721.

HARMON, L. D. 1961 Studies with artificial neuron I: properties and functions of an artificial neuron. *Kybernetik* **1**, 89–101.

HAYASHI, H., ISHIZUKA, S. & HIRAKAWA, K. 1983 Transition to chaos via intermittency in the onchidium pacemaker neuron. *Phys. Lett.* A **98**, 474–476.

HOFSTADTER, D. R. 1985 Metamagical themas: questing for the essence of mind and pattern. In *Mathematical Chaos and Strange Attractors*, pp. 364–395. Basic Books.

JACKSON, A. E. 1989 *Perspectives of Nonlinear Dynamics*, vol. 1 and 1990 vol. 2. Cambridge.

KOWALSKI, J. M., ALBERT, G. L., RHOADES, B. K. & EIOLL, G. W. 1992 Neural networks with bursting activity. *Neural Networks.* To appear.

LONGTIN, A., BULSARA, A. & MOSS, F. 1991 Time-interval sequences in bistable systems and the noise-induced transmission of information by sensory neurons. *Phys. Rev. Lett.* **67**, 656–659.

MALSBURG, C., VON DER & BIENENSTOCK, E. 1986 Statistical coding and short term synaptic plasticity: a chime for knowledge representation in the brain. In *Disordered System and Biological Organization* (ed. E. Bienenstock). NATO ASI Series, vol. F20, pp. 247–271.

NAGUMO, J. & SATO, S. 1972 On a response characteristic of a mathematical neuron model. *Kybernetik* **10**, 155–164.

PRIBRAM, K. 1971 *Languages of the Brain Experimental Paradoxes and Principles in Neuropsychology.* Prentice Hall.

PRINCIPE, J. C. & LO, P. 1991 Chaotic dynamics of time-delay neural networks. *IJCNN-91 San Diego* **II**, 403–409.

SCIENCE 1989 *Frontiers of Neurosciences. Science* special issue.

SOMPOLINSKY, H. & KANTER, I. 1986 Temporal association in asymmetric neural networks. *Phys. Rev. Lett.* **57**, 2861–2864.

SZU, H. 1989 A dynamic reconfigurable neural network (annotated by Walter Freeman). *J. Neural Network Computing* **1** (special issue), 3–23.

SZU, H. 1990 Neural networks based on Peano curves and Hairy neurons. *Telematics and Informatics* **7**, 403–430.

SZU, H. & ROGERS, G. 1992a Single neuron chaos. *IJCNN-92 Baltimore* **III**, 103–108.

SZU, H. & ROGERS, G. 1992b Generalized McCullouch–Pitts neuron model with threshold dynamics. *IJCNN-92 Baltimore* **III**, 535–540.

SZU, H., TELFER, B., ROGERS, G., LEE, K., MOON, G., ZAGHLOUL, M. & LOEW, M. 1992 Collective chaos in neural networks. In *Intl. Joint Conf. Neural Networks, IJCNN-92*, Beijing, China.

SZU, H., TELFER, B., ROGERS, G., GOBOVIC, D., HSU, C., ZAGHLOUL, M. & FREEMAN, W. 1993 Spatiotemporal chaos information processing in neural networks-electronic implementation. In *Proc. World Conference of Neural Network*, Oregon, pp. 719–734.

TSUDA, I. 1992 Dynamic link of memory-choatic memory map in nonequilibrium neural networks *Neural Networks* **5**, 313–326.

WANG, L., PICHLER, E. & ROSS, J. 1990 Oscillations and chaos in neural networks: an exactly solvable model. *Proc. Natl. Acad. Sci. USA* **87**, 9467–9471.

YAMAKAWA, T., MIKI, T. & UCHINO, E. 1992 A chaotic chip for analyzing nonlinear discrete dynamical network systems. In *Proc. 2nd Intl. Conf. Fuzzy Logic & Neural Networks*, Iizuka, Japan, pp. 563–566.

YAO, Y. & FREEMAN, W. 1990 Model of biological pattern recognition with spatially chaotic dynamics. *Neural Networks* **3**, 153–170.

YOSHIZAWA, S., OSADA, H. & NAGUMO, J. 1982 Pulse sequence generated by a degenerate analog neuron model. *Bio. Cyb.* **45**, 23–33.

See also

BAI-LIN, H., ED. 1985 *41 Reprints & 900 Author-indexed References in Relation to Chaos.* Reprint up to 1985, vol. 1. World Scientific Publishing.

BASTI, G., PERRONE, A., CIMAGALLI, V., GIONA, M., PASERO, E. & MORGAVI, G. 1991 A dynamic approach to invariant feature extraction from time-varying inputs by using chaos in neural nets. *IJCNN-91 San Diego* **III**, 505–510.

CHUA, L. O., ED. 1987 Special issue on *Chaotic Systems Proc. IEEE* **75**, No. 8. In it specifically A. Rodriguez-Vazquez *et al.*, Chaos from switched-capacitor circuits: discrete maps, pp. 1109–1106.

HOGG, T. & HUBERMAN, B. Controlling chaos in distributed systems. *IEEE Trans.* SMC-**21**, 1325–1332.

MANDELL, A. J. 1983 From intermittency to transitivity in neuropsychobiological flows. *Am. Physiol. Soc.* R484–R494.

MATSUMOTO, G., AIHRA, K., HANYU, Y., TAKAHASHI, N., YOSHIZAWA, S. & NAGUMO, J. 1987 Chaos and phase locking in normal squid axons. *Phys. Lett.* A **123**, 162–166.

MOON, G., ZAGHLOUL, M. E. & NEWCOMB, R. W. An enhancement mode MOS voltage-controlled linear resistor with large dynamic range. *IEEE Trans. on CAS* CAS-**37**, 1284–1288.

ZAK, M. 1991 An unpredictable-dynamics approach to neural intelligence. *IEEE Expert*, 4–10.

Phase transitions and learning in neural networks

By C. VAN DEN BROECK AND G. J. BEX

Limburgs Universitair Centrum, Diepenbeek, Belgium

We show how networks that learn by examples can undergo abrupt transitions to partial or perfect generalization. These transitions are reminiscent of the 'Eureka' transition that we know from our daily experience. We explain the origin of these transitions in terms of concepts from statistical mechanics, as arising from the competition between entropic effects and an 'error' contribution representing the information contained in the training examples. We show how, in order to display such transitions, the network has to operate on the basis of a hierarchy of hypotheses. This hierarchy arises naturally from the architectural rules by which the network is constructed.

1. Introduction

It is difficult, if not impossible, to give a crisp and clear definition of intelligence. But one aspect that should certainly be included in such a definition is the ability to learn from examples. In order to get more insight into this mechanism of learning, several explicit scenarios and learning schemes have been proposed and studied over the last decade. Our purpose here is to review some results obtained in the statistical mechanics of learning, see Györgyi & Tishby (1990), Hertz *et al.* (1991), Seung *et al.* (1992), Watkin *et al.* (1993), Opper & Kinzel (1994), Van den Broeck (1994), Engel (1994) and Bouten *et al.* (1995). Phenomena reminiscent of those occurring in human intelligence are observed, for example 'Eureka' transitions and the presence of a hierarchy of hypotheses.

To keep matters simple, we will mainly focus on the supervised learning of classification tasks. Classification corresponds to a large class of problems in which one is asked to produce the output classification for a given input pattern. Examples are medical diagnosis (input: symptoms, output: illness), prediction (input: past, output: future), speech and image recognition (input: sound/picture, output: word/object), addition (input: numbers, output: their sum), etc. Usually, the output class has a much smaller cardinality than the input one, so that the classification corresponds to a clustering or projection of the input patterns into output classes.

The object that performs the classification is called a classifier and it has the following structure (see figure 1). Patterns are presented at the input channel, and processed by the classifier to produce an output class. In order to be able to learn from examples, the classifier which is characterized by a specific architecture, possesses internal degrees of freedom $\mathbf{J}$ that can be adapted during the learning process following some learning algorithm. The resulting performance can be quantified by the so-called generalization error, defined as the probability for missclassification of a new pattern.

2. Boolean networks

One of the first classifiers in which the process of learning by examples was studied in some detail, is the feedforward Boolean network introduced by Carnevali & Patarnello (1987), see figure 2. It consists of 16 logic 0–1 input channels, 8 logic 0–1 output channels, and a number of binary logic gates (AND, OR, XOR, etc.) ordered and connected together in a random feedforward way. The purpose was to train this system to learn

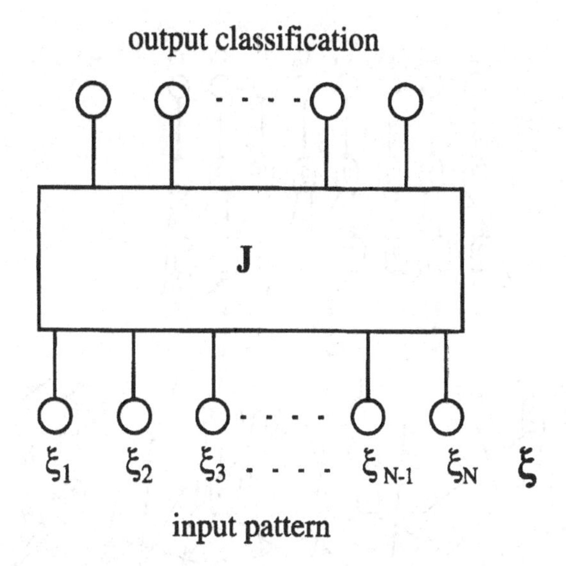

FIGURE 1. Schematic representation of a classifier, with N input channels, a number of output channels, and internal degrees of freedom **J**.

the addition (output = sum of two 8-digit binary numbers) on the basis of examples. The training algorithm was a simulated annealing technique searching for a network configuration that reproduces the training examples with zero error. It was found that the resulting network would operate as a pure memory device, without generalization, if training consisted of a relatively small number of training examples in an architecture with a large number of gates. On the other hand, for a relatively large number of training examples and a small number of gates, no configuration could be found that stored all the training examples correctly. However, for a moderate number of gates (e.g. 200), it was found that the system would undergo a sharp transition from perfect memorization to perfect generalization when the number of training examples exceeded a critical threshold (e.g. 300), which is much smaller than the total number of entries in the addition table (namely 2^{16}).

In order to investigate this surprising result, we turn to a simple variant of the problem, see Van den Broeck & Kawai (1990). We consider a feedforward Boolean network with 4 logic inputs and a single logic output, and consider the task of learning the addition modulo 2, or the so-called parity problem. Before considering the learning problem itself, we construct the *a priori* hierarchy of hypotheses for this architecture. We do this by generating a large number of configurations at random, and by counting how frequently any of the 2^{16} possible input–output tables occur. The result is reproduced in figure 3 for an architecture with 25 gates. In about 20% of the cases, one finds that a random construction will lead to a configuration that returns a logic 0 on any of the inputs. Less frequently, one finds configurations that give as output one of the inputs, or their negation. Next come outputs that are functions of two inputs, etc. The parity problem is found to be realized by only 2 out of 10000 configurations. Incidently, we mention that the resulting hierarchy possesses a rather clear mono-fractal structure, as can be deduced from the observed Zipf law (linear relation between log(frequency) and log(rank)).

The learning algorithm is defined as follows. We choose at random a set of p input

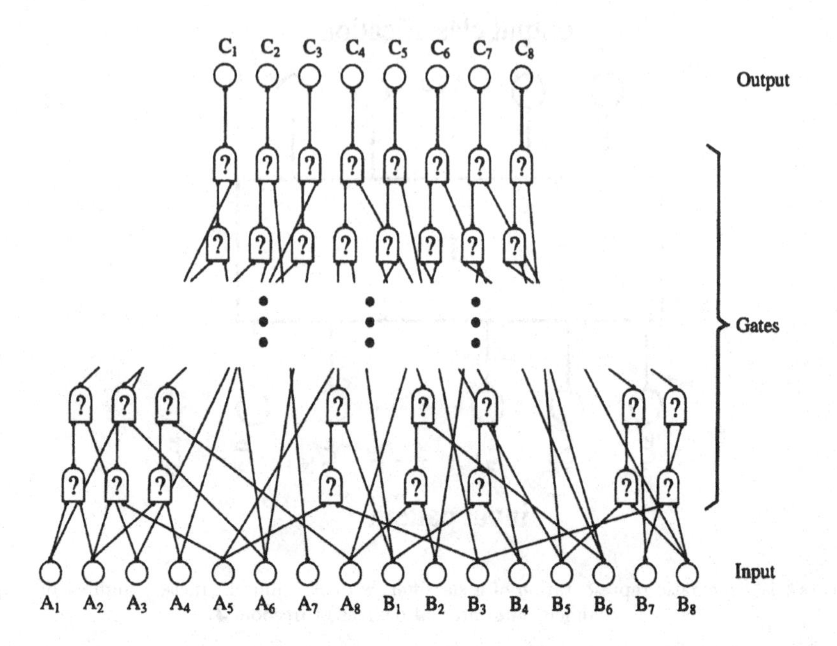

FIGURE 2. Feedforward Boolean net consisting of 16 logic input units, 8 logic output units and a set of feedforward binary gates.

patterns with the correct parity output, and search for a compatible configuration by simulated annealing (changing randomly one gate and one connection). In figure 4, we plot the resulting learning curves, including cases with 4, 6 and 8 logic input gates. As the number of inputs increases, one observes a sharp Eureka-like transition from pure memorization to perfect learning at a value $\alpha = p/N$ of approximately 3.2, corresponding to an exponentially small fraction of the total logic table for the parity.

To explain the mechanism by which this remarkable phenomenon occurs, we plot schematically in figure 5 the space Ω_0 of all configurations. Every point corresponds to a specific choice of the logic gates and a specific way to connect them, and can be called a micro-state. The macro-states then correspond to the ensemble of points that return a specific input–output table.

In figure 5, we also represent the subspace Ω_∞ corresponding to the macro-state of the target problem at hand, in the above case the parity problem. Finally, we focus on the space Ω_p of compatible configurations, constituted by all the macro-states that make no error on p training examples. Clearly, Ω_p is a random variable depending on the specific choice of these examples. But its average can easily be evaluated as follows. Let $\Omega_0(\varepsilon)$ be the sum of the subspaces of all macro-states that have a generalization error ε with respect to the target, i.e. their output tables differ in a fraction ε of all places. This quantity can be constructed on the basis of the *a priori* landscape reproduced in figure 3. For the parity with 4, 6 and 8 inputs, the resulting curves are reproduced in figure 6. Note that $\Omega_0(\varepsilon)$ typically has a maximum at $\varepsilon = \frac{1}{2}$, because tables that differ in half of their output from the target are the most numerous.

Now, it is clear from the definition of ε that, on the presentation of a training example, only a fraction $(1-\varepsilon)$ of the configurations in the space $\Omega_0(\varepsilon)$ will on average agree with

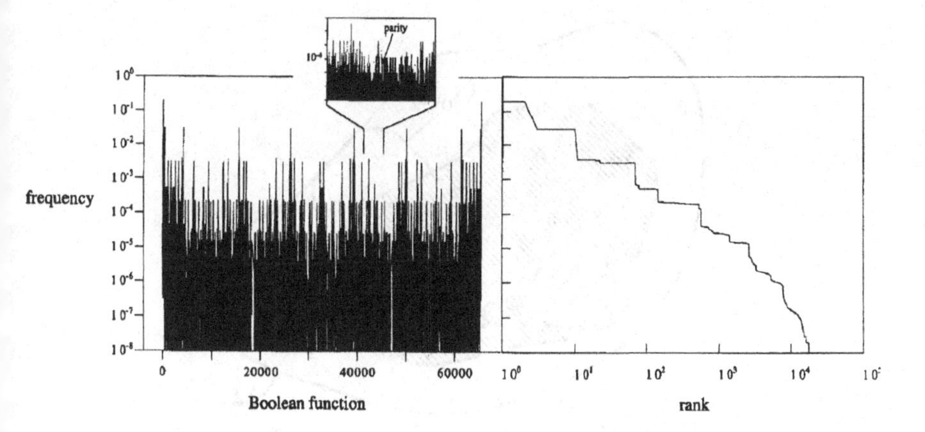

FIGURE 3. The frequency of occurrence of any of the 2^{16} possible input–output tables, for randomly generated feedforward Boolean networks with 25 gates. By plotting log(frequency) versus log(rank), we find that the hierarchy obeys Zipf's law, indicating the existence of an underlying fractal.

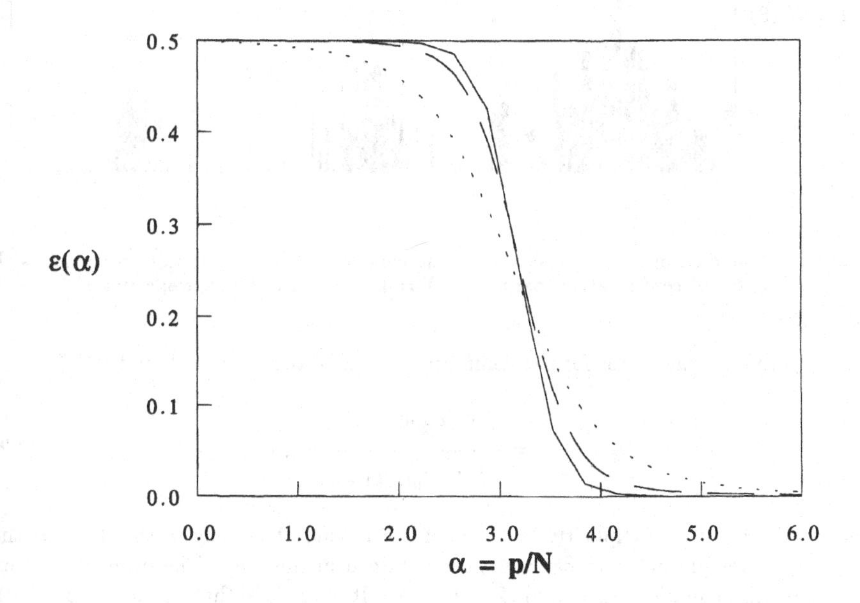

FIGURE 4. The generalization error ε versus the number of training examples for the parity problem with $N = 4$, 6 and 8 inputs.

the classification provided by the target. Since the examples are supposed to be chosen at random and independent of one another, one finds that the average $\langle \Omega_p(\varepsilon) \rangle$ of such configurations compatible with p training examples is given by

$$\langle \Omega_p(\varepsilon) \rangle = \Omega_0(\varepsilon)(1 - \varepsilon)^p. \tag{2.1}$$

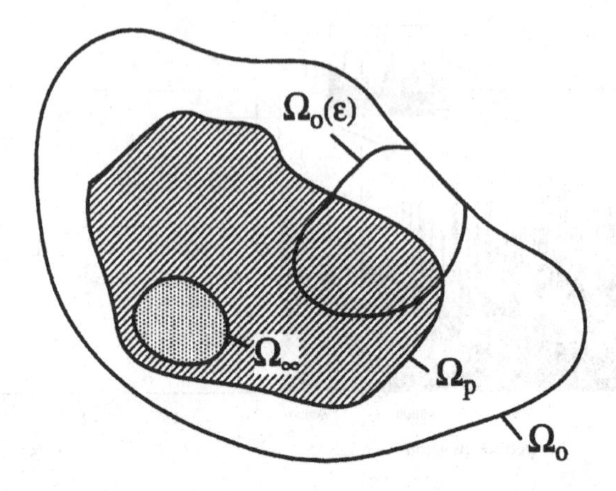

FIGURE 5. Phase space Ω_0 of all micro-configurations, with the space of the target configurations Ω_∞, and the phase space $\Omega_0(\varepsilon)$ of all configurations with a generalization error ε.

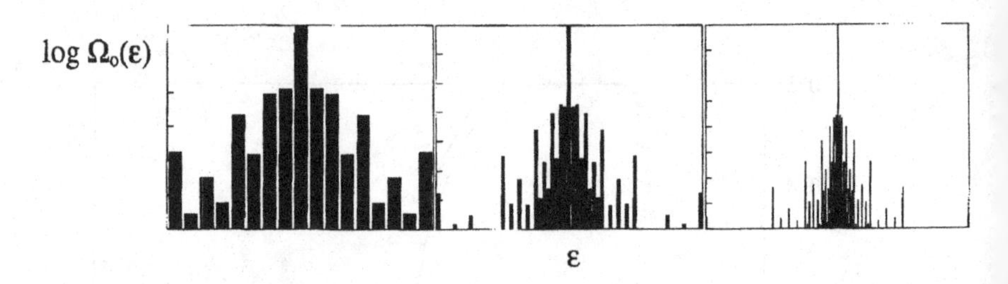

FIGURE 6. The distribution of the errors of randomly generated configurations compared with the target configuration (parity) for $N = 4$, $N = 6$ and $N = 8$, respectively.

The resulting average error for the compatible configurations is then found to be

$$\langle \varepsilon_p \rangle = \frac{\int d\varepsilon \, \varepsilon \, \Omega_0(\varepsilon)(1-\varepsilon)^p}{\int d\varepsilon \, \Omega_0(\varepsilon)(1-\varepsilon)^p}, \tag{2.2}$$

and can be calculated explicitly in terms of the *a priori* hierarchy in Ω_0. In fact, the learning curves presented in figure 4 were obtained in this way. The intuitive picture that is behind equations (2.1) and (2.2) is clear. It is unlikely that configurations with large generalization error make no errors on the training data, and they are eliminated quite easily. As more training examples become available, configurations with a smaller generalization error are eliminated and the generalization error decreases until it eventually reaches zero. This process can be quite abrupt if the configurations with low but non-zero generalization error are very rare. This is precisely what happens for the parity problem in the Boolean network, see figure 6. Finally we would like to stress that good learning properties for some targets imply that the system will not learn well for other targets. The architecture of the network and the resulting hierarchy of hypotheses determines which problems will be easy to learn and which ones will not.

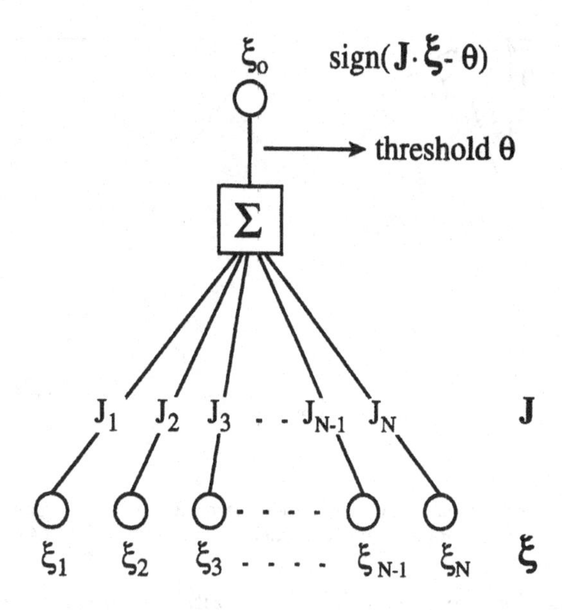

FIGURE 7. A perceptron with a synaptic vector **J** which classifies an input signal $\vec{\xi}$ to an output ξ_0.

3. Perceptrons

The Boolean network, discussed in the previous section, nicely illustrates the phenomenon of learning by examples, operating through the elimination of wrong hypotheses in a fractal hierarchy, and leading ultimately to a Eureka type of transition to perfect generalization. The analytic treatment of Boolean networks is, however, very difficult. We therefore turn to another architecture arising in the important context of artificial neural networks. We consider the simplest neural network, namely that consisting of a single neuron. This 'net', called the perceptron, was popularized by Rosenblatt (1962), the founding father of artificial neural networks. The perceptron is a multi-input, single output threshold sensitive device, vaguely modeling the operation of a physical neuron. We consider it here under its simplest form only (see figure 7): an input signal $\boldsymbol{\xi}$ is classified as $\xi_0 = \text{sgn}(\mathbf{J}\cdot\boldsymbol{\xi})$, where $\mathbf{J}$ is the 'synaptic' vector that describes the internal state of the perceptron. In other words, the perceptron is a hyperplane classifier returning $+1$ on all patterns that make an angle smaller than $90°$ with its $\mathbf{J}$-vector and -1 for the other patterns. We will focus, as far as training is concerned, on the same simple algorithm as considered for Boolean networks, namely the random search for a $\mathbf{J}$ vector that correctly reproduces the classification ξ_0^μ for a set of p training examples $\boldsymbol{\xi}^\mu$, $\mu = 1, \ldots, p$. Since only the angles between pattern vectors and synaptic vector matter for the classification, one may also restrict oneself to the case in which these vectors lie on a sphere. For reasons that will become clear later, one chooses the radius of this sphere to be $\sqrt{N}$, where N is the dimensionality of the input patterns.

We now specify the classification task that we will submit to the perceptron, see Gardner & Derrida (1989). We suppose that the data are generated by a target perceptron with unknown synaptic vector $\mathbf{T}$, i.e. $\xi_0^\mu = \text{sgn}(\mathbf{T}\cdot\boldsymbol{\xi}^\mu)$. Furthermore, we assume that its components are of the Ising type $T_i = \pm 1$, and we search for student perceptrons $\mathbf{J}$ that are also of the Ising type. In order to construct the quantity $\Omega_0(\varepsilon)$ that was introduced

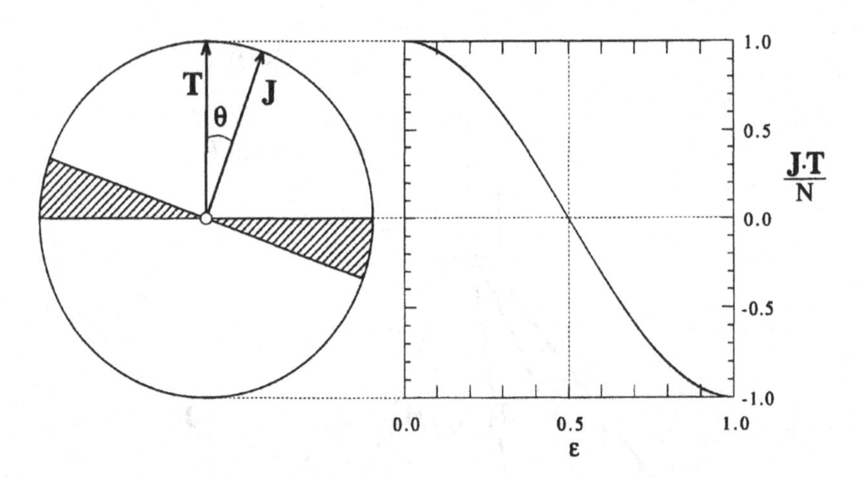

FIGURE 8. Student perceptron **J** and teacher perceptron **T** with their region of disagreement. The corresponding generalization error ε is given by (3.4).

in the previous section, we proceed in 2 steps. First we observe that the number of perceptrons **J** with binary components that differ in exactly Nx legs from the teacher, is given by

$$\binom{N}{Nx} \sim \exp\Big(N\big[-x\ln x-(1-x)\ln(1-x)\big]\Big) \quad \text{as } N\to\infty. \tag{3.3}$$

The generalization error is given by the probability that $\mathbf{T}\cdot\boldsymbol{\xi}$ and $\mathbf{J}\cdot\boldsymbol{\xi}$ have different signs for a random choice of the pattern $\boldsymbol{\xi}$. These expressions are the sum of N terms, with Nx terms having the opposite sign. In the limit $N\to\infty$, we can apply the central limit theorem and conclude that the probability for obtaining a different sign is given by

$$\varepsilon = \frac{1}{\pi}\arccos(1-2x) = \frac{1}{\pi}\arccos\left(\frac{\mathbf{J}\cdot\mathbf{T}}{N}\right). \tag{3.4}$$

The same result can be obtained by the simple geometrical argument that ε is equal to twice the angle between **T** and **J**, divided by 2π, see figure 8. By combining (3.3) and (3.4), we obtain

$$\Omega_0(\varepsilon) \sim \exp\left\{N\ln\left(-\frac{1-\cos(\pi\varepsilon)}{2}\ln\frac{1-\cos(\pi\varepsilon)}{2}-\frac{1+\cos(\pi\varepsilon)}{2}\ln\frac{1+\cos(\pi\varepsilon)}{2}\right)\right\} \tag{3.5}$$

as $N\to\infty$. To obtain a nontrivial result from (2.1) in the limit $N\to\infty$, it is clear that one has to let the number of training examples grow proportionally, $p=\alpha N$, with α fixed. Furthermore, it is convenient to switch to an 'intensive' quantity

$$s_\alpha(\varepsilon) = \lim_{N\to\infty}\frac{1}{N}\ln\langle\Omega_p(\varepsilon)\rangle\Big|_{p=\alpha N},$$

very similar to the entropy from the usual statistical mechanics. The fact that this quantity is of order N^0 is a result of our choice for the radius of the sphere equal to $\sqrt{N}$. One thus finds:

$$s_\alpha(\varepsilon) = -\frac{1-\cos(\pi\varepsilon)}{2}\ln\frac{1-\cos(\pi\varepsilon)}{2}-\frac{1+\cos(\pi\varepsilon)}{2}\ln\frac{1+\cos(\pi\varepsilon)}{2}+\alpha\ln(1-\varepsilon). \tag{3.6}$$

This quantity represents the relative average abundance of **J** vectors with generalization

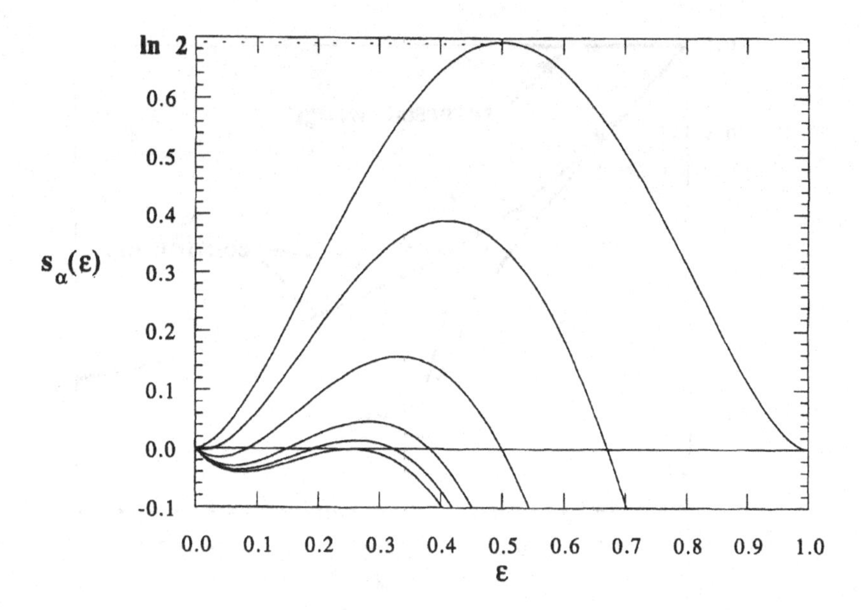

FIGURE 9. Plots of $s_\alpha(\varepsilon)$ given by (3.6) for several values of α; from top to bottom: $\alpha = 0,\ 0.5,\ 1.0,\ 1.3,\ 1.4$ and 1.45.

error ε after eliminating all those incompatible with the training set of size α. It is represented in figure 9 for several values of α. Clearly, the observed generalization error will be that of the, on average, exponentially dominant configurations, i.e. $\varepsilon(\alpha) = \max_\varepsilon \arg s_\alpha(\varepsilon)$. From figure 9, it follows that $\varepsilon(\alpha)$ will start at $\varepsilon = \frac{1}{2}$ corresponding to the *a priori* most numerous configurations and move continuously towards smaller values as α increases. However, at $\alpha = 1.448$, a first order Eureka phase transition from $\varepsilon \approx .276$ to $\varepsilon = 0$ takes place, with all $\mathbf{J}$ vectors at finite ε becoming exponentially unlikely to survive the selection by the training set. The $\varepsilon(\alpha)$-curve is represented in figure 10.

It is straightforward to formulate variants of the above scenario. The simplest one consists of considering $\mathbf{J}$ and $\mathbf{T}$ vectors with continuous components. One then finds

$$s_\alpha(\varepsilon) = \tfrac{1}{2}\ln\left(1 - \cos^2(\pi\varepsilon)\right) + \alpha\ln(1-\varepsilon).$$

In this case, there is no phase transition to perfect generalization, but rather a smooth decay of $\varepsilon(\alpha)$ from $\varepsilon(\alpha) \sim \frac{1}{2} - O(\sqrt{\alpha})$ for α small to $\varepsilon(\alpha) \sim 1/\alpha$ for α large. It may be interesting to point out that these results are rather typical. According to a general argument put forward in Van den Broeck & Parrondo (1993), the fastest decrease of the generalization error for small α is conjectured to be of the form $\varepsilon(\alpha) - \frac{1}{2} \sim \sqrt{\alpha}$, while a general theorem based on considerations from statistics, see Haussler *et al.* (1990) and Parrondo & Van den Broeck (1993), implies that $\varepsilon(\alpha) < (\ln\alpha)/\alpha$ for α large for any architecture and learning rule, provided the task is learnable.

We now turn to another variant, namely the Ising reversed wedge perceptron described in Bex *et al.* (1995), see figure 11. This perceptron is characterized by an Ising vector $\mathbf{T}$, but the classification of a pattern $\boldsymbol{\xi}$ is now a non-monotonic function of the overlap $\lambda = \mathbf{T}\cdot\boldsymbol{\xi}/\sqrt{N}$ (remember that we follow the convention $\boldsymbol{\xi}^2 = \mathbf{T}^2 = N$), namely $\xi_0 = \mathrm{sgn}((\lambda - K)(\lambda + K)\lambda)$; K is a control parameter that can be varied at will. We consider

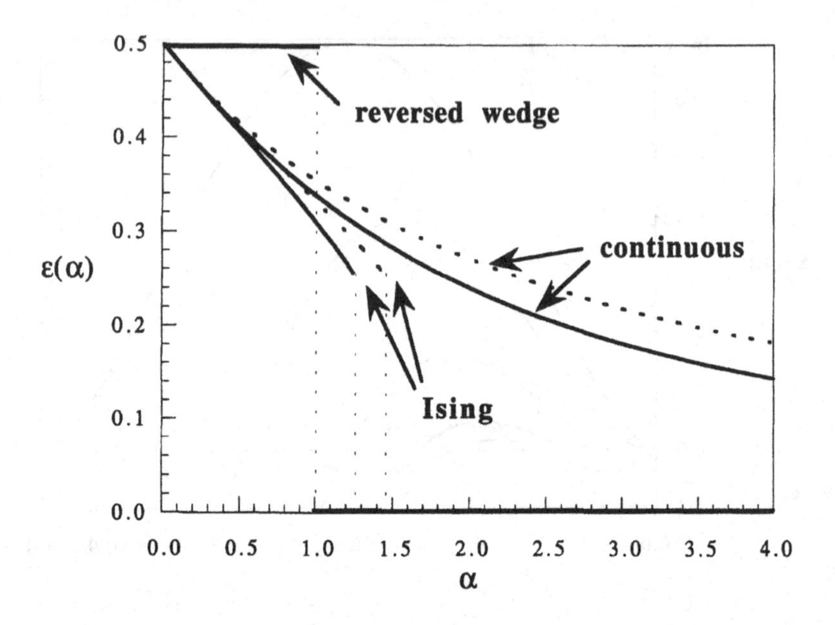

FIGURE 10. Generalization error for the Ising perceptron, perceptron with continuous components, and reversed wedge Ising perceptron ($K = \sqrt{2 \ln 2}$). The full line corresponds to the exact results obtained through replica calculations.

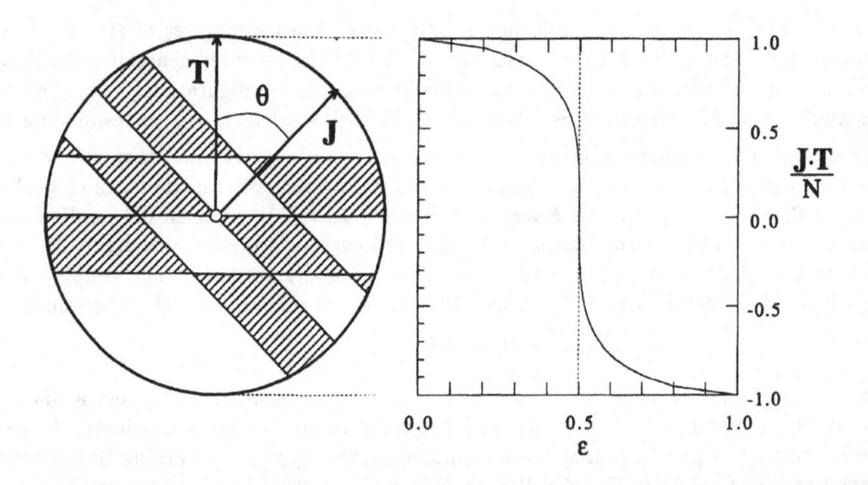

FIGURE 11. Reversed wedge student perceptron **J** and teacher perceptron **T** with their regions of disagreement and the corresponding generalization error ε as a function of $\mathbf{J}\cdot\mathbf{T}/N$

again the teacher–student scenario, with the students **J** of the Ising reversed wedge type being selected on the basis of compatibility with the training set $(\boldsymbol{\xi}^\mu, \xi_0^\mu; \mu = 1, \ldots, p)$.

At this point, a short discussion about the information carried by the training set is in order. Every training example μ carries information of a binary valued quantity ξ_0^μ corresponding to at most one bit of information. Specifying the teacher vector requires N bits. Since one expects that there is overlap between the information carried by the

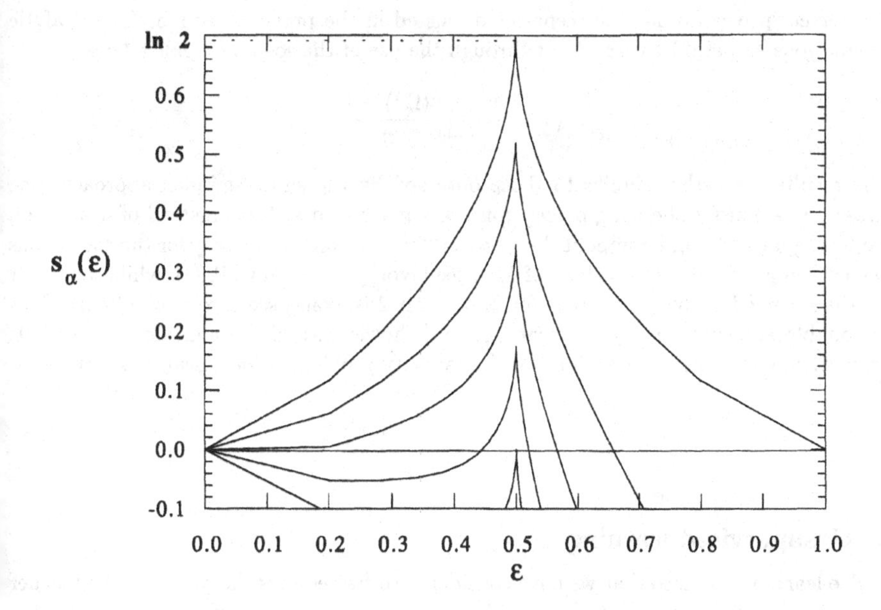

FIGURE 12. The entropy $s_\alpha(\varepsilon)$ for $K = \sqrt{2\ln 2}$ B for several values of α; from top to bottom: $\alpha = 0,\ 0.25,\ 0.5,\ 0.75$ and 1.

different examples, a first order phase transition to perfect generalization should take place for $\alpha = p/N \geq 1$. This is indeed the case for the Ising perceptron (corresponding to $K = 0$) discussed before. However, it is found that as K increases, the Eureka transition shifts to lower values of α and exactly attains the $\alpha = 1$ lower bound at the value $K = \sqrt{2\ln 2}$! For larger values of K, the phase transition again shifts to larger values of α. Furthermore, at $K = \sqrt{2\ln 2}$, one finds that before the transition, $\alpha < 1$, the generalization error is equal to $\frac{1}{2}$ so that the training set is memorized without any generalization, see figure 10. An explanation for this remarkable behaviour can be found by plotting ε as a function of $\mathbf{J}\cdot\mathbf{T}/N$. It is found that ε possesses a horizontal inflection point at $\mathbf{J}\cdot\mathbf{T} = 0$, with the result that $s_\alpha(\varepsilon)$ always possesses a local cusp-like maximum at $\varepsilon = \frac{1}{2}$ (see figure 12).

4. Statistical mechanics of learning and replica theory

The learning theory presented in the previous sections was derived independently in Van den Broeck & Kawai (1990) and Schwartz *et al.* (1990), and corresponds to the annealed approximation in the statistical mechanical approach. It is simple and intuitively appealing. Unfortunately, it only gives an approximate description of the learning process. The reason is that the quantities $\Omega_\alpha(\varepsilon)$ are random variables which are typically not self-averaging, so that $\langle\Omega_\alpha(\varepsilon)\rangle$ can be quite different from its typical value. This problem is accentuated by the fact that we are usually interested in the limit of a large number N of input channels in which case $\Omega_\alpha(\varepsilon)$ becomes exponentially large in N. The problem that arises can be illustrated by the following example: $P(\Omega) = \delta(\Omega - e^N)(1 - e^{-N}) + \delta(\Omega - e^{2N})e^{-N}$. Clearly $\Omega_{\text{typ}} = e^N$ is the typical value, yet $\langle\Omega\rangle = 2e^N$. Note that $\Omega_{\text{typ}} = \exp\langle\ln\Omega\rangle$, where the application of the logarithm kills the contribution of very large but unlikely events. In fact, it turns out that this is also the correct procedure for the scenarios discussed in the previous section. The analytic calculations can still be carried out through the use of the so-called replica trick

$$\langle \ln \Omega \rangle = \lim_{n \to 0} \frac{\langle \Omega^n \rangle - 1}{n} .$$

The results are rather similar to those obtained by the simple-minded approach: the phase transition for the Ising perceptron takes place at $\alpha \approx 1.245$ instead of $\alpha \approx 1.448$, see Györgyi (1990) and Sompolinsky *et al.* (1990), the long time decay for the continuous perceptron goes as $0.62/\alpha$ instead of $1/\alpha$, see Györgyi & Tishby (1990), while the result for the reversed wedge perceptron for $K = \sqrt{2 \ln 2}$ is exact, see Bex *et al.* (1995). This reasonable agreement is by no means general, however, and the annealed theory fails dramatically in other cases such as for the capacity problem or for unlearnable problems.

5. Unsupervised learning

The learning scenario that we have considered so far requires the presence of a teacher who provides the correct classification. It is therefore an example of supervised learning. A brief glance at figure 13 will convince us that learning without a teacher is possible if the patterns or data that we want to distinguish are clustered. The corresponding learning scheme is called unsupervised. To briefly illustrate some of its features, we give a few examples. Suppose that the patterns $\boldsymbol{\xi}^\mu$ lie on the equator orthogonal to an unknown direction $\mathbf{B}$. In order to detect this symmetry breaking orientation, one has to solve the set of equations $\boldsymbol{\xi}^\mu \cdot \mathbf{B} = 0$, $\mu = 1, \ldots, p$. Clearly the direction of $\mathbf{B}$ can only be found for $p \geq N$. In fact, in the limit, $p \to \infty$, $N \to \infty$ with $\alpha = p/N$ fixed, one finds that randomly chosen patterns $\boldsymbol{\xi}^\mu$ orthogonal to $\mathbf{B}$ are linearly independent with probability 1. Hence, one can identify $\mathbf{B}$ perfectly for $\alpha > 1$, while for $\alpha < 1$, the large majority of compatible orientations (i.e. those orthogonal on the patterns) are orthogonal to $\mathbf{B}$. We thus observe a first order phase transition similar to the one discussed in the context of the Ising perceptrons. A more detailed analysis, based on a statistical mechanics description of the problem, see Reimann & Van den Broeck (1996) and Van den Broeck & Reimann (1996), reveals that the delayed detection of a symmetry breaking orientation and the presence of a phase transition of both first and second order are in fact generic if the patterns are unbiased.

A revealing interpretation of the delay is suggested by an analysis of a simple competitive learning algorithm, that searches for the existence of two clusters. We have performed a detailed analysis of this algorithm including the case where the data are not clustered, but are in fact completely random, see Lootens & Van den Broeck (1995). It is found that the algorithm will nevertheless identify two clusters with a sharp and well defined structure, whenever the sampling from the random distribution is not exhaustive, i.e. $\alpha = p/N$ finite. This implies that the sampling from a high dimensional uniform distribution typically generates pronounced non-random like structure that will be extracted by algorithms that search for such structure. This type of structure by chance will compete with a genuine underlying structure, e.g. non-uniformity, of the pattern distribution. As a result, a critical amount of sampling, corresponding to a sufficiently large value of α, will be required to detect the genuine structure of clusters in the pattern distribution.

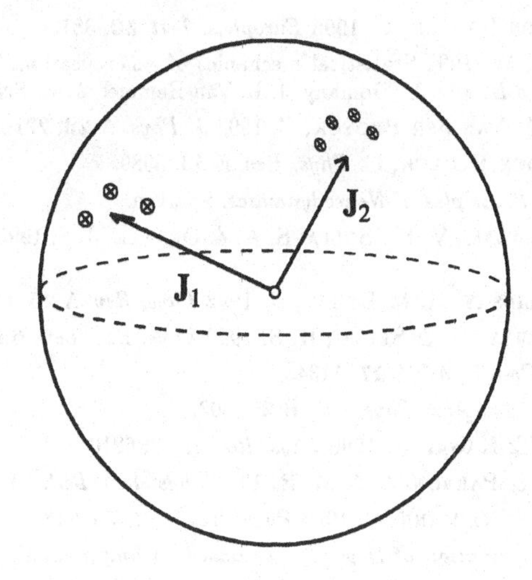

FIGURE 13. Schematic drawing of data grouped in two clusters.

6. Discussion

We have illustrated how the process of learning by examples can be formalized, studied and understood in terms of simple models. One encounters ingredients and phenomena reminiscent of the way we operate, such as the elimination of incompatible hypotheses from a hierarchy determined by the underlying architecture of the network and the presence of first and second order Eureka-like phase transitions. The learning process can be described as a competition between entropy (= log(number of configurations)) and the error (of a given configuration with respect to the target). It is thus not surprising that a technical formalism, similar to that of equilibrium statistical mechanics, can be set up. The existence of a formal framework also allows us to study the influence of special effects such as various learning algorithms, the effect of noise, dilution of synapses, rejection, queries, inclusion of surprise and self-evaluation, etc.. The results obtained through the study of these specific models can be compared with the general results obtained within the context of a statistical mechanical description, or derived from other fields such as statistics, see Vapnik (1982), or computational learning theory, see Valiant (1984).

REFERENCES

BEX, G. J., SERNEELS, R. & VAN DEN BROECK, C. 1995 *Phys. Rev.* E **51**, 6309.

BOUTEN, M., SCHIETSE, J. & VAN DEN BROECK, C. 1995 *Phys. Rev.* E **52**, 1958.

ENGEL, A. 1994 *Mod. Phys. Lett.* B **8**, 1683.

CARNEVALI, P. & PATARNELLO, S. 1987 *Europhys. Lett.* **4**, 1199.

GARDNER, E. & DERRIDA, B. 1989 *J. Phys.* A **22**, 1983.

GYÖRGYI, G. 1990 *Phys. Rev.* A **41**, 7097.

GYÖRGYI, G. & TISHBY, N. 1990 *Neural Networks and Spin Glasses* (ed. W. K. Theumann & R. Koberle), pp. 3–36. World Scientific.

HAUSSLER, D., LITTLESTONE, N., & WARMUTH, M. K. 1990 Predicting {0,1}-functions on randomly drawn points. Technical report, University of California, Santa Cruz.

HERTZ, J., KROGH, A. & PALMER, R. G. 1991 *Introduction to the Theory of Neural Computing.* Addison-Wesley.

LOOTENS E. & VAN DEN BROECK, C. 1995 *Europhys. Lett.* **30**, 381.

OPPER, M. & KINZEL, W. 1994 Statistical mechanics of generalisation. To appear in *Physics of Neural Networks III* (ed. E. Domany, J. L. Van Hemmen & K. Schulten).

PARRONDO, J. M. R. & VAN DEN BROECK, C. 1993 *J. Phys.* A **26**, 2211.

REIMANN, P. & VAN DEN BROECK, C. *Phys. Rev.* E **53**, 3989.

ROSENBLATT, F. 1962 *Principles of Neurodynamics.* Spartan.

SCHWARTZ, D. B., SAMALAM, V. K., SOLLA, S. A. & DENKER, J. S. 1990 *Neural Computation* **2**, 374.

SEUNG, H. S., SOMPOLINSKY, H. & TISHBY, N. 1992 *Phys. Rev.* A **45**, 6056.

SOMPOLINSKY, H., TISHBY, N. & SEUNG, H. S. 1990 *Phys. Rev. Lett.* **65**, 1683.

VALIANT, L. G. 1984 *Comm. ACM* **27**, 1134.

VAN DEN BROECK, C. 1994 *Acta Phys. Pol.* B **25**, 903.

VAN DEN BROECK, C. & KAWAI, R. 1990 *Phys. Rev.* A **42**, 6210.

VAN DEN BROECK, C. & PARRONDO, J. M. R. 1993 *Phys. Rev. Lett.* **71**, 2355.

VAN DEN BROECK, C. & REIMANN, P. 1996 *Phys. Rev. Lett.* **76**, 2188.

VAPNIK, V. N. 1982 *Estimation of Dependences Based on Empirical Data.* Springer.

WATKIN, T. L. H., RAU, A. & BIEHL, M. 1993 *Rev. Mod. Phys.* **65**, 499.

See also

BIEHL, M. & MIETZNER, A. 1993 Statistical mechanics of unsupervised learning. *Europhys. Lett.* **24**, 421; see also 1994 *J. Phys.* A **27**, 1885.

WATKIN, T. L. H. & NADAL, J.-P. 1994 Optimal unsupervised learning. *J. Phys.* A **27**, 1899.

Synthesis of chaos

By ANTONÍN VANĚČEK AND SERGEJ ČELIKOVSKÝ

Institute of Information Theory and Automation, Academy of Sciences of the Czech Republic, Prague, Czech Republic

The virtues of chaotic behaviour are listed, and the problem of synthesis of chaos stated. They are then demonstrated on examples provided first in R^3 by Chua, and then in R^n, $n \geq 3$, by the present authors. A scenario for the synthesis of chaos is given. The classical root locus is useful for both approaches to chaos synthesis. The trajectory root locus is introduced to both analyse and control chaotic behaviour.

1. The virtues of chaos

Until recently, chaotic behaviour was perceived negatively. In the rare cases in which analysis was extended, the goal of synthesis was an introduction of a control which would lead to a decay of such behaviour, see Baillieul *et al.* (1980) and Aleksejev & Loksutov (1987).

We take as fundamental the change of the paradigm from a negative to a positive perception of chaos as the outcome of a behaviour, see Babloyantz *et al.* (1985), Goldberger *et al.* (1986), Garfinkel (1987), Scarda & Freeman (1987), Freeman (1988) and Ottino (1989).

The reasons for such a new paradigm are:

(*a*) Chaos makes possible better absorption of energy and motion, see Garfinkel (1987).

(*b*) Chaos makes mixing possible, see Ottino (1989).

(*c*) Chaos is distinguished by spectral reserve with a spectrum of type $\frac{1}{f}$. This is resistive with respect to resonance, see Goldberger (1987).

(*d*) Chaos in biological systems is substantially connected with healthy activity – contrary to periodic behaviour and the point attractor, connected with the end of all activity, see Goldberger (1987), Goldberger (1989) and Scarda & Freeman (1987).

(*e*) Chaos is very closely connected with new fundamental tools for the description of Nature – fractals, self-similarity and renormalization groups: *chaos acts mainly as an organizing principle*, see Freeman (1988).

(*f*) Chaos is useful for a new encryptic modulation scheme, using at the sender end a synthetized chaos which modulates a signal, and at the receiver end a modulator identical with chaos. After synchronization, or after observing the state of the second chaos generator, the signal is deciphered, see Pecora & Carroll (1991).

(*g*) Chaos is a bounded behaviour which can be a reasonable substitute for the stabilization to a single stable equilibrium point, see Čelikovský & Vaněček (1992).

2. Synthesis of chaos as a control on a chaotic attractor

Contrary to the linear case, nonlinear systems may have complicated structure of their invariant sets: several equilibria, robust limit cycles, tori (quasiperiodic motion) or chaotic attractors.

Seven years ago, the first author formulated the problem of stabilization of a chaotic attractor as a new part of nonlinear control systems theory, i.e. the stabilization to some special bounded subset of the state space, see Vaněček (1989).

We model a general control system as a dynamical system with parameters

$$\dot{x} = f_p(x), \quad x \in R^n, \quad p \in R^m \text{ (or } p \in C^q).$$

Our task is to design such $f_\cdot(\cdot)$, and later find such p that, after some transient behaviour, the trajectories of $\dot{x} = f_p(x)$ for some large set of initial conditions

$$x(0) \in \{\alpha\}$$

approach the chaotic attractor. As the working definition of the chaotic attractor we use the following: it is the trajectory after some transient behaviour, defined for $t \in [t_a, \infty)$, which is bounded and has at least one positive Lyapunov exponent defined on the interval in question. This set does not cover all R^n, counter-examples being stable submanifolds near all equilibria.

In a narrower sense, $f_p(\cdot)$ is realizable by passive and active electronic components with given technical tolerancies.

3. The synthesis of chaos in R^3 according to Chua and Matsumoto

We quote Leon O. Chua (Berkeley) in Chua (1992) about his synthesis of chaos: 'Prior to 1983, the conspicious absence of a reproductible functioning chaotic circuit or system seems to suggest that chaos is a pathological phenomenon that can exist only in mathematical abstractions, and in computer simulation of contrived equations. Consequently, electrical engineers in general, and nonlinear circuit theorists in particular, have heretofore paid little attention to a phenomenon which many had regarded as an esotheric curiosity. ... Suddently it dawned to upon me, that since the main mechanism which gives rise to chaos, in both the Lorenz and the Roessler Equations, is the presence of at least two *unstable* equilibrium points – 3 for the Lorenz Equations and 2 for the Roessler Equations – it seems only prudent to design a simpler and more robust circuit having these attributes.

Having identified this alternative approach and strategy, it becomes a simple exercise in elementary circuit theory ... to enumerate systematically *all* such circuit candidates, of which there were only 8, and then systematically eliminate these that, for one reason or another, cannot be chaotic. This simple exercise quickly led to two contenders, which ... finally led me to the circuit' – later called Chua's circuit.

Chua's circuit gave rise to hundreds of papers, two books, Parker & Chua (1989) and Matsumoto *et al.* (1994), and one journal, Chua (from 1991).
Chua's equations read:

$$\begin{bmatrix} \dot{x}_1 \\ \dot{x}_2 \\ \dot{x}_3 \end{bmatrix} = \begin{bmatrix} -\alpha & \alpha & 0 \\ 1 & -1 & 1 \\ 0 & \beta & 0 \end{bmatrix} \begin{bmatrix} x_1 \\ x_2 \\ x_3 \end{bmatrix} + \begin{bmatrix} \varphi(x_1) \\ 0 \\ 0 \end{bmatrix},$$

where the scalar nonlinearity $\varphi(x_1)$ is the piecewise linear function:

$$\varphi(x_1) = \begin{cases} bx_1 - a + b, & x_1 \le -1, \\ ax_1, & |x_1| < 1, \\ bx_1 + a - b, & x_1 \ge 1, \end{cases}$$

and $\alpha, \beta, a, b \in R$ are parameters.

Figure 1 shows that in each of three equilibria some poles are stable, some unstable i.e. every equilibrium is unstable. During the trajectory evolution, the root locus of Chua's system is switched between the pole configurations.

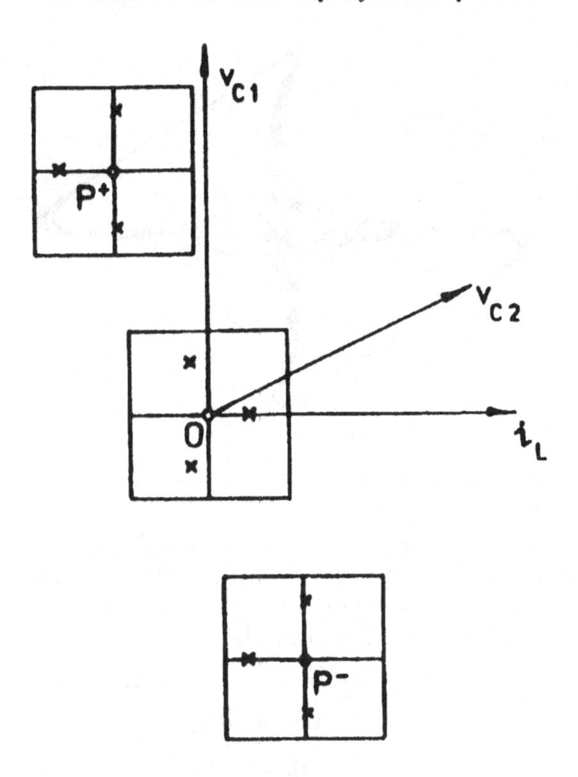

FIGURE 1. Three roots for each of three equilibria of Chua's circuit.

4. Synthesis of chaos in R^3 according to Čelikovský and Vaněček

To synthetize the chaotic behaviour, we started from a working knowledge of bilinear systems, i.e. the following control systems with state x and input u:

$$\dot{x} = Ax + Bxu, \quad u = Cx, \quad x \in R^3, \quad u \in R.$$

Using elementary geometric arguments, the following two systems were derived from the bilinear system:

$$\frac{dx}{dt} = \begin{bmatrix} \mu & 0 & 0 \\ 0 & \rho & \omega \\ 0 & -\omega & \rho \end{bmatrix} x + \left([\, c_1 \quad c_2 \quad c_3 \,]x\right)^2 E\, x,$$

$$\frac{dx}{dt} = \begin{bmatrix} \lambda_1 & 0 & 0 \\ 0 & \lambda_2 & 0 \\ 0 & 0 & \lambda_3 \end{bmatrix} x + \left([\, c_1 \quad c_2 \quad c_3 \,]x\right) E\, x,$$

where:

$$E = \begin{bmatrix} 0 & -c_3 & c_2 \\ c_3 & 0 & -c_1 \\ -c_2 & c_1 & 0 \end{bmatrix} \in or_3(R),$$

$\mu > -\rho > 0, \quad -\lambda_2 > \lambda_1 > -\lambda_3 > 0,$
$or_3(R)$ is the matrix algebra of infinitesimal rotations in R_3;

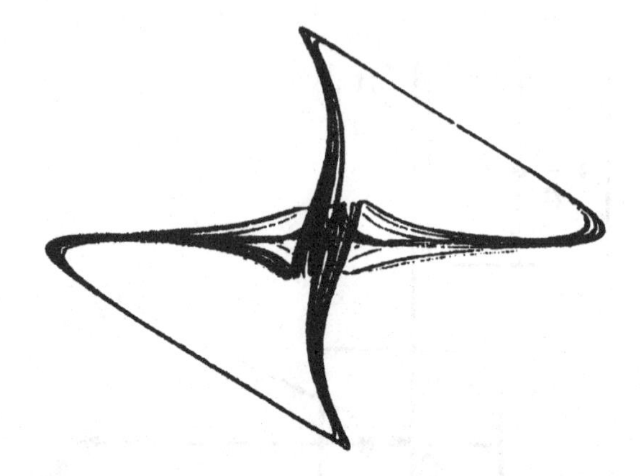

FIGURE 2. New type of attractor for $\rho = -0.8$, $\omega = 1$, $\mu = 1$, $\alpha = 1.20437$.

$c_1 = \cos\alpha, \quad c_2 = \sin\alpha, \quad c_3 = 0;$
$\alpha \in [0, \pi/2]$ is the angle between the x_1-axis (the unstable eigenvector of the approximate linearization) and the c-axis of the infinitesimal rotation.
The solution near equilibrium is approximated by:

$$x_i^{\mathrm{ne}}(t) = x_i^{\mathrm{ne}}(0)\exp(s_i t), \quad i = 1, 2, 3.$$

The solution far from equilibrium is approximated by:

$$x^{\mathrm{fe}}(t) = cx_0^{\mathrm{fe}} \exp(Et)x_0^{\mathrm{fe}},$$

where $Et \in Or_3(R)$ and $Or_3(R)$ is the orthogonal matrix group of rotations in R^3. It acts in R^3 as the rotation around the c'-axis by the angle t.

The parameter $\alpha \in [0, \pi/2]$ has a clear geometrical interpretation: it is the angle between the x_1-axis and the axis of the infinitesimal rotation defined by the skew-symmetric matrix of the nonlinear part of our equation.

Orbits of the purely nonlinear part (i.e. described by the equation with deleted linear part) are periodic. They are the circles around c'-axis, perpendicular to it. The angular velocity along these circles increases linearly along the rotation axis.

It is clear (due to the quadratic character of the purely nonlinear part) that sufficiently far from the origin in the direction of the rotation axis, the influence of the linear part is negligible as compared to the purely nonlinear part. As a consequence, we may assume that far from the origin along the c'-direction, the orbits may be qualitatively approximated by the above computed orbits of the purely nonlinear part.

The synergy of these two types of behaviour is very complicated, since the influence of the linear behaviour decreases only gradually.

The parameter causing the homoclinicity is the parameter α.

The system makes it possible to give a natural geometrical explanation of its behaviour. There is an approximate linearization (or linear part) that expands along the x_1-axis and contracts along the $x_{2,3}$-axes. Locally, around the origin, the system behaves in such a way that all trajectories are pushed away along the x_1-axis far from the vicinity of the origin sooner or later.

The main strongly nonlinear effect is the following: since the rotation axis lies in the semistable (x_1, x_2)-plane, the nonlinear part rotates the trajectories initially going along the antistable x_1-axis so that after some time they get closer to the (x_2, x_3)-plane (formed

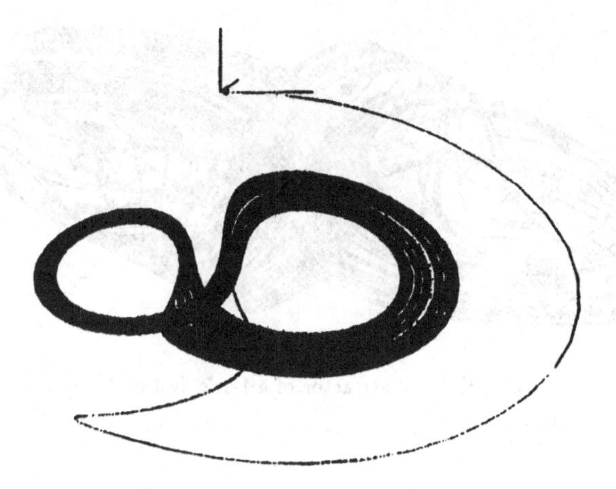

FIGURE 3. New type of attractor for $\lambda_1 = 8$, $\lambda_2 = -16$, $\lambda_3 = -1$, $\alpha = 1.5707 \to \pi/2$.

by the stable directions of the system). For a carefully selected $\alpha \in [0, \pi/2]$ this causes trajectories to approach the origin along this stable plane, which in turn leads to the existence of the homoclinic orbit. For nearby values of the parameter α, a complicated behaviour may then be expected.

See Čelikovský & Vaněček (1994) for more details.

5. Synthesis of chaos in R^n $(n \geq 3)$ according to Vaněček and Čelikovský

A second class of control systems considered by the authors were the so-called Lur'e control systems, single input, single output linear systems with static nonlinear scalar feedback. In the time domain:

$$\dot{x} = Ax + B\varphi(Cx), \quad x \in R^n, \quad n \geq 3, \quad \varphi(Cx) \in R.$$

The feedback $\varphi(\cdot)$ is to be nonlinear, static, odd, strictly monotonous, smooth (or at least piecewise smooth). The simplest such nonlinearity (up to the calibration) is $\varphi(y(t)) = -y^3(t)/3$. (The quintic φ was also considered as well the feedbacks $\arctan(y) - y$ and $\tanh(y) - y$.)

In the complex frequency domain:

$$Y(s) = \frac{g(s - z_1)(s - z_2) \cdots (s - z_{n-1})}{(s - s_1)(s - s_2) \cdots (s - s_n)} U(s), \quad U(s) = \varphi(Y(s)) = -\varphi(-Y(s)).$$

For the state space realization of the nonlinear circuit being an interconnection of the linear dynamical part and the static nonlinear feedback, the canonical linear Frobenius observable form was used:

$$F_1 = (-1)^{n-1} s_1 s_2 \cdots s_n, \ \ldots, \ F_n = s_1 + s_2 + \cdots + s_n,$$

$$G_1 = (-1)^{n-2} g\, z_1 \cdots z_{n-1}, \ \ldots, \ G_{n-1} = -g\,(z_1 + \cdots + z_{n-1}), \ G_n = g,$$

$$\begin{bmatrix} \dot{x}_1 \\ \dot{x}_2 \\ \vdots \\ \dot{x}_n \end{bmatrix} = \begin{bmatrix} 0 & \ldots & 0 & F_1 \\ 1 & & & F_2 \\ & \ddots & & \vdots \\ 0 & \ldots & 1 & F_n \end{bmatrix} \begin{bmatrix} x_1 \\ x_2 \\ \vdots \\ x_n \end{bmatrix} - \begin{bmatrix} G_1 \\ G_2 \\ \vdots \\ G_n \end{bmatrix} \frac{x_n^3}{3}.$$

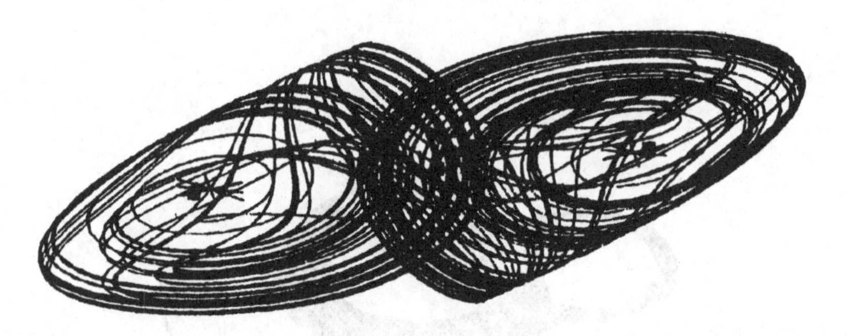

FIGURE 4. Chaotic attractor of a Lur'e system for $n = 6$.

We used the classical Root Locus Method of Bode and Evans which originated from the the stability analysis with respect to the change of gain, taken as the slope of some differentiable nonlinear static characteristic with the change of the working point. Here, the working points are not only three equilibria but also any of other solutions, parametrized by the slope of the characteristic, i.e. by $-x_n^2$. The equation of the root locus, parametrized by x_n^2, is

$$(s-s_1)(s-s_2)\cdots(s-s_n)+x_n^2\,g\,(s-z_1)(s-z_2)\cdots(s-z_{n-1})=0.$$

For $x_n = 0$, the roots are $s_1, s_2, \ldots, s_n$; for x_n improper, the roots are $z_1, z_2, \ldots, z_{n-1}$ and the improper root, i.e. the zeros and the improper zero of the transfer function. The condition of dissipativity, or the condition of the divergence of the vector field to be negative, gives

$$s_1+s_2+\cdots+s_n-x_n^2\,g<0.$$

Figure 4 shows the chaotic trajectories for $n = 6$.

The synthesis of chaos for the Lur'e system may be described as follows.

Scenario – Let us take:

(i) The linear system with single input and output, at least of the third order, having semistable poles (i.e. some stable, some unstable), hyperbolic (i.e. no poles on the imaginary axis), dissipative (i.e. the sum of poles is negative), and nonpotential (i.e. some poles having nonzero imaginary parts).

(ii) The feedback from the output to the input being nonlinear, static, odd, and strictly monotonous, giving rise, in addition to the central equilibrium, to the off-central equilibria.

(iii) The linear system zeros being attracting, according to the rules of the Root Locus Method (the gain being the slope of the nonlinearity), the central equilibrium poles and the off-central equilibria poles chosen in such a way that these off-central equilibria are again semistable, hyperbolic, nonpotential and dissipative.

Figure 5 shows the root locus used for synthesis of chaos for $n = 3, 4, 5, 6$. The reader is invited to synthetize other types of chaos, e.g. hyperchaos with two positive Lyapunov exponents.

See Vaněček & Čelikovský (1994a) and Vaněček & Čelikovský (1994b) for more details.

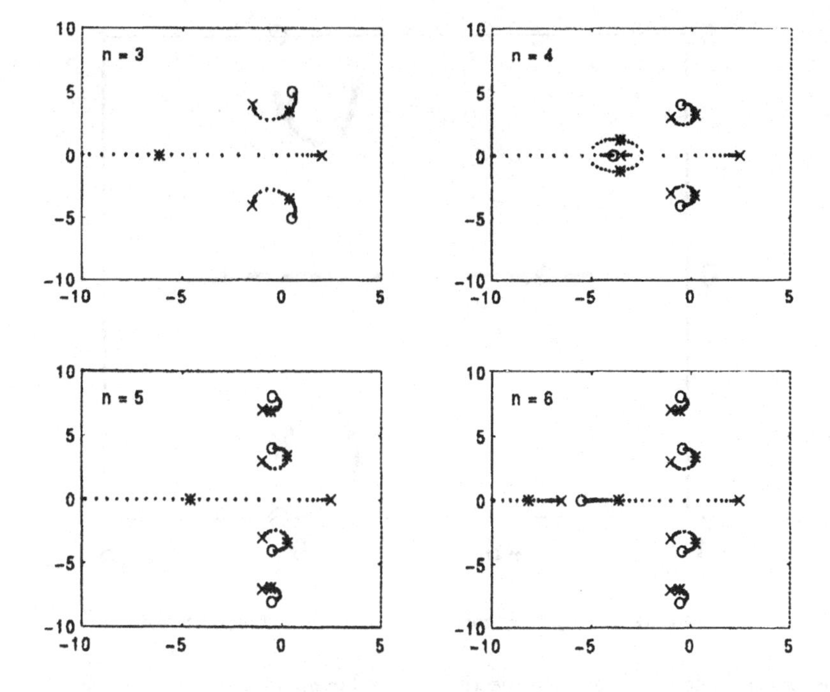

FIGURE 5. Root locus of a Lur'e system for $n = 3, 4, 5, 6$.

6. Trajectory root locus: from trajectories in R^n to trajectories in C

A new object, the trajectory root locus, will be introduced in this section. First, introduce the nonlinear systems described by

$$\dot{x} = f(x) = L(x)\,x + [\,f(x) - L(x)\,x\,],$$

$$L(x) = \mathrm{D}_x f(x), \quad x \in R^n,$$

$$s_i(x) = i\text{th eigenvalue of } L(x), \quad i = 1, 2, \ldots, n, \quad s_i(x) \in C.$$

The set $\{s_i(x(t)),\ i = 1, 2, \ldots, n,\ t \in R_+\}$ is called the root locus along the trajectory $\{x(t),\ t \in R_+\}$ or, briefly, the trajectory root locus. Notice that the equation for the fundamental matrix $\Psi(t)$ of the system $\dot{x} = f(x)$ linearized near the trajectory $x(t)$:

$$\dot{\Psi} = L(x)\Psi, \quad \Psi(0) = I,$$

$$p_i(x(t)) = i\text{th eigenvalue of } \Psi(x(t)) = \mu_i(x(t)) + \mathrm{i}\omega_i(x(t)), \quad i = 1, 2, \ldots, n$$

can also be used for computation of the Lyapunov exponents:

$$m_i = \lim_{t \to \infty} \frac{1}{t} \ln |\mu_i(t)|, \quad i = 1, 2, \ldots, n.$$

Nevertheless, the relation of our root locus to the Lyapunov coefficients of the trajectory is not straightforward, see e.g. Bylov *et al.* (1966).

Figure 6 shows the trajectory root locus for a Lur'e system for $n = 3$. It is part of the root locus used for the synthesis, see figure 4. During the trajectory evolution, the stability boundary, $\mathrm{Re}\, s = 0$, is systematically crossed. The segments are the multiple folds.

Figure 7 shows the trajectory root locus for the Lorenz system, see Anosov (1995).

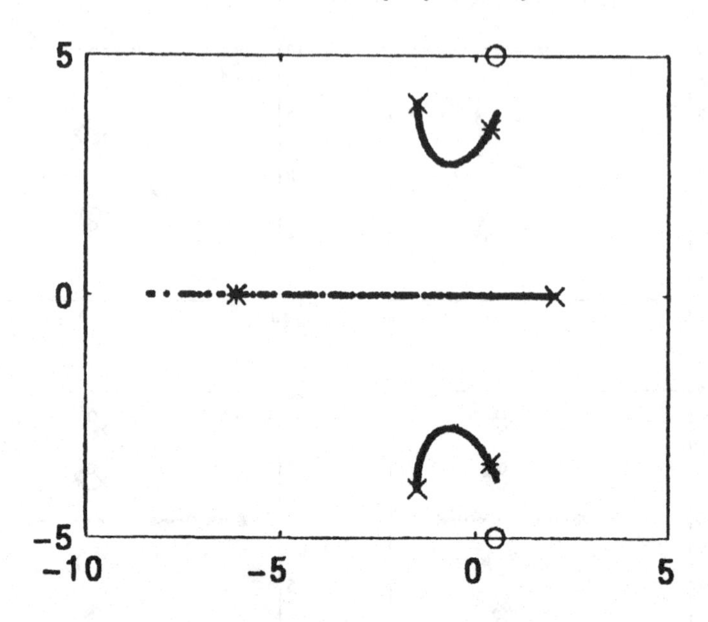

FIGURE 6. Trajectory root locus of a Lur'e system for $n = 3$.

(The bottom left is the usual projection of the Lorenz system trajectory from R^3 to $R^2 = C$.) Near the center equilibrium, in contrast to Chua's and Lur'e chaotic systems, the eigenvalues are real. In the vicinity of this equilibrium the trajectory root locus is located on the real axis. Notice again that the stability boundary is systematically crossed.

In what follows, the trajectory root locus will be used to extend the linear control theory to a strongly nonlinear one.

7. From linear to strongly nonlinear control theory – preliminaries

By the control of chaos one usually means the 'art' of using some feedback to transform the chaotic trajectories to the cyclic ones, see Infeld *et al.* (1996). For this purpose some *ad hoc* dynamic feedback is used as the derivative of output (in the presence of noise – *sic*!) or of time-delayed output (increasing the state space dimension from R^n to R^∞ – *sic*!).

To start some theory, we will try to extend the now classical state space theory of single input (dually single output) linear systems with the dyadic feedback with the trajectory root locus concept. For the linear system with a single input $u \in R$

$$\dot{x} = Fx + Gu, \quad u = Kx,$$

or eliminating u,

$$\dot{x} = (F + GK)\, x,$$

the accessibility condition

$$\operatorname{rank}\,[\,G \quad FG \quad \ldots \quad F^{n-1}G\,] = n$$

is the condition of shiftability of the eigenvalues

$$s_i(F)$$

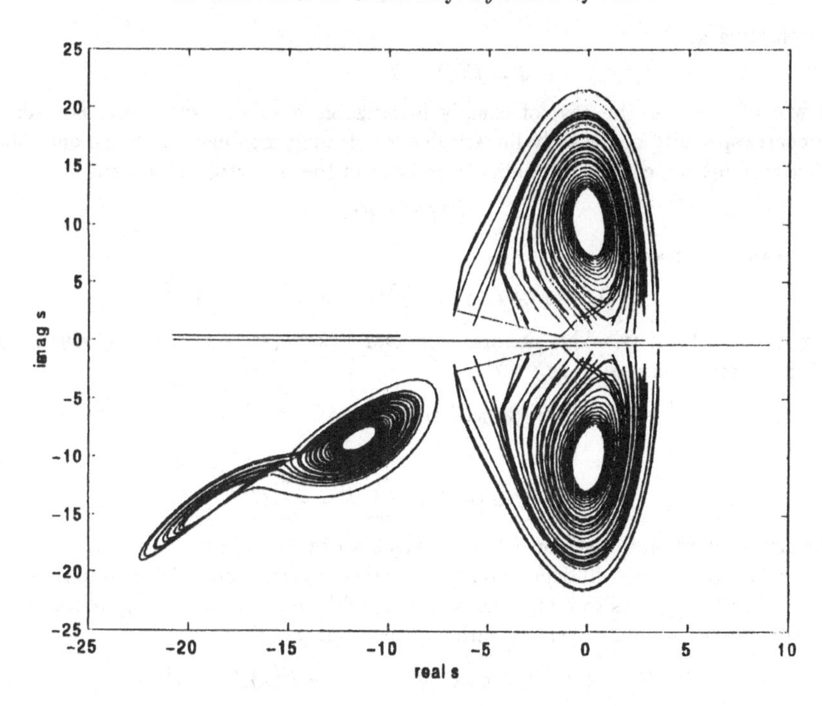

FIGURE 7. Trajectory root locus of a Lorenz system. Bottom-left: the trajectory.

to arbitrally chosen eigenvalues

$$s_i(F+GK), \quad i=1,2,\ldots,n.$$

The latter eigenvalues are chosen in such a way that the trajectories of $F+GK$ are stable and suitably damped. Both the former and the latter eigenvalues are the fixed points of the trajectory root locus. This suggests that the control of chaotic trajectories should be some extension of the fixed points to some more general fixed points of the trajectory root locus.

Dually, for the linear system with single output $y \in R$

$$y = Hx, \quad \dot{x} = Fx + Ly \quad \text{or} \quad \dot{x} = (F+LH)x$$

the observability condition

$$\operatorname{rank}\begin{bmatrix} H \\ HF \\ \vdots \\ HF^{n-1} \end{bmatrix} = n$$

is the condition of shiftability of the eigenvalues

$$s_i(F)$$

to arbitrally chosen eigenvalues

$$s_i(F+LH), \quad i=1,2,\ldots,n.$$

Now the nonlinear extension with a single input $u \in R$ is

$$\dot{x} = F(x) + G(x)u, \quad u = K(x),$$

or eliminating u,

$$\dot{x} = F(x) + G(x)K(x).$$

(We will not discuss the case of exactly linearizable nonlinear systems.) For general nonlinear, especially not exactly linearizable, or strongly nonlinear systems, one should furthermore introduce the shiftability conditions of the trajectory root locus

$$s_i(\mathrm{D}_x F(x))(t)$$

to the trajectory root locus

$$s_i(\mathrm{D}_x[F(x) + G(x)K(x)])(t), \quad i = 1, 2, \ldots, n.$$

Using the results of (weakly) nonlinear control theory (see e.g. Isidori (1989)), these conditions can be expressed as

$$\operatorname{rank}[\,G(x) \quad \mathrm{ad}_F G(x) \quad \ldots \quad \mathrm{ad}_F^{n-1} G(x)\,] = n,$$

where

$$\mathrm{ad}_F G = [F, G] = \frac{\partial G}{\partial x} F - \frac{\partial F}{\partial x} G$$

is the so-called adjoint operator of the Lie algebra of vector fields.

Now, the shiftability will depend on the initial state. Typically, the chaotic trajectory starting at $x(0) = \alpha$ has to be tranformed to a stable limit cycle starting from α.

Dually, for the nonlinear system with single output $y \in R$ we have

$$y = H(x), \; \dot{x} = F(x) + L(x)y \quad \text{or} \quad \dot{x} = F(x) + L(x)H(x).$$

For general nonlinear systems one should furthermore introduce the shiftability conditions of trajectory root locus

$$s_i(\mathrm{D}_x F(x))(t)$$

to the trajectory root locus

$$s_i(\mathrm{D}_x[F(x) + L(x)H(x)])(t), \quad i = 1, 2, \ldots, n.$$

Dually, again using the results of (weakly) nonlinear control theory (see e.g. Isidori (1989)), these conditions can be expressed as

$$\operatorname{rank}\begin{bmatrix} \mathrm{d}H(x) \\ \mathrm{dL}_F H(x) \\ \vdots \\ \mathrm{dL}_F^{n-1} H(x) \end{bmatrix} = n,$$

where

$$\mathrm{dL}_F H = \langle\, \mathrm{d}H, F\,\rangle$$

is the so-called adjoint operator of the Lie algebra of vector fields, and $\langle\,\cdot\,,\,\cdot\,\rangle$ denotes the scalar product.

Clearly, an example of the control of chaos is the reverse of our chaos synthesis. First put in the inverse output static feedback to compensate the cubic nonlinearity. Second apply the linear static state feedback to shift the unstable poles.

The generalization of a part of the above Frobenius form of § 5 is given in Isidori (1989). It would be as useful for trajectory root locus shift as has been the classical Frobenius form for the linear case, as well as the duals of both forms.

As far as the control of chaos is concerned, there is another problem, solved again as an 'art'. This is the problem of chaos synchronization, which is closely related to the construction of the nonlinear observer. Even here, the art of synchronization, in fact the observer of the state of a chaotic system without the knowledge of the input (now the message to be decoded) should be put on some theoretical basis.

This work was supported by the Grant Agency of the Academy of Sciences of the Czech Republic, grant number 275403.

REFERENCES

ALEKSEJEV, V. V. & LOKSUTOV, A. JU. 1987 Control of system with strange attractor by means of periodical parametric control. *Dokl. Akad. Nauk SSSR.* **293**, 1346–1348.

ANOSOV, D. V., ED. 1995 *Dynamical Systems IX.* Encyclopaedia of Mathematical Sciences, vol. 66. Springer-Verlag. VINITI, 1991.

BABLOYANTZ, A., SALAZAR, J. M. & NICOLIS, C., 1985 Evidence of chaotic dynamics of brain activity during the sleep cycle. *Phys. Lett.* A **111**, 152–156.

BAILLIEUL, J., BROCKETT R. W. & WASHBURN, R. B. 1980 Chaotic motion in nonlinear feedback systems. *IEEE Trans. on Circuits Syst.* **CAS-27**, 990–997.

BYLOV, B. F. *et al.* 1966 *Theory of Lyapunov Exponents and its Applications to Stability Problems.* Nauka. In Russian.

ČELIKOVSKÝ, S. & VANĚČEK, A. 1992 Bilinear systems as strongly nonlinear systems. In *Systems Structure and Control,* pp. 264–267. Pergamon.

ČELIKOVSKÝ, S. & VANĚČEK, A. 1994 Bilinear systems and chaos. *Kybernetika* **30**, 403–424.

CHUA, L. O. 1992 The genesis of the Chua circuit. *AEÜ* **46**, 250–257.

CHUA, L. O., ED., From 1991 *International Journal of Bifurcations and Chaos in Applied Sciences and Engineering.*

FREEMAN, W. J. 1988 Strange attractors that govern mammalian brain dynamics shown by trajectories of EEG potential. *IEEE Trans. on Circuits Syst.* **CAS 35**, 791–783.

GARFINKEL, A. 1987 The virtues of chaos. *Beh. Brain Sci.* **10**, 178–179.

GOLDBERGER, A. R., BHARGAVA, V., WEST, B. J. & MANDELL A. J. 1986 Some observations on the question: Is ventricular fibrilation 'chaos'? *Physica* D **19**, 282–289.

GOLDBERGER, A.R. 1987 Nonlinear dynamics, fractals, cardiac physiology and sudden death. In *Temporal Disorder in Human Oscillatory Systems.* Springer.

GOLDBERGER, A. 1989 Why a steady heart may not be healthy. (Review of address given by A. Goldberger of the Harvard Medical School to the American Association for the Advancement of Science.) *New Scientist* **21**, 31.

INFELD, E. *et al.*, EDS. 1996 *Proc. Intl. Conf. on Nonlinear Dynamics, Chaotic and Complex Systems.* Zakopane, Poland, November 1995. Contributed papers. *J. Technical Phys.* **37**, **38**. In press.

ISIDORI, A. 1989 *Nonlinear Control Systems.* Springer.

MATSUMOTO, T., KOMURO, M., KOKUBU, H. & TOKUYNAGA, R. 1994 *Bifurcations. Sights, Sounds and Mathematics.* Springer.

OTTINO, J. M. 1989 The mixing of fluids. *Sci. Amer.* 56–67.

PARKER, T. S. & CHUA, L. O. 1989 *Practical Numerical Algorithms for Chaotic Systems.* Springer.

PECORA, L. M. & CARROLL, T. L. 1991 Synchronization in chaotic systems. *Phys. Rev. Lett.* **64**, 821–824.

SCARDA, A. & FREEMAN, W. J. 1987 How brains make chaos in order to make sense of the world. *Beh. Brain Sci.* **10**, 161–195.

VANĚČEK, A. 1989 Double scroll – chaotic attractor of nonlinear electrical circuit of the third order. *Fyzika a synergetika,* 178–187. In Czech.

VANĚČEK, A. & ČELIKOVSKÝ, S. 1994a Chaos synthesis via root locus. *IEEE Trans. on Circuits Syst.-I: Fundamental Theory and Applications* **41**, 59–60.

Vaněček, A. & Čelikovský, S. 1994b Synthesis of chaotic systems. *Kybernetika* **30**, 537–542.

See also

Chua, L. O., Komuro, M. & Matsumoto, T. 1986 The double scroll family. *IEEE Trans. on Circuits Syst.* **CAS-33**, 1073–1118.

Matsumoto, T. 1984 A chaotic attractor from Chua's circuit. *IEEE Trans. Circuits Syst.* **CAS-31**, 1055–1058.

Matsumoto, T., Chua, L. O. & Komuro, M. 1985 The double scroll. *IEEE Trans. Circuits Syst.* **32**, 798–812.

Sparrow, C. 1982 *The Lorenz Equations: Bifurcations, Chaos and Strange Attractors.* Springer.

Vaněček, A. 1991 Strongly nonlinear and other control systems. *Probl. Contr. Inf. Th.* **20**, 2–12.

Vaněček, A. & Čelikovský, S. 1993 Wrapped eigenstructure of chaos. *Kybernetika* **29**, 73–79.

Vaněček, A. & Čelikovský, S. 1996 *Control Systems: From Linear Analysis to Synthesis of Chaos.* Prentice.

Computational complexity of continuous problems

By HENRYK WOŹNIAKOWSKI

Department of Computer Science, Columbia University, New York, USA
and
Institute of Applied Mathematics, Warsaw University, Warsaw, Poland

Computational complexity studies the intrinsic difficulty of solving mathematically posed problems. Information-based complexity is a branch of computational complexity that deals with continuous problems defined on spaces of multivariate functions. For such problems, computations can only lead to approximate solutions. The complexity is defined as the minimal cost needed to compute an approximation with error at most ε. Error and cost can be defined in different settings, such as the worst case, average case, and probabilistic or randomized settings.

We survey recent results on complexity of linear multivariate problems and on path integration. In particular, we show that multivariate integration and approximation are (strongly) tractable in the average case setting, for the class of continuous functions equipped with the Wiener sheet measure. This means that their complexity is a polynomial in ε^{-1}.

We consider path integration for the Wiener measure in the worst case, and randomized settings. For the class of r times Frechet differentiable functions, the problem is intractable in the worst case setting, whereas it is tractable in the randomized setting, and the classical Monte Carlo algorithm is optimal. On the other hand, for the specific class of entire functions, the problem is tractable in the worst case setting, and its complexity is proportional to $\varepsilon^{-2/3}$.

1. Introduction

The goal of this paper is to introduce the reader to computational complexity. This is a relatively new and fast developing area of theoretical computer science. Computational complexity studies the intrinsic difficulty of solving mathematically posed problems.

To study complexity we must first define a model of computation. The model states which operations are allowed, what the cost of each operation is, and how computation is performed. Not surprisingly, complexity results depend on the model of computation, and sometimes an apparently innocent change of the model leads to a completely different complexity result.

In discrete computational complexity, the Turing machine model is usually assumed. Roughly speaking, in this model we operate on bits, the cost depends on the size of numbers, and we count how many bit operations are necessary to solve the problem. The Turing machine model is used for discrete problems, and there is a deep theory culminating in the famous question whether P$\neq$NP, see e.g. Garey & Johnson (1979).

In continuous computational complexity, we study continuous problems. Many scienific phenomena correspond to continuous problems. They are usually solved using fixed precision floating point arithmetic. The cost of floating point operations is independent of the size of the numbers. Furthermore, all arithmetic operations cost about the same to execute. If we ignore rounding errors, floating point arithmetic corresponds to the *real number* model of computation. That is why, for continuous problems, we usually choose the real number model and study computational complexity in this model. For the precise definition of the real number model the reader is referred to Blum *et al.* (1989) and Novak (1995).

Continuous computational complexity may be split into two branches. The first branch deals with problems for which the information is *complete.* Informally, information may be complete for problems which are specified by a finite number of inputs. Examples include matrix multiplication, and the solution of linear algebraic systems or systems of polynomial equations.

To illustrate this branch of continuous computational complexity, consider the problem of solving linear systems $Ax = b$, where the $n \times n$ matrix A and $n \times 1$ vector b are given. If n is not too large and the matrix A is dense, we input $n^2 + n$ data (the system coefficients, i.e. elements of A and b). Information is then complete.

What is the complexity of solving systems of linear equations? That is, what is the minimal number of arithmetic operations needed to solve $Ax = b$ for an arbitrary non-singular $n \times n$ matrix A and an arbitrary $n \times 1$ vector b? We do not know exactly the complexity. We only know bounds on it. The lower bound is given by the total number of data and is proportional to n^2. The upper bound is given by the cost of an algorithm that solves the problem. Classical algorithms compute the solution vector $x = A^{-1}b$ using $\Theta(n^3)$ arithmetic operations.† Examples of such algorithms include Gaussian elimination and Householder's method. However, we can do better. Strassen (1969) found an algorithm which computes the solution using $\Theta(n^{\log_2 7})$ arithmetic operations. Since $\log_2 7 = 2.81\ldots$, this yields a better upper bound, at least for large n. Today, the best known upper bound is due to Coppersmith & Winograd (1987), equal to $\Theta(n^{2.376})$. The constant in the theta notation of the latter bound is, unfortunately, huge.

We stress that problems with complete information may be very hard in the real number model. The first NP-complete problem over the reals was established in Blum *et al.* (1989). This is the problem of deciding whether a real polynomial of degree 4 in n variables has a real root. Hence, modulo the conjecture $\mathrm{P} \neq \mathrm{NP}$, but this time over the reals, the complexity of the latter problem is *not* polynomial in n.

The second branch of continuous computational complexity is *information-based complexity*, denoted for brevity as IBC. It deals with problems for which the information is *partial.* Typically, IBC studies problems whose input is an element of an infinite-dimensional space. Examples of such problems include multivariate integration or approximation, solution of ordinary or partial differential equations, integral equations, optimization, and solving non-polynomial equations. The input of such problems is often a multivariate function on the reals. Information is usually supplied by a subroutine which computes function values. Using this subroutine finitely many times, we know only partial information about the function. Typically, this partial information is *contaminated* with errors such as round-off errors or measurement errors. Thus, the available information is partial and/or contaminated. Therefore, the original problem can only be solved *approximately.* The goal of IBC is to compute such an approximation at minimal cost. The error and the cost of approximation can be defined in different settings, including the worst case, average case, probabilistic, randomized and mixed settings. The ε-complexity is then defined as the minimal cost of computing an approximation with error at most ε. The reader who wants to find more about IBC is referred to books and recent surveys: Heinrich (1993a), Novak (1988), Novak (1994) Plaskota (1996), Traub *et al.* (1988), Traub & Woźniakowski (1993), Werschulz (1991).

We believe that the readers of this proceedings are mainly interested in solving scientific problems for which only partial information is available. That is why in the rest of this paper we restrict ourselves to IBC issues. To make this paper self-contained, we presen

† By $\Theta(n^3)$ we mean a function which can be bounded from below and from above by a multiple of n^3.

an abstract formulation of IBC in § 2. This abstract formulation is illustrated by a simple example of scalar integration.

We then briefly survey recent results on complexity of linear multivariate problems in § 3. Many problems in science, engineering, economics and finance are modeled by multivariate problems involving functions of d variables with large or even huge d. For path integration, we even have $d = +\infty$; the approximation of path integrals yields multivariate integration with huge d.

We are interested in the complexity of linear multivariate problems in various settings. In particular, the complexity depends on the error parameter ε, and on the number of variables d.

In the worst case setting, it is known that many problems are intractable. More specifically, for many problems the complexity is an exponential function of d. This means that for large d the complexity is so huge that it is impossible to solve the problem. This is sometimes called the *curse of dimension.*

We stress that the exponential dependence on d is a complexity result and it is impossible to get around it by designing efficient algorithms. The only way to break the curse of dimension is to weaken the notion of error and/or cost. This can sometimes be done by switching from the worst case setting to another setting. Hence, we wish to examine how the complexity depends on ε and d in other settings. If the dependence is polynomial in d and ε^{-1}, the curse of dimension is broken.

For a given setting, we say that a linear multivariate problem is *tractable* if its complexity depends polynomially on d and ε^{-1}. It is called *strongly tractable* if its complexity is independent of d and depends polynomially on ε^{-1}. There are some general results characterizing which linear multivariate problems are tractable or strongly tractable, see Woźniakowski (1994a). In particular, multivariate integration and approximation are strongly tractable in the average case setting for the class of continuous functions equipped with the Wiener sheet measure. Specific complexity bounds are given in § 3.

The final section deals with path integration, see Wasilkowski & Woźniakowski (1996a) and Wasilkowski & Woźniakowski (1996b). Usually Monte Carlo algorithms are used to approximate path integrals. We study deterministic algorithms in the worst case setting. Then path integration is tractable (i.e. its complexity is polynomial in ε^{-1}) if the class of integrands consists of entire functions. Finite smoothness of integrands is not enough if the measure of the path integration problem is supported on an infinite dimensional subspace. In this case, the classical Monte Carlo algorithm is almost optimal in the randomized setting. We conclude with a remark on Feynman–Kac path integrals.

2. Basic concepts of IBC†

In this section we present an abstract formulation of IBC and illustrate it by a simple example. A proof technique which leads to tight complexity bounds for some problems will also be indicated. Let

$$S : F \to G,$$

where F is a subset of a linear space and G is a normed linear space over the real or complex field. We wish to approximate $S(f)$ for all f from F.

Let $U(f)$, where $U : F \to G$, denote a computed approximation to $S(f)$ for $f \in F$. We now explain how the approximation U can be constructed. To do this we first need to discuss the concept of information.

The basic assumption of IBC is that, in general, we do not have full knowledge of

† This section is based on § 2 of Traub & Woźniakowski (1991).

an element f since typically f is a multivariate function and it cannot be represented exactly on a digital computer. Instead, it is assumed that we can gather some knowledge about f by computations of the form $L(f)$, where $L : F \to H$ for some set H.

Let Λ denote a class of permissible information operations L. That is, $L \in \Lambda$ if and only if $L(f)$ can be computed for each f from F. For example, if F is a set of functions then Λ is often taken as a set of L consisting of function evaluations, $L(f) = f(x)$, $\forall f \in F$, for some x from the domain of f. Such Λ is denoted by Λ_{std}. If the class Λ is taken as a set of all linear functionals L then it is denoted by Λ_{all}. Let

$$N(f) = \left[L_1(f), L_2(f), \dots, L_n(f)\right], \quad L_i \in \Lambda, \quad \forall f \in F, \tag{2.1}$$

be the computed information about f. We stress that the L_i as well as the number n can be chosen adaptively. That is, the choice of L_i may depend on the already computed $L_1(f), L_2(f), \dots, L_{i-1}(f)$. The number n may also depend on the computed $L_i(f)$. (This permits arbitrary termination criteria.)

$N(f)$ is called the information about f, and N the information operator. In general, N is many-to-one, and thus knowing $y = N(f)$ it is impossible to recover the element f. For this reason, the information N is called *partial*.

The approximation $U(f)$ is constructed by combining the computed information $N(f)$. That is, $U(f) = \phi(N(f))$, where $\phi : N(F) \to G$. A mapping ϕ is called an algorithm. The approximation U can thus be identified with the pair (N, ϕ), where N is an information operator and ϕ an algorithm that uses the information N.

We illustrate these concepts by an example.

Example. Integration.
Let F be a class of functions $f : [0,1] \to \mathbb{R}$ that satisfy the Lipschitz condition with constant q,

$$|f(x) - f(y)| \le q\,|x - y|, \quad \forall\, x, y \in [0,1].$$

Let $G = \mathbb{R}$ and

$$S(f) = \int_0^1 f(t)\,\mathrm{d}t.$$

The class Λ is a collection of $L : F \to \mathbb{R}$, such that for some x from $[0,1]$, $L(f) = f(x)$, $\forall f \in F$. The information N is given by

$$N(f) = \left[f(x_1), f(x_2), \dots, f(x_n)\right]$$

with the points x_i and the number n adaptively chosen. The approximation U is now of the form $U(f) = \phi(N(f)) = \phi(f(x_1), f(x_2), \dots, f(x_n))$. An example of an algorithm ϕ is a quadrature given by $U(f) = \phi(N(f)) = \sum_{i=1}^{n} a_i\, f(x_i)$ for some numbers a_i.

We now present a model of computation. It is defined by two postulates:

• We are charged for each information operation. That is, for every $L \in \Lambda$ and for every $f \in F$, the computation of $L(f)$ costs c, where c is positive and fixed, independent of L and f.

• Let Ω denote the set of permissible combinatory operations including the addition of two elements in G, multiplication by a scalar in G, arithmetic operations, comparison of real numbers, and evaluations of certain elementary functions. We assume that each combinatory operation is performed exactly with unit cost.

In particular, this means that we use the real number model, where we can perform operations on real numbers exactly and at unit cost.

We now discuss the cost of the approximations $U(f) = \phi(N(f))$. Let $\text{cost}(N, f)$ denote the cost of computing the information $N(f)$. Note that $\text{cost}(N, f) \ge c\,n$, and the inequality may occur since adaptive selection of L_i and n may require some combinatory operations. If $N(f)$ cannot be computed by using n information operations and a finite number of operations from Ω, then $\text{cost}(N, f) = +\infty$.

Knowing $y = N(f)$, we compute $U(f) = \phi(y)$ by combining the information $L_i(f)$. Let $\text{cost}(\phi, y)$ denote the number of combinatory operations from Ω needed to compute $\phi(y)$. If $\phi(y)$ cannot be computed by using a finite number of operations from Ω, then $\text{cost}(\phi, y) = +\infty$.

The cost of computing $U(f)$, $\text{cost}(U, f)$, is given by

$$\text{cost}(U, f) = \text{cost}(N, f) + \text{cost}(\phi, N(f)).$$

We now define the concepts of error and cost of the approximation U. The definitions of error and cost depend on the setting. We first discuss three settings: worst case, average case and probabilistic. Then we turn to a randomized setting.

In the worst case setting, the error and cost of U are defined as

$$e(U) = \sup_{f \in F} \|S(f) - U(f)\|,$$
$$\text{cost}(U) = \sup_{f \in F} \text{cost}(U, f).$$

In the average case and probabilistic settings, we assume that the set F is equipped with a probability measure μ. In the average case setting the error and cost of U are defined as

$$e(U) = \left(\int_F \|S(f) - U(f)\|^2 \mu(\mathrm{d}f) \right)^{1/2},$$
$$\text{cost}(U) = \int_F \text{cost}(U, f)\, \mu(\mathrm{d}f).$$

In the probabilistic setting, we assume that we are given a number $\delta \in [0, 1]$, and the error and cost of U are defined as

$$e(U) = \inf \left\{ \sup_{f \in F - A} \|S(f) - U(f)\| : A \text{ such that } \mu(A) \leq \delta \right\}.$$
$$\text{cost}(U) = \sup_{f \in F} \text{cost}(U, f).$$

We now discuss a randomized setting. In this setting the approximation U is defined by a random selection of information and algorithm. More precisely, let ρ be a probability measure on a set T. Then for each $t \in T$ we select information N_t and an algorithm ϕ_t, and compute $U_t(f) = \phi_t(N_t(f))$. Here t is a random variable distributed according to the measure ρ. Random information N_t is of the form (2.1) with randomly chosen L_i and n. A random algorithm is $\phi_t : N_t(F) \to G$. The approximation U can now be identified as the 4-tuple, $U = (N, \phi, T, \rho)$.

The error of U in the randomized setting is defined as

$$e(U) = \sup_{f \in F} \int_T \|S(f) - U_t(f)\|\, \rho(\mathrm{d}t),$$

In the randomized setting, the cost of $U_t(f) = \phi_t(N_t(f))$ is defined as above, and then the cost of U is defined as

$$\text{cost}(U) = \sup_{f \in F} \int_T \text{cost}(U_t, f)\, \rho(\mathrm{d}t).$$

We illustrate the randomized setting by continuing the integration example.

Example. (continued) Consider the classical Monte Carlo algorithm

$$U_t(f) = \frac{1}{n}\sum_{i=1}^{n} f(t_i),$$

with uniformly distributed points t_i. That is, $t = [t_1, t_2, \ldots, t_n] \in T = [0,1]^n$ and ρ is the uniform distribution over the unit n dimensional cube. In this case,

$$N_t(f) = \big[f(t_1), f(t_2), \ldots, f(t_n)\big]$$

is random information with randomly chosen points t_i and deterministically chosen n. The algorithm ϕ_t is deterministic and equal to $\phi_t(y_1, y_2, \ldots, y_n) = \frac{1}{n}\sum_{i=1}^{n} y_i$. The error of U is proportional to $n^{-1/2}$ and the cost of U is proportional to n.

We are ready to define the computational complexity of IBC problems. The basic notion is the ε-complexity, which is defined as the minimal cost of *all* U with error at most ε,

$$\text{comp}(\varepsilon) = \inf\big\{\text{cost}(U) : \; U \text{ such that } e(U) \le \varepsilon\big\}.$$

Here we use the convention that the infimum of the empty set is infinity.
Depending on how $e(U)$ and $\text{cost}(U)$ are specified, this defines ε-complexity in each of the four settings discussed above.

We stress that we take the infimum over *all* possible U for which the error does not exceed ε. In the worst case, average case and probabilistic settings, U can be identified with the pair (N, ϕ), where N is the information and ϕ is the algorithm that uses that information. This means that we take the infimum over *all* information N consisting of information operations from the class Λ, and over *all* algorithms ϕ that use N such that (N, ϕ) computes approximations with error at most ε. In the randomized setting, U can be identified with the 4-tuple (N, ϕ, T, ρ) and we take the infimum over *all* random information N_t and *all* random algorithms ϕ_t, where $t \in T$ is distributed accordingly to an arbitrary probability measure ρ. Sometimes we write

$$\text{comp}_{\text{wor}}(\varepsilon), \quad \text{comp}_{\text{avg}}(\varepsilon), \quad \text{comp}_{\text{prob}}(\varepsilon, \delta), \quad \text{comp}_{\text{ran}}(\varepsilon)$$

to emphasize the setting and the dependence on the parameter δ in the probabilistic setting. If we want to stress that we use one of the deterministic settings we then say, for example, 'the worst case deterministic setting' or 'the average case deterministic setting' and write $\text{comp}_{\text{wor,det}}(\varepsilon)$ or $\text{comp}_{\text{avg,det}}(\varepsilon)$.

Example. (continued) For the integration problem, the model of computation assumes that one function evaluation costs c, and each arithmetic operation, comparisons of real numbers, and evaluations of certain elementary functions can be performed exactly at unit cost. Usually $c \gg 1$.
The worst case ε-complexity for the class F of Lipschitz functions with constant q is

$$\text{comp}_{\text{wor}}(\varepsilon) = \Theta(\varepsilon^{-1} c q) \quad \text{as } \varepsilon \to 0.$$

For the average case and probabilistic settings, assume that μ is a truncated classical Wiener measure placed on the first derivatives. Then in the average case setting we have

$$\text{comp}_{\text{avg}}(\varepsilon) = \Theta(\varepsilon^{-1/2} c) \quad \text{as } \varepsilon \to 0.$$

In the probabilistic setting, for $q \gg \ln(1/\delta)$, we have

$$\text{comp}_{\text{prob}}(\varepsilon, \delta) = \Theta\Big(\varepsilon^{-1/2} c \sqrt[4]{\ln(1/\delta)}\Big) \quad \text{as } \varepsilon \to 0.$$

Finally, in the randomized setting we have

$$\text{comp}_{\text{ran}}(\varepsilon) = \Theta(\varepsilon^{-2/3} c) \quad \text{as } \varepsilon \to 0.$$

The complexity of integration in different settings has been studied for various classes of functions by many researchers, see Novak (1988) and Traub *et al.* (1988) for a list of references.

One of the main goals of IBC is to find or estimate the ε-complexity, and to find an ε-complexity optimal U, or equivalently, an ε-complexity optimal pair (N, ϕ). In the randomized setting, we want to find an ε-complexity optimal 4-tuple (N, ϕ, T, ρ). By *ε-complexity optimality* of U we mean that the error of U is at most ε and the cost of U is equal to, or not much greater than, the ε-complexity. For a number of problems this goal has been achieved due to the work of many researchers.

We briefly indicate a proof technique often used to obtain tight bounds on computational complexity of IBC problems. In what follows, we restrict ourselves to the worst case setting although a similar approach can be used in other settings.

As already explained, the approximation $U(f)$ is computed by combining information operations from the class Λ. Let $y = N(f)$ denote this computed information. In general, the operator N is many-to-one, and therefore the set $N^{-1}(y)$ consists of many elements from F which are indistinguishable from f. Then the set $S(N^{-1}(y))$ consists of all elements from G which are indistinguishable from $S(f)$. Since $U(f)$ is the same for any f from the set $N^{-1}(y)$, the element $U(f)$ must serve as an approximation to any element g from the set $SN^{-1}(y)$. It is clear that the quality of the approximation $U(f)$ depends on the 'size' of the set $SN^{-1}(y)$.

The intuitive notion of size can be formalized by using the concept of radius. The radius of the set $A = SN^{-1}(y)$ is defined as the smallest radius of the ball which contains A,

$$\operatorname{rad}(A) = \inf_{g \in G} \sup_{a \in A} \|a - g\|.$$

The *radius of information* $r(N)$ is then defined as the maximal radius of the set $SN^{-1}(y)$ for $y \in N(F)$,

$$r(N) = \sup_{y \in N(F)} \operatorname{rad}\Big(SN^{-1}(y)\Big).$$

Clearly, the radius of information $r(N)$ is a sharp lower bound on the worst case error of any U. We can guarantee an ε-approximation if and only if $r(N)$ does not exceed ε (modulo a technical assumption that the corresponding infimum is attained).

The cost of computing $N(f)$ is at least $c\,n$, where n, called the *cardinality* of N, denotes the number of information operations in N. By the *ε-cardinality number* $m(\varepsilon)$ we mean the minimal number n of information operations for which the information N has radius $r(N)$ at most ε,

$$m(\varepsilon) = \min\big\{n : \text{ there exists } N \text{ of cardinality at most } n \text{ such that } r(N) \le \varepsilon\big\}.$$

From this we obtain a lower bound on the ε-complexity,

$$\operatorname{comp}_{\text{wor}}(\varepsilon) \ge c\, m(\varepsilon).$$

It turns out that for many problems it is possible to find an information operator N_ε consisting of $m(\varepsilon)$ information operations, and a mapping ϕ_ε such that the approximation $U(f) = \phi_\varepsilon(N_\varepsilon(f))$ has error at most ε and $U(f)$ can be computed with cost at most $(c+2)\,m(\varepsilon)$. (For examples of such problems see Traub *et al.* (1988), chapters 5 and 7). This yields an upper bound on the ε-complexity,

$$\operatorname{comp}_{\text{wor}}(\varepsilon) \le (c+2)\,m(\varepsilon).$$

Since usually $c \gg 1$, the last two inequalities yield almost the exact value of the ε-

complexity,

$$\mathrm{comp}_{\mathrm{wor}}(\varepsilon) \simeq c\,m(\varepsilon).$$

This also shows that the pair $(N_\varepsilon, \phi_\varepsilon)$ is almost ε-complexity optimal.

In each setting of IBC one can define a radius of information such that we can guarantee an ε-approximation if and only if $r(N)$ does not exceed ε. This permits one to sometimes obtain tight complexity bounds in other settings.

The essence of this approach is that the radius of information as well as the ε-cardinality number $m(\varepsilon)$ and the information N_ε do not depend on particular algorithms, and they can often be expressed entirely in terms of well known mathematical concepts. Therefore we can sometimes obtain tight complexity bounds by drawing on powerful mathematical results.

3. Linear multivariate problems†

In this section we discuss complexity of linear multivariate problems. By a linear multivariate problem we mean an approximation of a linear operator defined on functions f of d variables. More precisely, let F_d be a class of functions $f: [0,1]^d \to \mathbb{R}$, and let

$$S_d : F_d \to G_d,$$

where G_d is a normed linear space.

We wish to approximate $S_d(f)$ for $f \in F_d$. Two primary examples are *multivariate integration*,

$$S_d(f) = \int_{[0,1]^d} f(t)\,dt \quad \text{with } G_d = \mathbb{R},$$

and *multivariate approximation*,

$$S_d(f) = f \quad \text{with } G_d = L_2\Big([0,1]^d\Big),$$

with F_d being a class of functions that are continuously differentiable r times.

As in § 2, the cost of one function evaluation (or one evaluation of $L(f)$) is denoted by c. To stress the dependence on the number d of variables, we write $c = c(d)$.

We are particularly interested in the complexity for large d and/or in large ε^{-1}. To stress the dependence on the error parameter ε and on the number of variables d, we denote the complexity by $\mathrm{comp}(\varepsilon, d)$.

Many multivariate problems are *intractable* and their complexity grows exponentially with the number of variables d. This is sometimes called the *curse of dimension*. Typically,

$$\mathrm{comp}(\varepsilon, d) = \Theta\Big(\varepsilon^{-d/r} c(d)\Big) \quad \text{as } \varepsilon \to 0,$$

where r stands for the smoothness of the functions in the class F_d.

Problems which suffer the curse of dimension in the worst case setting include integration, approximation, global optimization, integral and partial differential equations for classes of functions whose rth derivatives are uniformly bounded in L_∞, see Bakhvalov (1959), Heinrich (1993b), Nemirovsky & Yudin (1983), Novak (1988), Pereverzev (1989), Traub *et al.* (1988) and Werschulz (1991).

In the average case and randomized settings, the curse of dimension is present for approximation over the class of functions with r continuous derivatives which is equipped

† This section is based on § 3 of Woźniakowski (1995).

with the folded isotropic Wiener measure, see Ritter & Wasilkowski (1996) and Wasilkowski (1993) for the average case, and Mathé (1993), Novak (1992) and Wasilkowski (1989) for the randomized setting.

For some problems we can break the curse of dimension by switching to a different setting. For example, in the randomized setting, it is well known that the classical Monte Carlo algorithm breaks the curse of dimension for multivariate integration. In the average case setting, the curse of dimension is broken for multivariate integration no matter what probability measure is given on the class of functions. However, in general, the proof is not constructive. For the Wiener sheet measure, the proof is constructive and we know almost optimal algorithms, see Wasilkowski & Woźniakowski (1995), Woźniakowski (1991). For multivariate approximation, the curse of dimension is broken only for some probability measures. For instance, it is broken for the Wiener sheet measure, see Wasilkowski & Woźniakowski (1995), Woźniakowski (1992), however, as already mentioned, it is not broken for the isotropic Wiener measure, see Ritter & Wasilkowski (1996), Wasilkowski (1993).

It seems natural to characterize which multivariate problems are *tractable* or *strongly tractable* in various settings. More precisely, we say that the multivariate problem is *tractable* if there exist nonnegative numbers K, p and q such that

$$\mathrm{comp}(\varepsilon, d) \leq K\, c(d)\, d^q\, \varepsilon^{-p}, \quad \forall\, d, \forall\, \varepsilon \leq 1. \tag{3.2}$$

If $q = 0$ then we say that the multivariate problem is *strongly tractable*. For strongly tractable problems, the only dependence of the complexity on d is through the cost $c(d)$.

Tractability and strong tractability of linear multivariate problems have been studied in Woźniakowski (1994a) for the information classes Λ_{std} and Λ_{all}. In the worst case and randomized settings we assume that the domain F_d and the range of S_d are Hilbert spaces. In the average case and probabilistic settings we assume that F_d is a Banach space equipped with a Gaussian measure μ_d and that the range of S_d is a Hilbert space.

For the class Λ_{all}, necessary and sufficient conditions for tractability and strong tractability can be obtained by using known IBC results on complexity of linear problems. They are expressed in terms of singular values of S_d or in terms of eigenvalues of the covariance operator of the measure $\mu_d S_d^{-1}$. Roughly speaking, tractability and strong tractability hold if the singular values tend to zero sufficiently fast.

Tractability and strong tractability in the randomized setting and the worst case setting are equivalent, and the corresponding complexities differ only by constants. This follows easily from Novak (1992). Similarly, tractability and strong tractability in the probabilistic setting and the average case setting are equivalent due to relations between these two settings for linear problems, see Traub *et al.* (1988).

We stress that for the class Λ_{all} the construction of an ε-approximation with minimal cost is easy since we know the optimal choice of linear functionals, and that linear algorithms are optimal.

We now turn to the class Λ_{std}. Under mild assumptions, we prove in Woźniakowski (1994a) that tractability and strong tractability in the classes Λ_{std} and Λ_{all} are equivalent. In particular, we prove that the exponents in ε^{-1} may differ by at most two. The proof of this equivalence is, however, *not* constructive.

One may suspect that only trivial problems are strongly tractable. However, even in the worst case setting, this is not true. More precisely, if F_d is a unit ball of a reproducing kernel Hilbert space and the linear problem is suitably normalized, then there exists a constant K such that

$$\mathrm{comp}(\varepsilon, d) \leq K\, c(d)\, \varepsilon^{-p},$$

where $p = 2$ for the class $\Lambda_{\rm all}$, and $p = 4$ for the class $\Lambda_{\rm std}$, see Woźniakowski (1994a). It is also known that $p = 2$ for the class $\Lambda_{\rm all}$ is sharp, whereas it is open whether $p = 4$ for the class $\Lambda_{\rm std}$ can be improved.

As before, the proof for the class $\Lambda_{\rm std}$ is not constructive. A construction is known for linear multivariate problems that are defined by tensor products, see Wasilkowski & Woźniakowski (1995a), Woźniakowski (1994b). For tractable tensor product problems and for the class $\Lambda_{\rm std}$, we construct polynomial-time algorithms, see Wasilkowski & Woźniakowski (1995a). This construction is based on Smolyak's algorithm, see Smolyak (1963). More precisely, in the worst case and average case settings, we present linear algorithms that compute an ε-approximation for the multivariate tensor product problem with cost

$$\mathrm{cost}(\varepsilon, d) \le (c(d) + 2)\beta_1 \left(\beta_2 + \beta_3 \frac{\ln(1/\varepsilon)}{d-1}\right)^{\beta_4(d-1)} \left(\varepsilon^{-1}\right)^{\beta_5}.$$

The coefficients β_i do not depend on d; they are determined by the properties of the problem for $d = 1$.

Note the intriguing dependence of the cost bound on d. The leading term $\varepsilon^{-\beta_5}$ does not depend on d, whereas $\ln(1/\varepsilon)$ is divided by a multiple of $d - 1$ and then raised to a multiple of $d - 1$. If the tensor product problem is tractable then the cost bound does not exceed $c(d)K\varepsilon^{-p}$ for some numbers K and p, both independent of d.

We illustrate the results for multivariate approximation and integration for the class $\Lambda_{\rm std}$ in the average case setting for the class of continuous functions $f : [0,1]^d \to \mathbb{R}$ equipped with the Wiener sheet measure. For the approximation problem, we know a linear algorithm, see Wasilkowski & Woźniakowski (1995), that computes an ε-approximation with cost

$$\mathrm{cost}(\varepsilon, d) \le c(d)\, 0.8489 \left(2.9974 + 4.3869\, \frac{-0.9189 + \ln(1/\varepsilon)}{d-1}\right)^{2(d-1)} \varepsilon^{-2}.$$

This algorithm has optimal powers of ε^{-1} and $\ln(1/\varepsilon)$ since

$$\mathrm{comp}(\varepsilon, d) = \Theta\left(\varepsilon^{-2} \left(\ln(1/\varepsilon)\right)^{2(d-1)}\right),$$

see Woźniakowski (1992). This approximation problem is strongly tractable since

$$\mathrm{cost}(\varepsilon, d) \le c(d)\, 2.37632\, \varepsilon^{-5.672}.$$

The exponent 5.672 seems to be too high; however, no smaller exponent has been found so far.

We would like again to add that the choice of the Wiener sheet measure is essential. It is known, see Wasilkowski (1993), that if we replace the Wiener sheet measure by the isotropic Wiener measure then the approximation problem is intractable since $\mathrm{comp}(\varepsilon, d) = \Theta\left(c(d)\,\varepsilon^{-2d}\right)$.

Consider now the integration problem $S_d f = \int_{[0,1]^d} f(x) dx$ in the average case setting for the class of continuous functions $f : [0,1]^d \to \mathbb{R}$ equipped with the Wiener sheet measure. Then we know a linear algorithm, see Wasilkowski & Woźniakowski (1995), which computes an ε-approximation with $\mathrm{cost}(\varepsilon, d)$ bounded by

$$\mathrm{cost}(\varepsilon, d) \le c(d)\, 3.304 \left(1.77959 + 2.714\, \frac{-1.12167 + \ln(1/\varepsilon)}{d-1}\right)^{1.5(d-1)} \varepsilon^{-1}.$$

The power of ε^{-1} is optimal and the power of $\ln(1/\varepsilon)$ is too large since

$$\mathrm{comp}(\varepsilon, d) = \Theta\left(\varepsilon^{-1} \left(\ln(1/\varepsilon)\right)^{(d-1)/2}\right),$$

see Woźniakowski (1991). This integration problem is strongly tractable since

$$\mathrm{cost}(\varepsilon, d) \le c(d)\, 7.26\, \varepsilon^{-2.454}.$$

The exponent 2.454 is too high. There exists an algorithm with an exponent at most 1.4788..., see Wasilkowski & Woźniakowski (1996c). The proof of this latter fact is, however, *not* constructive.

This integration problem is related to discrepancy in the L_2-norm, see Woźniakowski (1991). Using this relation we obtain an upper bound, which is independent of d, for the number $n(\varepsilon, d)$ of points for which discrepancy (with unequal weights) is at most ε,

$$n(\varepsilon, d) \le 7.26\, \varepsilon^{-2.454}, \quad \forall\, d, \forall\, \varepsilon \le 1.$$

4. Path integration

Path integrals occur in many applied fields including quantum physics and chemistry, differential equations, and financial mathematics, as well as average case complexity. The path integration problem is defined as the approximation of

$$S(f) = \int_X f(x)\, \mu(\mathrm{d}x), \quad \forall\, f \in F.$$

Here, X is a separable infinite dimensional Banach space and μ is a zero mean Gaussian measure on X. The class F is a class of (Borel) measurable real functions defined on X.

A typical approach is to approximate the path integral by high dimensional integrals and apply a Monte Carlo (randomized) algorithm. Do we really need to use randomized algorithms for path integrals? Perhaps we can find an effective *deterministic* algorithm that approximates path integrals with small error. To answer this question, we study the *worst case complexity* of path integration in the class Λ^{std}. Path integration is considered with respect to different Gaussian measures μ and different classes F of integrands.

Tractability of path integration means that the complexity depends polynomially on ε^{-1}. For the class F of integrands that are r times Frechet differentiable, tractability of path integration holds iff the covariance operator of the Gaussian measure μ has finite rank. Hence, if the Gaussian measure μ is supported on an infinite dimensional space then path integration is intractable. In this case, there exists no effective deterministic algorithm, and the use of randomized algorithms is reasonable. In fact, for this class of integrands, the classical Monte Carlo algorithm is optimal and the complexity in the randomized setting is proportional to ε^{-2}, see Wasilkowski & Woźniakowski (1996a).

On the other hand, for a particular class F of entire integrands, the worst case complexity of path integration is at most of order ε^{-p} with p depending on the Gaussian measure μ. Hence, path integration is now tractable. Furthermore, for any Gaussian measure μ, the exponent p is less than or equal to 2. For the Wiener measure we have $p = 2/3$. For this class of entire integrands, we provide effective deterministic algorithms that solve the path integration problem with (worst case) cost that is usually much less than the (randomized) cost of the classical Monte Carlo algorithm, see Wasilkowski & Woźniakowski (1996a).

In Wasilkowski & Woźniakowski (1996b) we consider a class of functions related to the Feynman–Kac formula. More precisely, this is the class of potential and initial conditions functions that define the heat equation. Although these functions do not need to be very smooth, we prove tractability of path integration, and in many cases, the worst case complexity is substantially smaller than ε^{-2}.

We thank L. Plaskota and A. G. Werschulz for useful comments on this paper.

REFERENCES

BAKHVALOV, N. S. 1959 On approximate calculation of integrals. *Vestnik MGU, Ser. Mat. Mekh. Astron. Fiz. Khim.* **4**, 3–18. In Russian.

BLUM, L., SHUB, M. & SMALE, S. 1989 On a theory of computation and complexity over the real numbers: NP-completeness, recursive functions and universal machines. *Bull. Amer. Math. Soc.* **21**, 1–46.

COPPERSMITH, D. & WINOGRAD, S. 1987 Matrix multiplication via arithmetic progression. In *Proc. of the Nineteenth ACM Symp. on Theor. of Comp.* 1–6.

GAREY, M. R. & JOHNSON, D. S. 1979 *Computers and Intractability: A Guide to the Theory of NP-Completeness.* Freeman.

HEINRICH, S. 1993a Random approximation in numerical analysis. In *Proc. Functional Analysis Conference,* Essen, 1991. Lecture Notes in Pure and Applied Mathematics (ed. K.D. Bierstedt *et al.*) vol. 150, pp. 123–171. Marcel Dekker.

HEINRICH, S. 1993b Complexity of integral equations and relations to s-numbers. *J. Complexity* **9**, 141–153.

MATHÉ, P. 1993 Random approximation of Sobolev embedding. *J. Complexity* **7**, 261–281.

NEMIROVSKY, A. S. & YUDIN, D. B. 1983 *Problem Complexity and Method Efficiency in Optimization.* Wiley.

NOVAK, E. 1988 *Deterministic and Stochastic Error Bounds in Numerical Analysis.* Lectures Notes in Math., vol. 1349. Springer.

NOVAK, E. 1992 Optimal linear randomization methods for linear operators in Hilbert spaces. *J. Complexity* **8**, 22–36.

NOVAK, E. 1994 Algorithms and complexity for continuous problems. In *Geometry, Analysis, and Mechanics* (ed. J. M. Rassias), pp. 96–128. World Scientific.

NOVAK, E. 1995 The real number model in numerical analysis. *J. Complexity* **11**, 57-73.

PEREVERZEV, S. V. On the complexity of the problem of finding solutions of Fredholm equations of the second kind with differentiable kernels. *Ukrain. Mat. Sh.* **41**, 1422–1425. In Russian

PLASKOTA, L. 1996 *Noisy Information and Computational Complexity.* Cambridge University Press.

RITTER, K. & WASILKOWSKI, G. W. 1996 Integration and L_2-approximation: average case setting with isotropic Wiener measure for smooth functions. *Rocky Mount. J. Math.* To appear.

SMOLYAK, S. A. 1963 Quadrature and interpolation formulas for tensor products of certain classes of functions. *Dokl. Akad. Nauk SSSR*, pp. 240–243.

STRASSEN, V. 1969 Gaussian elimination is not optimal. *Numer. Math.* **13**, 354–356.

TRAUB, J. F., WASILKOWSKI, G. W. & WOŹNIAKOWSKI, H. 1988 *Information-Based Complexity.* Academic Press.

TRAUB, J. F. & WOŹNIAKOWSKI, H. 1991 Theory and applications of information-based complexity. In *1990 Lectures in Complex Systems* (ed. L. Nadel & D. Stein). Santa Fe Institute Lect., vol. III, pp. 163–193. Addison-Wesley.

TRAUB, J. F. & WOŹNIAKOWSKI, H. 1993 Recent progress in information-based complexity. *Bulletin of EATCS* **51**, 141–154.

WASILKOWSKI, G. W. 1989 Randomization for continuous problems. *J. Complexity* **5**, 195-218.

WASILKOWSKI, G. W. 1993 Integration and approximation of multivariate functions: average case complexity with isotropic Wiener measure. *Bull. Amer. Math. Soc. (N.S.)* **28**, 308–314.

WASILKOWSKI, G. W. & WOŹNIAKOWSKI, H. 1995 Explicit cost bounds of algorithms for multivariate tensor product problems. *J. Complexity* **11**, 1–56.

WASILKOWSKI, G. W. & WOŹNIAKOWSKI, H. 1996a On tractability of path integration. *J. Math. Physics.* To appear.

WASILKOWSKI, G. W. & WOŹNIAKOWSKI, H. 1996b Worst case complexity of Feynman–Kac path integration. To appear.

WASILKOWSKI, G. W. & WOŹNIAKOWSKI, H. 1996c The exponent of discrepancy is at most 1.4778.... To appear.

WERSCHULZ, A. G. 1991 *The Computational Complexity of Differential and Integral Equations: An Information-Based Approach.* Oxford University Press.

WOŹNIAKOWSKI, H. 1991 Average case complexity of multivariate integration. *Bull. Amer. Math. Soc. (N.S.)* **24**, 185–194.

WOŹNIAKOWSKI, H. 1992 Average case complexity of linear multivariate problems: Part I. Theory, Part II. Applications. *J. Complexity* **8**, 337–372, 373–392.

WOŹNIAKOWSKI, H. 1994a Tractability and strong tractability of linear multivariate problems. *J. Complexity* **10**, 96–128.

WOŹNIAKOWSKI, H. 1994b Tractability and strong tractability of multivariate tensor product problems. *J. Computing and Information* **4**, 1–19.

WOŹNIAKOWSKI, H. 1995 Overview of information-based complexity. In *Lectures in Applied Mathematics* (ed. J. Renegar, M. Shub & S. Smale). To appear.

Stochastic differential geometry in finance studies

By VLADIMIR G. MAKHANKOV

Center for Nonlinear Studies, Los Alamos National Laboratory, Los Alamos, USA

A geometrical theory has been developed to describe the term structure of interest rates. The theory is based on arbitrage-free pricing of zero-coupon bonds in the case of a complete market and the equations of Stochastic Differential Geometry. By using properties of the internal space of the system spanned by a variety of bond maturities, the drift term in the pricing equation for bonds is shown to define a curvature of the state space.The volatility matrix defines its metric. So that finance markets are to be classified with respect to the curvature of the state space they generate:

(a) Euclidean (pure diffusional) and

(b) non-Euclidean (risk-adverse) markets.

A set of models is proposed using a martingale Euclidean fiber bundle (a bundle of frames). An exact solution for a one-factor discrete maturity model was found and used for checking out a computer experiment facility. The theory predicts, and computer experiments already perfomed reveal, all possible types of term-structure existing in reality, along with a possible instability of the financial market.

---

1. Objectives

Our aim is to restrict the freedom in description of the interest rate term-structure and estimate the parameters of models applied to the evolution of interest rates. A principally new approach is suggested to evolve in time rates, prices, and their volatilities which is based on exploiting the geometrical structure of the multi-dimensional state space (internal space of the system) *generated by the variety of bond maturities*. In this way a new arbitrage-free theory of interest rate term structure is developed. This theory incorporates the approach of Heath, Jarrow & Morton (HJM), see Heath *et al.* (1992) and the geometric picture of stochastic motion on Riemannian manifolds, and allows one to construct *nonlinear models* with a different number of factors (Wiener generators).

By implementing the arbitrage free requirement, HJM succeeded in reducing the freedom in Ito's description of interest rate dynamics to one two-dimensional function, the volatility $\sigma(t, T)$. Our theory provides equations for volatility dynamics that make it, to the author's knowledge, the first known term structure theory which is self-consistently complete. The freedom is reduced to a specification of initial and boundary conditions.

Stochastic differential equations are obtained which describe the stochastic behaviour of the interest rates along with the volatility matrix in the multi-factor approach. We construct a multi-dimensional state space with curvature, expressed through the volatility matrix and the drift term, such that the volatility structure acquires a sound geometrical meaning. The drift term in the Ito equation will be shown to fall into two parts. The first, called an 'external drift', is determined by external forces, and the second one, a 'geometrical drift' is inherent in the system and defined solely by its volatility structure. The external drift always defines some connexion curvature. The geometrical drift can be either Euclidean or non-Euclidean. The nature of the geometry can be determined by calculating the connexion curvature. If the curvature vanishes, we deal with a Euclidean state space which implies martingale properties of the process considered. In the absence of the external drift, the volatility matrix defines the geometry of the phase space of the system entirely. As a result, the problem of obtaining dynamical equations is reduced to purely geometric considerations.

In the first part of the paper, we derive the stochastic equations on a manifold defined by the volatility. The second part illuminates the mathematics of the Stochastic Differential Geometry (SDG). In the third part, we give an exact solution for the single factor discrete model and finally discuss results of some preliminary computer experiments, see Makhankov *et al.* (1995).

2. Current status of the problem

The explosive growth of the over-the-counter (OTC) derivatives and several nearly disastrous events occurring in the last few years have raised questions as to the stability of the market. The very instruments' nature makes them flexible tools for managing risk and cash flows. Since, unlike listed financial instruments traded on an exchange and hence having a real and known price, they do not trade publicly, their fair value is difficult to evaluate without relying on complex computational models. Research of the theory and methods that determine the pricing and market parameters to which these instruments are sensitive, is needed to manage risk as well as managing portfolios of these instruments. Derivatives exist across all market segments: equity, fixed income and commodities. They take the form of future and forward contracts, options and swaps. Often the final instrument is really a combination of elements from each of these groups. The global nature of these instruments raises the possibility that a failure in one economy may propagate to another. Cross-border structured contracts also involve a certain degree of political risk exposure. In these cases, unilateral action by a sovereign state can disrupt the integrity of a structured products enlarging risk. Probably the most encouraging feature of derivatives, allowing their thorough study, is that complex ones can be decomposed into a set of elemental instruments (contracts) which are more easily studied and standardized.

However, the current models *available to practitioners* are of unknown quality and accuracy. This causes any current scheme to evaluate derivatives and analyze, measure and neutralize risk in derivative portfolios to become an art on the verge of a science. We underline that the level of opaqueness and uncertainty surrounding derivatives presents a threat to the stability of the economy. A need exists for a better understanding of this threat.

The recent theoretical understanding allows one to develop a generalized pricing methodology for these instruments. What is needed are techniques for modeling the underlying processes of the various markets. The necessary approaches in this direction lead to solving problems that are nonlinear and complex in nature, requiring computationally intensive models. The nature and principles arrived at in this way have deep parallels in physics.

The area that needs the most research is the identification of the underlying stochastic processes and their interdependence. This research impacts not only the evaluation of derivatives, but also that of more elemental securities such as bonds and stocks and thus shed light not only on the derivative markets, but also on emerging markets and their interaction with established ones. In this way the most difficult problem to solve is the description of the so-called term structure of interest rates, i.e. the dependence of interest rates on the term of the bond. Determination of correct interest rates is an important consideration in both swap and option valuation.

There are several 'conventional' (econometric and arbitrage-free) term-structure models of bond pricing which use interest rates as underlying (proxy) variables, e.g. the one-factor models of Cox *et al.*, Vasicek and Heath *et al.*, see Hull (1993), and the two-factor model of Brennan & Schwartz (1979). These have their own advantages and shortcomings, see Ingber (1990). These models are supposed to be valid under certain assumptions, in particular: (i) 'efficiency of the market' and (ii) bond prices are functions of a number of state variables that follow Markov processes. Here we have to stress that, in recent years, there has been a 'hot' discussion on what is the source of randomness in financial markets: a low-dimensional dynamic chaos, conventional noise (Markovian processes), or both. A number of contradictory works have been published on this subject. However, recent results seem to have pointed to the absence, in the time series studied, of any underlying dynamic chaos bearing low-dimensional structure (of the kind of a low-dimensional dynamical system). At least no reliable evidence of dynamical chaos has been obtained yet, see Alexander & Giblin (1993) and also Ingber (1990). It is probably too early to make the final point, for some dynamical correlations could appear when considering some models for certain markets and short range epochs. Anyway, this makes us quite certain that Markov process techniques are now most appropriate.

Below we constuct a theory of interest rate term-structure on the basis of two known approaches:

(*a*) Risk-neutral arbitrage-free pricing of the discount bond, see Heath *et al.* (1992) and Hull (1993).

(*b*) Equations of Stochastic Differential Geometry, see, e.g. Kendal (1987).

3. Financial structures and mathematical counterparts

Let us first consider the financial issue. In order to construct a mathematical description of a financial phenomenon, we need to establish a kind of correspondence between certain financial categories and notions and their mathematical counterparts in the form of functions and equations. It is now believed that such a correspondence exists for at least a number of entities and categories such as a risk-neutral pricing process for simple derivatives (Black–Scholes equation, see Hull (1993)), interest rate term structure efficiency of the market, its stochasticity, completeness and arbitrage-free property (see Table 1). We also give a short survey of the models. There are two kinds: econometric (also known as equilibrium models) and arbitrage-free models.

DEFINITION 3.1. *The instantaneous forward rate $F(t,T)$ $(t \le T)$ is defined by*

$$F(t,T) = -\frac{\partial \ln P(t,T)}{\partial T}, \quad 0 < t \le T, \tag{3.1}$$

where $P(t,T)$ is the zero-coupon bond price at time t with principal \$1, maturing at $t = T$ so that $P(t,t) = 1$.

DEFINITION 3.2. *The spot rate of interest $R(t,T)$ is defined as*

$$R(t,T) = -\frac{\ln P(t,T)}{T-t}, \quad 0 < t \le T, \tag{3.2}$$

so that

$$P(t,T) = \mathrm{e}^{-R(t,T)(T-t)} \tag{3.3}$$

and hence $R(t_0,T)$ is a yield curve.

Furthermore we denote by $r(t)$ the *short term interest rate*, and by $\nu_p(t,T)$ the *volatility* of $P(t,t)$ corresponding to the pth component of an n-dimensional vector *Wiener process* $\mathrm{d}\vec{W} = \{\mathrm{d}W^1, \mathrm{d}W^2, \dots, \mathrm{d}W^n\}$.

FINANCIAL (market property)	MATHEMATICAL
1. efficiency of the market	stochastic differential equation: $\mathrm{d}X = a(X,t)\,\mathrm{d}t + \sigma(X,t)\,\mathrm{d}W$
2. arbitrage-freedom	restriction on the functions $a(X,t)$ and $\sigma(X,t)$
3. completeness	$F(t,T_i) = \sum_{j=1}^{N} C_j F(t,T_j)$
4. fair game rule	$\delta\sigma(X,t) = 0$

TABLE 1. Interplay of financial and mathematical entities and relations

Definitions 3.1 and 3.2 imply the following relation:

$$F(t,T) = R(t,T) + (T-t)\frac{\partial R(t,T)}{\partial T} \tag{3.4}$$

or

$$R(t,T) = \frac{1}{T-t}\int_t^T F(t,\tau)\,\mathrm{d}\tau,$$

with boundary condition at $t = T$:

$$F(t,t) = R(t,t) = r(t). \tag{3.5}$$

3.1. *Econometric models*

A basic variable is the short range interest rate $r(t)$. A general description is an Ito process (efficiency of the market assumed):

$$\mathrm{d}r(t) = m(r,t)\,\mathrm{d}t + s(r,t)\,\mathrm{d}W$$

(one-factor models).

The freedom in the description is given by two 2-dimensional unknown functions: a drift $m(r,t)$ and a deviation $s(r,t)$. The pricing formula is as follows

$$P(t,T) = E\{\mathrm{e}^{-r(t)(T-t)}\},$$

where $E\{f(X)\}$ is the expectation of a function $f(X)$ of a random variable X with probability density $\rho(x)$:

$$E\{f(X)\} = \int f(x)\rho(x)\,\mathrm{d}x.$$

Here we mention four models.

1. *Vasicek*, see Hull (1993): $m = a(b-r)$, $s = \sigma$; a, b and σ are constants. A solvable one-factor model, rates can be negative.

2. *Rendlemann and Bartter*, see Hull (1993): $m = Mr$, $s = Sr$; M and S are constants. The simplest solvable one-factor model, rates can be negative.

3. *Cox, Ingersoll and Ross*, see Hull (1993): $m = a(b-r)$, $s = \sigma\sqrt{r}$ is a solvable one-factor model with positive rates.

4. *Brennan and Schwartz*, see Brennan & Schwartz (1979): coupled equations for a short term interest rate $r(t)$ and a long term interest rate $l(t)$. A two-factor model (two Wiener generators involved).

3.2. *Arbitrage-free models*

A basic variable is the forward rate $F(t,T)$ described by the multifactor equation:

$$\mathrm{d}F(t,T) = \sum_p (\sigma_p \nu_p\,\mathrm{d}t + \sigma_p\,\mathrm{d}W^p)$$

with

$$\nu_p = \int_t^T \sigma_p(t,\tau)\,\mathrm{d}\tau,$$

i.e. the freedom is now up to a twice reduced set of 2-dimensional functions, $\sigma_p(t,T)$. The pricing formula is

$$P(t,T) = \mathrm{e}^{-\int_t^T F(t,\tau)\,\mathrm{d}\tau} \equiv \mathrm{e}^{-R(t,T)(T-t)},$$

and the models are:

1. *Ho and Lee*, see Hull (1993): $\sigma_p = \sigma = \text{const}$. Integrable case:

$$\mathrm{d}r = \big(F_t(0,t) + \sigma^2 t\big)\mathrm{d}t + \sigma\,\mathrm{d}W,$$
$$m(t,T) = \theta(t) = F_t(0,t) + \sigma^2 t.$$

2. *Hull and White*, see Hull (1993):

$$m(t,T) = \theta(t) - ar(t)$$

with a constant.

3. *Black, Derman and Toy*, see Hull (1993):

$$\mathrm{d}\ln r = \big[\theta(t) - a\ln r\big]\,\mathrm{d}t + \sigma \mathrm{d}W.$$

4. *Heath, Jarrow and Morton's* two-factor model (includes short range r_1 and long range r_2 rates):

$$\sigma_1(t) = \text{const} > 0,$$
$$\sigma_2(t) = \sigma_2(0)\,\mathrm{e}^{-\lambda(T-t)} > 0, \quad \lambda = \text{const}.$$

Here we give a short derivation of the arbitrage-free equation for $F(t,T)$ after HJM. Let us consider a general-form-pricing equation for bonds,

$$\frac{\mathrm{d}P(t,T)}{P(t,T)} = \mu(t,T)\,\mathrm{d}t + \nu(t,T)\,\mathrm{d}W,$$

and rewrite it in the form

$$\frac{\mathrm{d}P(t,T)}{P(t,T)} = r(t)\,\mathrm{d}t + \nu(t,T)\,\mathrm{d}W + \big(\mu(t,T) - r(t)\big)\mathrm{d}t,$$

then the arbitrage-free conditions mean that the equation

$$\mu(t,T) - r(t) = \lambda(t,T)\nu(t,T)$$

has a solution $\lambda(t)$ which does not depend on the maturiry T:

$$\frac{\partial\lambda(t)}{\partial T} = 0.$$

Here $\lambda(t)$ is a market price of risk. Whereby

$$\frac{\mathrm{d}P(t,T)}{P(t,T)} = r(t)\,\mathrm{d}t + \nu(t,T)\,\mathrm{d}\tilde{W} \tag{3.6}$$

with

$$\mathrm{d}\tilde{W} = \mathrm{d}W + \lambda\,\mathrm{d}T.$$

In Heath *et al.* (1992), it was shown that there is a universal martingale measure with respect to which all $\mathrm{d}\tilde{W}^p$ are independent martingale type Wiener generators, so that in what follows we can drop the tilde over W.

By applying Ito's lemma to (3.6) we have

$$\mathrm{d}\ln P(t,T) = \Big(r - \tfrac{1}{2}\sum_{q=1}^{m} \nu_q^2\Big)\mathrm{d}t + \sum_{q=1}^{m} \nu_q(t,T)\,\mathrm{d}W^q(t),$$

whence

$$\mathrm{d}F(t,T) = \sum_{p=1}^{n}\big(-\sigma_p(t,T)\nu_p(t,T)\,\mathrm{d}t + \sigma_p(t,T)\,\mathrm{d}W^p\big) \tag{3.7}$$

describes an n-factor model with a set of n unknown 2-dimensional functions:

$$\sigma_p(t,T) = -\frac{\partial \nu_p(t,T)}{\partial T}.$$

STATEMENT 3.1 (HJM). *Equation* (3.6) *is a master equation for arbitrage-free theory.*

3.3. *Internal space*

A variety of bond prices over their maturity

$$P(t,T_1),\ldots,P(t,T_N)$$

from a mathematical point of view constitute an N-dimensional metric space with respect to the expectation $E\{\cdot\}$ as the metric. Similarly, we have for forward rates a state vector space

$$F(t,T_1) \equiv X_1(t),\ldots,F(t,T_N) \equiv X_N(t),$$

so that a point in this space corresponds to a portfolio (a set) of different bonds.

The dynamics of a portfolio is a movement of the corresponding point in the state space (e.g. on a sphere). Is there a financial market feature that can help one to mathematically describe this movement?

The answer is *yes*.

These are already familiar concepts of market efficiency, arbitrage-freedom and, as a result, the so-called 'fair game' law. The last, in mathematical language, means that the sensitivity (volatility structure) of the portfolio *does not alter* while moving from point to point due to market impacts. Consider the equation for forward rate dynamics:

$$\mathrm{d}X_i(t) = b_i(t)\,\mathrm{d}t + \sum_p \sigma_p^i\,\mathrm{d}W^p.$$

On the curved manifold we have, for the set of vectors σ_p^i, see Dubrovin *et al.* (1984):

$$\delta\sigma_p^i = \mathrm{d}\sigma_p^i + \sum_{j,k}\Gamma_{jk}^i\sigma_p^i\,\mathrm{d}X^k.$$

This is zero due to the fair game condition, or

$$\mathrm{d}\sigma_p^i = -\sum_{j,k}\Gamma_{jk}^i\sigma_p^i\,\mathrm{d}X^k,$$

the equation wanted to describe volatility dynamics.

Here Γ_{jk}^i is the connexion of the space. It describes its geometrical features, in particular its curvature.

4. Some mathematical features of the construction

Here we recall that there are two kinds of stochastic calculus: after Ito, and after Stratonovich, see Stratonovich (1968), Pathria (1972) and Gardiner (1994). The first clearly gives statistical properties of the process, but it has definite difficulties in finding differentials and integrals, the second hides some of the statistical properties and instead gives conventional rules for derivatives and integrals. Naturally, a connection exists between the two representations of the *same* stochastic process, namely the Ito and Stratonovich differentials are related as follows

$$\mathrm{d}_{\mathrm{I}} Y^i = b^i\,\mathrm{d}t + \sum_{q=1}^{m} \sigma_q^i \mathrm{d}W^q(t), \quad i = 1, \ldots, N, \tag{4.8a}$$

$$\mathrm{d}_{\mathrm{S}} Y^i = b_{\mathrm{S}}^i\,\mathrm{d}t + \sum_{q=1}^{m} \sigma_q^i \mathrm{d}W^q(t), \quad i = 1, \ldots, N, \tag{4.8b}$$

where

$$b_{\mathrm{S}}^i = b^i - \tfrac{1}{2} \sum_{j,q} \sigma_q^j \frac{\partial \sigma_q^i}{\partial Y^j}, \tag{4.8c}$$

so that

$$\mathrm{d}_{\mathrm{I}} Y^i = \mathrm{d}_{\mathrm{S}} Y^i,$$

when $\partial \sigma_j^i / \partial Y_j = 0$, i.e. when *the volatility does not depend on the stochastic variable* Y^i.

Now we face a problem of how to obtain the coefficients of (4.8). There are at least two possible ways for doing this:

(*a*) Conventional method: assuming a certain time dependence of $b(t,T)$ and $\sigma(t,T)$ and from some additional, exogenous, considerations fitting them to historical market data, see Heath *et al.* (1992) and Hull (1993).

(*b*) Alternative method: find (endogenously) a mathematical reason to construct these dependences, e.g. a geometrical approach.

We discuss the latter method in this paper. Our task is to continue in time a forward rate curve $F(0,T)$ given at time $t = 0$, see figure 1.

For the sake of simplicity, we assume trivial boundary conditions on the 'left end' $T = t$. Later on we consider a more realistic and complicated situation. Normally the curve is continuous and we have an infinite dimensional system with respect to the parameter T. To pursue our goal we consider its discrete (chain) approximation

$$T = j\Delta T, \quad j = 0, \ldots, N, \tag{4.9}$$

and

$$T_{\mathrm{b}} = N\Delta T \tag{4.10}$$

is a boundary magnitude of the bond maturity: $T \in (0, T_{\mathrm{b}}]$. Therefore

$$F(0,T) = F(0, j\Delta T) \equiv F_j(0). \tag{4.11}$$

Each point of this chain, j, is supposed to evolve due to its own Wiener process $\mathrm{d}W^j$ though correlations are possible. So we have an N-dimensional Wiener process $\mathrm{d}\vec{W}$ governing the stochastic behaviour of

$$X^j(t) = F_j(t), \tag{4.12}$$

a vector in the state space of the system,

$$\mathrm{d}X^j(t) = b^j(t)\,\mathrm{d}t + \sum_q \sigma_q^j(t)\,\mathrm{d}W^q(t). \tag{4.13}$$

These equations describe the time evolution of the forward rate chain F_j. Now our goal is to describe the geometrical structure of the phase space and the trajectories in this space as given by (4.13).

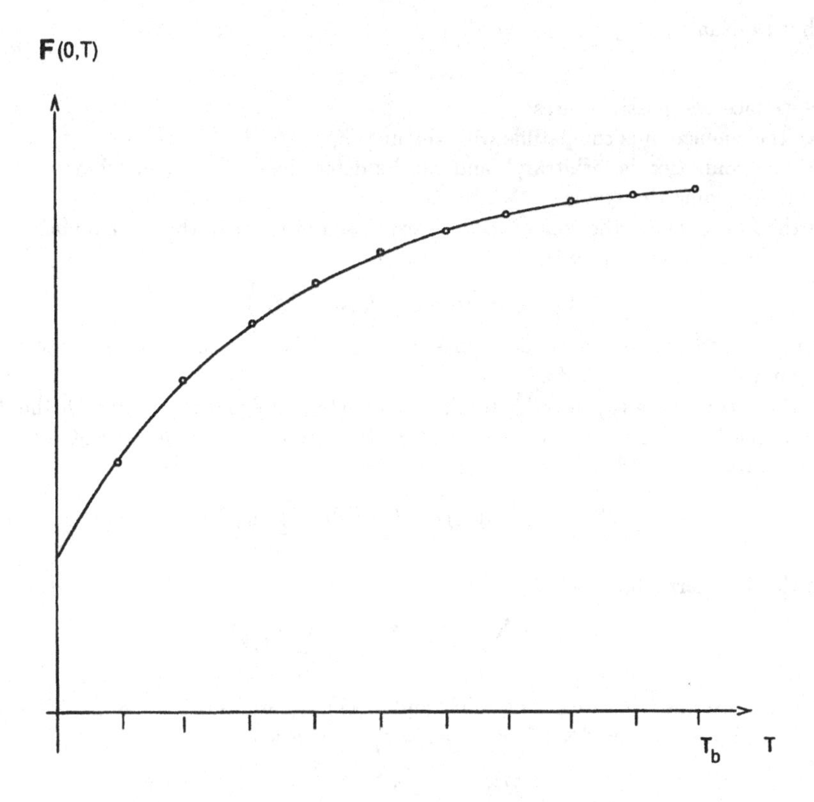

FIGURE 1. Forward rates curve in arbitrary units.

4.1. *Equations of stochastic differential geometry and the term structure models*

Iere we come to the equations of Stochastic Differential Geometry (SDG). Our construc-ion is similar to one well-known in differential geometry, i.e. the fibre bundle. SDG gives he generalization of this notion to stochastic processes, see Kendal (1987). Its equations n the invariant form and Stratonovich differentials are (qualitative derivation is given in ppendix A)

$$\mathrm{d_S}\vec{X} = \hat{\sigma}\,\mathrm{d}\vec{W}, \tag{4.14}$$

$$\mathrm{d_S}\hat{\sigma} = -\Gamma\hat{\sigma}\,\mathrm{d_S}\vec{X}, \tag{4.15}$$

r in the coordinate representation

$$\mathrm{d_S}X^i = \sum_q \sigma^i_q \mathrm{d}W^q, \tag{4.16}$$

$$\mathrm{d_S}\sigma^i_q = -\sum_{j,k} \Gamma^i_{jk}\sigma^j_q\,\mathrm{d_S}X^k, \tag{4.17}$$

here Γ^i_{jk} are Christoffel's symbols (the connexion).

If we assume the phase space to be a Riemannian manifold, then the inverse of the iemannian metric is given by

$$g^{ij} = \sum_q \sigma^i_q \sigma^j_q, \tag{4.18}$$

so that as usual

$$g_{ij}g^{jk} = \delta_i^k .$$

Here we face two possible cases:

(a) the connexion is compatible with the metric,

(b) the connexion is 'arbitrary', and can be determined, e.g. by matching some Ito process governing the system.

In the first case (a) the connection Γ is expressed in terms of the Riemannian metric g_{ij}, see Dubrovin *et al.* (1984),

$$\Gamma^i_{jk} = \tfrac{1}{2} g^{il}(\partial_j g_{kl} + \partial_k g_{jl} - \partial_l g_{jk}), \tag{4.19}$$

and we come to a closed system of stochastic differential equations describing a stochastic path on a Riemannian manifold.

In the second case (b) in order to close our system of equations we match the drift term of the Ito process for the forward rates with the drift term in the SDG equation (4.16). Rewriting (4.16) in the Ito representation, one obtains

$$\mathrm{d}X^i = (\tilde{\sigma}^i_q + \tfrac{1}{2}\mathrm{d}\tilde{\sigma}^i_q)\,\mathrm{d}W^q \equiv \tilde{b}^i \mathrm{d}t + \sum_q \tilde{\sigma}^i_q\,\mathrm{d}W^q, \tag{4.20}$$

with the drift term due to (4.17) being

$$\tilde{b}^i = -\tfrac{1}{2}\sum_{j,k,q} \tilde{\sigma}^j_q \Gamma^i_{kj} \tilde{\sigma}^k_q = -\tfrac{1}{2}\sum_{k,j} \Gamma^i_{kj} g^{kj}. \tag{4.21}$$

In order to use the SDG equations (4.16) and (4.17), equations (4.20) and (4.13) have to be identical, i.e. we must have $\tilde{b}^i = b^i$, $\tilde{\sigma}^i_q = \sigma^i_q$. This leads to

$$b^i(t) = -\tfrac{1}{2}\sum_{k,j} \Gamma^i_{k,j} g^{kj},$$

or

$$\Gamma^i_{kj} = -\frac{2}{m}\, g_{kj} b^i(t). \tag{4.22}$$

This gives us relations between the connexion Γ, the drift of the interest rate $b^i(t)$, and the volatility structure σ^i_q, through the space metric g^{ij}.

If the drift only consists of the geometrical part,

$$b^i(t) = f(\sigma^j_q),$$

the system of stochastic differential equations (4.16) and (4.17) becomes closed and self consistent.

For example, in the case of forward rates, process (4.13), we have

$$\sum_{q=1}^{m} \nu^i_q \sigma^i_q = \tfrac{1}{2}\sum_{k,j} \Gamma^i_{kj} g^{kj},$$

or

$$\Gamma^i_{kj} = \frac{2}{m}\, g_{kj} \sum_{q=1}^{m} \nu^i_q \sigma^i_q, \tag{4.23}$$

so that we have obtained the direct relation between the connexion Γ, and the volatility structure of the interest rate σ^i_q, and that of the bond price ν^i_q, thereby again closing the system of stochastic differential equations.

Substituting (4.23) into (4.17), we come to the equations for the volatility and stat vector trajectories:

$$\mathrm{d}_S\sigma^i_q(t) = -2\nu^i_q \sum_p \sigma^i_p \,\mathrm{d}W^p(t), \tag{4.24a}$$

$$\mathrm{d}_S X^i(t) = \sum_p \sigma^i_p \,\mathrm{d}W^p(t), \tag{4.24b}$$

thereby again obtaining a closed system of equations.

The problem of initial states and boundary conditions goes beyond the scope of the scheme considered here. However, those states are, in fact, available from the market data. By adjusting the initial states and boundary conditions, we can obtain a number of mathematically self-consistent models of the term structure. Several examples are given below.

4.2. *The connexion curvature and classification of markets*

It is known from differential geometry that one of the most important characteristics of a manifold is its connexion curvature tensor

$$-R^i_{jkl} = \partial_k \Gamma^i_{jl} - \partial_l \Gamma^i_{jk} + \Gamma^i_{pk}\Gamma^p_{jl} - \Gamma^i_{pl}\Gamma^p_{jk}. \tag{4.25}$$

Namely this tensor defines invariant geometrical properties of a manifold and vanishes in the Euclidean case:

$$R^i_{jkl} = 0.$$

Consider, for example, a Euclidean plane $\mathbf{R}^2$ with polar coordinate system:

$$\rho = \sqrt{x^2 + y^2},$$
$$\phi = \arctan(y/x),$$

$$g_{ij} = \begin{bmatrix} 1 & 0 \\ 0 & \rho^2 \end{bmatrix}, \quad g^{ij} = \begin{bmatrix} 1 & 0 \\ 0 & \rho^{-2} \end{bmatrix}.$$

Then since

$$\Gamma^i_{jk} = \tfrac{1}{2} g^{il}(\partial_j g_{kl} + \partial_k g_{jl} - \partial_l g_{jk}),$$

we have

$$\Gamma^1_{22} = -\tfrac{1}{2} g^{11} \partial_\rho g_{22} = -\rho,$$
$$\Gamma^2_{12} = \Gamma^2_{21} = \tfrac{1}{2} g^{22} \partial_\rho g_{22} = 1/\rho,$$

the rest are zero. The drift terms are

$$a^1 = -\tfrac{1}{2}\Gamma^1_{22} g^{22} = \frac{1}{2\rho}$$

and the curvature tensor components are

$$R^1_{112} = R^1_{212} = R^2_{212} = 0,$$

and

$$R^2_{112} = -\partial_1 \Gamma^2_{12} - \Gamma^2_{12}\Gamma^2_{21} = \frac{1}{\rho^2} - \frac{1}{\rho^2} = 0.$$

In the spherical case we have

$$R_{1212} = g_{1k} R^k_{212} = \sin^2(\rho/R_0) \neq 0,$$

with R_0 being the sphere radius and $\rho = R_0\theta$.

So in the case of polar coordinates on a plane, the nonvanishing drift (and hence the connexion) is defined by fictitious ('inertial') forces and can be removed by the appropriate coordinate transformation with the connexion vanishing. This is impossible in the spherical case.

The above means that a nonzero connection does not imply a curved space. A nonzero curvature tensor surely does, implying essential drift (and forces) present in the system studied.

Sometimes, instead of the curvature tensor R^i_{jkl}, it is easier to calculate the so-called scalar curvature

$$R = \sum_{i,j,l} g^{jl} R^i_{jil}$$

that gives us a clear understanding of the state space nature as well. For instance, in the case of forward rates, equation (4.23), we have the scalar curvature

$$-R = g^{ql}\partial_i(b^i g_{ql}) + (m-1)b_i b^i \equiv g^{ql}\operatorname{div}(\vec{b}g_{ql}) + (m-1)(\vec{b}\cdot\vec{b})$$

expressed through the drift vector $\vec{b}$ given by

$$b^i(t) = -\sum_{p=1}^{n} \sigma^i_p(t)\nu^i_p(t),$$

and the metric of the state space g_{ij} (m stands for its dimensions). This formula says that the state space of zero-coupon bonds is in general non-flat (non-Euclidean) and hence dynamics of their rates and prices cannot be reduced to pure independent diffusions.

In the spherical example

$$R = \det(g_{ij})\, R_{1212} = 2/R_0^2.$$

Now we come to the conclusion:

Processes with internal state space can be classified by the connexion curvature of this space.

1. Zero curvature means a martingale property of the market: no drifts, no correlations just independent diffusion processes.

2. Nonzero curvature implies a richer picture for the market with drifts, correlations and so forth.

4.3. *Solutions*

Let us consider a single-factor model. In this case we can obtain an exact analytical solution, namely we integrate the stochastic differential equations for the volatility matrix $\sigma^i_j(t)$ in the discrete maturity case. Knowledge of this structure along with the arbitrage free equation for forward rates of discount bonds enables one to price (and hedge) all contingent claimes on the term structure such as options, swaps, swaptions, and so forth.

In Stratonovich's differentials, equations (4.24) can be represented back in the continuous form as follows

$$\mathrm{d}\sigma_q(t,T) = 2\left(\int_t^T \sigma_q(t,\tau)\,\mathrm{d}\tau\right)\sigma_p(t,T)\,\mathrm{d}W^p, \quad 0 \le t \le T,$$

$$\mathrm{d}F(t,T) = \sigma_p(t,T)\,\mathrm{d}W^p, \quad p = 1,\ldots,m,$$

where $F(t,T)$ is the forward rate, and $\sigma_p(t,T)$ its volatility structure matrix. In the case of a single-factor model we can put

$$\mathrm{d}W^p = \mathrm{d}W\delta^{pi}$$

and drop the index i taking $\sigma_i \equiv \sigma$ such that

$$\mathrm{d}\sigma(t,T) = 2\left(\int_t^T \sigma(t,\tau)\,\mathrm{d}\tau\right)\sigma(t,T)\,\mathrm{d}W, \tag{4.26}$$

$$\mathrm{d}F(t,T) = \sigma(t,T)\,\mathrm{d}W. \tag{4.27}$$

Consider (4.26) and (4.27). Let us proceed to the model with continuous time and discrete maturity defined as follows

$$F(t,T) \to F(t, jT_1) \equiv F_j(t), \quad t \in \big((i-1)T_1, iT_1\big]$$

and the same for $\sigma(t,T)$, then

$$\mathrm{d}\sigma_i(t) = 2\Big((T_1 - t)\sigma_1(t) + T_1 \sum_{j=2}^{i} \sigma_j(t)\Big)\sigma_i(t)\,\mathrm{d}W, \quad 0 \le t \le T_1 \tag{4.28}$$

$$\mathrm{d}F_i(t) = \sigma_i(t)\,\mathrm{d}W. \tag{4.29}$$

Take initial 'curves' $F_i(0)$ and $\sigma_i(0)$ (in fact, a two-line matrix of initial data).
Consider (4.28) for σ_1:

$$\mathrm{d}\sigma_1(t) = 2(T_1 - t)\sigma_1^2(t)\,\mathrm{d}W, \quad t \in (0, T_1]. \tag{4.30}$$

First we should stress that our system is defined on the (t,T) plane and its evolution goes under the diagonal: t=T, step by step:

$$t_1 \in (0, T_1],\ t_2 \in (T_1, 2T_1],\ \ldots,\ t_i \in \big((i-1)T_1, iT_1\big].$$

Final values of F_i and σ_i of the preceding step serve as initial conditions for the current evolution with an extra bond of the maximal maturity appearing at the end of each time interval $\Delta t = T_1$, see figure 2.

Equation (4.30) is self-consistent and easy to solve

$$\sigma_1(t) = \frac{\sigma_1(0)}{1 - 2\sigma_1(0)\displaystyle\int_0^t (T_1 - \tau)\,\mathrm{d}W_\tau}, \quad \sigma_1(0) = \sigma_1(t=0). \tag{4.31}$$

The behaviour of this solution is defined by the stochastic integral

$$I_1 = \int_0^t (T_1 - \tau)\,\mathrm{d}W_\tau. \tag{4.32}$$

This integral is easy to evaluate:

$$\langle I_1 \rangle = 0,$$

$$D_{I_1} \equiv \langle I_1^2 \rangle - \langle I_1 \rangle^2 = \int_0^t (T_1 - \tau)^2 \mathrm{d}\tau = \tfrac{1}{3} t(t^2 + 3T_1(T_1 - t)),$$

with $D_{I_1}|_{\max}$ at the point $t_{\max} = T_1$:

$$D_{I_1}|_{\max} = \tfrac{1}{3} T_1^3.$$

The standard deviation is as follows

$$\sqrt{D_{I_1}|_{\max}} = \frac{1}{\sqrt{3}} T_1^{3/2}, \tag{4.33}$$

the value we can use for estimations due to the form of the curve $D_{I_1}(t)$.

Let us consider the denominator in (4.31):

$$d_1 \equiv 1 - 2\sigma_1(0) I_1. \tag{4.34}$$

For the solution $\sigma_1(t)$ to be well defined (finite deviations of σ_1) d_1 must not vanish, i.e.

$$1 - 2\kappa\sigma_1(0) I_1 > 0$$

or

$$\kappa \frac{2}{\sqrt{3}} \sigma_1(0) T_1^{3/2} < 1,$$

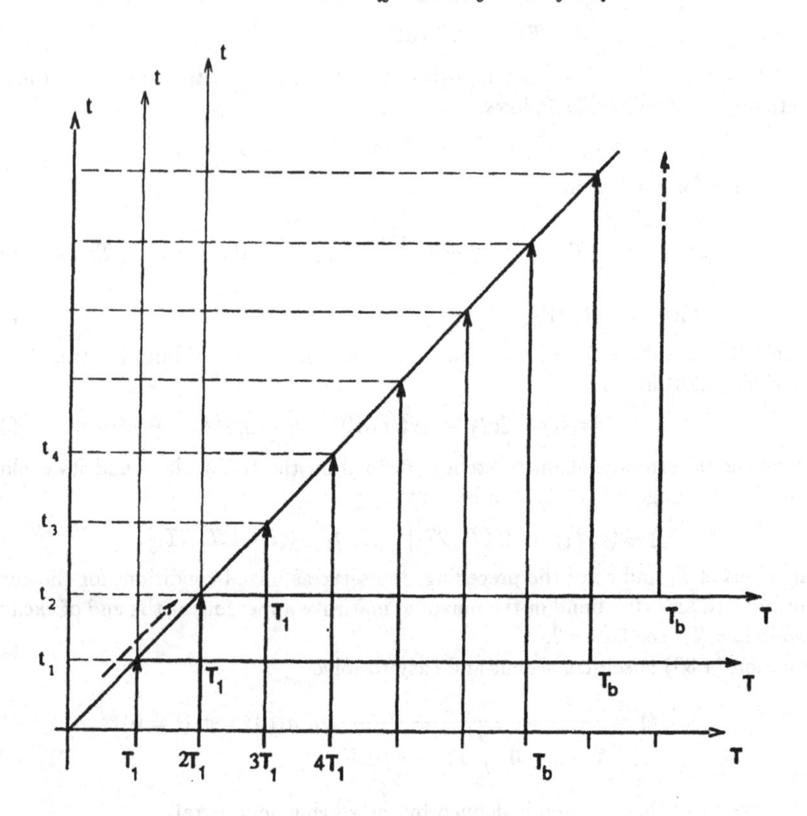

FIGURE 2. Time evolution of the discrete maturity model: short range, $0 < t_1 < T_1$; long-range, $t \gg T_1$.

hence

$$\kappa\sigma_1(0) < \frac{\sqrt{3}}{2}\tau_1^{-3/2} \tag{4.35}$$

with τ_1 being the annualized maturity and $\kappa \lesssim 3$ a numerical factor. It is easy to check the dimensions of F and σ via their definitions:

$$F = -\frac{\partial \ln P}{\partial T}, \quad \text{i.e. } [F] = [T^{-1}],$$

$$[\nu] = [T^{-1/2}], \quad \text{volatility of } P,$$

$$\sigma = -\frac{\partial \nu}{\partial T}, \quad \text{i.e. } [\sigma] = [T^{-3/2}].$$

Consider now the equation for $\sigma_2(t)$:

$$\mathrm{d}\sigma_2(t) = 2\left[(T_1 - t)\sigma_1(t) + T_1\sigma_2(t)\right]\sigma_2(t)\,\mathrm{d}W. \tag{4.36}$$

Since (4.36) is given in Stratonovich's form, we treat the related equation

$$\dot{\sigma} = \alpha_2(t)\sigma + \sigma^2,$$

where

$$\mathrm{d}t = 2T_1 \mathrm{d}W$$

and

$$\alpha_2 = \sigma_1(t)\frac{T_1 - t}{T_1}$$

with the solution

$$\sigma \equiv \sigma_2(t) = \frac{\sigma_2(0)\exp\left(2T_1\int_0^t \alpha_2(\tau, W_\tau)\mathrm{d}W_\tau\right)}{1 - 2T_1\sigma_2(0)\int_0^t \mathrm{d}W_x \exp\left(2T_1\int_0^x \alpha_2(\tau, W_\tau)\mathrm{d}W_\tau\right)}.$$

The reccurent formula is straightforward:

$$\sigma_i(t) = \frac{\sigma_i(0)\exp\left(\int_0^t \alpha_i(\tau, W_\tau)\,\mathrm{d}\tilde{W}_\tau\right)}{1 - \sigma_i(0)\int_0^t \mathrm{d}\tilde{W}_\tau \exp\left(\int_0^\tau \alpha_i(\tau_1, W_{\tau_1})\,\mathrm{d}\tilde{W}_{\tau_1}\right)}, \tag{4.37}$$

where

$$\alpha_i(t, W) = \frac{T_1 - t}{T_1}\sigma_1(t, W) + \sigma_2(t, W) + \cdots + \sigma_{i-1}(t, W) + \frac{\nu_0(t)}{T_1}, \tag{4.38a}$$

$$\mathrm{d}\tilde{W} = 2T_1\mathrm{d}W. \tag{4.38b}$$

While deriving (4.37) and thereafter, we assume that at the left end of the system, at $T = \Delta t \ll T_1$, new short term bonds with price volatility $\nu_0(t)$ are issued constantly, so that we have a singular behaviour of the system there. In hydrodynamics, for example, such a phenomenon is known as a boundary layer and can be treated as effective boundary conditions.

For small time intervals, the average values are

$$\begin{aligned}\langle\sigma_1(t)\rangle = \sigma_1(0)\big[1 &+ \tfrac{4}{3}\sigma_1^2(0)(t^2 + 3T(T - t))t\\ &+ 6\sigma_1(0)\nu_0(0)(T - \tfrac{1}{2}t)t + 2\langle\delta\nu_0\rangle t + 2\nu_0^2(0)t\big],\end{aligned} \tag{4.39}$$

$$\begin{aligned}\langle\sigma_i(t)\rangle = \sigma_i(0)\Big[1 &+ 4\sigma_i(0)\nu_{i-1}(0)Tt + 2\nu_{i-1}^2(0)t + 2\sigma_i(0)^2T^2t\\ &+ 2Tt\sum_{k=2}^{i}\sigma_k(0)\left[\sigma_k(0)T + \nu_{k-1}(0)\right] + \sigma_1(0)^2\left(\tfrac{4}{3}t^2 + 2T - Tt\right)t\\ &-3\,\sigma_1(0)\nu_i(0)t^2 + 2\sigma_1(0)\nu_0(0)Tt + 2\langle\delta\nu_0\rangle t\Big], \quad i \neq 1,\end{aligned} \tag{4.40}$$

where

$$\nu_i(0) = \nu_0(0) + T\sum_{k=1}^{i}\sigma_k(0),$$

$$\nu_i(t) = \nu_0(t) + \sum_{k=1}^{i}\Delta T_k\sigma_k(t),$$

$$\delta T_1 = T - t, \quad \delta T_i = T, \quad i > 1,$$

$$\langle\delta\nu_0\rangle = \left\langle\int_0^t \delta\nu(W)\,\mathrm{d}W\right\rangle.$$

This result means that the forward rate volatility of a given maturity is formed by all preceding ones. The same is valid for the forward rates themselves $F(t, T)$ by virtue of the equation

$$\mathrm{d}_S F(t, T) = \sigma(t, T)\,\mathrm{d}W.$$

By integrating this equation and

$$R(t, T) = \frac{1}{T - t}\int_t^T F(t, \tau)\,\mathrm{d}\tau,$$

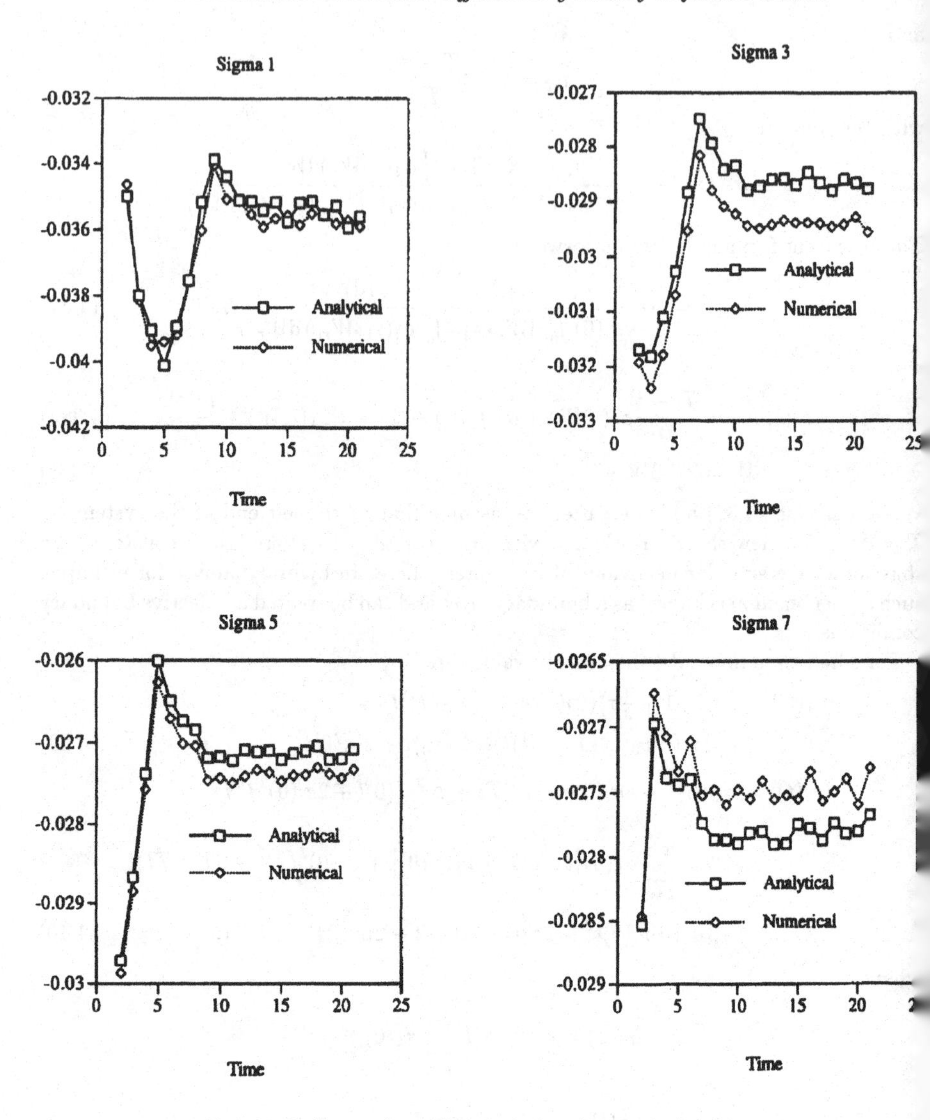

FIGURE 3. Comparison of analytical and numerical results.

we solve exactly the problem of finding the interest rate term structure (its dynamics) in the single factor, arbitrage free, risk neutral evaluation approach. In fact, we construct a theory based on these three cornerstones. Note that the specific behaviour of the system now depends on either the initial condition (for short range dynamics) or boundary conditions (for long range epochs and steady states).

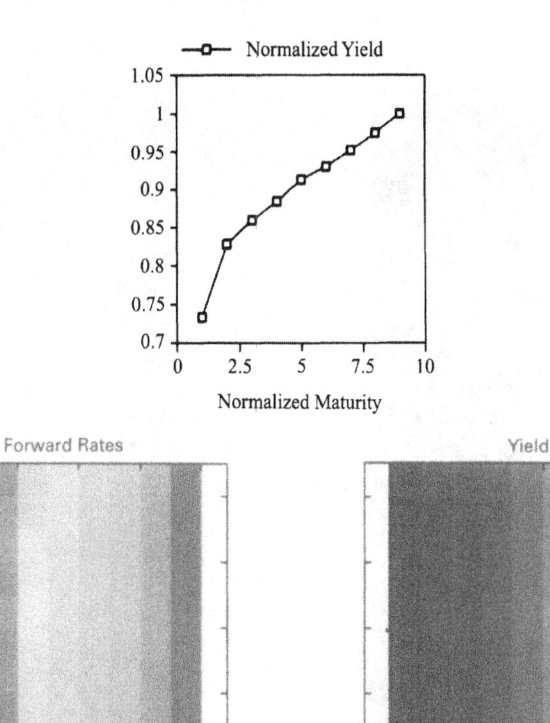

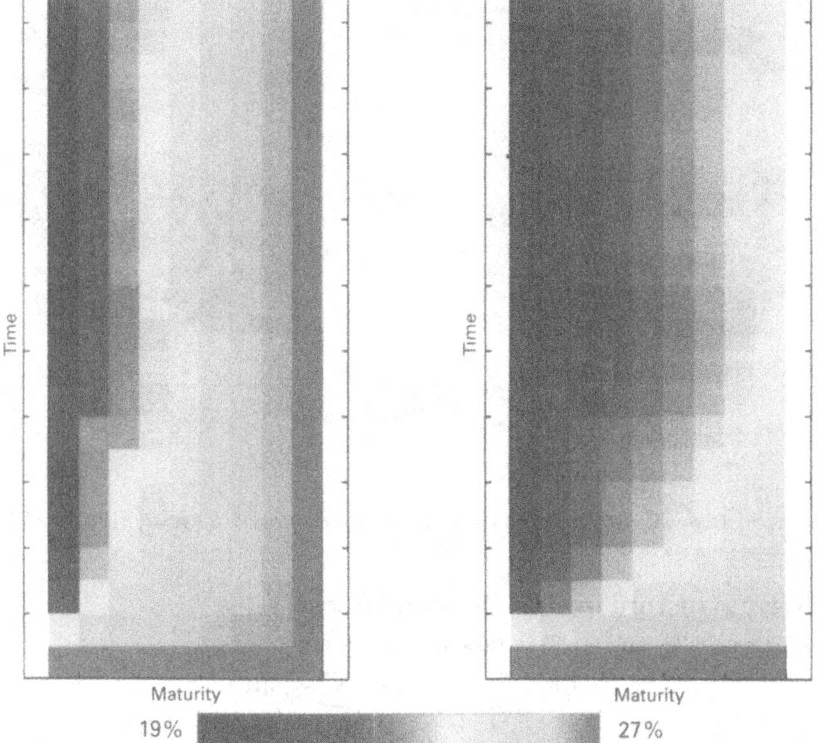

FIGURE 4. Upward sloping yield curve. Maximum maturity is 30 years, $\sigma(T_{N_m}) = -0.7\%$, $\nu_0 = 52\%$. A line plot at the top shows a steady state yield curve.

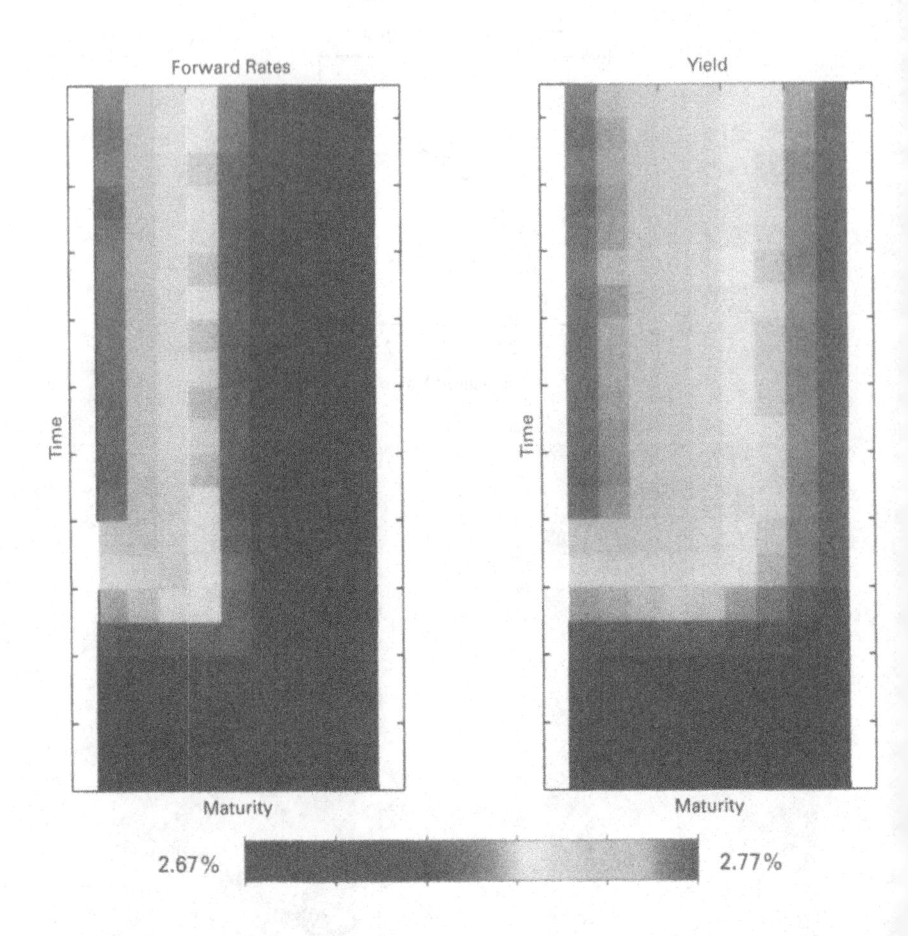

FIGURE 5. Downward sloping yield curve. Maximum maturity is 30 years, $\sigma(T_{N_m}) = 0.07\%$, $\nu_0 = 52\%$.

5. Term structure models for bond interest rates

Equations (4.28) and (4.29) in Ito's differentials are

$$\mathrm{d_I} F_i = \sum_q \nu_{i,q}\sigma_{i,q}\,\mathrm{d}t + \sum_q \sigma_{i,q}\,\mathrm{d}W^q, \tag{5.41}$$

$$\mathrm{d_I}\sigma_{i,q} = 2\sum_p \sigma_{i,p}\Big(\nu_{i,q}\nu_{i,p} + \sum_{l\le i}\sigma_{l,p}\nu_{l,q}\Delta T_l\Big)\mathrm{d}t + 2\nu_{i,q}\sum_p \sigma_{i,q}\,\mathrm{d}W^q, \tag{5.42}$$

where

$$\nu_{i,q} = -\sum_{j=1}^{i}\sigma_{j,q}\Delta T_j,$$

and ΔT_j is the time interval between maturities for $p > 1$ and

$$\Delta T_1 = T_1 - t.$$

5.1. *Initial and boundary conditions*

In order to numerically evolve the term structure, we should specify the initial and boundary conditions, as well as possible sources and sinks. In other words, we assume that, in general, issues of new bonds or their calls are not determined by the present model but rather are due to *external economic conditions.* However, it is the volume of the bond issued that decdes if it is determined by the market or else conversely it determines the market. In the case of limited issues, prices are driven by the market and as such can be determined in the framework of the present theory.

For the sake of clarity and simplicity, zero coupon bonds are assumed to be issued with two maturities only: (i) short term bonds maturing at $T_0 = t + \Delta t$, issued every Δt, where Δt is a small trading interval that determines spot interest rates (e.g. Δt is one day), and (ii) long term bonds that are issued with a time interval $\Delta T \gg \Delta t$ and which have a maturity of $N_m T_1$.

We specify the issuing of short term bonds with interest rate F_0 (i.e. spot interest rate) and volatility $\nu_{0,q}$. We specify issuing of long term bonds with interest rate F_{N_m} and volatility of that interest rate $\sigma_{N_m,q}$.

We note that at $t = T_{N_m}$, we find it more reasonable to specify the volatility of the forward interest rate rather than the volatility of the bond itself.

5.2. *Numerical experiments*

In Makhankov *et al.* (1995) a series of numerical experiments has been performed with the aim to study possible time dependent and stationary solutions of equations (5.41) and (5.42). The numerical algorithm was based on multi-process Monte Carlo modelling of the equations checked against an exact solution, see figure 3. Different initial and boundary conditions have been considered. As a result, it was shown that under certain *boundary conditions* rather than initial ones, various forms of stationary solutions occur revealing all known shapes of interest rate structures (yield curves). In what follows we give a brief survey of the results of Makhankov *et al.* (1995).

(i) Depending on the sign of the drift term in the interest rates (5.41) at least two basic yield curves (upward sloping and downward sloping) have been obtained. This sign is determined by the boundary conditions on both ends (short range epochs, $t \to T$ and long range epochs, $T \gg t$).

An example of an upward sloping yield curve is shown in figure 4. The initial distribution of the interest rates and volatilities was flat. The upward sloping term structure developed over time. The boundary conditions are specified in the plot.

An example of the downward sloping term structure development is shown in figure 5. The initial distribution of the interest rates and volatilities was flat. The boundary conditions are specified in the plot.

From experiments with the boundary conditions it seems that the upward sloping curves are much more robust than the downward sloping curves. The downward sloping curves become unstable (explode) as we increase values of volatilities at the boundary that would bring a larger difference in short and long term interest rates.

(ii) By playing with the number of Wiener processes involved, it was shown that the simplest approach for analyzing interest rate dynamics (to match only initial volatilities of forward interest rates) would be that using a single Wiener process. If we follow a typical assumption used by many researchers, and assume that volatilities of interest rates are constant, then forward rate dynamics of interest rates is fully described by two independent Wiener proceses. However, in the general case, long time dynamics requires more than two Wiener processes, since according to our theory, volatilities do change and solutions (4.37)–(4.40) can give us simple estimates of the time needed to change them. On some occasions, the system exhibited explosive behaviour. This result allows us to conclude that, in general, long term dynamics of the market is described by multi-dimensional Wiener processes. However, since some degeneration of parameters is possible (i.e. one can get the same results in terms of averaged interest rate values) we should try to find the minimum number of processes. We expect that only a statistical analysis of the market data can give us an answer about the minimum number of processes operating under different market conditions.

6. Conclusions

The theory developed above allows one to conclude that:

(i) By implementing a natural requirement of the fair game rule (which in our theory is in fact a consequence of the arbitrage-free feature of the market) we close the system of equations governing the common dynamics of forward rates and their volatilities. Therefore we reduce a freedom hidden in a set of two-dimensional functions $\sigma_p(t,T)$ to that contained in a set of one-dimensional functions. These are defined by boundary and initial conditions. The latter can be extracted from market data on the basis of conventional techniques.

(ii) Analytical estimates and computer experiments reveal a variety of possible developments depending on initial and boundary conditions:

(*a*) explosive instability of the solution which leads to unpredictability of the bonc market,

(*b*) a number of stationary solutions among which are well known upward sloping downward sloping and humped term structures. The form of the solution i completely determined by boundary conditions and hence it is an exogeneously defined feature of the system rather than an intrinsic one.

(iii) Stationary states occur thanks to the balance of four different effects:

(*a*) movement of bonds along the maturity T-axis (bonds get older, approachin their maturities),

(*b*) emission of new long term bonds at $T = T_b$ with time step T_1,

(*c*) emission of new short term bonds at $T = \Delta t < T_1$,

(*d*) increase/decrease in bond value due to the bond price drift process.

Two of these processes are given by boundary conditions, and their relative strengt determines the stationary structure.

(iv) Our theory is based on three crucial assumptions: efficiency of the bond marke (no short range dynamical structures, just a Markov stochastic description), arbitrage free requirement (HJM, reduction of freedom from two 2D functions to one 2D function and fair game law. Violation of any one of these assumptions makes the predictions c the theory invalid.

First of all my gratitude goes to the organizers of the Conference, in particular to Pro ssors R. Żelazny and J. Zagrodziński and Dr A. Gałkowski, for giving me the opportunit to visit Poland and participate in the meeting. I am also indebted to my colleagues an co-authors, R. Jones, C. Gomez and Y. Taranenko, for numerous valuable discussion and help.

Appendix A. Qualitative derivation of the SDG equations

In this section we give a brief qualitative derivation of (4.16) and (4.17) and henc expose the assumptions that should be made for their validity. Our goal is to describ

Brownian motion on a curved manifold. For the sake of simplicity, we consider the most elementary example of a two-dimensional phase space. To be more concrete, we will use the $\mathbf{S}^2$ sphere, although we could have used either a hyperboloid or indeed any other curved surface.

Why a curved surface?

Recent developments in physics show that all types of interactions (and hence, correlations) admit geometrization. For example, two free particles on a curved surface are seen interacting with each other, and the curvature plays the role of the coupling constant (interaction rate).

Many modern physical theories have been constructed via implemention of this principle (e.g. general relativity, strings, QCD and so on). In its most transparent form, it can be seen in the theory of so-called $(1+1)$-dimensional integrable systems, where in a consistent manner the relation is established between geometry of the internal (isotopic) space and the kind of interaction. This relation, based on so-called gauge equivalence, is discussed in detail in Makhankov & Pashaev (1992).

From figure 6 we see that deviation from Euclidicity automatically gives rise to an nteraction (or correlation). Let us now consider a 'pure' Brownian motion on the sphere, and first build a frame bundle of the sphere.

Consider a point X_1 on $\mathbf{S}^2$ and a patch of a tangential plane in this point, we denote t $T_{X_1}\mathbf{S}^2$. It is called a fibre at the point X_1. Then we proceed to a neighbouring point X_2, and, doing the same, we get $T_{X_2}\mathbf{S}^2$. In such a way, we can cover the sphere with such patches, sticking them along the lines of their intersections like on a soccer ball, obtaining a polyhedron.

Now we have an example of a fiber bundle with the sphere as the base and the polyhedron a bundle of fibres (really a bundle of frames in our case). We call this polyhedron a 'covering' of the sphere.

Crucial points of the construction are:

(*a*) the sphere is curved (a manifold),

(*b*) the covering is a Euclidean space.

A pure Wiener process (a martingale), satisfying the equations

$$\mathrm{d}W^q \mathrm{d}W^p = \delta^{qp}\mathrm{d}t,$$

occurs in the Euclidean world. This equation for vectors is only valid in a Euclidean space. The same takes place for semimartingales (approximately), for only in a Euclidean space it is possible to represent a stochastic process in the semimartingale form,

$$\mathrm{d}\tilde{W}^q = \alpha^q \mathrm{d}t + \mathrm{d}W^q. \tag{A 43}$$

This means that Wiener processes can only appear on the covering while a particle is moving on the sphere. Now we should adjust both phenomena. Let us consider a covering 'boiling' with fluctuating forces (Wiener processes) and a particle at a point X_1 on the phere. This point also belongs to the fibre $T_{X_1}\mathbf{S}^2$. Hence the particle undergoes a andom shock

$$\mathrm{d}\vec{X}_1 = \hat{\sigma}_1 \mathrm{d}\vec{W}_1,$$

umping onto a point X_2 on the sphere. At this new point, it again undergoes a shock

$$\mathrm{d}\vec{X}_2 = \hat{\sigma}_2\, \mathrm{d}\vec{W}_2,$$

nd so forth. The matrix $\hat{\sigma}$ defines particle mobility (sensitivity). Here we should mphasize that all differentials considered in this section are of the Stratonovich type, which allows us to use standard differential calculus.

Assumption 1. All $\mathrm{d}\vec{W}_i$ are the same in distribution (pure Brownian motion).

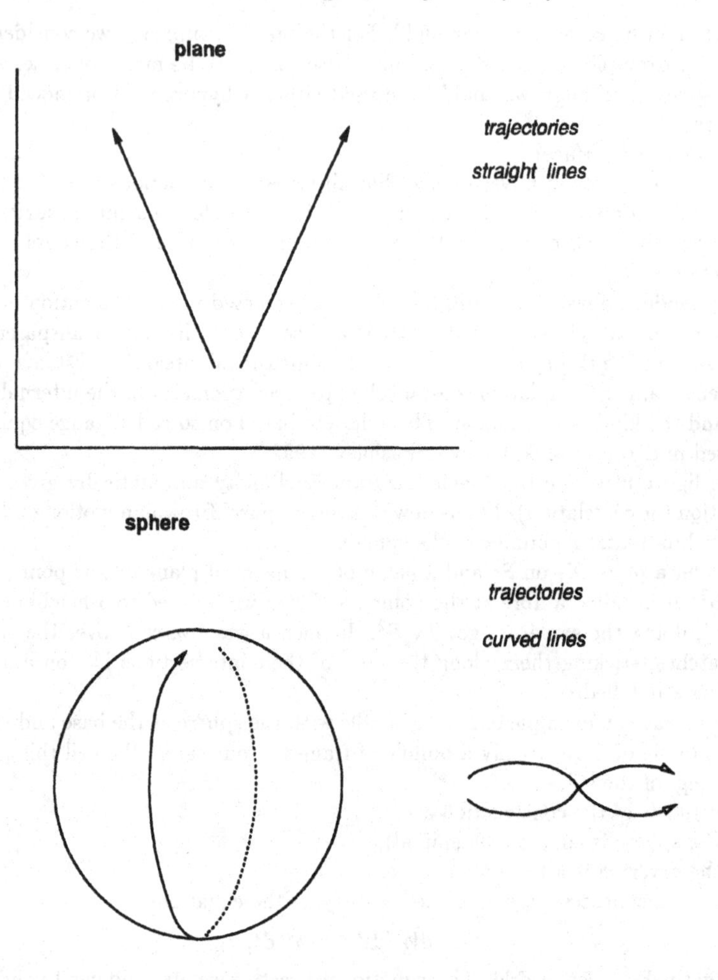

FIGURE 6. Curvature and interactions.

Now our task is to connect $\hat{\sigma}_1$ and $\hat{\sigma}_2$. Note that the matrix $\hat{\sigma}$, being in fact a rotatin operator, can be constructed from two vectors $\vec{j}_1$ and $\vec{j}_2$

$$\hat{\sigma} = \begin{bmatrix} j_{1x} & j_{2x} \\ j_{1y} & j_{2y} \end{bmatrix},$$

which gives a natural frame on the patch considered.

While moving from one patch to another, this frame changes its orientation. The tota change of a vector $\vec{B}$, due to moving from one point to another, consists of two parts Dubrovin *et al.* (1984),

$$\delta\vec{B} = \mathrm{d}\vec{B} + \Gamma\vec{B}\,\mathrm{d}\vec{X}, \tag{A 44}$$

where the first term is the differential along the path,

$$\mathrm{d}\vec{B} = \frac{\partial\vec{B}}{\partial X^i}\mathrm{d}X^i(t), \tag{A 45}$$

and the second allows for a change of the frame orientation. Since the matrix $\hat{\sigma}$ consist of two vectors $\vec{j}_1$ and $\vec{j}_2$, it is transformed following the same rule,

$$\delta\hat{\sigma} = \mathrm{d}\hat{\sigma} + \Gamma\hat{\sigma}\,\mathrm{d}\vec{X}(t). \tag{A 46}$$

Assumption 2 (of 'fair game'). 'Rules must not change from play to play' means that the total change of $\hat{\sigma}$ should vanish

$$\delta\hat{\sigma} = 0,$$

or

$$\mathrm{d}\hat{\sigma} = -\Gamma\hat{\sigma}\,\mathrm{d}\vec{X}(t). \tag{A 47}$$

This along with the equation for the elementary shock,

$$\mathrm{d}\vec{X}(t) = \hat{\sigma}\,\mathrm{d}\vec{W}, \tag{A 48}$$

gives us the equations of *Stochastic Differential Geometry* on the sphere.

Generalization to other curved manifolds is straightforward.

In order to clarify the geometrical meaning of the geometrical drift term, we consider here probably the simplest solvable example of diffusion on a sphere. The intimate relation between the two processes, namely Brownian motion and diffusion was revealed by Einstein and Smoluchovski long ago (see e.g. Pathria (1972)), and can be easily seen from the Ito stochastic differential equation for the pure Brownian process

$$\mathrm{d}X = \sigma(X,t)\,\mathrm{d}W.$$

A.1. *Diffusion*

Take Ito's stochastic differential equation for the pure Brownian process given in Cartesian coordinates

$$\begin{aligned} \mathrm{d}x_1 &= a_1\mathrm{d}t + \sigma_1\mathrm{d}W_1, \\ \mathrm{d}x_2 &= a_2\mathrm{d}t + \sigma_2\mathrm{d}W_2. \end{aligned}$$

Consider a function $f(x_1, x_2, t)$. Then by Ito's lemma we have

$$\mathrm{d}f = \left(\partial_t f + a_1\partial_{x_1} f + a_2\partial_{x_2} f + \tfrac{1}{2}\sigma_1^2\partial_{x_1}^2 f + \tfrac{1}{2}\sigma_2^2\partial_{x_2}^2 f\right)\mathrm{d}t + \sigma_1\partial_{x_1} f\,\mathrm{d}W_1 + \sigma_2\partial_{x_2} f\,\mathrm{d}W_2.$$

Take

$$\begin{aligned} x_1 &= \rho\cos\phi, \\ x_2 &= \rho\sin\phi, \end{aligned}$$

$$\begin{aligned} f_1 &\equiv \rho = \sqrt{x_1^2 + x_2^2}\,, \\ f_2 &\equiv \phi = \arctan\frac{x_2}{x_1}\,. \end{aligned}$$

A pure diffusion process takes place when $a_1 = a_2 = 0$. We can also put $\sigma_1 = \sigma_2 = \sigma$, and then

$$\mathrm{d}\rho = \frac{1}{2\rho}\sigma^2\mathrm{d}t + \sigma(\cos\phi\,\mathrm{d}W_1 + \sin\phi\,\mathrm{d}W_2), \tag{A 49a}$$

$$\rho\,\mathrm{d}\phi = \sigma(-\sin\phi\,\mathrm{d}W_1 + \cos\phi\,\mathrm{d}W_2), \tag{A 49b}$$

such that

$$\hat{\sigma} = \sigma\begin{bmatrix} \cos\phi & \sin\phi \\ -\sin\phi & \cos\phi \end{bmatrix} \in O(2).$$

It is well-known (see e.g. Gardiner (1994) and Doering (1991)) that the stochastic equation in the Ito form,

$$dx(t) = a(x,t)\,dt + \sigma(x,t)\,dW,$$

generates a Fokker–Planck equation for the probability density $p(x,t)$,

$$\partial_t p(x,t) = \left[-\partial_x a(x,t) + \tfrac{1}{2}\partial_x^2 \sigma^2(x,t)\right] p(x,t), \tag{A 50}$$

in the one-dimensional case.

The multi-dimensional process

$$dx^i(t) = a^i(\vec{x},t)\,dt + \sigma_q^i(\vec{x},t)\,dW^q, \quad i,q = 1,\ldots,m,$$

gives rise to

$$\partial_t p(\vec{x},t) = \left[-\frac{\partial}{\partial x^i} a^i(\vec{x},t) + \tfrac{1}{2}\frac{\partial}{\partial x^i}\frac{\partial}{\partial x^j}\sum_q \sigma_q^i \sigma_q^j\right] p(\vec{x},t).$$

So we see that the diffusion coefficient

$$D^{ij} = \sum_q \sigma_q^i \sigma_q^j = g^{ij}$$

is nothing but the inverse metric in our consideration.

Now in the case of a cylindrically symmetric process, we have the following diffusion type equation for the cylindrical density $p_c(\rho,t)$

$$\partial_t p_c = -\partial_\rho\left(\frac{\sigma^2}{2\rho}p_c\right) + \tfrac{1}{2}\partial_\rho^2\left(\sigma^2 p_c\right)$$

and equation

$$\partial_t p_{car} = \frac{\sigma^2}{2}\left(\frac{\partial^2}{\partial x^2} + \frac{\partial^2}{\partial y^2}\right) p_{car} = \frac{\sigma^2}{2\rho}\frac{\partial}{\partial\rho}\left(\rho\frac{\partial}{\partial\rho}p_{car}\right), \tag{A 51}$$

for the Cartesian density $p_{car}(\rho,t)$.

These two densities are connected through

$$p_{car}(x,y,t)dxdy = p_{car}(\rho,\phi,t)\rho d\rho d\phi \equiv p_c(\rho,\phi,t)d\rho d\phi,$$

or

$$p_c(x,y,t) = \rho p_{car}(\rho,\phi,t).$$

Substituting the last equation into (A 50) we obtain

$$\partial_t p_{car} = \frac{\sigma^2}{2}\left(\frac{1}{\rho}\partial_\rho + \partial_\rho^2\right) p_{car} \equiv \frac{\sigma^2}{2\rho}\frac{\partial}{\partial\rho}\left(\rho\frac{\partial}{\partial\rho}p_{car}\right),$$

which coincides with (A 51) for the centrally symmetric case, $\partial\phi = 0$.

It means that the drift term in the process (A 49)

$$b(\rho) = \frac{\sigma^2}{2\rho},$$

has a purely geometrical character, and allows for the fact that in polar coordinates $\rho \geq 0$ and the diffusion results in increasing the mean value of ρ, a phenomenon that was absent in Cartesian coordinates invariant under the transformation $(x,y) \to (-x,-y)$.

We now see that the self-consistent drift term is geometrical in nature and in the Euclidean case considered, it describes an expansion of the corresponding mean coordinate $\langle\rho\rangle$, due to the diffusion process itself.

Sometimes the geometrical drift can be negative and lead to decreasing the mean value of the coordinate of the process studied. This is so as in the case of geometrical Brownian motion, where the equation (see Hull (1993))

$$\frac{\mathrm{d}S}{S} = a\,\mathrm{d}t + \sigma\,\mathrm{d}W$$

gives for the return

$$\mathrm{d}\ln S = \left(b - \frac{\sigma^2}{2}\right)\mathrm{d}t + \sigma\,\mathrm{d}W,$$

and the drift can be negative when

$$b < \frac{\sigma^2}{2}.$$

Now look at the sphere. The centrally symmetric diffusion process around the north pole of the sphere is defined by the Laplacian

$$\partial_\rho^2 + \frac{1}{R_0}\cot(\rho/R)\partial_\rho,$$

hence by the Ito process

$$\mathrm{d}\rho = \frac{1}{2R_0}\cot(\rho/R)\mathrm{d}t + \sigma_q^\rho\,\mathrm{d}W^q, \tag{A 52a}$$

$$\mathrm{d}\phi = \sigma_q^\phi\,\mathrm{d}W^q, \tag{A 52b}$$

where drift term is

$$b(\rho) = \frac{b^2}{2R_0}\cot(\rho/R)\,\mathrm{d}t. \tag{A 53}$$

This drift term must naturally occur in the framework of the SDG equations on the sphere:

$$\mathrm{d}\vec{X} = \hat{\sigma}\mathrm{d}\vec{W},$$
$$\mathrm{d}\hat{\sigma} = -\Gamma\hat{\sigma}\mathrm{d}\vec{X}.$$

On the sphere we have (see e.g. Dubrovin *et al.* (1984))

$$\mathrm{d}l^2 = \mathrm{d}\rho^2 + R_0^2\sin^2(\rho/R_0)\,\mathrm{d}\phi^2, \tag{A 54}$$

or

$$g_{ij} = \begin{bmatrix} 1 & 0 \\ 0 & R_0^2\sin^2(\rho/R_0) \end{bmatrix},$$

nd

$$g^{ij} = \begin{bmatrix} 1 & 0 \\ 0 & R_0^{-2}\sin^{-2}(\rho/R_0) \end{bmatrix}.$$

The sphere is a Riemannian manifold with connexion compatible with the metric. Equation (4.19) holds, whence

$$\Gamma_{12}^2 = \Gamma_{21}^2 = \tfrac{1}{2}g^{22}\frac{\partial g_{22}}{\partial X_1} = \frac{1}{R_0}\cot(\rho/R_0),$$

$$\Gamma_{22}^1 = -\tfrac{1}{2}g^{11}\frac{\partial g_{22}}{\partial X_1} = -R_0\sin(\rho/R_0)\cos(\rho/R_0),$$

nd the other components vanish. The process (in Stratonovich's form) is given by the equations

$$\mathrm{d}X^1 = \sigma_1^1\,\mathrm{d}W^1 + \sigma_2^1\,\mathrm{d}W^2,$$
$$\mathrm{d}X^2 = \sigma_1^2\,\mathrm{d}W^1 + \sigma_2^2\,\mathrm{d}W^2,$$
$$\mathrm{d}\sigma_1^1 = -\Gamma_{11}^1\sigma_1^1\,\mathrm{d}X^1 - \Gamma_{22}^1\sigma_1^2\,\mathrm{d}X^2.$$

The Ito drift term is defined by the third equation and the equation for the drift, by virtue of

$$\sum_q \sigma_q^i \sigma_q^j = g^{ij},$$

is as follows

$$\tilde{b}^1 = -\tfrac{1}{2}\sum_{jk}\Gamma^1_{jk}g^{jk} = -\tfrac{1}{2}\Gamma^1_{22}g^{22} = \frac{1}{2R_0}\cot(\rho/R_0), \tag{A 55}$$

$$\tilde{b}^2 = -\tfrac{1}{2}\Gamma^1_{12}g^{12} - \tfrac{1}{2}\Gamma^1_{21}g^{21} = 0, \tag{A 56}$$

since $g^{12} = g^{21} = 0$. As a result we obtain the following Ito process

$$\mathrm{d}\rho = \frac{1}{2R_0}\cot(\rho/R_0)\,\mathrm{d}t + \sigma_1^1\,\mathrm{d}W^1 + \sigma_2^1\,\mathrm{d}W^2,$$

$$\mathrm{d}\phi = \sigma_1^2\,\mathrm{d}W^1 + \sigma_2^2\,\mathrm{d}W^2.$$

We see that (A 55) and (A 56) give the correct drift term $b(\rho) = \tilde{b}^1$, which now appears due to the curvature of the sphere.

REFERENCES

ALEXANDER, C. & GIBLIN, I. 1993 Searching for chaos in financial markets. Univ. of Sussex Math. Research Report, pp. 93–24; see also 1994 Chaos in the system? Can chaos theory provide a way of predicting movements in financial markets? *Risk* **7**, 71–76.

BRENNAN, M. J. & SCHWARTZ, E. S. 1979 A continuous time approach to pricing bonds. *J. Banking and Finance* **3**, 133–155; see also 1982 An equilibrium model of bond pricing and a test of market efficiency. *J. Finan. Quant. Anal.* **17**, 301–329; SCHAEFER, S. M. & SCHWARTZ, E. S. 1984 A two-factor model on the term structure: an approximate analytical solution. ibid. **19**, 413–424.

DOERING, CH. R. 1991 Modeling complex systems: stochastic processes, stochastic differential equations, and Fokker–Planck equations. In *1990 Lectures in Complex Systems* (ed. L. Nadel & D. Stein). SFI Studies in the Sciences of Complexity, vol. 3. Addison–Wesley.

DUBROVIN, B., FOMENKO, A. & NOVIKOV, S. 1984 *Modern Geometry. Methods and Applications. Part I. The Geometry of Surfaces, Transformations Groups and Fields.* Springer.

GARDINER, C. W. 1994 *Handbook of Stochastic Methods for Physics, Chemistry and the Natural Sciences.* Springer.

HEATH, D., JARROW, R. & MORTON, A. 1992 Bond pricing and the term structure of interest rates: a new methodology. *Econometrica* **60**, 77–105.

HULL, J. C. 1993 *Options, Futures and other Derivative Securities.* Prentice-Hall.

INGBER, L. 1990 Statistical mechanics aids to calculating term structure models. *Phys. Rev. A* **42**, 7057–7064; see also INGBER, L., WEHNER, M. F., JABBOUR, G. M. & BORNHILL, T. M. 1991 Application of statistical mechanics methodology to term-structure bond-pricing models. *Math. Comput. Modelling* **15**, 77–98.

KENDAL, W. S. 1987 Stochastic differential geometry: an introduction. *Acta Applicandae Mathematica* **9**, 29–60.

MAKHANKOV, V. & PASHAEV, O. 1992 *Integrable Pseudospin Models in Condensed Matter.* Harwood Acad. Publishers.

MAKHANKOV, V., TARANENKO, YU., GOMEZ, C. & JONES, R. 1995 Stochastic differential geometry and the term structure of interest rates. Preprint LA-UR-95-449, Los Alamos Nat. Laboratory. Submitted for publication.

PATHRIA, R. K. 1972 *Statistical Mechanics*, p. 451. Pergamon.

STRATONOVICH, R. L. 1968 *Conditional Markov Processes and their Application to the Theory of Optimal Control.* Elsevier.

Where will the future go?

By MITCHELL J. FEIGENBAUM

Rockefeller University, New York, USA

This is the first after dinner talk I have ever given other than for fund raising. Maybe that's different. A long time ago, shortly after I arrived at Los Alamos, this is approximately 1975, there was the dedication banquet for a something called LAMPF, which was the *Los Alamos Meson Physics Factory*, and the new head of the theory division at Los Alamos, Peter Carruthers decided upon who to invite to be the after dinner speaker. The person he chose, I won't mention the name, gave a talk entitled *Academic integrity*. As the talk proceeded, all that one could make out amongst the snoring was something like agready–agready, and this produced something of a scandal, and I'll try the best that I can to appear to be sober. So I think my task is to say something about *where the future will go*; I'm not very sure about such things, and start off by saying where the past will go. So let me tell you something very quickly about my background.

As an undergraduate I was an electrical engineer at the City College of New York; at the same time I also took all of the physics courses and graduate courses and all the mathematics courses and graduate courses. One of the things that I learned from that is that I was a member of so to speak three communities, and the most extraordinary part is that there was no communication whatsoever between them. In fact there was something that one might term racism, although that's probably too harsh a comment, and so the engineers felt that the physicists didn't know anything about the real world, and the phrase is always the real world, and they regarded them as rather foolish dreamers, and they never said much about the mathematicians. The physicists, conversely, thought the engineers were rather stupid, and more to the point, they thought that mathematicians had no intuition whatsoever, and so their presumed contributions were unimportant. Probably the fairest of the lot were the mathematicians; they always felt, a great number of them, that physics was potentially interesting, but they couldn't make any sense out of it. And so, as I remember back to a long time ago, there were really quite discordant worlds, and what I'll quickly lead up to is that things are decidedly better than that now. (I guess I could give another speech in which I would say that they are the same. It is one of these curiosities of the world where it depends upon the vantage point.) So in any case from my education, what struck me is that the world was still with all these complicated things, for one thing wonderful machines, for another thing all of the things of nature, fluids and whatnot that were immensely complicated. It was easy to learn in high school very simple comments when something is completely symmetric. You just write down 1/r, and you know how to calculate it. It's a different story when there are very many parts. And what it seemed all of this knowledge was leading to it is that there is a possibility to understand a decidedly grander world. Of the physics things that I learned, the subject that impressed me the most as an undergraduate was statistical mechanics. I thought this was a natural thing that I might want to do, but it was very carefully pointed up to me in 1964 – that subject was dead. And this will be a recurrent theme in what I want here to say. Well, I went to graduate school. I studied electrical engineering and quickly turned to the physics department. My initial interest was in general relativity, and, to immediately return to the previous comment, after studying for one year it was pointed out to me – general relativity was a dead subject. And as a matter of fact, and it was, this was now the mid to end of the sixties, the period that I am talking about, and that subject was dead. And it was pointed out by the person who then became my theses adviser, that first of all there are no jobs there, and second of all, however much I found general relativity interesting, there wasn't any particular reason I shouldn't find high energy physics interesting, and any way I didn't have a choice. So I became a high energy theorist. One of the curiosities and something that was a profound disappointment to me was that with all of this dreaming about all these complicated configurations the problem at hand, the problem to gain insight into nature at very small distances, was conceptually, as the geometry of arrangement of particles, the most boring thing in the world: the entire discussion was the scattering of one particle against another particle. And to my mind that was always dreadful. It was a very wonderful education growing up as a high energy theorist, to learn a very large amount of stuff from papers mostly wrong that would have a half life of only a few months, but nevertheless shot with all sorts of ideas, was a rather interesting way to realize you can very quickly pick up varieties of knowledge. However, I certainly always felt dissatisfied and I didn't feel I was learning something about the heart of the world, whatever that might mean. In any case, as I went on, it was increasingly clear to me I wasn't finding what I cared about in high energy physics; not that the theory wasn't interesting, not that the mathematics wasn't interesting, but what I always cared about was what this bigger world looked like, and so starting in the very early seventies I had an impression that there was some sort of a subject to be made that for want of a name I called big systems. And I presumed that with time there would be a chance to do something about the existence. At the end of my graduate school career I also spent a large amount of time with a neurophysiologist Jerome Lition and that also had profound consequences in coloring my thoughts and things that I cared about. It's worth pointing out that when the physicists at MIT discovered I was talking to him it was pointed out that it was dangerous and I shouldn't. That is there is something unfortunate in the world that it is a little too easy, while things are happening, that people feel uncomfortable with things that are different. And I think it's one of the greatest consequences of this subject that embraces this meeting, that some of these barriers have at least become much more diffused.

Well, as a post doctoral person, the most interesting thing that arose, and this was just about 1970, was Ken Welthon's work on resurrecting the renormalization group, which opened up at least conceptually the possibility that one can start saying something about somewhat complicated problems. And many of you probably remember (might know) that in Wilson's very first paper on the renormalization group he immediately pointed out that of course this was methodology that was aimed at problems not with the separational scale, but with the continuum of them and what problem was more natural that had never been understood, and that of course was the problem of fluids, the problem of fluid turbulence. Among the things that one might call big systems, was certainly all these things one sees in a storm, watching the waves not only breaking, but almost dashing and destroying things in front of them, and this extraordinary spectacle of nature which didn't look anything like the symmetric solutions to some particular problem. And whatever was to happen with that subject I'll make a few little comments in the course of my comments. When I went to Los Alamos after two other post docs in 1974, the head of the theory division then felt it was altogether appropriate to see if Wilson's ideas (was actually) might say something about the turbulence. That obviously was a topical thing to do, Los Alamos had gone in some decline theoretically and with the resurrection of theory in 1973, there was definitely some feeling of unhappiness amongst people, the people in the laboratory, that here should be these mandarins who aren't doing directed research, and so it was obviously useful if we could show off that the stuff that we knew could actually be of benefit and beyond with other of the people there to do

So Carruthers told me why shouldn't I start thinking about using the renormalization group in turbulence. Well, I'm going to truncate the story a whole lot. I simply started thinking quickly about nonlinear oscillations; it was natural not to look at differential equations but to look at some appropriate maps. I did that for a few months. It got very very hard. The problem was sort of a toy anyway, and so after a while I gave it up and I did other things. About a year later, a little less than a year later, I returned to the problem and I realized that there was a very pretty conformal picture which has to do with cutting up the complex plane, as there are successive instabilities, and I produced some version of an analytic theory of what this stuff looks like. However, the theory was only contingent on the fact that there would be more and more complications which the theory couldn't say very much about, but it produced some appropriate sort of equations.

When I retuned to Los Alamos I took out my calculator. It was very hard to solve these equations that were at stake for this conformal theory. And so I quickly checked out if one couldn't find the places where the doubling of periodicity occurred on the calculator and then in doing that, because the calculator was so slow, I discovered the fact it was an asymptotic geometric convergence, and that made it very easy to go through the entire sequence, and understand what was happening. It was interesting to me. A geometric convergence is an absolute number. If you change coordinates the space changes; if you change the parameter the parameter changes. For a nonlinear problem you change coordinates, the parameter mixes up. Nevertheless, unless something is profoundly wrong, if it's a geometric convergence, there obviously is an invariant. So I paid some attention to the number. I didn't have any more thoughts about it a few months later. I realized that another problem, which would not have subscribed to my conformal thoughts, was supposed to do the same thing. I checked it out on the calculator, indeed it did the same thing. And moreover I very quickly realized that again this was geometrically convergent, and by my efforts of thinking about the number a few months before, I immediately noticed it was the same number, and so now three weeks ago, and twenty years ago, I discovered that this number was some four and two thirds, was universal over dynamics. So this meeting is in some sense appropriately timely. It's now the twentieth anniversary (which is a long time ago). In any case having done that was it clear how to precede? The answer is no. The three months I had this curious fact, I learned how to use a computer to really check out if the number was the same, had this curious fact that dynamics didn't make any difference; pick any equation you wanted to, you always get the same thing, no matter where did such a thing come from. And then some three months later I finally realized, more or less, what would give rise to such a theory, that that was a certain functional business, there was some sort of a fixed point, and then I started using a computer. Wholesale, it was not clear how any of this would exactly go out. This was one of the most interesting periods in my life. This was a period of some three months where I worked approximately 22 hours a day, seven days a week, until I became very sick. But nevertheless it was very amusing business of an interchange between thinking and using a computer. I'm going to make some comments about that. I've made a few comments before about computers. Computers were different then from what they are now. The attitude towards them was different then from what they are now. So going back to 1975, 1976 in Los Alamos, the serious computers were first of all behind defense. That meant they were classified and there was not easy means to use them for some purposes. If you wanted to compute at home, there was only one machine, machine 'zero', which later became machine 'M'; there was great originality in the naming of these machines, which I think is vastly superior to the foolish names that one now encounters, but nevertheless there was this one machine; the secretaries would use it; it was the only machine that you could call up, and so if you would connect during the day, first of all if you didn't do something after three minutes, they would disconnect you. If you were disconnected after around 10 in the morning, you could forget about ever getting on to the machine again until the evening. Now that sets up quite a tyranny. So imagine you produce a 1500 line program, you get some results, you have to look at them very quickly, see what in it was interesting to now go ahead, and figure out the next things to do. If you paused too long you were dead. And so there was this machine buzzing at you, a so called Silon 700, and it was an extraordinary tension that it set up. So this was partly what it meant to compute in the past. Now there is a myth about Los Alamos. The myth is that there was great graphical capabilities. The right answer is there were virtually none whatsoever. The only terminals that existed, which had these great thick pictures, were in the weapons division, there were no others, they were illegal out of the cleared area, and so there was fundamentally no graphics that was available. And so in olden days what we had was the so called Silon 700. How many people here remember Silon 700? One. Probably only one. So what this was, it was a sort of a typewriter in a box. It had two rubber cups and you stuffed the telephone into it, and it had thermal paper. How did you make graphics?

The idea was real simple. You had 80 characters. You would blank out a line, so you had to become a master in Fortran to encode the code statements in Fortran, which we all did, you would blank out a line, you would pick where the axis would be, say put a colon over there, so as you would keep running it out you would get a line, this was one abscissa, and as the thing would keep running out, each line was one abscissa unit, and you would keep generating it. And then in the place where there was a point on a graph say, you would put an asterisk there. That proved to be very problematic Los Alamos was very strong in computation, and that meant that there were constantly system upgrades. At some point, all of a sudden, none of my programs worked, because it turned out for a week colon was illegal for anyone but the system to use. And there were more amusing things that happened. Well, I was busily trying to solve some very complicated functional equations that have monstrously complicated graphs, and so this meant, in order to use this wonderful method, that there were tens and thirties of these printouts and I still own most of this. Somehow I suppose as a museum piece to recall truly what it meant to use graphics in the past. So this is simply as an historic comment computation was something very different then from what it is now. Well, at some point there was a finished theory about this period doubling, the theory of course says that the dynamics don't matter at least for a one dimensional map, whatever they are, by the time you see an onset of chaotic behavior, there is something universal. Well, this was a one dimensional map. Now the question is where is it going to go, and what do people think about it.

Well, the answer is very clear. The condensed matter community was absolutely uninterested. The question was, is this a toy, or is this really saying something say about a fluid. Well, it ostensively wasn't saying anything that anyone knew about a fluid. It was only sitting as a discrete time problem, so it was clearly regarded as a toy. So at some point, say in 1978, Mark Katz made a comment to me, that you know it's brilliant café table conversation, but until it really turns into physics, no one will really be interested. That of course was among physicists. During the same period mathematicians started getting intensely interested, and so for a number of years my contact was much more with the mathematicians. What I am trying to say is how does one guess along what lines in what communities are things going to develop, and obviously one has limited foresight.

The whole story obviously changed in 1979 when Libchaber did his rather famous experiment and discovered indeed that these things literally and precisely happened i the fluid. Then it became a part of fluid physics. Indeed one then learned that classical thinking that when you see a broad spectrum of noise, people thought that meant that each mode, each spectral line, meant another degree of freedom, one would only see that when an infinite number of degrees of freedom were excited. One of the things that of course we learned from that is that that was ill thinking. And so one learned a lot of things, and in the course of it one learned to discover that there was now wind turbulence. Perhaps it was temporal turbulence as opposed now to genuine turbulence. And I think that's an immediate prediction to the future. With each piece of knowledge one will discover the full problem of turbulence is always something different. And I offer that up as a prediction of the future. That's not meant facetiously. It's meant truly. This is a very complicated problem and it will own parts by the time we understand it, that are different from all that we know. Does this applause mean that I said enough? I'm perfectly happy to stop. So let me go on a little bit. So this whole story is one doing physics, is one doing mathematics? The one thing that was clear is that after Libchaber's experiment, since the mathematicians had already become involved, there was suddenly open a new discussion. I think one of the best things that happened from this early work on chaotic dynamics is that all of the sudden there was a more serious dialogue between fluid physicists, already some engineers, certainly between mathematicians and then amongst physicists. And indeed after 1980 it was altogether conceivable that a person doing nonlinear dynamics might be hired in a physics department. So one of the things that occurred was definitely a first breaking down the barriers of communication, and I think this meeting very well indicates that one is now doing something rather different from a meeting that would have happened twenty years ago, in which there would be no such spectrum of people present. Well, let me make some quick comments about that.

As I told you my interest was understanding these very big systems. Did one understand the big systems? Well, it turns out that one understood something in some context. And so the first impediment was that we've not been able to get out of the small number of dimensions. It makes no difference if we are talking about a fluid or whatever, what makes the difference is that we have something like some ultimate states in which we have comprehension, say a renormalization group theory, that one at the moment still has to have a very low dimensionality, this already partly presses our knowledge, and in general we don't have very much knowledge beyond that. We've obviously learned, various amongst us, how to capitalize upon these insights in order to do some interesting new data analysis. And of course the work of Procaccia comes to mind. In thinking about that, again one of the ways in which the subject has bloomed and enabled to embrace more people, is the possibility of new ways of looking at signals, looking at unknown data, to say something or other about it with attendant new methods of modeling. It's nevertheless an impediment to the progress of understanding much modern problems that we really don't know how to go very much further than a few dimensions. And one might hope one day we can better the map, but I think it is a warning. The warning is when you have an idea, and the people who have done it, they have usually saturated what can handily be done, and usually a new advance requires, at least initially, a completely different idea. So, however these things work out, we will simply see. Well, let me say something further about fluids now. As I said, the progenitor, certainly within the physics community, of looking at complicated systems has been the issue of fluids. Fluids certainly, seem to us (fluid turbulence), as the prototype, as the icon of these very complicated systems. As I told you, in 1964 statistical mechanics was dead. Because of Wilson's work it had an extraordinary rebirth in the very early seventies. One of the reasons that nonlinear dynamics did very well in the physics world was that a period doubling theory has a renormalization group theory, provided natural things for experts in a renormalization group to do, especially after they had already worked out all of the other phase transitions that they could think about. So, one of the things that the renormalization group provides are exponents. And that leads to the question: is the solution of turbulence, is the meaning of a fluid exponents?

Let's change the question a little bit. We've embraced even more fields. We quietly have done, or are at the very beginnings of to employ physics, to employ analytic thoughts in the understanding of biology. It is another complicated problem. Is the solution to biology a set of exponents? The answer to the second question is absolutely and overwhelmingly no. It's very easy to see why that is. When you look at a cell, the first thing a cell does is to use up all physics that we know as fast as possible. What a cell does is to use physics to build machines, and thereafter it has no interest in the laws of the physics, it made itself something different. When I say no interest, that's obviously an exaggeration. Nevertheless, one of the profound parts of molecular biology is that it's the first process which forgets about the fusion. A cell refuses to have anything to do with the fusion. It's simply not important to its functioning. It's anathema to it. Most of the things that we know from statistical physics are dealt with very early in the construction of the processes in the cell, instead one makes machines. Perhaps then a hundred different machines now. Machines that read DNA, machines that replicate it, machines that of course start crafting proteins. And one after another the most annoying discovery of modern molecular biology is that the idea that one day, when we know enough statistical mechanics, and you shake up the bag with all kinds of chemicals, that's out. That's certainly not how biology works. So what can we say about fluids? Exponents, to know how to get them well, is a wonderful enterprise; we've not been able to do it yet, we're beginning to make progress, some of us, have made progress over the years, it's constantly a new work, there will be more of that work, it will be wonderful when one understands the statistical behavior of the fluid. But as a matter of fact a fluid has form, it has texture, it has excitations in it. These are things that we care about. We don't only care about fluids, we care about them even more than anything else as an icon, as a prototype where *what* are the complicated things that are perhaps within our reach. And so if we think of texture, do we think of fractals? What are fractals? Fractals are exponents. When we look at textures, are they fractals? The answer is unambiguously no! These are things that we know very well. We know from the theory of period doubling already that it is not a fractal in any simple sense, in exponent sense, we already know it's much more information than that, one has structural information. I've mentioned this because the theory of period doubling is very special. In the right world it had so much symmetry in it that in the end even though it gives rise to very complicated things, it's much too special a problem, and the attempt to try to metamorphose from it to more complicated things has been partly very valuable in crafting newer thoughts, for example thoughts about how to start measuring unknown signals, more than that, but nevertheless we have open to us a large variety of questions. How will we succeed, where will the future go, we're going to work on all of these problems with all of the tools we know, some of us will prove to be brilliant and they will deliver results that no one expected. And along the line there will be totally new ideas (unless my subject dies) and I think that's really the future.

Thank you very much.